INDUSTRIAL BIOTECHNOLOGY

Sustainable Production and Bioresource Utilization

INDUSTRIAL BIOTECHNOLOGY

Sustainable Production and Bioresource Utilization

Edited by
Devarajan Thangadurai, PhD
Jeyabalan Sangeetha, PhD

Apple Academic Press Inc. | Apple Academic Press Inc.
3333 Mistwell Crescent | 9 Spinnaker Way
Oakville, ON L6L 0A2 | Waretown, NJ 08758
Canada | USA

ISBN-13: 978-1-77463-582-7 (pbk)
ISBN-13: 978-1-77188-269-9 (hbk)

Library and Archives Canada Cataloguing in Publication

Industrial biotechnology : sustainable production and bioresource utilization / edited by Devarajan Thangadurai, PhD, Jeyabalan Sangeetha, PhD.

Includes bibliographical references and index.
Issued in print and electronic formats.
ISBN 978-1-77188-269-9 (hardcover).--ISBN 978-1-77188-262-0 (pdf)

1. Biotechnology industries. 2. Biotechnology--Environmental aspects.
I. Thangadurai, D., editor II. Sangeetha, Jeyabalan, editor

HD9999.B442I54 2016   338.4'76606   C2016-901490-8   C2016-901491-6

Library of Congress Cataloging-in-Publication Data

Names: Thangadurai, D., editor. | Sangeetha, Jeyabalan, editor.
Title: Industrial biotechnology : sustainable production and bioresource utilization / editors, Devarajan Thangadurai, PhD, Jeyabalan Sangeetha, PhD.
Description: Toronto ; [Hackensack?] New Jersey : Apple Academic Press, 2016.
| Includes bibliographical references and index.
Identifiers: LCCN 2016009307 | ISBN 9781771882699 (hardcover : alk. paper)
Subjects: LCSH: Biotechnology--Industrial applications. | Industrial microbiology. | Green chemistry.
Classification: LCC TP248.2 .I54 2016 | DDC 660--dc23
LC record available at http://lccn.loc.gov/2016009307

CONTENTS

LIST OF CONTRIBUTORS

Mohd Azmuddin Abdullah
Institute of Marine Biotechnology, Universiti Malaysia Terengganu, 21030 Kuala Terengganu, Terengganu, Malaysia

Huma Ajab
Department of Chemistry, COMSATS Institute of Information Technology, 22060 Abbottabad, Pakistan

Ricardo Alfán-Guzmán
School of Biotechnology and Biomolecular Sciences, The University of New South Wales, Sydney, NSW, 2052, Australia

Sakinatu Almustapha
Department of Chemical Engineering, Universiti Teknologi Petronas, 32610 Seri Iskandar, Perak, Malaysia

Shradha Bashetti
Department of Microbiology, Modern College of Arts, Science and Commerce, Shivajinagar, Pune, Maharashtra, India

Suchitra B. Borgave
Microbial Sciences Division, MACS-Agharkar Research Institute, G.G. Agharkar Road, Pune, 411004, India

Fernanda Bravim
Núcleo de Biotecnologia, Centro de Ciências da Saúde, Universidade Federal do Espírito Santo, Vitória, ES, Brazil

Safoura Daneshfozoun
Department of Chemical Engineering, Universiti Teknologi Petronas, 32610 Seri Iskandar, Perak, Malaysia

Ajay K. Dalai
Department of Chemical and Biological Engineering, University of Saskatchewan, Saskatchewan, Canada

Siddharth V. Deshmukh
Interactive Research School for Health Affairs (IRSHA), Bharati Vidyapeeth Deemed University, Pune Satara Road, Pune, 411043, India

Prashant K. Dhakephalkar
Microbial Sciences Division, MACS-Agharkar Research Institute, G.G. Agharkar Road, Pune, 411004, India

Melina Campagnaro Farias
Núcleo de Biotecnologia, Centro de Ciências da Saúde, Universidade Federal do Espírito Santo, Vitória, ES, Brazil

Patricia Machado Bueno Fernandes
Núcleo de Biotecnologia, Centro de Ciências da Saúde, Universidade Federal do Espírito Santo, Vitória, ES, Brazil

Sonali Joshi
Department of Biotechnology, Fergusson College, Shivajinagar, Pune, Maharashtra, 411004, India

Vaishnavi S. Joshi
Life Science Research Centre, Department of Biotechnology, Modern College of Arts, Science and Commerce, Pune, Maharashtra, 411005, India

Pradnya P. Kanekar
Department of Biotechnology, Modern College, Pune 411005, Maharashtra, India

Sagar P. Kanekar
Microbial Sciences Division, MACS-Agharkar Research Institute, G.G. Agarkar Road, Pune, 411004, India

Meghana S. Kulkarni
Microbial Sciences Division, MACS-Agharkar Research Institute, G.G. Agharkar Road, Pune, 411004, India

Janusz A. Kozinski
Lassonde School of Engineering, York University, Ontario, Canada

Matthew Lee
School of Biotechnology and Biomolecular Sciences, The University of New South Wales, Sydney, NSW, 2052, Australia

D. R. Majumder
Department of Microbiology, Abeda Inamdar Senior College, 2390-KB Hidayatullah Road, Azam Campus, Camp, Pune 411001, Maharashtra, India

Nirupama Mallick
Agricultural and Food Engineering Department, Indian Institute of Technology, Kharagpur, West Bengal, India

Michael Manefield
School of Biotechnology and Biomolecular Sciences, The University of New South Wales, Sydney, NSW, 2052, Australia

Padmanabh Mishra
Molecular Biophysics Unit, Indian Institute of Science, Bangalore, Karnataka, 560012, India

Pravakar Mohanty
Science and Engineering Research Board, Department of Science and Technology, Government of India, New Delhi, India

Shilpa Mujumdar
Department of Microbiology, Modern College of Arts, Science and Commerce, Shivajinagar, Pune, Maharashtra, India

Dattatraya G. Naik
Chemistry Group, MACS-Agharkar Research Institute, G.G. Agharkar Road, Pune, 411004, India

Sonil Nanda
Department of Chemical and Biological Engineering, University of Saskatchewan, Saskatchewan, Canada

Muhammad Shahid Nazir
Department of Chemical Engineering, COMSATS, Institute of Information Technology, Lahore, 54000 Punjab, Pakistan

Smita S. Nilegaonkar
Microbial Sciences Division, Agharkar Research Institute, Pune, Maharashtra, 411004, India

Radhika S. Oke
Department of Biotechnology, Modern College of Arts, Science and Commerce, Shivajinagar, Pune, 411005, Maharashtra, India

Sheetal Pardeshi
Department of Microbiology, Modern College of Arts, Science and Commerce, Shivajinagar, Pune, Maharashtra, India

Oeber De Freitas Quadros
Núcleo de Biotecnologia, Centro de Ciências da Saúde, Universidade Federal do Espírito Santo, Vitória, ES, Brazil

Prabhakar K. Ranjekar
Interactive Research School for Health Affairs (IRSHA), Bharati Vidyapeeth Deemed University, Pune Satara Road, Pune, 411043, India

Akhilesh Kumar Singh
Amity Institute of Biotechnology, Amity University, Lucknow, Uttar Pradesh, India

S. K. Singh
Biodiversity and Palaeobiology Group, Agharkar Research Institute, Pune, Maharashtra, 411004, India

Rebecca S. Thombre
Department of Biotechnology, Modern College of Arts, Science and Commerce, Pune, Maharashtra, 411005, India

Vasudeo P. Zambare
Center for Bioprocessing Research and Development, South Dakota School of Mines and Technology, E. Saint Joseph Street, Rapid City, SD, USA

LIST OF ABBREVIATIONS

[Amim]$^+$	1-Allyl-3-Methylimidazolium
[Ammim]$^+$	1-Allyl-2,3-Dimethylimidazolium
[Bu$_4$P]$^+$	Tetrabutylphosphonium
[C$_4$mP$_y$]$^+$	1-Butyl-3-Methylpyridinium
[Ch]Ac	Cholineacetate
[C$_n$mim]$^+$	1-Alkyl-3-Methylimidazolium
[C$_n$mmim]$^+$	1-Alkyl-2,3-Dimethylimidazolium
[TBMA]Cl	Tributyl Methyl Ammonium Chloride
°C	Degree Celsius
2DGE	Two-Dimensional Gel Electrophoresis
3,4-DCI	3,4-Dichloroisocoumarin
3D	3-Dimensional
3HHD	3-Hydroxyhexadecanoate
3HOD	3-Hydroxyoctadecanoate
3HV	3-Hydroxyvalerate
4HB	4-Hydroxybutyrate
5-HMF	5-Hydroxymethyl-2-furaldehyde
ACP	Acyl Carrier Protein
ADH6	Alcohol Dehydrogenase 6 Gene
ADY	Active Dry Yeast
AFEX	Ammonia Fiber Expansion
Ag	Silver
AIDS	Acquired Immune Deficiency Syndrome
ALD5	Aldehyde Dehydrogenase Gene
Ald6p	Cytosolic Aldehyde Dehydrogenase
ALK	Alkaliphilic Bacteria
AmpR	Ampicillin Resistance
Arg1p	Arginosuccinate Synthetase
ARS	Autonomously Replicating Sequence
ATP	Adenosine Triphosphate
Au	Gold
BACs	Bacterial Artificial Chromsomes
BmimCl	1-Butyl-3-Ethyl-Imidazolium Chloride

BOD	Biological Oxygen Demand
bR	Bacteriorhodopsin
BRE	B Recognition Element
BTU	British Thermal Units
Bya	Billion Years Ago
c-DCE	*cis*-Dichloroethene
CAGR	Compounded Annual Growth Rate
CB	Chlorobenzoate
CBP	Consolidated Bioprocessing
CD's	Cyclodextrins
CDase	Cyclodextrinase
cDNA	Complementary DNA
Cen	Centromers
CF	Chloroform
CGTase	Cyclodextrin Glycosyl Transferase
CHEMEX	Changhae Ethanol Multi Explosion
CLPs	Chitinase Like Proteins
CMC	Critical Micelle Concentration
CNW	Cellulose Nanowhiskers
CO_2	Carbon Dioxide
CoA	Coenzyme A
COD	Chemical Oxygen Demand
CP	Chlorophenol
CSTR	Continuous Stirred-Tank Reactor
CT	Carbon Tetrachloride
Cys4p	Cystathionine beta-Synthase
Da	Daltons
DCA	Dichloroethane
DCB	Dichlorobenzene
DCM	Dichloromethane
DCP	Dichloropropane
DESs	Deep Eutectic Solvents
DFP	Diisopropylfluorophosphate
DNA	Deoxyribonucleic Acid
DNSA	3,5-Dinitrosalicylic Acid
EBI	Electron Beam Irradiation
EFB	Empty Fruit Bunches
EISA	Energy Independence and Security Act

EOR	Enhanced Oil Recovery
EPS	Exopolysaccharides
ER	Endoplasmic Reticulum
ESR	Environmental Stress Response
FBA	Flux Balance Analysis
FDH1	Formate Dehydrogenase 1 Gene
FDH2	Formate Dehydrogenase 2 Gene
g/L	Gram per Liter
G+C	Guanine + Cytosine Content
G6PD	Glucose-6-Phosphate Dehydrogenase
GAPN	Glyceraldehyde-3-Phosphate Dehydrogenase
GC rich	Guanosine Cytosine rich
GC-MS	Gas Chromatography-Mass Spectrometry
GDH1	NADP(+)-Dependent Glutamate Dehydrogenase
GHz	Giga Hertz
GLN1	Glutamine Synthetase
GLT1	Glutamate Synthase
GPD1	NAD-Dependent Glycerol-3-Phosphate Dehydrogenase
GPD2	Glycerol-3-Phosphate Dehydrogenase
GSMT	Glycine Sarcosine Methyl Transferase
H_2O_2	Hydrogen Peroxide
HA	Hyaluronic Acid
HAIB	Horizontal Anaerobic Immobilized Bioreactors
HBD	Hydrogen Bond Donor
HCB	Hexachlorobenzene
HCH	Hexachlorocyclohexane
HDF	High-Density Fiberboard
HHP	High Hydrostatic Pressure
HHV	Higher Heating Value
HIV-I	Human Immunodeficiency Virus I
Hom6p	Homoserine Dehydrogenase
HRT	Hydraulic Retention Time
HSF1	Heat Shock Transcription Factor
HSR	Heat Shock Response
HSV	Herpes Simplex Virus
HTS	High Throughput Screening
HXK2	Hexokinase Gene

HXT4	Hexose Transporter Gene
ILs	Ionic Liquids
ITS	Intermediate Temperature Stability
K+	Potassium Ion
kDa	KiloDaltons
LC-MS	Liquid Chromatography-Mass Spectrometry
LCL	Long-Chain-Length
LDL	Low Density Lipoprotein
LPS	Lipopolysaccharide
LR-CD's	Large Ring Cyclodextrins
LTTMs	Low Transition Temperature Mixtures
mboe/d	Million barrels of oil equivalent per day
MBR	Membrane Bioreactor
MCB	Monochlorobenzene
MCCs	Microcrystalline Celluloses
MCL	Medium-Chain-Length
MELS	Mannosylerythritol Lipids
MEOR	Microbial Enhanced Oil Recovery
Met22p	Bisphosphate-3′-nucleotidase
Min	Minute
Mm	Millimeter
Mpa	Mega Pascal
MRSA	Methicillin-Resistant *Staphylococcus aureus*
NAD⁺	Aldehyde Dehydrogenase
NADH	Nicotinamide Adenine Dinucleotide
NADP⁺	Nicotinamide Adenine Dinucleotide Phosphate
NAG/(GlcNAc)	N-Acetyl-D-Glucosamine
nm	Nanometer
NMR	Nuclear Magnetic Resonance
O_2	Oxygen
OLE1	Delta(9) Fatty Acid Desaturase
OLR	Organic Loading Rate
OPF	Oil Palm Frond
ORB	Organohalide Respiring Bacteria
Ori	Origin of Replication
P(3HB-*co*-3HV)	3-Hydroxybutyrate and 3-Hydroxyvalerate
PAD1	Phenylacrylic Acid Decarboxylase Gene
PAH	Polycyclic Aromatic Hydrocarbon

PCB	Pentachlorobenzene
PCD	Programmed Cell Death
PCE	Perchloroethene
PCMB	ρ-Chloromercuribenzoate
PCP	Pentachlorophenol
PCR	Polymerase Chain Reaction
PEA	Policy Energy Act
PHA	Polyhydroxyalkonate
PHB	Polyhydroxybutyrate
PLA	Polylactic Acid
PMSF	Phenylmethanelsulfonyl Fluoride
RD	Reductive Dechlorination
RFS	Renewable Fuel Standard
RNA	Ribonucleic Acid
ROS	Reactive Oxygen Species
SAR	Silica-Alumina Ratio
SCL	Short-Chain-Length
SCP	Single Cell Protein
SCV	Small Colony Variants
SDMT	Sarcosine Dimethylglycine Methyl Transferase
SDS-PAGE	Sodium Dodecyl Sulphate-Polyacrylamide Gel Electrophoresis
SEM	Scanning Electron Microscopy
Sti1p	Hsp90 Cochaperone
TAT	Twin Arginine Translocation
TCA	Trichloroethane
TCB	Trichlorobenzene
TCE	Trichlorethene
TCs	Terminal Complexes
TDS	Total Dissolved Solids
TeCB	Tetrachlorobenzene
TEL	Telomeres
TetR	Tetracycline Resistance Gene
TLCK	Tosyl-L-Lysine Chloromethyl Ketone
TLP	Thermolysin Like Protease
TPS1	Trehalose-6-P Synthase
TPS2	Trehalose-6-P Synthase/Phosphatase Complex
UASB	Upflow Anaerobic Sludge Blanket

UMB	Ultramicrobacteria
UNICA	Brazilian Sugarcane Industry Association
US-EPA	United States Environmental Protection Agency
UV	Ultraviolet
VC	Vinyl Chloride
VCX1	Vacuolar H^+/Ca^{2+} Exchanger Gene
VFA	Volatile Fatty Acids
WAS	Waste Activated Sludge
YACs	Yeast Artificial Chromosomes
YCp	Yeast Centromeric Plasmid
Yep	Yeast Episomal Plasmid
Yip	Yeast Integrating Plasmid

PREFACE

Industrial biotechnology, also known as *white biotechnology*, is the application of biotechnology for industrial purposes, technology that makes use of cells and its biomolecules in the manufacturing of various industrially important products. This book covers broad areas such as bioenergy, biomining and biomaterials, including protein engineering, metabolic engineering, synthetic biology, systems biology, and downstream processing. It is evident from the past that industrial biotechnology is a form of a relatively older technology with more advanced scientific research and innovations, particularly in the fields of biosciences and chemical sciences. The importance of industrial biotechnology and the future prospects being realized these days can offer new opportunities for fossil fuel substitution and carbon sequestration, while at the same time providing a more sustainable foundation for the developing world. New biotechnological methods and processes holds a great promise in solving global challenges, including pollution prevention, resource conservation, and cost reduction. As a consequence it yields more viable solutions for our environment by reducing greenhouse gas emissions and helping to fight global warming. Thus, it has added benefits for both our climate and economy. Ultimately, it is worthwhile to mention that advancement in biotechnological research may play a significant role in the success of industrial biotechnology in the near future. It is expected that industrial biotechnology will be increasingly adopted by all production sectors of chemical, pharmaceutical, food, agro- and environment-based industries.

This book aims to deal with recent advancements and technologies of industrial biotechnology and the application of various biomolecules in industrial production, cleaning, and environmental remediation sectors in a comprehensive manner. The book starts with the chapter on the production of exopolysaccharides from halophilic microorganisms, a polymer which is normally very useful in various production sectors of food, pharmaceutical, and petroleum industries. Subsequently, production of antimicrobial compounds from alkaliphilic bacteria, thermophilic actinomycetes, food, agro and pharmaceutical potential and biotechnological applications of biosurfactants, halophiles, cyclodextrin glycosyltransferase, fungal chitinase, proteases, yeasts and yeast products are discussed in the following chapters. In addition, environmental aspects of industrial biotechnology as genetic enhancement for biofuel production, production of biodegradable thermoplastics, advancement in the synthesis of bio-oil, ecofriendly treatment of agrobased lignocelluloses and anaerobic bioreactors for hydrocarbon remediation are also reviewed.

The authors of the individual chapters have been chosen for their renowned expertise and contribution to the various fields of industrial biotechnology. Their willingness to contribute to this book is gratefully acknowledged. The editors are indebted to Mr. Ashish Kumar, Apple Academic Press and AAP staff members of for their foresight, valuable and diligent support throughout the editorial task. This book is an excellent source of innovation, production and application in industrial biotechnology sector. This book is equally very suitable to biology students, chemists, biotechnologists from research institutes, academia and for industry to advance the industrial biotechnology to the future decades for the benefit of mankind and the health of the globe.

—Devarajan Thangadurai, PhD
Jeyabalan Sangeetha, PhD

ABOUT THE EDITORS

Devarajan Thangadurai, PhD

Devarajan Thangadurai is Assistant Professor at Karnatak University in South India. He is also President of the Society for Applied Biotechnology; General Secretary of the Association for the Advancement of Biodiversity Science; and Journal Editor-in-Chief of *Biotechnology, Bioinformatics and Bioengineering*; *Acta Biologica Indica*; *Biodiversity Research International*; and *Asian Journal of Microbiology*. He received his PhD in Botany from Sri Krishnadevaraya University in South India (2003). During 2002–2004, he worked as CSIR Senior Research Fellow with funding from the Ministry of Science and Technology, Government of India. He served as a Postdoctoral Fellow at the University of Madeira (Portugal), University of Delhi (India), and ICAR National Research Centre for Banana (India) from 2004 to 2006. He is the recipient of a Best Young Scientist Award with a Gold Medal from Acharya Nagarjuna University (2003) and the VGST-SMYSR Young Scientist Award of Government of Karnataka, Republic of India (2011). He has edited and authored 15 books, including *Genetic Resources and Biotechnology* (3 vols.), *Genes, Genomes and Genomics* (2 vols.), and *Mycorrhizal Biotechnology*, with publishers of national and international reputation.

Jeyabalan Sangeetha, PhD

Jeyabalan Sangeetha earned a BSc in Microbiology (2001) and PhD in Environmental Sciences (2010) from Bharathidasan University, Tiruchirappalli, Tamil Nadu, India. She holds an MSc in Environmental Sciences (2003) from Bharathiar University, Coimbatore, Tamil Nadu, India. Between 2004 and 2008, she was the recipient of a Tamil Nadu Government Scholarship and the Rajiv Gandhi National Fellowship of University Grants Commission, Government of India, for doctoral studies. She served as a Dr. D.S. Kothari Postdoctoral Fellow and UGC Postdoctoral Fellow at Karnatak University, Dharwad, South India, during 2012-2016 with funding from the University Grants Commission, Government of India, New Delhi. Her research interests are in environmental microbiology and environmental biotechnology and her scientific/community leadership have included serving as editor of an international journal, *Acta Biololgica Indica*. Presently, she is an Assistant Professor in Central University of Kerala at Kasaragod, South India.

CHAPTER 1

EXOPOLYSACCHARIDES OF HALOPHILIC MICROORGANISMS: AN OVERVIEW

PRADNYA P. KANEKAR,[1] SIDDHARTH V. DESHMUKH,[2] SAGAR P. KANEKAR,[3] PRASHANT K. DHAKEPHALKAR,[3] and PRABHAKAR K. RANJEKAR[2]

[1]Department of Biotechnology, Modern College, Pune 411005, Maharashtra, India

[2]Interactive Research School for Health Affairs (IRSHA), Bharati Vidyapeeth Deemed University, Pune Satara Road, Pune, 411043, India

[3]Microbial Sciences Division, MACS-Agharkar Research Institute, G.G. Agarkar Road, Pune, 411004, India

CONTENTS

1.1   INTRODUCTION

Every living organism on the earth finds its own way to protect itself from the surrounding environmental conditions. Microorganisms are not exception to that. One of the mechanisms of microbial cells to protect them from outer environment is production of extracellular polysaccharides also called as exopolysaccharides (EPS). EPS are high molecular weight, carbohydrate-containing biopolymers produced by certain microorganisms exterior to the cell surface as capsule or loosely associated as slime. They are involved in cell adhesion, dehydration, protection of cell from freezing, biofilm formation, pathogenicity and virulence, quorum sensing, storage of carbon sources, etc. They were extensively studied in the past as virulence factor in pathogenic organisms (e.g., *Streptococcus pneumoniae*). EPS are now looked upon more as industrially important macromolecules as gelling agents, antimicrobial compounds, immunomodulatory substances, etc. EPS occur widely specially among microorganisms, microalgae, a few yeasts and fungi, and plants. The diversity of EPS produced by plants is considerably less than that produced by microorganisms. The number of different sugars found in EPS is an indicator of diversity of structure of EPS. In natural environment, many microorganisms occur in aggregates whose structural and functional integrity is based on the presence of a matrix of extracellular polymeric substances. The production of EPS appears to be necessary for their survival (Wingender et al., 1999). The EPS are important in microbial interaction and emulsification of various hydrophobic substances (Maki et al., 2000; Yim et al., 2005).

1.2   STRUCTURE AND COMPOSITION OF EPS

EPs are found to be of two types, viz. homopolysaccharides which contain only one type of monosaccharide and heteropolysaccharides that contain more than one. Many of them are neutral glucans which contain the monosaccharide component D-glucose. Three types of homopolysaccharides are known. In single linkage type, many are neutral glucans, for example, curdlan, some are polyanionic homopolymers and some may contain acyl groups. Under side chain type, scleroglycan is a classical example which has tetrasaccharide-repeating units due to 1,6-β-D-glucosyl-side chain on every third main chain residue. In branched types, dextrans composed entirely of 2-linked glycosyl residues are classified (Jankins and Hall, 1997).

Microbial heteropolysaccharides are mostly composed of repeating units of 2–8 monosaccharides. The unit primarily contains D-glucuronic acid and short side-chains of 1 to 4 residues. The side chains may vary in different heteropolysaccharides. Xanthan is the classical example of heteropolysaccharide. The anionic nature of Xanthan is due to presence of glucuronic acid and pyruvate (Jankins and Hall, 1997). The unique physical properties of microbial EPS are based upon their molecular conformation which is decided by the primary structure and from associations between molecules in solution. The shape of EPS is determined by the angle of bonds which governs the relative orientations of adjacent sugar residues in the chair. EPS in solution have single, double or triple helical conformation in an orderly manner. Xanthan has a double or triple helix. These are stabilized by intermolecular hydrogen bonds. The helical conformation makes the EPS semi rigid. The molecules can move large volumes of solution which overlap even at low concentrations of EPS leading to relatively highly viscosity. The interaction between molecules stabilizes the helical structure and influences the properties of EPS in solution such as solubility, viscosity and gel formation. The insolubility of EPS is due to a strong interaction between molecules while a poor interaction leads to solubility of EPS. The side chains influence interaction between molecules. By introduction of a 3-monosaccharide side chain into the cellulose chain, soluble Xanthan is obtained although cellulose is insoluble (Jankins and Hall, 1997).

EPS are primarily composed of carbohydrates. Different types of sugars are found in microbial EPS, for example, D-glucose, D-galactose, D-mannose, L-rhamnose and L-fucose. Pentoses like D-xylose and D-ribose occur rarely. The presence of uronic acids in EPS impart them polyanionic nature. Microbial EPS contain pyruvate ketals and different ester-linked organic substituents. Anionic nature of EPS is attributed to pyruvate ketals which are generally present in stoichiometric ratios with the carbohydrate component. Pyruvate is normally attached to the neutral hexoses. Acetate is the common ester-linked component of EPS while propionate, glycerate, succinate may occur sometimes. The presence of organic acid substituents in EPS increases lipophilicity of the molecule. Various amino acids have also been found in EPS, for example, serine and L-glutamic acid. Some EPS contain inorganic substituents like phosphate (in EPS of *Escherichia coli*) and sulphate (in EPS of cyanobacteria) (Jankins and Hall, 1997).

1.3   PROPERTIES OF EPS

In case of microbial EPS, mostly rheological properties like viscosity, gel formation and stability are studied. EPS show pseudoplastic flow in solution, which is also known as shear thinning. When solutions of EPS are sheared, the molecules align in the shear field leading to reduction in the effective viscosity which is not due to degradation. The viscosity recovers immediately when the shear rate is decreased. The combined viscous and elastic behavior recognized as viscoelasticity differentiates microbial viscosifiers from solutions of other thickeners. Xanthan gum, Succinoglycan, Scleroglucan are the examples of viscosifiers. The viscosity of a solution of EPS is defined as a function of the shear rate. Although the rate of flow may be very low in very viscous solutions, there is no yield stress. Some EPS exhibit yield stress characteristics. The pseudoplastic flow behavior and elasticity are important characteristics of microbial EPS. High viscosity at low shear rate is essential to inhibit particle sedimentation or to form a film of EPS whereas reduced viscosity under conditions of high shear rate maintains good pumping and spraying characteristics (Jankins and Hall, 1997).

1.4   ROLE AND FUNCTION OF EPS

EPS play many important roles and functions based upon their structure and composition. Production of EPS enhances growth and survival of microbes and the communities in which they live (Wolfaardt et al.,

1999). EPS strengthens ability of microorganisms to compete and survive in altered environmental conditions by making changes in physical and biogeochemical microenvironment surrounding the microbial cells (Costerton, 1974). EPS are essential in the production of aggregate (Harris and Mitchell, 1973; Biddanda, 1985; Alldredge and Silver, 1988), biofilm formation (Sutherland, 1999; Sutherland, 2001), providing protection (Bitton and Friehofer, 1978; Decho and Lopex, 1993) and adhesion to surfaces (Fletcher and Floodgate, 1973; Paerl, 1975, 1976; Marshall, 1985; Vincent et al., 1994; Holmstrom and Kjelleberg, 1999). Decho (1990) and Wolfaardt et al. (1999) have reviewed role of EPS in ocean. Nichols et al. (2005) have reviewed bacterial EPS from extreme marine environments, their biosynthesis, structure-function relationship, role in marine environment, etc. EPS have significant role in biofilm formation by virtue of their properties like adhesion, water retention, cohesion of biofilms, sorption of organic compounds and inorganic ions, binding of enzymes, export of cell components, etc. They serve as nutrient source and protective barrier (Nwodo et al., 2012).

1.5 MICROORGANISMS PRODUCING EPS

Microorganisms synthesizing EPS widely occur in nature. Patel and Prajapati (2013) have reviewed applications of EPS produced by lactic acid bacteria. Dextran is the first industrial polysaccharide produced by *Leuconostoc mesenteroides* (Crescenzi, 1995) while xanthan gum produced by plant pathogen, *Xanthomonas campestris* is the second microbial EPS approved for use in food (Sutherland, 1998). Nwodo et al. (2012) have reviewed functionality of capsular polysaccharides as capsular antigens of human pathogens *Escherichia coli, Haemophilus influenza, Neumococcus meningitides, Klebsiella pneumonia, Streptococcus pneumonia, Staphylococcus aureus* and *Pseudonomonas aeruginosa* in biofilms associated with cystic fibrosis pneumonia. Agriculturally important bacteria like *Azotobacter vinelandii* (Pindar and Bucke, 1975) and *Rhizobium* sp. (Duta et al., 2006) have been studied extensively for their exopolysaccharides. Several *Bacillus* species have been reported for production of EPS (Patil et al., 2009; Sayem et al., 2011; Orsod et al., 2012; Chen et al., 2013; Razack et al., 2013). Kanekar et al. (2014) have reviewed alkaliphilic bacteria for production of biomolecules including EPS. Joshi and Kanekar (2011) have described alkaliphilic strain of *Vagococcus carniphilus* producing EPS. Bales et al. (2013) have reported extraction and purification of EPS from human pathogenic bacteria namely *Staphylococcus*, *Klebsiella*, *Acinetobacter*, *Pseudomonas* and *Escherichia*. The information is compiled in Table 1.1.

1.6   EXTENSIVELY STUDIED MICROBIAL EPS

A variety of EPS are produced by microorganisms. These include xanthan, dextran, gellan, curdlan, alginate, etc. Xanthan is popularly known as 'xanthan

TABLE 1.1   Microorganisms Known to Produce EPS

Name of the organism	References
Streptococcus thermophilus	Levander et al. (2001); Broadbent et al. (2003)
Bacillus cereus	Orsod et al. (2012)
Brachybacterium sp.	Orsod et al. (2012)
Bacillus licheniformis	Sayem et al. (2011)
Bacillus subtilis	Patil et al. (2009)
Bacillus amyloliquefaciens	Chen et al. (2013)
Gluconacetobacter hansenii	Valepyn et al. (2012)
Rhizobium sp.	Duta et al. (2006)
Bacillus subtilis	Razack et al. (2013)
Lactococcus lactis ssp. *cremoris* strains	Yang et al. (1999)
Lactobacillus delbrueckii subsp. *bulgaricus* 291	Faber et al. (2001)
Lactobacillus kefiranofaciens WT-2 B (T)	Maeda et al. (2004)
Bifidobacterium sp.	Salazar et al. (2008)
Pediococcus parvulus 2.6	De Palencia et al. (2009)
Alteromonas infernus	Colliec-Jouault et al. (2001)
Leuconostoc mesenteroides	Crescenzi (1995)
Xanthomonas campestris	Sutherland (1998)
Azotobacter vinelandii	Pindar and Bucke, 1975
Escherichia coli	Silver et al. (1998)
Vagococcus carniphilus	Joshi and Kanekar (2011)
Salmonella sp.	Crawford et al. (2008)
Pseudomonas aeruginosa	Matsukawa and Greenberg (2004)
Staphylococcus aureus	Cramton et al. (1999)
Klebsiella pneumonia	Elasser-Beile et al. (1978)

gum' and is produced by plant pathogenic bacterium *Xanthomonas campestris*. The primary structure of xanthan consists of a repeating unit of β-D-glucose. It has a β-(1–4)-linked glucan main chain with alternating residues substituted on the 3-position with a trisaccharide chain containing two mannose and one glucuronic acid residue making it a charged polymer. This EPS is a large molecule with molecular weight Mw in the range ~6 × 10^6 g/mole. Xanthan gives very viscous solution making it ideal as a thickening and suspending agent. Xanthan has been approved by FDA as a food grade product. It is used in sauces, syrups and salad dressings. Dextran is produced by lactic acid bacterium namely *Leuconostoc mesenteroides*. It is a polymer of mostly α (1–6) D-glucopyranosyl unit and of molecular weight Mw in the range ~ 1 × 10^6 g/mole. With reduced molecular weight as 6 × 10^4 g/mole, dextran is used in medicine as a blood extender. Its solubility in both aqueous and non-aqueous media makes dextran a good membrane former using electrospinning technique. Gellan is produced by *Pseudomonas elodea* and has a linear structure with repeating unit of tetrasaccharide each with one carboxyl group and one acetyl group in the native state. Gellan is one of the many film-forming polysaccharides likely to be used in implants for insulin in the treatment of diabetes.

Curdlan is an EPS with a repeat unit of three β-(1–3)-linked D-glucose residues and is not charged. It is produced mostly in Japan using mutant strains of *Agrobacterium*. It has several applications in food industry. Bacterial Hyaluronic acid (HA) is produced by *Staphylococcus* and *Streptococcus* sp. and has high molar mass (1–3 × 10^6 g/mole). HA is a linear polymer consisting of alternate units of β-(1–4) N-acetyl-D-glucosamine and β-(1–3) D-glucuronic acid. HA has been shown to have mucoadhesive and anti-inflammatory effects.

EPS available in market are produced, for example, dextran by lactic acid bacteria, xanthan by *Xanthomonas campestris*, gellan by *Pseudomonas elodea*, hyaluronic acid by *Staphylococcus* and *Streptococcus* sp. Some of these organisms are pathogenic while in case of lactic acid bacteria, the yield is low To overcome these problems, exploration of other microorganisms continues globally. Extremophilic microorganisms including halophiles are considered new sources of bioactive molecules. Halophiles find important place in biotechnology.

1.7   HALOPHILES

Halophiles are the salt loving microorganisms that thrive in hypersaline environments originated mostly by evaporation of seawater. They include both prokaryotic and eukaryotic microorganisms having ability to balance the osmotic pressure of the surrounding environment and resist denaturing effects of salt. Their metabolic types represent oxygenic and anoxygenic phototrophs, aerobic heterotrophs, sulfate and nitrate reducing organisms, etc. They are classified based upon their requirement of salt as halotolerant (0%, salt tolerance upto 15%), slight halophiles (2–5%), moderate halophiles (5–15%) and extreme halophiles (20–30%) (Kushner, 1993). Moderately halophilic bacteria constitute a heterogenous physiological group of microorganisms belonging to different genera while extreme halophiles are the archaea belonging to family Halobacteriaceae. Halotolerant organisms grow in absence of salt as well as in high salinity condition. Halophilic microorganisms have been isolated from environments including aquatic habitats of both high-and low salinity, salty marshy places, saline lakes (Javor, 1989), from Dead Sea (Oren, 2002). They are reported from ancient salt sediments as reviewed by Grant et al. (1998) and McGenity et al. (2000). *Halococcus salifodinae* was the first strain isolated from ancient rock salt (Permian salt sediment) by Denner et al. (1994). Haloarchaea were further reported from ancient rock salt (Stan Lotter et al., 1999; Stan Lotter et al., 2002; Vreeland et al., 2002; Gruber et al., 2004). Halophilic microorganisms have been isolated from a number of hypersaline ecosystems, for example, the Great Salt Lake, Utah, USA, the Dead Sea, the extremely alkaline brines of the Wadi Natrun, Egypt and Lake Magadi, Kenya (Oren, 1994; Kamekura, 1998).

Halophilic microorganisms are cultivated using different nutrient media containing high concentration of salt. Various techniques like SDS gel electrophoresis. 16S rRNA gene sequencing, BIOLOG system, DGGE, etc. were used for isolation and characterization of halophiles. Ventosa et al. (1998) have made in-depth review of halophilic aerobic bacteria with respect to their habitats, taxonomy, amazing physiological features, biotechnological potential in production of enzymes, polymers, etc. They have described a number of species of *Halomonas*, other moderately halophilic bacteria namely *Salinivibrio costicola, Pseudomonas halophila, Flavobacterium gondwanense, Arhodomonas aquaeolei*, etc.

Taxonomic, physiological and biotechnological aspects of halophiles have also been investigated by Das Sarma and Das Sarma (2012). Production of EPS using halophilic bacteria from extreme marine habitats and their biological activities have been reviewed by Poli et al. (2010). Kanekar et al. (2012) have reviewed literature on taxonomy, diversity, physiology and applications of halophiles. They have also compiled information on new reports of halophilic microorganisms from different saline environments during the last decade. Oren (2013) has predicted and discussed on the occurrence of acidic proteomes in halophiles of different physiology and phylogenetic affiliation. Indeed, halophilic microorganisms attract researchers all over the world by virtue of their amazing features, unusual physiology and prospective biotechnological potential.

1.8 EXOPOLYSACCHARIDES (EPS) PRODUCED BY HALOPHILIC MICROORGANISMS

There occurs a wide diversity in the EPs produced by halophiles. Ruiz-Ruiz et al. (2011) have studied sulfated EPS by a novel halophilic bacterium *Halomonas stenophila* strain B100. The EPS when over sulfated exhibited antitumoural activity on T cell lines derived from acute lymphoblastic leukemia. The tumor cells were susceptible to apoptosis induced by the sulfated EPS while T cells were resistant. The authors predict the possible use of EPS from halophiles in search of new antineoplastic drugs. Bacterial exopolymers may be of great value in enhanced oil recovery processes because of their surfactant activity and bioemulsifying properties. Since the conditions existing in oil deposits are often saline to hypersaline, use of salt resistant surfactants may be advantageous. Pfeiffer et al. (1986) isolated over 200 bacterial strains from oil well, soils in the vicinity of oil wells, brine injection water, etc., which were found to produce EPS anaerobically in a sucrose mineral salt medium with 10% NaCl. The EPS showed pseudoplastic behavior, e.g., showing a decreased viscosity with increase in shear rate and was resistant to shear and thermal degradation. The polymer is a charged heteropolysaccharide. Isolates of *Halomonas* from Morocco were found to produce EPS having applications as emulsifiers in oil industry. The strains were further identified as *Halomonas maura* sp. nov. (Bouchotroch et al., 2001). Mancuso et al. (2004) have reported production of EPS by Antarctic marine bacteria.

Satpute et al. (2010) have reviewed work on EPS of marine microorganisms, for example, *Bacillus*, *Halomonas*, *Planococcus*, *Enterobacter*, *Alteromonas* and cyanobacteria. *Halomonas eurihalina* strain F2–7 was found to produce large amounts of polyanionic EPS. The polymer is a potent emulsifier and shows pseudoplastic behavior. The EPS F2–7 may have a range of applications in pharmaceuticals, food industry and biodegradation (Quesada et al., 1993; Calvo et al., 1995; Bejar et al., 1998). The polymer also has immunomodulatory activity *in vitro*. It enhances the proliferative effect of human lymphocytes as a response to the presence of anti-CD3 monoclonal antibody in blood (Perez-Fernandez et al., 2000). EPS produced by *Halomonas ventosae* and *Halomonas anticariensis* were found to form solutions of low viscosity and have pseudoplastic behavior (Mata et al., 2006). They also have capacity to bind with cations suggesting their applications in bioremediation. Carmen et al. (2008) isolated moderately halophilic bacteria from saline soils in Spain. The organisms were designated as *Halomonas cerina* sp. nov. and found to produce EPS. Poli et al. (2007) have reported *Halomonas alkaliantarctica* strain CRSS from Salt lake in Cape Russell in Antarctica which produces EPS having high viscosity and different chemical composition on various substrates utilized as nutrients.

Halophilic cyanobacteria namely *Aphanothece halophytica*, *Aphanocapsa halophytica*, *Cyanothece*, etc. have been investigated to produce interesting EPS (De Philippis et al., 1993; Sudo et al., 1995; Morris et al., 2001). The EPS of *Aphanothece* is found to have Xanthan like properties indicating its biotechnological potential. EPS produced by *Pseudoalteromonas* strain 721 isolated from deep sea hydrothermal vent is composed of octasaccharide repeating unit with two side chains and exhibits gelling properties (Rougeaux et al., 1999; Guezennec, 2002). *Pseudoalteromonas* strain SM 9913 was isolated from deep-sea sediment in the Bohai Gulf, gulf of the Yellow sea, China by Qin et al. (2007). The organism produced EPS having linear arrangement of α-(1–6) linkage of glucose with a high degree of acetylation. Its flocculation behavior and biosorption capacity was further investigated by Li et al. (2008).

Alteromonas macleodii subsp. *figiensis* isolated from deep sea hydrothermal vent in North Fijian Basin is reported to produce EPS which is sulfated heteropolysaccharide and has high uronic acids with pyruvate. The repeating unit of the EPS is a branched hexasaccharide containing glucose, mannose, galactose, glucuronic acid, galacturonic acid and pyruvated mannose (Costaouec et al., 2012). The EPS has wide range of applications such as thickening agent in food processing industry, bioremediation of hazardous waste waters, bone healing and treatment of cardiovascular diseases (Raguenes et al., 1996; Rougeaux et al., 1998; Colliec et al., 2001). EPS produced by *Geobacillus* sp. isolated from sediment in marine hot spring near the sea shore of Maronti, Ischia Island, Italy, has a pentasaccharide repeating unit, two of them with a gluco-galacto configuration and three with a mannose configuration. The EPS has pharmaceutical application (Nicolaus et al., 2002).

Bacillus licheniformis isolated from water of a shallow marine hot spring, Vulcano Island, Italy was reported by Maugeri et al. (2002) to produce a novel EPS having mannose as the main monosaccharide, tetrasaccharide repeating unit and a manno-pyranosidic configuration. The EPS was found to have antiviral and immunoregulatory effect (Arena et al., 2006). Marx et al. (2009) have reported EPS producing halophilic organism *Colwella psychrerythraea* strain 34 H from Arctic marine sediments. The EPS has role in cryoprotection. A marine bacterium, *Hahella chejuensis* was investigated by Lee et al. (2001) from marine sediment sample collected from Marado, Cheju Island, Republic of Korea. The organism produced EPS named as EPS-R having glucose:galactose in the proportion 0.68:1.0. The EPS-R acts as biosurfactant and thus has application in detoxification of polluted areas from petrochemical oils.

Among the haloarchaea, *Haloferax* and *Haloarcula* are studied extensively for production of EPS. *Haloferax mediterranei* was the first haloarchaeon to be reported for EPS. The organism is isolated from Mediterranean Sea. The EPS has complex chemical composition and can be used in oil recovery especially in oil deposits with high salinity (Anton et al., 1988; Parolis et al., 1996). Acidic EPS produced by *Haloferax denitrificans* has been described by Parolis et al. (1999) and EPS of *Haloferax gibbonsii* by Paramonov et al. (1998). Nicolaus et al. (1999) have described EPS producing *Haloarcula japonica* strain T5, *Haloarcula* sp. strain T6 and strain T7. The information on halophiles producing EPS is summarized in Table 1.2.

TABLE 1.2 Halophilic Microorganisms Reported to Produce EPS

Name of the organism	Source	References
Bacillus strain B3–15	Marine hot spring (Eolian Islands, Italy)	Maugeri et al. (2002)
Bacillus strain B3–72	Marine hot spring (Eolian Island, Italy)	Nicolaus et al. (2000)
Bacillus licheniformis	Volcano island	Arena (2004)
Bacillus thermoantarcticus	Sea sand in Ischia island	Nicolaus et al. (2000)
Bacillus licheniformis	Marine hot spring of volcano Island, Italy	Arena et al. (2006)
Planococcus maitriensis	Coastal sea water of Bhavnagar, India	Kumar et al. (2007)
Enterobacter cloacae	Marine sediment from Gujarat coast, India	Iyer et al. (2005)
Alteromonas sp.	*Alvinella pompejana* of deep sea hydrothermal vent of East Pacific Rise	Vincent et al. (1994)
Alteromonas sp.	South sea of Korea	Yim et al. (2005)
Alteromonas haloplanktis	Marine	Gorshkova et al. (1993)
Pseudoalteromonas sp.	Sea water and sea ice in Southern Ocean	Nichols et al. (2004)
Pseudomonas sp. strain S9	Marine	Wrangstadh et al. (1990)
Rhodococcus erythropolis	Marine	Urai et al. (2007)
Zoogloea sp.	Marine	Kwon et al. (1994)
Cyanothece strain 16S om2	Somaliland salt pan	De Philippis et al. (1993)
Cyanothece sp. 113	Salt lakes in China	Chi et al. (2007)
Vibrio alginolytics	Marine	Jayaraman and Seetharaman (2003)
Desulfovibrio sp.	Marine	Zinkevich et al. (1996)
Geobacillus thermodenitrificans	Vent of Vulcano Island, Italy	Arena, 2009
Halomonas maura		Arias et al. (2003)
Halomonas almeriensis	Marine	Llamas et al. (2012)
Aphanothece halophytica	Guangrao Salt work, China	Morris et al. (2001)
Halomonas smyrnensis	Saline pond soil, Turkey	Poli et al. (2013)
Alteromonas macleodii	Deep sea hydrothermal vent	Costaouec et al. (2012)
Pantoea sp.	Sea sediment	Silvi et al. (2013)
Haloferax mediterranei	Marine solar saltern	Anton et al. (1988); Parolis et al. (1996)
Haloferax gibbonsii	Marine solar saltern	Paramonov et al. (1998)
Haloferax denitrificans	Marine solar saltern	Parolis et al. (1999)
Haloarcula japonicum	Marine	Nicolaus et al. (1999)

1.9   BIOSYNTHESIS OF EPS

It is interesting to study how the microorganisms synthesize and secrete EPS. Nichols et al. (2005) have reviewed biosynthesis of EPS in marine bacteria. When the substrate enters the microbial cell either as it is or after phosphorylation, the biosynthesis of EPS is initiated (Sutherland, 1977). With the help of activated precursors and carrier molecules, EPS is synthesized near the cytoplasmic membrane. In many microorganisms, for production of a precursor of EPS, uridinediphosphate glucose phosphorylase is the main enzyme involved. Sutherland (1977, 1982) has described in detail the synthesis of EPS. The construction of the repeating units is dependent on the transfer of the appropriate monosaccharides from sugar nucleotides to a carrier lipid-isoprenoid alcohol phosphate. After polymerization, the polysaccharide chain may be hydrolyzed from the isoprenoid carrier lipid by a very specific enzyme to form EPS. Simultaneously, the polysaccharide is transported through the inner and outer membranes.

The building blocks for EPS are usually sugar nucleotide diphosphates with some monophosphates. The sugar nucleotide diphosphates provide energy for the synthesis of the oligosaccharides and are readily interconverted. The energy released from the sugar nucleotide diphosphates is greater than that released by sugar monophosphates and provides 70% of the energy requirement for the synthesis (Jankins and Hall, 1997). Other groups are added to modify the basic structure of the polysaccharide, for example, acetate, pyruvate, succinate, phosphates and sulfate. Acetyl CoA is the likely source of acetyl groups in most of the EPS. Biosynthesis of heteropolysaccharides involves synthesis of the sugar nucleotide diphosphates, assembly of the repeat unit on lipid carrier, pyruvylation and addition of other substitutes and transfer of the polysaccharide to the new subunit. Thus formed EPS is then transported to the cell surface through membrane adhesion zones (Jankins and Hall, 1997).

1.10   PRODUCTION OF EPS

The structure, composition and yield of EPS vary with the microbial strains and the fermentation conditions. Microorganisms differ in critical factors like carbon and nitrogen sources, temperature, pH, requirement of minerals, etc. for production of EPS. Kumar et al. (2007) have stated that that the nutritional and environmental conditions influence the yield and quality of EPS. When marine bacteria were grown on limited nutrients like nitrogen, phosphorous, sulfur and potassium, the yield of EPS increased (Sutherland, 1982). The selection of carbon source is an important step in optimization of production of EPS. Lee et al. (2001) found that a marine organism *Hahella chejuensis* isolated from Cheju Island, Republic of Korea, produced the highest EPS when the growth medium was amended with sucrose. *Halomonas alkaliantartica* strain CRSS, an haloalkaliphilic bacterium isolated from saline lake in Antartica produced maximum EPS with acetate as the carbon source (Poli et al., 2007). To improve yield of EPS by an *Alteromonas* strain 1644, isolated from deep-sea hydrothermal vent, Samain et al. (1997) employed fed-batch method and nitrogen limitation by using ammonium chloride, an inorganic nitrogen source.

Sutherland (1982) has suggested that production of EPS is not growth associated. Most of the bacteria secrete the highest amount of EPS in the stationary phase, for example, *Alteromonas* strain 1644 (Samain et al., 1997); *Geobacillus*, a thermophilic bacterium from sea sand in Ischia Island, Italy (Schiano et al., 2003); *Bacillus licheniformis* from marine hot spring at Vulvano Island (Maugeri et al., 2002). In *Alteromonas macleodii* subsp. *fijiensis* isolated from a deep sea vent, the production of EPS initiated at the end of exponential phase and continued throughout the stationary phase (Raguenes et al., 1996). Poli et al. (2010) have described rheology of the growth media as an important parameter in EPS production. The culture broth develops non-Newtonian characteristics acting as a pseudoplastic fluid where the viscosity decreases with the increasing shear rate. Such change may be due to presence of EPS, their metabolic products as well as the lack of homogeneity in terms of mixing, mass and oxygen. Under these conditions, the organisms may produce heterogenous EPS in terms of molecular weight, etc. The rheological shifts during the fermentation process may be used as a parameter to monitor the consistency in quality and yield of EPS produced. The rate of aeration also can influence production of EPS. Lee et al. (2001) reported that high aeration rates enhanced production of EPS and increased the viscosity of the culture broth of *Hahella chejuensis*. Thus, the production of EPS by microorganisms is influenced by a variety of process parameters.

1.11   APPLICATIONS OF EPS

Microbial EPS with their unique properties are exploited for a wide range of industrial applications, as biomedical agents and remedy for environmental pollution (Sutherland, 1990; Sutherland, 1998; Tombs and Harding, 1998). Among the industrial applications, EPS find important place in food industry and petroleum oil industry.

1.11.1   FOOD INDUSTRY

EPS are used to alter the food texture by virtue of their thickening or gelling property. The gellan type of EPS are used as adhesive in icing and glazes; as binding agents in pet foods; as gelling agent in pastry, filling, jelly; as thickening agent in jams and sauces and as stabilizer in ice cream. The alginate type of EPS are used as binding agents in pet foods; as emulsifying agent in salad dressing; as inhibitor of crystal formation in frozen foods; as stabilizer in ice cream; as thickening agent in jams and sauces and as foam stabilizer in beer and dough. Gellan is stable to heat and gives a very clear, thermoreversible gel which sets at lower concentrations and more rapidly than most other EPS. Gellan also has superior flavor release than most other EPS. Alginates form non-thermoreversible gel and are useful when shape retention on heating is desired (Jenkins and Hall, 1997).

1.11.2   OIL RECOVERY IN PETROLEUM INDUSTRY

EPS are of great value as enhancers of oil recovery by virtue of their surfactant activity and bioemulsifying properties. Since the oil deposits occur in saline conditions, EPS of salt resistant bacteria may be advantageous. Poli et al. (2010) have reviewed biological activities of EPS produced by halophilic bacteria of marine origin. The extremely halophilic archaeon *Haloferax mediterranei* was the first haloarchaeon to be reported for production of EPS having a high viscosity at low concentration, excellent rheological properties and resistance to extreme pH and temperature. The extreme salt tolerance of the organism and its EPS showed that the organism is valuable candidate for the recovery of oil, especially in oil deposits with high salinity (Anton et al., 1988; Poli et al., 2010). Among the halophilic bacteria, *Halomonas* has been the potential candidate for production of EPS having efficient emulsifying activity. EPS of *Halomonas alkalian-tartica* (Bouchotroch et al., 2001; Poli et al., 2007), *H. ventosae, H. anti-cariensis* (Bouchtroch et al., 2001; Mata et al., 2006), *H. maura* (Arias et al., 2003) has been proposed to have role as emulsifying agent in oil recovery.

1.11.3   BIOMEDICAL APPLICATIONS

Microbial EPS can be exploited as immunogenic, anti-tumor and anti-viral agents and for their functional properties based on rheology and gel formation. The extensively studied EPS, Dextrans are explored as plasma substitute. About 6% dextran with 50,000–100,000 relative molecular weight has equivalent viscosity and colloid-osmotic properties to blood plasma. Dextran can also be used as non-irritant absorbent wound dressings (Jenkins and Hall, 1997). EPS are used in lotions and in encapsulated drugs because of their gel formation property. The EPS curdlan exhibits antitumor activity. The polysaccharides can be used for preparation of vaccines but pose some practical problems, for example, poor immune response elicited by polysaccharide antigens. This can be overcome by chemical modification. The antigenicity of the EPS can be increased by coupling with proteins (Jenkins and Hall, 1997). The EPS of *Halomonas maura* has been found to work as an immunomodulatory (Bejar et al., 1998; Arias et al., 2003); *H. eurihalina* also has immunomodulating activity (Quesda et al., 1993; Perez-Fernandez et al., 2000). EPS of *H. stenophila* has been shown to have antitumor activity (Ruiz-Ruiz et al., 2011). The EPS produced by *Bacillus thermodenitrificans* has been reported to have immunomodulatory and antiviral activity (Arena et al., 2006). Antiviral and immunoregulatory activity is also observed in EPS of *Bacillus licheniformis* (Arena et al., 2006). A wide range of applications as thickening agent in blood processing industry, bone healing, treatment of cardiovascular diseases are reported for EPS of *Alteromonas macleodii* subsp. *fijiensis* (Raguenes et al., 1996; Rougeaux et al., 1998).

1.11.4   BIOREMEDIATION OF ENVIRONMENTAL POLLUTANTS

The ability of EPS to accumulate metals has been known (Bitton and Friehofer, 1978; Brown and Lester, 1979; Loaec et al., 1997, 1998). By the virtue of metal biosorption capacity, the EPS could be employed in heavy metal bioremediation (Brown and Lester, 1979). At the pH of ambient temperature (pH around 8), anionically charged EPS can remove over 99% of Zn and Ag (Harvey and Luoma, 1985). Complexation of EPS with trace metals may impact on availability of these micronutrients to microorganisms in marine environment. This phenomenon may be important in the downward transport of trace metals and micronutrients in the ocean (Decho, 1990). EPS of *Halomonas almeriensis* finds application as biodetoxifier and in metal chelation (Llamas et al., 2012), *H. stenophila* and *Planococcus maitriensis* in bioremediation (Kumar et al., 2007; Llamas et al., 2011). The EPS produced by *Alteromonas macleodii* subsp. *fijiensis* has been reported for biodetoxification of waste waters (Raguenes et al., 1996; Rougeaux et al., 1998), *Pseudoalteromonas* strain SM9913 for flocculation and biosorption (Qin et al., 2007; Li et al., 2008). Various applications of EPS produced by halophilic bacteria and archaea have been described in Table 1.3. One can see through the table a vast range of applications of EPS produced by halophiles. Indeed, an array of biological activities is exhibited by halophilic microorganisms which could be explored for their applications in industries, biomedical field and in environmental pollution control.

1.12 CONCLUSION AND FUTURE PROSPECTS

A lot of research has been carried out on EPS of halophilic microorganisms. In extreme environments, EPS mainly serves for survival of the halophiles. Due to large diversity in the structure, composition and other properties, EPS from halophiles show biotechnological promise. They have wide range of applications, for example, textile, dairy, cosmetic industries, medicine and pharmaceuticals and bioremediation covering industry, health and environment sectors. Since the EPS produced on commercial scale are from pathogenic organisms, for example, *Streptococcus*, *Xanthomonas*, *Haemophilus*, etc., exploring halophilic microorganisms would be a novel approach. EPS produced by *Alteromonas* sp. has been extensively studied and should be further developed for commercialization. Likewise, EPS from haloarchaea, for example, *Haloferax* sp., *Haloarcula* also possess enormous potential. Production of EPS using lactic acid bacteria suffers

TABLE 1.3   Applications of EPS Produced by Halophilic Microorganisms

Organism	Application	References
Haloferax mediterranei	Oil recovery from oil deposits with high salinity	Anton et al. (1988); Parolis et al. (1996)
Halomonas alkaliantarctica	Oil recovery	Poli et al. (2007); Bouchotroch et al. (2001)
Halomonas almeriensis	Bio-detoxifier emulsifier, heavy metal chelation	Llamas et al. (2012)
Halomonas eurihalina	Immunomodulating activity, emulsifier	Quesada et al. (1993); Perez-Fernandez et al. (2000); Davis (1974)
Halomonas stenophila	Gelling agent, bioremediation, antitumour activity	Llamas et al. (2011); Bouchotroch et al. (2001); Ruiz-Ruiz et al. (2011)
Halomonas ventosae	Emulsifier	Mata et al. (2006); Bouchotroch et al. (2001)
Halomonas anticariensis	Emulsifying agent	Mata et al. (2006); Bouchotroch et al. (2001)
Halomonas maura	Emulsifier in oil industry	Arias et al. (2003)
Alteromonas infernus	Anticoagulant	Colliec et al. (2001)
Hahella chejuensis	Biosurfactant	Lee et al. (2001)
Planococcus maitriensis	Emulsifying agent, bioremediation	Kumar et al. (2007)
Bacillus thermdenitrificans	Immunomodulatory and antiviral	Arena et al. (2009)
Bacillus licheniformis	Antiviral	Maugeri et al. (2002); Arena et al. (2006)
Geobacillus sp.	Pharmaceutical	Nicolaus et al. (2002)
Alteromonas macleodii subsp. *fijiensis*	Thickening agent in blood processing industry, bone healing, treatment of cardiovascular diseases, biodetoxification of waste waters	Raguenes et al. (1996); Rougeaux et al. (1998); Bouchotroch et al. (2001); Colliec et al. (2001)
Pseudoalteromonas sp.	Gelling agent	Rougeaux et al. (1999); Bouchotroch et al. (2001); Guezennec (2002)
Pseudoalteromonas strain SM 9913	Flocculation and biosorption	Qin et al. (2007); Bouchotroch et al. (2001); Li et al. (2008)
Colwellia psychrerythraea	Cryoprotection	Marx et al. (2009)

from low yield, expensive nutrient media and specific growth conditions. Hence search for microorganisms that would overcome these problems is required. Extremophilic microorganisms including halophiles are looked upon as new source of bioactive molecules, for example, EPS, enzymes, etc. Initially applications of EPS were restricted to as gelling and flocculating agents and as emulsifiers. New dimensions of applications of EPS especially from halophiles encompass biomedical, environmental and health areas. Several EPS have exhibited antiviral, antitumor, immunomodulatory activities which can be further exploited. Environmental pollution due to toxic heavy metals is known since long and EPS has demonstrated its potential in bioremediation of toxic waste waters containing metals. The demand of hydrocarbon oil continues and EPS from halophiles is a boon to recovery of oil from deep sea sediments. For use of EPS in food, health and biomedical areas, toxicity testing of EPS becomes necessary. They have to pass several animal trials followed by human subjects. A lot of research is therefore needed in transforming the investigations of new microbial resources to commercial production of EPS and their applications.

KEYWORDS

- **Antitumor**
- **Antiviral**
- **Biofilms**
- **Bioremediation**
- **Emulsifier**
- **Exopolysaccharides**
- **Gelling agents**
- **Halophiles**
- **Immunomodulatory**
- **Immunoregulatory**
- **Metal biosorption**
- **Oil recovery**

REFERENCES

Alldredge, A., Silver, M. V. Characteristics, dynamics and significance of marine snow. *Prog Oceanogr* 1988, 20, 41–82.

Anton, J., Meseguer, I., Rodriguez-Valera, F. Production of an extracellular polysaccharide by *Haloferax mediterranei*. *Appl Environ Microbiol* 1988, 54(10), 2381–2386.

Arena, A., Pavone, B., Gugliandolo, C., Maugeri, T. L., Bisignano G., Messina, I. Exopolysaccharides from marine thermophilic bacilli induce a Thl cytokine profile in human PBMC. *Clin Microbiol Infect* 2004, 10(3), 366.

Arena, A., Gugliandolo, C., Stassi, G., Pavone, B., Iannello, D., Bisiqnano, G. An exopolysaccharide produced by *Geobacillus thermodenitrificans* strain B 3–72: antiviral activity on immunocompetant cells. *Immunol Lett* 2009, 123(2), 132–137.

Arena, A., Maugeri, T. L., Pavone, B., Iannello, D., Gugliandolo, C., Bisiqnano, G. Antiviral and immunoregulatory effect of a novel exopolysaccharide from a marine thermotolerant *Bacillus licheniformis*. *Int Immunopharmacol* 2006, 6(1), 8–13.

Arias, S., del Moral, A., Ferrer, M. R., Tallon, R., Quesada, E., Bejar, V. M. An exopolysaccharide produced by the halophilic bacterium *Halomonas maura*, with a novel composition and interesting properties for biotechnology. *Extremophiles* 2003, 7, 319–326.

Bales, P. M., Renke, E. M., May, S. L., Shen, Y., Nelson, D. C. Purification and characterization of biofilm-associated EPS exopolysaccharides from ESKAPE organisms and other pathogens. *PLoS One* 2013, 8(6), e67950.

Bejar, V., Llamas, I., Calco, C., Quesada, E. Characterization of exopolysaccharides produced by 19 halophilic strains of the species *Halomonas eurihalina*. *J Biotechnol* 1998, 61, 135–141.

Biddanda, B. A. Microbial synthesis of macroparticulate matter. *Mar Ecol Prog Ser* 1985, 20, 241–251.

Bitton, G., Friehofer, V. Influence of extracellular polysaccharide on the toxicity of copper and cadmium toward *Klebsiella aerogenes*. *Microb Ecol* 1978, 4, 119–125.

Bouchotroch, S., Quesada, E., del Moral, A., Llamas, I., Bejar, V. *Halomonas maura* sp. nov., a novel moderately halophilic exopolysaccharide – producing bacterium. *Int J Syst Evol Microbiol* 2001, 51, 1625–1632.

Broadbent, J. R., McMahon, D. J., Welker, D. L., Oberg, C. J., Moineau, S. Biochemistry, genetics and applications of exopolysaccharide production in *Streptococcus thermophilus*: a review. *J Dairy Sci* 2003, 86, 407–423.

Brown, M. J., Lester, J. N. Metal removal in activated sludge: the role of bacterial extracellular polymers. *Water Res* 1979, 13, 817–837.

Calvo, C., Ferrer, M. R., Martinez-Checa, F., Bejar, V., Quesada, E. Some rheological properties of the extracellular polysaccharide produced by *Volcaniella eurihalina* F2–7. *Appl Biochem Biotechnol* 1995, 55, 45–54.

Gonzalez-Domenech, C. M., Martinez-Checa, F., Quesada, E., Bejar, V. *Halomonas cerina* sp. nov., a moderately halophilic, denitrifying, exopolysaccharide-producing bacterium. *Int J Syst Evol Microbiol* 2008, 58(4), 803–809.

Chen, Y. T., Yuan, Q., Shan, L. T., Lin, M. A., Cheng, D. Q., Li, C. Y. Antitumor activity of bacterial exopolysaccharides from the endophyte *Bacillus amyloliquefaciens* sp. isolated from *Ophiopogon japonicus*. *Oncol Lett* 2013, 5, 1787–1792.

Chi, Z., Su, C. D., Lu, W. D. A new exopolysaccharide produced by marine *Cyanothece* sp. 113. *Biores Technol* 2007, 98(6), 1329–1332.

Colliec, J. S., Chevalot, L., Helley, D., Ratiskol, J., Bros, A., Sinquin, C., Roger, O., Fischer A. M. Characterization, chemical modifications and *in vitro* anticoagulant properties of an exopolysaccharide produced by *Alteromonas infernus*. *Biochem Biophys Acta* 2001, 1528, 141–151.

Costaouec, T. L., Cerantola, S., Ropartz, D., Ratiskol, J., Sinquin, C., Jouault, S. C., Boisset, C. Structural data on a bacterial exopolysaccharide produced by a deep sea *Alteromonas macleodii* strain. *Carbohyd Polym* 2012, 90(1), 49–59.

Costerton, J. W. Structure and function of the cell envelope of Gram negative bacteria. *Bacteriol Rev* 1974, 38, 87–110.

Cramton, S. E., Gerke, C., Schnell, N. F., Nichals, W. W., Gotz, F. The intercellular adhesion (ica) locus is present in *Staphylococcus aureus* and is required for biofilm formation. *Infect Immun* 1999, 67, 5427–5433.

Crawford, R. W., Gibson, D. L., Kay, W. W., Gunn, I. S. Identification of a bile-induced exopolysaccharide required for *Salmonella* biofilm formation on gallstone surfaces. *Infect Immun* 2008, 76, 5341–5349.

Crescenzi, V. Microbial polysaccharides of applied interest: ongoing research activities in Europe. *Biotechnol Prog* 1995, 11, 251–259.

Das Sarma, S., Das Sarma, P. Halophiles. In *Encyclopedia of Life Sciences*, John Wiley and Sons Ltd.: Chichester, 2012, pp. 1–11.

Davis, J. S. Importance of microorganisms in solar salt production. In *Proceedings of the 4th Symposium on Salt, Vol. 1,* Coogan, A. L. (Ed.), Northern Ohio Geological Society: Cleveland, 1974, pp. 369–372.

De Palencia, F. P., Weming, M. L., Sierra-Filardi, E., Duenas, M. T., Irastorza, A. Probiotic properties of the 2-substituted (1, 3)-D-glucan-producing bacterium *Pedioccusparvulus* 2.6. *Appl Environ Microbiol* 2009, 75, 4887–4891.

De Philippis, R., Margheri, M. C., Pelosi, E., Ventura, S. Exopolysaccharide production by a unicellular cyanobacterium isolated from a hypersaline habitat. *J Appl Phycol* 1993, 5, 387–394.

Decho, A. W. Microbial exopolymer secretions in ocean environments: their roles in food webs and marine processes, In *Oceanography and Marine Biology Annual Review*, Barnes, M. (Ed.), Aberdeen University Press: Scotland, 1990, pp. 73–153.

Decho, A. W., Lopez, G. R. Exopolymer microenvironments of microbial flora – multiple and interactive effects on trophic relationships. *Limnol Oceanogr* 1993, 38, 1633–1645.

Denner, E. B. M., McGenity, T. J., Busse, H. J., Grant, W. D., Wanner, G., Stan-Lotter, H. *Halococcus salifodinae* sp. nov., an archaeal isolate from an Austrian salt mine. *Int Syst Bacteriol* 1994, 44, 774–780.

Duta, F. P., Franca, F. P., Lopes, L. M. A. Optimization of culture conditions for exopolysaccharides production in *Rhizobium* sp. using the response surface method. *Electronic J Biotechnol* 2006, 9(4), 391–399.

Elsasser-Beile, U., Frieholin, H., Stirm, S. Primary structure of *Klebsiella* serotype 6 capsular polysaccharide. *Carbohydr Res* 1978, 65, 245–249.

Faber, E. J., Kamerling, J. P., Vlieganthart, J. F. Structure of the extracellular polysaccharide produced by *Lactobacillus delbrueckii* subsp. *bulgaricus* 291. *Carbohydrate Res* 2001, 331, 183–194.

Fletcher, M., Foodgate, G. D. An electron microscopic demonstration of an acidic polysaccharide involved in adhesion of a marine bacterium to solid surfaces. *J Gen Microbiol* 1973, 74, 325–334.

Gorshkova, R. P., Nazarenko, E. L., Zubkov, V. A., Ivanova, E. P., Ovodov, Y. S., Shash-kov, A. S., Knirel, Y. A. Structure of the repeating link of the acid polysaccharides of *Alteromonas haloplanktis* KMM 156. *Bioorg Khim* 1993, 19(3), 327–336.

Grant, W. D., Gemmell, R. T., McGenity, T. J. Halobacteria: the evidence for longevity. *Extremophiles* 1998, 2, 279–287.

Gruber, C., Legat, A., Pfaffenhuemer, M., Radax, C., Weidler, G., Busse, H. J., Stan-Lotter, H. *Halobacterium noricense* sp. nov., an archaeal isolate from a bore core of an alpine Perm-ian salt deposit, classification of *Halobacterium* sp. NRC-1 as a strain of *H. salinarum* and emended description of *H. salinarum. Extremophiles* 2004, 8(6), 431–439.

Guezennec, J., Pignet, P., Raguenes, G., Rougeaux, H. Marine bacterial strain of the genus *Vibrio*, water soluble polysaccharides produced by said strain and their uses. 2002, US Patent 6436680.

Harris, R. H., Mitchell, R. The role of polymers in microbial aggregation. *Ann Rev Microbiol* 1973, 27, 27–50.

Harvey, R. W., Luoma, S. N. Effect of adherent bacteria and bacterial extracellular polymers upon assimilation by *Macoma baltica* of sediment bound Cd, Zn and Ag. *Mar Ecol Prog Ser* 1985, 22, 281–289.

Holmstrom, C., Kjelleberg, S. Marine *Pseudoalteromonas* species are associated with higher organisms and produce biologically active extracellular agents. *FEMS Microbiol Ecol* 1999, 30, 285–293.

Iyer, A., Mody, K., Jha, B. Characterization of an exopolysaccharide produced by marine *Enterobacter cloacae. Indian J Exp Biol* 2005, 43, 467–471.

Javor, B. Hypersaline Environments: Microbiology and Biogeochemistry. In *Brock/Springer Series in Contemporary Bioscience*, Brock T. D. (Ed.), Springer: Berlin, 1989, pp. 1–328.

Jayraman, M., Seetharaman, J. Physiochemical analysis of the exopolysaccharides produced by a marine biofouling bacterium, *Vibrio alginolytics. Process Biochem* 2003, 38(6), 841–847.

Jenkins, R. O., Hall, J. F. Production and applications of microbial exopolysaccharides. In *Biotechnological Innovations in Chemical Synthesis*, Currell, B., Mieras Van Dam, R. C. (Ed.), Butterworth Heinemann: Oxford, 1997, pp. 193–230.

Joshi, A. A., Kanekar, P. P. Production of exopolysaccharide by *Vagococcus carniphilus* MCM B–1018 isolated from alkaline Lonar lake, India. *Annals of Microbiol* 2011, 61(4), 733–740.

Kamekura, M. Diversity of extremely halophilic bacteria. *Extremophiles* 1998, 2(3), 289–295.

Kanekar, P., Joshi, A., Kulkarni, S., Borgave, S., Sarnaik, S., Nilegaonkar, S., Kelkar, A., Thombare, R. Biotechnological potential of alkaliphilic microorganisms. In *Biotech-nology and Bioinformatics – Advances and Applications for Bioenergy, Bioremediation and Biopharmaceutical Research*, Thangadurai, D., Sangeetha, J. (Eds.), CRC Press: Boca Raton, USA, 2014, pp. 249–279.

Kanekar, P. P., Kanekar, S. P., Kelkar, A. S., Dhakephalkar, P. K. Halophiles, Taxonomy, Diversity, Physiology and Applications. In *Microorganisms in Environmental Manage-ment: Microbes and Environment*, Satyanarayana, T. (Ed.), Springer Science: London, 2012, pp. 1–34.

Kumar, A. S., Mody, K., Jha, B. Evaluation of biosurfactant/bioemulsifier production by a marine bacterium. *Bull Environ Contam Toxicol* 2007, 79, 617–621.

Kushner, D. J. Growth and nutrition of halophilic bacteria. In *The Biology of Halophilic bacteria*, Vreeland, R. H, Hochstein, L. I. (Eds.), CRC Press Inc.: Boca Raton, 1993, pp 87 103.

Kwon, K. J., Park, K. J., Kim, J. D., Kong, J. Y., Kong, I. S. Isolation of two different polysaccharides from halophilic *Zooglea* sp. *Biotechnol Lett* 1994, 16, 783–788.

Lee, H. K., Chun, J., Moon, E. J., Ko, S. H., Lee, D. S., Lee, H. S., Bae, K. S. *Hahella chejuensis* gen. nov. sp. nov., an extracellular-polysaccharide producing marine bacterium. *Int J Syst Evol Microbiol* 2001, 51, 661–666.

Levander, F., Svensson, M., Rådström, P. Small-scale analysis of exopolysaccharides from *Streptococcus thermophilus* grown in a semi-defined medium. *BMC Microbiol* 2001, 1, 23–27.

Li, W. W., Zhou, W. Z., Zhang, Y. Z., Wang, J., Zhu, X. B. Flocculation behavior and mechanism of an exopolysaccharide from the deep-sea psychrophilic bacterium *Pseudoalteromonas* sp. SM 9913. *Biores Technol* 2008, 99(15), 6893–6899.

Llamas, I., Amjres, H., Mata, J. A., Quesada, E., Bejar, V. The potential biotechnological applications of the exopolysaccharide produced by the halophilic bacterium *Halomonas almeriensis*. *Molecules* 2012, 17, 7103–7120.

Llamas, I., Bejar, V., Martinez-Checa, F., Martinez-Canovas, M. J., Molina, I., Quesada, E. *Halomonas stenophila* sp. nov., a halophilic bacterium that produces sulphate exopolysaccharides with biological activity. *Int J Syst Evol Microbiol* 2011, 61, 2508–2514.

Loaec M., Olier R., Gueennec J. Uptake of lead, cadmium and zinc by a novel bacterial exopolysaccharide. *Water Res* 1997, 31, 1171–1179.

Loaec, M., Olier, R., Guezennec, J. Chelating properties of bacterial exopolysaccharides from deep-sea hydrothermal vents. *Carbohydr Polym* 1998, 35, 65–70.

Maeda, H., Zhu, X., Suzuki, S., Suzuki, K., Kitamura, S. Structural characterization and biological activities of an exopolysaccharide kefiran produced by *Lactobacillus kefiranofaciens* WT-2B (T). *J Agri Food Chem* 2004, 52, 5533–5538.

Maki, J. S., Ding, L., Stokes, J., Kavouras, J. H., Rittschof, D. Substratum/bacterial interactions and larval attachment: films and exopolysaccharides of *Halomonas marina* (ATCC 25374) and their effect on barnacle cyprid larvae, *Balanus amphitrite* Darwin. *Biofouling* 2000, 16, 159–170.

Mancuso, N. C., Garon, S., Bowman, J. P., Raguenes, G., Guezennec, J. Production of exopolysaccharide by Antarctic marine bacterial isolates. *J Appl Microbiol* 2004, 96(5), 1057–1066.

Marshall, K. C. Mechanisms of bacterial adhesion at solid water interfaces. In *Bacterial adhesion*, Savage, D. C., Fletcher, M. (Eds.), Plenum Press: New York, 1985, pp. 133–161.

Marx, J. G., Carpenter, S. D., Deming, J. W. Production of cryoprotectant extracellular polysaccharide substances (EPS) by the marine psychrophilic bacterium *Colwellia psychrerythraea* strain 34 H under extreme conditions. *Can J Microbiol* 2009, 55(1), 63–72.

Mata, J. A., Bejar, V., Llamas, I., Arias, S., Bressollier, P., Tallon, R., Urdaci, M. C., Quesada, E. Exopolysaccharides produced by the recently described bacteria *Halomonas ventosae* and *Halomonas anticariensis*. *Res Microbiol* 2006, 157, 827–835.

Matsukawa, M., Greenberg, E. P. Putative exopolysaccharide synthesis genes influence *Pseudomonas aeruginosa* biofilm development. *J Bacteriol* 2004, 186, 4449–4456.

Maugeri, T. L., Gugliandolo, C., Caccamo, D., Panico, A., Lama, I., Gambacorta, A. Halophilic thermotolerant *Bacillus* isolated from a marine hot spring able to produce a new exopolysaccharide. *Biotechnol Lett* 2002, 24, 515–519.

McGenity, T. J., Gemmell, R. T., Grant, W. D., Stan-Lotter, H. Origins of halophilic microorganisms in ancient salt deposits. *Environ Microbiol* 2000, 2(3), 243–250.

Morris, G. A., Li, P., Puaud, M., Liu, Z., Mitchell, J. R., Harding, S. E. Hydrodynamic characterization of the exopolysaccharide from the halophilic cyanobacterium *Aphanothece halophytica* GR02: a comparison with xanthan. *Carbohydrate Polymers* 2001, 44, 261–268.

Nichols, C. A. M., Guezennec, J., Bowman, J. P. Bacterial exopolysaccharides from extreme marine environments with special consideration of the southern ocean, sea ice and deep sea hydrothermal vents: a review. *Marine Biotechnol* 2005, 7(4), 253–271.

Nicolaus, B., Lama, L., Esposito, E., *Haloarcula* sp. able to biosynthesize exo and endopolymers. *J Indus Microbiol Biotechnol* 1999, 23(6), 489–496.

Nicolaus, B., Lama, L., Panico, A., Schiano, M. V., Romano, I., Gambacorta, A. Production and characterization of exopolysaccharides excreted by thermophilic bacteria from shallow, marine hydrothermal vents of flegrean areas (Italy). *Syst Appl Microbiol* 2002, 25, 319–325.

Nicolaus, B., Panico, A., Manca, M. C., Lama, L., Gambacorta, A., Maugeri, T. L. A thermophilic *Bacillus* isolated from an Eolian shallow hydrothermal vent, able to produce exopolysaccharides. *Syst Appl Microbiol* 2000, 23, 426–432.

Nwodo, U. U., Ezeikel, G., Anthony, I. O. Bacterial exopolysaccharides: functionality and prospects. *Int J Mol Sci* 2012, 13, 14002–14015.

Oren, A. Diversity of halophilic microorganisms: environments, phylogeny, physiology and applications, *J Ind Microbiol Biotechnol* 2002, 28, 56–63.

Oren, A. The ecology of extremely halophilic archaea. *FEMS Microbiol Rev* 1994, 13, 415–440.

Oren, A. Life at high salt concentrations, intracellular KCl concentrations, and acidic proteomes. *Front Microbiol* 2013, 4, 1–6.

Orsod, M., Joseph, M., Huyop, F. Characterization of exopolysaccharides produced by *Bacillus cereus* and *Brachybacterium* sp. isolated from Asian Sea Bass (*Lates calcarifer*). *Malaysian J Microbiol* 2012, 8(3), 170–174.

Paerl, H. W. Microbial attachment to particles in marine and freshwater ecosystems. *Microb Ecol* 1975, 2, 73–83.

Paerl, H. W. Specific associations of blue-green algae Anabaena and Aphanizomenon with bacteria in freshwater blooms. *J Phycol* 1976, 12, 431–435.

Paramonov, N. A., Parolis, L. A. S., Parolis, H., Boan, I. F., Anton, J., Valera, F. R. The structure of the exocellular polysaccharide produced by the Archaeon *Haloferax gibbonsii* (ATCC 33959). *Carbohydrate Res* 1998, 309, 89–94.

Parolis, H., Parolis, L. A. S., Boan, I. F., Valera, F. R., Widmalm, G., Manta, M. C., Jansson, P. E., Sutherland, I. W. The structure of the exopolysaccharide produced by the halophilic archaeon *Haloferax mediterranei* strain R4 (ATCC 33500). *Carbohyd Res* 1996, 295, 147–156.

Parolis, L. A. S., Parolis, H., Paramonov, N. A., Boan, I. F., Anton, J., Valera, F. R. Structural studies on the acidic exopolysaccharide from *Haloferax denitrificans* ATCC 35960. *Carbohydrate Res* 1999, 319, 133–140.

Patel, A., Prajapati, J. B. Food and health applications of exopolysaccharides produced by lactic acid bacteria. *Adv Dairy Res* 2013, 1(2), 107.

Patil, V., Bathe, G. A., Patil, A. V., Patil, R. H., Sulunkea, B. K. Production of bioflocculant exopolysaccharide by *Bacillus subtilis*. *Advanced Biotech* 2009, 8, 14–17.

Perez-Fernandez, M. E., Quesada, E., Galvez, J., Ruiz, C. Effect of exopolysaccharide V2–7 isolated from *Halomonas eurihalina* on the proliferation *in vitro* of human peripheral blood lymphocytes. *Immunopharmacol Immunotoxicol* 2000, 22, 131–141.

Pfeiffer, S. M., McInerney, M. J., Jenneman, G. E., Knapp, R. M. Isolation of halotolerant, thermotolerant, facultative polymer-producing bacteria and characterization of the exopolymer. *Appl Environ Microbiol* 1986, 51, 1224–1229.

Pindar, D. F., Bucke, C. The biosynthesis of alginic acid by *Azotobacter vinelandii*. *J Biochem* 1975, 152, 617–622.

Poli, A., Anzelmo, G., Nicolaus, B. Bacterial exopolysaccharides from extreme marine habitats: production, characterization and biological activities. *Mar Drugs* 2010, 8, 1779–1802.

Poli, A., Esposito, E., Oriando, P., Lama, L., Giordano, A., de Appolonia, F., Nicolaus, B., Gambacorta, A. *Halomonas alkaliantarctica* sp. nov. isolated from saline lake Cape Russell in Antarctica, an alkalophilic moderately halophilic, exopolysaccharide-producing bacterium. *Syst Appl Microbiol* 2007, 30, 31–38.

Poli, A., Nicolaus, B., Denizei, A. A., Yavuzturk, B., Kazan, D. *Halomonas smyrnensis* sp. nov., a moderately halophilic exopolysaccharide producing bacterium. *Int J Syst Evol Microbiol* 2013, 63, 10–18.

Qin, K., Zhu, L., Chen, L., Wang, P. G., Zhang, Y. Structural characterization and ecological roles of a novel exopolysaccharide from the deep-sea psychrotolerant bacterium *Pseudoalteromonas* sp. SM 9913. *Microbiol* 2007, 153, 1566–1572.

Quesada, E., Bejar, V., Calvo, C. Exopolysaccharide production by *Volcaniella eurihalina*. *Experientia* 1993, 49, 1037–1041.

Raguenes, G., Pignet, P., Gauthier, G., Peres, A., Christen, R., Rougeaux, H., Barbier, G., Guezennec, J. Description of a new polymer-secreting bacterium from a deep-sea hydrothermal vent, *Alteromonas macleodii* subsp. *fijiensis* and preliminary characterization of the polymer. *Appl Environ Microbiol* 1996, 62(1), 67–73.

Razack, S. A., Velayutham, V., Thangavelu, V. Medium optimization for the production of exopolysaccharide by *Bacillus subtilis* using synthetic sources and agro wastes. *Turk J Biol* 2013, 37, 280–288.

Rougeaux, H., Quezennec, J., Carlson, R. W., Kervarec, N., Pichon, R., Talaga, P. Structural determination of the exopolysaccharide of *Pseudoalteromonas* strain HYD 721 isolated from a deep-sea hydrothermal vent. *Carbohydr Res* 1999, 315, 273–285.

Rougeaux, H., Talaga, P., Carlson, R. W., Guerzennec, J. Structural studies of an exopolysaccharide produced by *Alteromonas macleodii* subsp. *fijiensis* originating from a deep-sea hydrothermal vent. *Carbohydrate Res* 1998, 312, 53–59.

Ruiz-Ruiz, C., Srivastava, G. K., Carranza, D., Mata, J. A., Llamas, I., Santamaria, M., Quesada, E., Malina, I. J. An exopolysaccharide produced by the novel halophilic bacterium *Halomonas stenophila* strain B 100 selectively induces apoptosis in human T leukaemia cells. *Appl Microbiol Biotechnol* 2011, 89, 345–355.

Salazar, N., Gueimonde, M., Hemandez-Barranco, A. M., Ruas-Madiedo, P., delos Reyes-Gavilan, C. G. Exopolysaccharide produced by intestinal *Bifidobacterium* strains act as fermentable substrates for human intestinal bacteria. *Appl Environ Microbiol* 2008, 74, 4737–4745.

Samain, E., Milas, M., Bozzi, L., Dubreucq, M., Rinaudo, M. Simultaneous production of two different gel-forming exopolysaccharides by an *Alteromonas* strain originating from deep-sea hydrothermal vents. *Carbohydr Polym* 1997, 34, 235–241.

Satpute, S. K., Banat, I. M., Dhakephalkar, P. K., Banpurkar, A. G., Chopade, B. A. Biosurfactants, bioemulsifiers and exopolysaccharides from marine microorganisms. *Biotechnol Advances* 2010, 28, 436–450.

Sayem, S. M. A., Emiliano, M., Letizia, C., Annabella, T., Angela, C., Anna, Z., Maurilio, D., Mario, V. Anti-biofilm activity of an exopolysaccharide from a sponge associated strain of *Bacillus licheniformis*. *Microbial Cell Factories* 2011, 10, 74–86.

Schiano, M. V., Lama, L., Poli, A., Gugliandolo, C., Maugeri, T. L., Gambacorta, A., Nicolaus, B. Production of exopolysaccharides from a thermophilic microorganisms isolated from a marine hot spring inflegrean areas. *J Ind Microbial Biotechnol* 2003, 30, 95–101.

Silver, R. P., Aaronson, W., Vann, W. F. The K1 capsular polysaccharide of *Escherichia coli*. *Rev Infect Dis* 1998, 10, 282–286.

Silvi, S., Barghini, P., Aquilanti, A., Jimnez, B. J., Fenice, M. Physiologic and metabolic characterization of a new marine isolate (BM39) of *Pantoea* sp. producing high levels of exopolysaccharide. *Microbial Cell Factories* 2013, 12, 10.

Stan-Lotter, H., Pfaffenhuemer, M., Legat, A., Busse, H. J., Radax, C., Gruber, C. *Halococcus dombrowskii* sp. nov., an archaeal isolate from a Permian alpine salt deposit. *Int J Syst Evol Microbiol* 2002, 52, 1807–1814.

Stan-Lotter, H., McGenity, T. J., Legat, A., Denner, E. B. M., Glaser, K., Stetter, K. O., Wanner, G. Very similar strains of *Halococcus salifodinae* are found in geographically separated Permo-Triassic salt deposits. *Microbiol* 1999, 145, 3565–3574.

Sudo, H., Burgess, J. G., Takemara, H., Nakamora, M. Sulphated EPS production by the halophilic cyanobacterium *Aphanocopsa halophytica*. *Curr Microbiol* 1995, 30, 219–222.

Sutherland, I. W. Biosynthesis of microbial exopolysaccharides. *Adv Microbial Phys* 1982, 23, 79–150.

Sutherland, I. W. Biotechnology of microbial exopolysaccharides. In *Cambridge Studies in Biotechnology, Vol. 9*, Cambridge University Press: Cambridge, 1990, pp. 1–163.

Sutherland, I. W. Microbial exopolysaccharide synthesis. In *Extracellular Microbial Polysaccharides, Vol. 5*, Sanford, P. A., Laskin, A. (Ed.), American Chemical Society: Washington DC, 1977, pp. 40–57.

Sutherland, I. W. Novel and established applications of microbial polysaccharides. *Trends Biotechnol* 1998, 16, 41–46.

Sutherland, I. W. Biofilm exopolysaccharides: a strong and sticky framework. *Microbiol* 2001, 147, 3–9.

Sutherland, I. W. Microbial polysaccharides products. *Biotechnol Genet Eng Rev* 1999, 16, 217–229.

Tomb, M. P., Harding, S. E. In *An Introduction to Polysaccharide Biotechnology*, Bucke, C. (Ed.), Taylor and Francis: London, 1998, pp. 183.

Urai, M., Yoshizaki, H., Anzai, H., Ogihara, J., Iwabuchi, N., Harayama, S. Structural analysis of an acidic, fatty acid ester-bonded extracellular polysaccharide produced by a pristine assimilating marine bacterium, *Rhodococcus erythropolis* PR4. *Carbohydrate Res* 2007, 342, 933–942.

Valepyn, E., Berezina, N., Paquot, M. Optimization of production and preliminary characterization of new exopolysaccharides from *Gluconacetobacter hansenii* LMG1524. *Adv Microbiol* 2012, 2, 488–496.

Ventosa, A., Nieto, J. J., Oren, A. Biology of moderately halophilic aerobic bacteria. *Microbiol Mol Biol Rev* 1998, 62(2), 504.

Vincent, P., Pignet, P., Talmont, F., Bozzi, L., Fournet, B., Guezennec, J. G. Production and characterization of an exopolysaccharide excreted by a deep-sea hydrothermal vent bacterium isolated from the polychaete annelid *Alvinella pompejana*. *Appl Environ Microbiol* 1994, 60, 4134–4141.

Vreeland, R. H., Straight, S., Krammes, J., Dougherty, K., Rosenzweig, W. D., Kamekura, M. *Halosimplex carlsbadense* gen. nov., sp. nov., a unique halophilic archaeon with three 16S rRNA genes, that grows only in defined medium with glycerol and acetate or pyruvate. *Extremophiles* 2002, 6(6), 445–452.

Wingender, J., Neu, T. R., Flemming, H. C. What are bacterial extracellular polymer substances?. In *Microbial extracellular polymer substance*, Wingender, J., Neu, T. R., Flemming, H. C., Ed.; Springer: Berlin, Germany, 1999, pp. 1–19.

Wolfaardt, G. M., Lawrence, J. R., Korber, D. R. Function of EPS. In *Microbial Extracellular Polymeric substances: Characterization, Structure and Function*, Wingender, J., Neu, T. P., Flemming, H. C., Ed.; Springer-Verlag: New York, 1999, pp. 171–200.

Wrangstadh, M., Szewzyk, U., Ostling, J., Kjelleberg, S. Starvation specific formation of a peripheral exopolysaccharide by a marine *Pseudomonas* sp. *Appl Environ Microbiol* 1990, 56, 2065–2072.

Yang, Z., Huttunen, E., Staaf, M., Widmolm, G., Tenhu, H. Separation, purification and characterization of extracellular polysaccharides produced by slime-forming *Lactococcus-lactis* spp. *cremoris* strains. *Int Dairy J* 1999, 9, 631–638.

Yim, J. H., Ahn, S. H., Kim, S. J., Lee, Y. K., Park, K. J., Lee, H. K. Production of novel exopolysaccharide with emulsifying ability from marine microorganism, *Alteromonas* sp. strain 005511568. *Key Eng Mat* 2005, 277–279, 155–161.

Zinkevich, V., Bogdarina, I., Kang, H., Hill, M. A. W., Tapper, R., Beecht, I. B. Characterization of exopolymers produced by different isolates of marine sulfate reducing bacteria. *Int Biodeterior Biodegrad* 1996, 37, 163–172.

CHAPTER 2

ALKALIPHILIC BACTERIA AND THERMOPHILIC ACTINOMYCETES AS NEW SOURCES OF ANTIMICROBIAL COMPOUNDS

SUCHITRA B. BORGAVE,[1] MEGHANA S. KULKARNI,[1] PRADNYA P. KANEKAR,[2] and DATTATRAYA G. NAIK[3]

[1]*Microbial Sciences Division, MACS-Agharkar Research Institute, G.G. Agharkar Road, Pune, 411004, India*

[2]*Department of Biotechnology, Modern College, Pune 411005, Maharashtra, India*

[3]*Chemistry Group, MACS-Agharkar Research Institute, G.G. Agharkar Road, Pune, 411004, India*

CONTENTS

2.1   INTRODUCTION

Among the various groups of economically important biomolecules, antibiotics are perhaps the oldest and most widely studied compounds. Antibiotics spur a huge market that is constantly growing where demand for newer and better products far outstrips the supply. Antibiotics are low molecular weight natural substances produced by one microorganism that inhibits the growth of or kills other microorganisms. Antibiotics and similar natural products being secondary metabolites can be produced by almost all types of living things. They are prokaryotic and eukaryotic organisms belonging to the plant and animal kingdom. To survive in the environment and compete with other microorganisms for resources, many bacteria produce secondary metabolites to inhibit or kill other competing strains, including human and animal pathogens (Martin et al., 2003). The antibiotics may now be redefined as the secondary metabolites which in minimum concentration regulate growth processes, replications and exhibit some kind of responding action to the prokaryotic or eukaryotic cells at the biochemical level (Berdy, 2005).

Several classes or groups of antimicrobial compounds are known. The major classes of antibiotics and their examples are listed in Table 2.1. The 'Golden Age' of antibiotic research lasted from the 1940s to the late 1960s, such that by the late 1970s, the medical world proclaimed that the battle against infectious agents had been won. Unfortunately, the proclamations were all too premature as serious antibiotic resistance gradually emerged in pathogens such as *Staphylococcus aureus*, *Streptococcus pneumoniae* and *Enterococcus faecalis*. Today, methicillin-resistant *S. aureus*, multidrug resistant tubercle bacilli and vancomycin-resistant enterococci have become problematic in the hospital setting, and antibiotic

TABLE 2.1   Major Classes of Antibiotics

Name of the antibiotic class	Mode/Site of action	Examples
Beta lactams	Cell wall synthesis	Penicillins, Cephalosporins and Carbapenems
Macrolides	Protein synthesis inhibitors	Azithromycin, Erythromycin
Lincosamides	Protein synthesis inhibitors	Clindamycin
Streptogramins	Protein synthesis inhibitors	Pristinamycin
Aminoglycosides	Protein synthesis inhibitors	Gentamycin, Kanamycin
Quinolone	Inhibition of DNA synthesis	Nalidixic acid, Ciprofloxacin
Sulfonamides	Folic acid synthesis inhibitors	Trimethoprim, Sulfamethizole
Tetracyclines	Protein synthesis inhibitors	Tetracycline, Doxycycline
Others	RNA synthesis inhibitors, Peptidoglycan synthesis inhibitors	Glycopeptides, Polypeptides, Rifamycins, Lipoglycopeptides and Tuberactinamycins

resistance has entered the multidrug-resistant phase (Drlica, 2001). In spite of a large number of antimicrobials currently available, the need for the discovery of new antimicrobials and development of novel chemotherapeutic agents continues to exist due to development of drug resistance, emergence of new pathogens and resurgence of several diseases (Zahner and Fielder, 1995; Davies and Webb, 1998).

Although innumerable microbial cultures have been screened for their production of antimicrobial compounds and thousands of antibiotics have been listed, most of them have adverse properties that prevent their development as drugs. Only a few of these are therefore regularly employed in the therapy of diseases. This limiting number of actual chemotherapeutic agents sustains a constant pressure for the discovery of novel chemical agents either naturally occurring or of synthetic origin to assuage the need for the development of effective chemotherapeutic agents (Franklin and Snow, 2005). Research strategies for the discovery of new antimicrobial drugs are heavily influenced by the need to discover and develop new agents active against organisms resistant to earlier generations of drugs (Knowles, 1997; Hancock and Strohl, 2001). One strategy involves 'natural screening' for novel antibacterial drugs – ideally structurally novel inhibitors rather than analogues of existing compounds. Natural products make good drugs as they have existed in a biological system and offer unmatched diversity, providing many unique and patentable chemical structures. On the whole, however, natural product chemistry is largely unexplored, yet it is estimated that it has been the source of over 40% of all pharmaceutical compounds on the market today.

The continuing success of a biotechnologist in the search of microbial metabolites as antimicrobial compounds is in combating human, animal and plant diseases for stimulating the belief that microorganisms constitute an inexhaustible reservoir of compounds with pharmacological, physiological, medical or agricultural applications. Antibiotics continue to play a crucial role in the development of tissue culture techniques and basic screenings, primarily in biochemistry, molecular biology, microbiology and genetics including genetic engineering and in lesser extent, pharmacology and organic chemistry. The antibiotic research from the discovery of Fleming to our days has been a fascinating, exciting, continuously changing and developing adventure. As a result of the more than 50 years research, ten thousands of natural products derived from microbial sources are known. Interest towards the field were generally increasing, although sometimes declining and the whole story shows some cyclic features with successes and failures and evolved around changing clinical needs with new developing technology (Berdy, 2005).

2.1.1 SECONDARY METABOLITES

Secondary metabolites are defined as natural products, have a restricted taxonomic distribution, possess no obvious function in cell growth and are synthesized by cells that have stopped dividing. Secondary metabolites are not products of the primary metabolic pathways of the producing organisms. Secondary metabolites are accepted to be essential for the producing cell as inhibitors of other organisms that compete for the same food supply or as regulators of cellular differentiation processes (Demain, 1998; Demain, 1999). In addition, it is reported that, they are indeed products of biosynthetic pathways, which have evolved to give these types of advantages. The nature has provided a broad spectrum of structurally diverse secondary metabolites. Microbial secondary metabolites are synthesized from only a few precursors, in pathways with a relative small number of reactions, which branch from just a limited number of reactions of the primary metabolism (Demain and Fang, 2000). Its structural diversity is reflected in a variety of biological activities such as inhibitors of enzymes and antitumor, immunosuppressive and antiparasitic agents. Aside their medical relevance, these compounds are used in many industrial, agricultural and forest applications (Demain, 1999).

Different alternatives for improving production of secondary metabolites with different activities for biotechnological applications have been extensively investigated. Optimization of fermentation process and improvement of strains are the two methods used for increasing production of secondary metabolites (Parekh, 2000). The microbial production of secondary metabolites is extremely sensitive to environmental factors or culture conditions. Media optimization has been the standard procedure for optimizing antibiotic production. Overproduction of the secondary metabolites is a complex process and the successful development of improved strains requires knowledge of physiology of the microbial producer, pathway regulation and control of the product (Bunch and Harris, 1986).

2.1.2 FUNCTION AND IMPORTANCE OF SECONDARY METABOLITES

The use of antimicrobial agents is critical to successful treatment of infectious diseases. Although there are numerous classes of drugs that are routinely used to treat infections in humans, there are several reasons why the discovery and development of new antimicrobial agents are important. Over the past decade there has been an increased development of resistance in organisms that are typical pathogens in humans. These include methicillin/oxacillin-resistant *Staphylococcus aureus* (Hsueh et al., 2005), vancomycin resistant and intermediate *Staphylococcus aureus* (Hanaki et al., 2007), vancomycin-resistant *Enterococcus*, Gram-negative bacilli that produce extended spectrum betalactamases (Paterson and Bonomo, 2005), carbapenem-resistant *Klebsiella pneumoniae* and *Pseudomonas* and *Acinetobacter* strains that are resistant to all antibiotics typically used for treatment.

New antimicrobials are also needed for certain groups of organisms. Very limited number of antimicrobials is available to treat infections caused by fungi and mycobacterium. Chemotherapy for cancer treatment, immunosuppressive drugs for treatment of autoimmune diseases and organ transplant recipients, and infections (such as AIDS) that alter the effectiveness of the host immune system render individuals at high risk for fungal infections and certain mycobacterial infections. Often these infections are caused by environmental organisms that would not typically cause disease in a normal host. Extended length of treatment (sometimes up to one year) and adverse side effects of the drugs used for treatment contribute to the lack of recovery.

Although many infectious diseases have been known for thousands of years, over the past 30 years a number of new infectious diseases have been discovered. Some examples include Lyme disease caused by *Borrelia burgdorferi*, Legionnaires' disease caused by *Legionella pneumophila*, peptic ulcers caused by *Helicobacter pylori*, antibiotic associated diarrhea caused by *Clostridium difficile*, and AIDS caused by Human Immunodeficiency Virus. In addition, microorganisms are constantly changing, finding new places to live and new ways to survive, and adapting to new situations. During this process, harmless organisms may turn deadly or deadly strains may move from their normal host to humans. With the continuing discovery of new infectious diseases and the development of new disease processes of existing pathogens (i.e., necrotizing fasciitis caused by *Streptococcus pyogenes*), it is important to continue to find anti-infective agents that can be used to treat these infections (Northern, 2007).

Development of novel classes of drugs, drugs with fewer side effects, and drugs with shorter lengths of treatment are key in continuing the fight against infectious disease. An interesting property common to most potent antibiotics is that at very low concentrations, they act as growth stimulants (Demain, 1999). This raises an interesting possibility that some secondary metabolites may actually stimulate growth of other microorganism antagonistic to the growth of the competitors of the secondary metabolite producing organisms. At present several arguments support the hypothesis that secondary metabolites improve the survival of the producer in competition with other living species. These arguments are as follows (Demain and Fang, 2000). Secondary metabolites act as an alternative defense mechanism, because only the organisms lacking an immune system are prolific producers of these compounds. They have sophisticated structures, mechanisms of action, and complex and energetically expensive pathways. Secondary metabolites act in the competition between microorganisms, plant and animals. They are produced by biosynthetic genes clusters, which would only be selected if the product conferred a selective advantage. Some particularities of these genes clusters are the absence of non-functional genes and the presence of resistance and regulatory genes. The production of secondary metabolites with antibiotic activities is temporarily related with sporulation when the cells are particularly sensitive to competitors and requiring special protection when a nutrient runs out.

Furthermore, the wide diversity of secondary metabolites suggests a broad range of functions. Nevertheless, these functions could depend on the conditions, optimal or not, surrounding the producer microorganism. Finally, due to their crucial importance, the study and exploitation of secondary metabolites continue to progress despite the lack of agreement regarding why microbes produce such chemical diversity of antimicrobial compounds. The practical importance of antibiotics and other secondary metabolites is tremendous. They are widely used in the human therapy, veterinary, agriculture, scientific research and in countless other areas.

In general, secondary metabolites including antimicrobial compounds may be practically utilized in three different ways namely applying the natural/fermentation product directly in the medicine, agriculture or in any other fields, as starting material for subsequent chemical or microbiological modification (derivatization) and as lead compounds for chemical synthesis of new analogs (Lange, 1996).

2.1.3 PHYSIOLOGY OF SECONDARY METABOLITE PRODUCTION

Traditionally, secondary metabolites are considered to be synthesized as a response to a decrease in the growth rate of the culture as a result of nutrient limitation. A secondary metabolite is secondary only because it is not required for vegetative growth in pure culture, the timing of product formation being immaterial. Secondary metabolism is brought on by exhaustion of a nutrient, biosynthesis or addition of an inducer, and/or by a growth rate decrease. These events generate signals which affect a cascade of regulatory events resulting in chemical differentiation (secondary metabolism) and morphological differentiation (morphogenesis). In a number of secondary metabolite pathways, primary metabolites increase production of the final product. These effectors are often precursors and one has to determine whether the effect is merely due to an increase in precursor supply and/or includes induction of one or more synthases of the biosynthetic pathway. Antibiotics are small molecules whose synthesis often requires dozens of enzymes. Enzyme activities are of necessity, closely regulated in such complex pathways. It is therefore important to understand the physiology of the producing organisms in order to maximize the fermentative production of antibiotics. As antibiotics are secondary metabolites, they do not seem to play a central role in growth and metabolism of the organisms.

2.1.4 MICROORGANISMS PRODUCING SECONDARY METABOLITES

Secondary metabolites can be produced by almost all types of living things that lack an immune system thus are rarely produced by higher animals (Maplestone et al., 1992). The secondary metabolite producing ability is very uneven in the species of living world. Indeed, these compounds are mostly biosynthesized by bacteria, fungi, algae, corals, sponges, plants and lower animals. Actually filamentous microorganisms are the main source of secondary metabolites with nearly 75% of all described antibiotics being produced by actinomycetes and 17% by molds. Approximately 40% of the filamentus fungi and actinomycetes produce antibiotics when freshly isolated from nature. Some isolates of several *Streptomyces* species can produce more than 180 different secondary metabolites (Demain and Fang, 2000). Table 2.2 shows total bioactive metabolites with the main producer types and several specific producer species (Berdy, 2005).

2.1.5 DEVELOPMENT OF RESISTANCE TO ANTIMICROBIAL AGENTS

A few decades after the introduction of antibiotics into clinical practice, resistance by pathogenic bacteria has become a major health concern. Indeed, while in the mid 1970s infectious diseases were considered virtually conquered (Breithaupt, 1999), actually many Gram positive bacteria and Gram negative opportunistic pathogens are becoming resistant to virtually every clinically available drug (Greenberg, 2003). The use of antimicrobial drugs for prophylactic or therapeutic purposes in human and veterinary or for agricultural purposes, has provided the selective pressure favoring the survival and spread of resistant organisms. *Staphylococcus aureus*, for instance, a virulent pathogen that is responsible for a wide range of infections including pimples, pneumonia, osteomyelitis, endocarditis and bacteremia, has developed resistance to most classes of antibiotics. Methicillin-resistant *S. aureus* (MRSA) strains appeared in the hospital environment after introduction of the semi-synthetic penicillin and methicillin, being vancomycin the last

TABLE 2.2   Approximate Number of Bioactive Microbial Metabolites according to Their Producers and Bioactivities

Source	Total bioactive metabolites
Bacteria	5455
Eubacteriales	2750
Bacillus sp.	860
Pseudomonas sp.	795
Myxobacter	410
Cyanobacteria	640
Actinomycetales	10,100
Streptomyces sp.	7,630
Rare actinomycetes	2,470
Fungi	10,600
Microscopic fungi	6,450
Penicillium/Aspergillus	1950
Basidiomycetes	2000
Yeasts	140
Slime moulds	60
Total microbial metabolites	26,155
Protozoa	50

chance for MRSA treatment (Enright, 2003). Certainly, vancomycin is the last tool for the treatment of the infections caused by the resistant Gram positive microorganisms. Indeed, vancomycin resistance is difficult to acquire because it is a complex system involving up to 7 genes. However vancomycin-intermediately-sensitive *S. aureus* were first isolated in 1997 in Japan (Hiramatsu et al., 1997) and later in other countries (Fridkin, 2001). In fact, Vancomycin-resistant clinical isolates have been recently reported (Tenover et al., 2004). Thus currently, no antibiotic class is effective against multidrug-resistant *S. aureus* infections and new antibiotics or alternative chemotherapeutic strategies are urgently needed.

Enterococci are responsible for urinary tract, wound, intra abdominal, and pelvic infections (Barsby et al., 2001). The increase of vancomycin-resistant enterococci (VRE) as important agents of nosocomial infections is cause of great concern (Perl, 1999). It is thought that a selective pressure favoring the survival and spread of VRE was the consequence of the use of antibiotics in food and agricultural practices (Bax et al., 2000). Actually, there are no effective antibiotics currently available for such organisms. Even worst, vancomycin-resistance is often associated with multiple-drug resistance (Perl, 1999; Cetinkaya et al., 2000; Lautenbach et al., 2003).

Another cause of great concern is the Gram-negative antibiotic-resistant opportunistic pathogens. These bacteria, like *Pseudomonas aeruginosa*, are common environmental organisms, which act as opportunistic pathogens in clinical cases where the defense system of the patient is compromised. For instance, over 80% of cystic fibrosis (CF) patients become chronically infected with *P. aeruginosa*. In addition, other intrinsically antibiotic resistant organisms such as *Burkholderia cepacia* and *Stenotrophomonas maltophilia* (Saiman et al., 2002) are emerging as opportunistic pathogens. Interestingly, changes in the bacterial phenotype have been observed concomitant with the appearance or increase of antibiotic resistance. Indeed, in CF infections, initially, strains are non-mucoid, but over a time, a mucoid population showing slow growth phenotype with a increased capability to form biofilm, the small colony variants (SCV) develops. This ability is considered a major virulence trait because the bacteria are protected from adverse environmental conditions as well as from biological and chemical antibacterial agents (Haussler, 2004). Thus, new therapeutic drugs and/or approaches are needed to improve the management of these diseases and overcome these problems (Saiman et al., 2002).

The appearance of multidrug-resistant pathogenic strains has caused a therapeutic problem of enormous proportions. For instance, they cause substantial morbidity and mortality especially among the elderly and immunocompromised patients. In response, there is a renewed interest in discovering novel classes of antibiotics that have different mechanisms of action (Spizek and Tichy, 1995; Barsby et al., 2001). In spite of the ever-increasing difficulty faced in screening research, a steady increase has continued to date in the number of newly discovered bioactive compounds of microbial origin (Omura, 1992).

2.2   EXTREMOPHILES AND THE SEARCH FOR NOVEL ANTIMICROBIALS

The vast majority of antibiotics have come from non-extreme, terrestrial microorganisms. However, while these microorganisms continue to be studied extensively, the rate of discovery of novel metabolites from terrestrial microorganisms is decreasing (Bentley et al., 2002). Discovery and identification of new sources of natural products, therefore, plays an important role in the uncovering of novel drug candidates and drug development process. In recent years extremophiles have been looked upon as valuable sources of novel bioproducts and this may well include antimicrobials (Da Costa et al., 1988; Horikoshi, 1995).

Extremophiles thrive in extreme environments where no other microorganisms are found, including high temperature, pH, pressure and salt concentration; or low temperature, pH, nutrient concentration, or water availability. Some extremophiles can tolerate extremely high levels of radiation or toxic compounds. Extremophilic microorganisms have been explored for their ability to produce bioactive compounds. They have provided an interesting and challenging platform for researchers since the time of their discovery. Besides growth in harsh conditions, production of industrially valuable compounds such as enzymes, antibiotics, hormones, etc. have fascinated and focused attention in the present scenario.

In recent years extremophiles have been recognized as valuable sources of novel bioproducts and this may well include antimicrobials (Da Costa et al., 1988; Horikoshi, 1995), although it is not so clear whether extremophiles can actually produce antibiotics. Microorganisms have evolved to produce antimicrobials in response to external stimuli and stresses such as other microbial competitors. It is difficult to believe that extremophiles well produce them as in their extreme environments they are likely to have few competitors. However, microorganisms produce a huge range of chemicals for which there is no obvious role, and any large collection of chemicals will, contain molecules with high biological activity. Indeed, the ability to produce and retain a rich chemical diversity surely enhances a microorganism to produce the very rare compound that gives it enhanced fitness. Many microorganisms will make chemicals with potent antibiotic properties, but only some of these chemicals will give the maker increased fitness because of that antimicrobial activity.

In Japan, several research groups have attempted to exploit alkaliphiles and piezophiles for production of antibiotic compounds. In case of alkaliphiles, although many compounds have been screened, many of them are unstable during cultivation (Horikoshi, 1999). Piezophilic mutants of antibiotic-producing *Actinomyces* species have been shown to produce novel antibiotics when grown under high pressure (Abe and Horikoshi, 2001). Some researchers believe that the salt-loving halophiles provide the most potential as they are more easily manipulated 'cell factories' compared with other extremophiles. Haloarchaea *Haloferax mediterranii* and *Haloferax gibbonsi* are known to produce halocins, which are like bacteriocins, inhibit growth of closely related species (Prangishvili et al., 2000).

Indirect evidence that extremophiles are capable of producing antibiotics also exists. A novel thermophilic *Pseudomonas* species was demonstrated to synthesize an iron-binding compound, pyohelin that has activity against several species of *Candida* and *Aspergillus fumigatus* (Phoebe et al., 2001). A heat labile ß-lactamase has been purified from the cold-adapted psychrophile, *Psychrobacter immobilis*. The kinetic parameters of this enzyme for the hydrolysis for some ß-lactam antibiotics are comparable to those observed for the highly specialized cephalosporinases from pathogenic mesophilic bacteria (Feller et al., 1997).

The biological diversity of the marine environment offers enormous scope of the discovery of novel natural products, several of which are potential targets for biomedical development (Fenical, 1997). Some marine organisms such as marine sponges are known to contain symbiotic microbes including extremophilic archaea (Webster et al., 2001). In some cases, it may be the symbiont microorganism rather than the organism itself that is the source of the novel compounds. The marine bacterium *Streptomyces tenjimariensis*, a producer of an aminoglycoside istamycin, successfully grew at about 5–7% NaCl in culture medium (Hotta et al., 1980) and the production of the antibiotic aplasmomycin was best in the presence of 1–3% NaCl (Okami et al., 1976). In the recent research study, bacteria from saline soils of Egypt were screened for their activity against other microorganisms.

These examples emphasize the value of screening of novel natural products and derived products as lead compounds. Therefore, extremophilic microorganisms represent a relatively unexplored resource, and may provide valuable antimicrobials. Research strategies for the discovery of new antimicrobial drugs are heavily influenced by the need to discover and develop new agents active against organisms resistant to earlier generations of drugs (Hancock and Strohl, 2001). Natural screening for novel antibacterial drugs which are ideally and structurally novel inhibitors rather than analogues of existing compounds is needed.

2.2.1   ALKALIPHILIC BACTERIA

Alkaliphiles are the microorganisms that grow optimally or very well at pH values above 9, often between 10 and 12 but cannot grow or grow very slowly at the near neutral pH value of 6.5 (Horikoshi, 1999). Although alka-liphiles are widespread, it seems probable that these organisms especially those unique to hyper saline lakes, evolved separately within an alkaline environment. Commercial processes ranging from cement manufacturing to the preparation of indigo dye, mining operations, paper and pulp production and food processing effluents are the examples of highly alkaline environments brought about by man's activities. Naturally occurring alkaline environments may arise through biological activities such as ammonification, sulfate reduction and oxygenic photosynthesis (Langworthy, 1978). In volcanic areas, alkaline hot springs have been reported with pH 9.5 (Hensel et al., 1997). Soda lakes and soda deserts (pH->11.5) are the most stable alkaline environments and these are some examples of that - Ashanti lake, Bosumatai, Ghana, Africa, New Quebec, Labrador, Canada, North America, Canayan Diable, Arizona, USA, Lonar Lake, India and Asia.

One of the key features on alkaliphily is associated with the cell surface, which discriminates and maintains the intracellular neutral environment separate from the extracellular alkaline environment. In addition, surface located and excreted enzymes resistant to the effects of extreme pH, the pH gradient reversed to carry out ATP synthesis. Alkaliphilic bacteria compensate the reversal of the pH gradient by having a high membrane potential or by coupling Na+ expulsion to electron transport for pH homeostasis and energy transduction. Alkaliphilic bacteria are considered for production of enzymes such as protese, amylase, lipase, xylanase, catalase, pectinase, pullulanase, chitinase which have potential applications in various fields of biotechnology. Alkaliphiles also produce polyhydroxyalkanoates (PHAs) which are bacterial polyesters having biodegradable nature and properties close to plastics from petrochemical routes. Some alkaliphilic bacteria produce exopolysaccharides which find a wide range of applications in food, pharmaceutical, petroleum and other industries (Sutherland, 1990).

2.2.2 ALKALIPHILIC MICROORGANISMS AS A SOURCE OF ANTIMICROBIALS

Alkaliphilic microorganisms have many industrial applications. Many of them produce compounds of industrial interest and also they possess useful physiological properties which can facilitate their exploitation for commercial purposes (Ulukanli and Digrak, 2002). Microbial communities in natural alkaline environments have attracted attention because of possible biotechnological use of enzymes and metabolites from such organisms. The information on alkaliphilic bacteria producing antimicrobial compounds is summarized in Table 2.3.

Alkaliphilic microorganisms isolated from Lonar Lake, India have been reported for production of antimicrobial compounds (Kharat et al., 2009; Deshmukh and Puranik, 2010; Deshmukh et al., 2011; Tambekar and Dhundale, 2012). Recently, the production of naphthospironone A (1) from alkaliphilic *Nocardiopsis* sp. (YIM DT266) was reported by Ding et al. (2010). This metabolite showed cytotoxic and antibiotic activity. A halotolerant alkaliphile *Streptomyces aburaviensis* Kut-8, isolated from saline desert of Kutch, Western India is reported to secret antibiotic active against Gram positive bacteria (Thumar et al., 2010). *Bacillus halodurans* B20 isolated from samples collected in eastern Africa produced antibacterial compounds (Danesh, 2011). Facultative anaerobic halophilic and alkaliphilic bacteria *Marinilactibacillus psychrotolerance* ALK 9 and *Facklamia tabacinasalis* ALK 1 isolated from a natural smear ecosystem inhibit *Listeria* growth in early ripening stages (Roth et al., 2011). Borgave et al. (2012) studied alkaliphilic bacteria isolated from Lonar Lake for their antimicrobial activity against clinical pathogens (*Salmonella typhi*, *Pseudomonas aeruginosa*, *Escherichia coli*, *Klebsiella pneumoniae* and *Staphylococcus aureus*) and phytopathogenic fungi (*Fusarium oxysporum*, *Fusarium moniliforme*, *Aspergillus parasiticus*, *Rhizoctonia solani* and *Colletotrichum gloeosporioides*). Out of 78 alkaliphilic bacteria, 25 strains showed either antibacterial or antifungal activity. The culture supernatants of 25 isolates were tested using agar well diffusion method out of which, moderately alkaliphilic halotolerant *Halomonas campisalis*, *Planococcus maritimus* and *Paenibacillus* sp. L55 stood out inhibitor of nine of the 10 indicator microorganisms. An extensive exploratory screening of alkaliphiles for antibiotics is very important for newer antimicrobial compounds.

2.3 SIGNIFICANCE OF ACTINOMYCETES IN THE WORLD OF ANTIMICROBIALS

In the realm of antibiotic world, opening of the field of antibiotics from actinomycetes was marked by isolation of actinomycin. Waksman and Woodruff reported the isolation of actinomycin, the first antibiotic produced by an actinomycete and obtained in a crystalline form. After this

TABLE 2.3 Alkaliphilic Bacteria Producing Antimicrobial Compounds

Alkaliphilic bacteria	Antimicrobial compound produced	Inhibitory activity against	Reference
Paecilomyces lilacinus	Peptide antibiotic no. 1907-VIII	*Bacillus subtilis, Staphylococcus aureus, Pseudomonas aeruginosa, Escherichia coli, Candida albicans, Aspergillus niger*	Sato et al. (1983)
Nocardiopsis dassonvillei OPC-15	Antibiotics I and II	*Proteus mirabilis, Bacillus subtilis*	Tsujibo et al. (1988)
Corynebacterium sp. YUA 25	*N*-2-methylbutanoyl tyramine	Aldose reductase inhibitor	Bahn et al. (1998)
Streptomyces sp. AK409	Pyrocoll	*Arthrobacter* strains, Filamentous fungi, pathogenic protozoa, human tumor cell lines	Dieter et al. (2003)
Streptomyces sannanensis RJT-1	Antibiotic	*Staphylococcus aureus, Bacillus cereus, Bacillus megaterium, Bacillus subtilis*	Vasavada et al. (2006)
Bacillus halodurans C-125	Haloduracin	*Lactobacilli, Listeria, Streptococcus, Enterococcus, Bacillus, Pediococcus*	Lawton et al. (2008)
Streptomyces tanashiensis A2D	Bioactive metabolite	*Candida albicans, Fusarium moniliforme*	Singh et al. (2009)
Streptomyces sp.	Antibacterial substance	*Bacillus subtilis, Staphylococcus aureus, Escherichia coli, Proteus vulgaris, Salmonella typhi,* human lung carcinoma A549 cell line	Kharat et al. (2009)
Synechocystis aquatilis	Antimicrobial compound	*Staphylococcus aureus, Proteus vulgaris, Pseudomonas aeruginosa, Bacillus subtilis, Escherichia coli*	Deshmukh and Puranik (2010)

TABLE 2.3 Continued

Alkaliphilic bacteria	Antimicrobial compound produced	Inhibitory activity against	Reference
Nocardiopsis sp. YIM DT266	Naphthospironone A	*Bacillus subtilis, Staphylococcus aureus, Escherichia coli, Aspergillus niger* and HeLa, L929, AGZY cells	Ding et al. (2010)
Streptomyces aburaviensis Kut-8	Antibiotic	*Staphylococcus aureus, Bacillus cereus, Bacillus megaterium, Bacillus subtilis*	Thumar et al. (2010)
Bacillus halodurans B20	Antibacterial compound	*Staphylococcus aureus, Enterococcus faecium, Enterococcus faecalis, Streptococcus* sp.	Danesh (2011)
Marinilactibacillus psychrotolerance ALK 9, *Facklamia tabacinasalis* ALK	Antimicrobial compound	*Listeria* sp.	Roth et al. (2011)

discovery, the attention of other investigators all over the world turned towards detailed scrutiny of the actinomycetes in screening programs. It was followed by the discovery of streptothricin and streptomycin. The chemotherapy of tuberculosis was made possible with streptomycin. Soil actinomycetes, especially *Streptomyces* represent an important source of biologically active compounds with high commercial value and important applications in human and livestock medicine and agriculture. The biologically active compounds produced by actinomycetes are antibiotics, immunosuppressants, extracellular hydrolytic enzymes, plant growth promoters and siderophores.

Actinomycetes are Gram-positive bacteria but are distinguished from other bacteria by their morphology, high G+C content and on the basis of nucleic acid sequencing and pairing studies. Although some show pleomorphic and even coccoid elements, they characteristically have a filamentous mycelium and many produce spores that are easily detached and may become airborne when disturbed. They may thus be considered as the prokaryotic equivalent of fungi. Actinomycetes are well known for their ability to produce antibiotics and enzymes and for their ability to degrade complex and recalcitrant molecules, especially cellulose, lignocellulose and lignin, which makes them particularly important in composting.

Actinomycetes are the most economical and biotechnologically valuable class of prokaryotes producing bioactive secondary metabolites notably antibiotics (Blunt et al., 2006), anti-tumor agents, immunosuppressive agents (Mann, 2001) and enzymes (Strohl, 2004; Berdy, 2005; Cragg and Newman, 2005). Actinomycetes are the main source of clinically important antibiotics, most of which are too complex to be synthesized by combinatorial chemistry, making three quarters of all known products; the *Streptomyces* are especially prolific, producing around 80% of total antibiotic products (Stach and Bull, 2005; Hoa et al., 2009). *Micromonospora* is the runner up with less than one-tenth as many as *Streptomyces* (Lam, 2006). In addition to antibacterial components they also produce secondary metabolites with biological activities of which the *Streptomyces* spp. amounts for 80% of the total production by actinomycetes (Cragg and Newman, 2005). Since the introduction of antibiotics as therapeutic agents, no other group of microorganisms has contributed so much to this field of human and animal therapy as the actinomycetes. Beginning with actinomycin, announced by Walksman in 1940, a large number of chemical compounds have been isolated from the cultures of these organisms. Streptomycin soon followed and repeated the success story of penicillin. Actinomycetes are known to produce almost two thirds of all known antibiotics of microbial origin.

It is well known that terrestrial microorganisms, in particular the Streptomycetes, as well as Gram-positive soil bacteria of the order, Actinomycetales are the major source of antimicrobial agents. These are produced as complex secondary metabolites which result in billions of US dollars per year in sales. The soil-dwelling filamentous *Streptomyces coelicolor* is responsible for over half the naturally-derived antibiotics in current use today. Actinomycetes are found in a wide variety of habitats, including aquatic and terrestrial habitats and are especially abundant in soils of different types. Soil samples from various locations have been screened all over the world with the objective of isolating actinomycetes with the ability to produce bioactive compounds. However most of the screenings have concentrated on the isolation of mesophilic actinomycetes. The information on secondary metabolites produced by actinomycetes is compiled by Tabarez (2005, 2006). An immense variety of active secondary metabolites with different properties have been isolated from actinomycetes.

2.3.1   THERMOPHILIC ACTINOMYCETES

Among the extremophiles, the thermophilic bacteria and actinomycetes have been less explored due to difficulties in isolation and maintenance in pure culture. A thermophile may be defined as an organism having a growth temperature optimum of 50°C or higher. In the case of hyperthermophiles the optimum may be between 80°C and 110°C. Thermophilic bacteria and actinomycetes were described almost simultaneously towards the end of the nineteenth century and have been the subject of several historical reviews (Waksman, 1954). Oxygenic photosynthesizers evolved around 1.2 Ba. Aerobic forms of life including the thermophilic aerobes originated on the Earth's surface approximately 0.5 billion years ago after the accumulation of sufficient oxygen had taken place. Evolutionary studies have shown conclusively on the basis of rRNA homology that the last universal common ancestor linking the roots of the three kingdoms of life forms was most likely a thermophilic organism from a hydrothermal system. Thermophilic actinomycetes are believed to be of ancient origin. There is strong molecular and physiological evidence from present day microorganisms to indicate that the "universal ancestor" was capable of growing at high temperatures. Thus, the importance of thermophiles in the evolutionary process is well established (Reysenbach and

Cady, 2001; Lineweaver and Schwartzman, 2004). The volume of literature available on thermophilic bacteria is immense and contrasts sharply with the relative lack of literature available on thermophilic actinomycetes. In some ways this is surprising because they are true bacteria, possess typical bacterial cell wall, nuclear material and physiology and are similarly susceptible to phages. They differ morphologically being filamentous or mycelia and produce spores very similar to fungi. However there was an initial reluctance by bacteriologists to accept them wholeheartedly. The thermophilic actinomycetes have long been regarded as curiosities, organisms that turn up now and again on plates designed to trap more interesting bacteria and fungi. They have aroused the interest of a few bacteriologists/microbiologists because they seem to have no role to play in food spoilage and only very recently and rarely have been implicated in animal/human disease. The major challenge for thermophilic microorganisms is their survival and production of bioactive molecules at elevated temperatures.

Natural and manmade water bodies including geothermal vents and natural meteorite crater lakes, thermal springs are well known habitats of thermophilic actinomycetes. Soil as always, serves as an excellent source of a rich diversity of actinomycetes especially *Streptomyces* spp. Thermophilic actinomycetes can be isolated in high numbers from composts and overheated plant materials such as hay and bagasse. Actinomycetes, especially thermophilic species, are well known components of the microflora of composts. Composts for mushroom cultivation, prepared from animal manures and straw, have been most studied but actinomycetes may also colonise household and green waste composts. Thermophiles especially the thermophilic actinobacteria have immense biotechnological importance. They have been used as excellent producers of several enzymes including DNA polymerases, pullulanases, amylases, xylanases, lipases and proteases on an industrial level as well as the commercial production of other active biomolecules including hormones. In the industrial production of bioactive molecules thermophiles and hyperthermophiles have the added advantage of lesser contamination problems and faster growth rates.

The rapid growth rates of actinomycetes at elevated temperatures necessitate a shorter incubation period as compared to mesophilic actinomycetes. Germination from inoculated spores occurs within 8–10 hours and they exhibit extraordinarily high requirement of oxygen. Since the solubility of oxygen in water at 55°C is half of the solubility at 30°C, it poses a problem for the efficient cultivation of thermophilic actinomycetes under laboratory conditions and is often an elemental factor in limiting their growth (Allen, 1953). The most interesting problem posed by the thermophilic microorganisms is a physiological one. It is fascinating to know how these microbes live and grow at high temperature at which many proteins are coagulated and the existence of life appears as a biochemical anomaly. The obvious answer to this anomaly would be that the thermophilic microorganisms may gain more energy by utilizing the available nutrients than the life forms that live in more temperate environments (Allen, 1953).

Thermophilic organisms exhibit a variety of modifications and adaptations at the structural and molecular level imparting them with the ability to resist and repair thermal damage. Thermophiles are reported to contain specialized 'chaperonins' that are thermostable and resist denaturation by refolding the proteins to their native form and restoring their function (Kumar and Nussinov, 2001). The cell membrane of a thermophile contains abundant saturated fatty acids which provide a hydrophobic environment for the cell and keep the cell membrane rigid enough for the cell to survive at elevated temperatures. The presence of a reverse DNA gyrase that introduces positive supercoils in the DNA of thermophiles has been reported (Lopez, 1999). This results in raising the melting point of the DNA to at least as high as the organism's optimum temperature for growth. Thermophiles also tolerate high temperatures by using increased number of interactions as compared to their mesophilic counterparts in terms of electrostatic and hydrophobic interactions as well as presence of other stabilizing bonds such as disulphide bridges. The structural flexibility of a thermophilic protein is more than its mesophilic analogue. Thus, it has been assumed that a mechanism characterized by entropic stabilization is responsible for the higher thermostability of thermophilic biocatalysts.

2.3.2   HABITATS OF THE THERMOPHILIC ACTINOMYCETES

During the geological evolution of the Earth's surface, the temperature conditions in the geothermal vents were obviously conducive for the growth of thermophilic organisms. As the fundamental facts about the existence and evolutionary significance of the thermophiles came to be known it became the subject of a lively field of investigation and since then thermophiles have been isolated from a variety of habitats. Thermophilic actinomycetes are known to be the microflora of several aquatic and terrestrial habitats including geothermal vents, water bodies such as lakes and hot water springs as well as compost and irrigated lands (Bergey, 1919; Allen, 1953; Goodfellow, 1988; Lacey, 1997). Information on the various habitats commonly colonized by thermophilic actinomycetes is summarized in Table 2.4.

2.3.3 ANTIMICROBIAL COMPOUNDS FROM THE THERMOPHILIC ACTINOMYCETES

One of the first reports of production of antibiotic from thermophilic actinomycetes was published by Schone (1951). The antibiotic was produced by *Streptomyces thermophilus*, and was named thermomycin. Thermomycin was found to be the most active against *C. diphtheriae* and against *Listeria monocytogenes* to a much lesser extent. Production and isolation of thermoviridin, an antibiotic produced by *Thermoactinomyces viridis* was later reported (Schuurmans, 1955). Thermoviridin appeared to be an organic acid which was dialyzable. Polyketides such as thermorubin produced by *Thermoactinomyces vulgaris* have been reported. Production of the antibiotic granaticin from the thermophilic *Streptomyces thermoviolaceus* has been

TABLE 2.4   Habitats of Thermophilic Actinomycetes

Source	Microorganism
Hot springs	*Thermus* sp., *Bacillus* sp., *Streptomyces megaspores*
Hot springs	*Bacillus thermoleovocans*
Deep sea hydrothermal vents	*Staphilothermus marimus*, *Pyrococcus abyssi*
Marine solfatare	*Thermococcus litoralis*
Decomposed plant samples from lakes	*Clostridium absonum*
Compost of fermenting citrus fruits	*Bacillus* strain
Compost	*Bacillus stearothermophilus*
Korean salt fermented anchovy	*Bacillus* sp.
Sediments of hot springs	*Bacillus* sp.
Garbage dumps	*Bacillus circulans*
Compost treated with artichoke juice	*Bacillus* sp.
Soil	*Bacillus thermoproteolyticus*
Lonar lake silt and water sample	*Streptomyces megasporus*
Lonar lake	*Streptomyces thermoviolaceus*

Sources: Patke and Dey (1998); Chitte et al. (1999).

reported. *Thermoactinopolyspora coremiallis* ATCC 15974 is known to produce a polyketide antibiotic effective against Gram positive bacteria but not against Gram negative bacteria. However the commercial use of these antibiotics derived from thermophilic actinomycetes is not reported. The number of antibiotics reported from thermophilic actinomycetes is but a mere fraction of the antibiotics that have been reported from their mesophilic counterparts. Work done on some better studied thermophilic species of the actinobacteria however indicates the presence of novel properties and suggests that with intensive research and investigation it may lead to the discovery of novel natural products (Clive, 1993). The absence of antibiotics from thermophilic actinomycetes in the thriving worldwide antibiotic market necessitates a need to study this group of organisms with the objective of isolating novel natural antibiotic like compounds whose presence is indicated by the special physiological and genetic features of these organisms. The presence of antibiotic producing thermophilic actinomycetes has been reported from niche locations such as thermal springs, Lonar lake (meteorite crater lake) and alkaline Pashan Lake (Kulkarni and Kanekar, 2011).

2.4   WORLDWIDE MARKET FOR ANTIMICROBIALS

The demand for antibiotics is expected to reach US$44.68 billion by 2016. Since 2005 this market has grown at the rate of 6.6% until 2011. Very few categories of antimicrobials such as penicillins, cephalosporins, fluoroquinolones and macrolids dominate this market. It has been predicted that in the near future two factors will contribute and influence the growth of the antibiotic market worldwide namely generic competition and antibiotic resistance. The antibiotic drug market is also expected to witness restricted growth as compared to the earlier decade due to patents of major players. As the patents expire, new products are expected to emerge in the market and possibly new classes/categories of antibiotic molecules are expected to make their presence felt.

2.5   CURRENT TRENDS IN NEW DRUG DISCOVERY

Within the past few decades, the time and cost of drug development have soared. Today it typically takes about 15 years and costs up to $800 million to convert a promising new compound into a drug on the market. These costs reflect the complexity of the process. While they promise a cornucopia of new drugs, genomic methods alone will not reduce the cost and time of drug development. However, other new developments, many stemming from biotechnology, will help to improve the productivity of drug development. Such approaches as rational drug design, combinatorial chemistry, and *in silico* experimentation via computers have started to expedite the overall search for new drugs. New methods of data management complement those approaches. Increasing productivity means screening more samples in less time and with less labor. To accomplish this, manufacturers have developed high throughput screening (HTS) systems that range from semi automated work stations to fully automated robotic systems. Pharmaceutical companies have long tried to replace serendipity with logic in the effort to develop new drugs. In years past, rational drug design required medicinal chemists working at a lab bench to synthesize a relatively small number of compounds with the desired properties of a potential new drug. The step-by-step approach took many months or even years to complete.

Current methods of rational drug design accelerate discovery by removing some of the randomness from the process. The methods involve the design and optimization of small organic molecules based on either information derived from a protein structure or a small collection of hits from HTS. Another approach to generating many compounds that may interact with a target is combinatorial chemistry, a method that creates every possible variant of a parent compound. Combinatorial chemistry plays a major role in constructing chemical libraries. Effective as they are individually, rational drug design and combinatorial chemistry work even better in partnership to shorten the drug discovery process. Recent trends in the drug discovery area have seen a variety of companies focusing on quality rather than quantity of compounds, instrumentation and data, which means the market demands high-quality compound synthesis for focused libraries; robust and easy-to-use compound storage; reliable transportation and compound handling across multiple sites; and information-rich assays such as those used for cell-based screening.

Automated combinatorial chemistry can now synthesize a near-infinite number of compounds. The problem is narrowing down the possibilities to a reasonable subset to synthesize and screen. Many off-the-shelf options exist for creating library subsets, but these are often limited in scope. Improved integration interfaces mean that multiple organizations now offer solutions that combine existing products to automate parts of a laboratory process.

However, when processes change, existing products do not necessarily adapt and can be throughput limiting. Companies that provide custom automation and are able to develop technologies to fill gaps in the process are therefore seeing an increased demand for bespoke automation solutions. More flexible systems can be produced if the automation is tailored to the needs of the chemist, rather than the other way around, and throughput requirements can be built in from the start.

2.6 CONCLUSION AND FUTURE PERSPECTIVES

Antimicrobial producing genes are tightly clustered in the microbial chromosomes, both gene replacement and coregulation of all the genes necessary for antibiotic formation are facilitated. In fact, by applying DNA recombinant technologies, a large amount of genes coding for enzymes involved in synthesis of secondary metabolites have been cloned in order to improve expression levels. In addition, with the genomic sequences of several antibiotic-producing microbes becoming available, there has been much interest in combinatorial biosynthetic approaches. Indeed, by manipulating the genes of antibiotic biosynthesis pathways, hybrid or variant antibiotics with novel properties and efficacies against resistant pathogens could be generated. Nanotubes present a new class of antibiotics with several advantages over conventional drugs. They target the cellular membrane and the interaction is not specific to any one protein or site in the membrane. Therefore it is difficult to the microbes to develop the resistance to nanotubes. With eight members to a nanotube ring and 20 amino acids, over 25 billion combinations are possible and even that could be expanded by changing the size of the ring and using amino acids not found in nature. Nanotubes act very quickly, completely destroying viability in a matter of 200 seconds. They may find utility in treating infections even very late in the course of a disease.

KEYWORDS

- **Alkaliphilic bacteria**
- **Antibiotic resistance**
- **Antimicrobial compounds**

- **Drug discovery**
- **Extremophiles**
- **Geothermal vents**
- **Multidrug resistance**
- **Peptide antibiotics**
- **Secondary metabolites**
- **Soda lakes**
- **Thermal springs**
- **Thermophiles**
- **Thermophilic actinomycetes**

REFERENCES

Abe, F., Horikoshi, K. The biotechnological potential of piezophiles. *Trends Biotechnol.* 2001, 19, 102–108.

Allen, M. B. The thermophilic aerobic spore forming bacteria. *Bacteriol Rev.* 1953, 17(2), 125–173.

Barsby, T., Kelly, M. T., Gagne, S. M., Andersen, R. J. Bogorol A produced in culture by a marine *Bacillus* sp. reveals a novel template for cationic peptide antibiotics. *Org Lett.* 2001, 3(3), 437–440.

Bax, R., Mullan, N., Verhoef, J. The millennium bugs – the need for and development of new antibacterials. *Int J Antimicrob Ag.* 2000, 16(1), 51–59.

Behal, V. Mode of action of microbial bioactive metabolites. *Folia Microbiol.* 2006, 51(5), 359–369.

Bergey, D. H. Thermophilic bacteria. *J Bacteriol.* 1919, 4(4), 301–306.

Blunt, J. W., Copp, B. R., Munro, M. H. G., Northcote, P. T., Prinsep, M. R. Marine natural products. *Nat Prod Rep.* 2006, 23(1), 26–78.

Berdy, J. Bioactive microbial metabolites. *J Antibiot.* 2005, 58(1), 1–26.

Borgave, S. B., Joshi, A. A., Kelkar, A. S., Kanekar, P. P. Screening of alkaliphilic, haloalka-liphilic bacteria and alkalithermophilic actinomycetes isolated form alkaline soda lake of Lonar, India for antimicrobial activity. *Int J Pharma Bio Sci.* 2012, 3(4), 258–274.

Breithaupt, H. The new antibiotics. *Nat Biotechnol.* 1999, 17(12), 1165–1219.

Bunch, A. W., Harris, R. E. The manipulation of microorganisms for the production of sec-ondary metabolites. *Biotechnol Genet Eng.* 1986, 4, 117–144.

Cetinkaya, Y., Falk, P., Mayhall, C. G. Vancomycin-resistant enterococci. *Clin Microbiol Rev.* 2000, 13(4), 686–707.

Chitte, R. R., Nalawade, V. K., Dey, S. Keratinolytic activity from the broth of a feather degrading *Streptomyces thermoviolaceus* strain SD8. *Lett Appl Microbiol.* 1999, 28, 131–136.

Clive, E. Graniticin production by *Streptomyces thermoviolaceus*. *Appl Biochem Biotechnol*. 1993, 42(2–3), 161–179.

Cragg, G. M., Newman, D. J. *Anticancer Agents from Natural Products*. Taylor and Francis: USA, 2005.

Da Costa, M. S., Duarte, J. C., Williams, R. A. D. Microbiology of extreme environments and its potential for biotechnology. *FEMS Symposium No. 49*. Elsevier Applied Science: London, 1988.

Davies, J., Webb, V. Antibiotic resistance in bacteria. In *Emerging Infections*, Krause, R. M., (Ed.), Academic Press: New York, 1998, pp. 239–273.

Demain, A. L. Induction of microbial secondary metabolism. *Int Microbiol*. 1998, 1, 259–264.

Demain, A. L. Pharmaceutically active secondary metabolites of microorganisms. *Appl Microbiol Biot*. 1999, 52(4), 455–463.

Demain, A. L., Fang, A. The natural functions of secondary metabolites. *Adv Biochem Eng Biotechnol*. 2000, 69, 1–39.

Deshmukh, K. B., Pathak, A. P., Karuppayil, M. S. Bacterial diversity of Lonar Soda Lake of India. *Ind J Microbiol*. 2011, 51(1), 107–111.

Deshmukh, D. V., Puranik, P. R. Application of Plackett-Burman design to evaluate media components affecting antibacterial activity of alkaliphilic cyanobacteria isolated from Lonar Lake. *Turk J Biochem*. 2010, 35(2), 114–120.

Dieter, A., Hamm, A., Fiedler, H. P., Goodfellow, M., Mueller, W. E. G., Brun, R., Beil, W., Bringmann, G. Pyrocoll, an antibiotic, antiparasitic and antitumor compound produced by a novel alkaliphilic *Streptomyces* strain. *J Antibiot*. 2003, 56(7), 639–646.

Ding, Z. G., Li, M. G., Zhao, J. Y., Ren, J., Huang, R., Xie, M. J., Cui, X. L., Zhu, H. J., Wen, M. L. Naphthospironone a: an unprecedented and highly functionalized polycyclic metabolite from an alkaline mine waste extremophile. *Chem*. 2010, 16(13), 3902–3905.

Drlica, K. Antibiotic resistance: can we beat the bugs? *Drug Discov Today*. 2001, 6, 714–715.

Enright, M. C. The evolution of a resistant pathogen - free case of MRSA. *Curr Opin Pharmacol*, 2003, 3(5), 474–479.

Feller, G., Zekhnini, Z., Lamotte-Brasseur, J., Gerday, C. Enzymes from cold-adapted microorganism. The class C β-lactamase from the antarctic psychrophile *Psychrobacter immobilis* A5. *Eur J Biochem*. 1997, 244, 186–191.

Fenical, W. New pharmaceuticals from marine microorganisms. *Trends Biotechnol* 1997, 15, 339–341.

Franklin, T. J., Snow, G. A. Biochemistry and molecular biology of antimicrobial drug action. Springer: New York, 2005, pp. 5–30.

Fridkin, S. K. Vancomycin-intermediate and resistant *Staphylococcus aureus*: what the infectious disease specialist needs to know. *Clin Infect Dis*. 2001, 32(1), 108–115.

Greenberg, E. P. Bacterial communication and group behavior. *J Clin Invest*. 2003, 112(9), 1288–1290.

Goodfellow, M., Williams, S. T., Modraski, M. *Actinomycetes in Biotechnology*, Academic Press: London, 1988.

Gordon, M. C., David, J. Newman drug discovery and development from natural products. In *The Way Forward, 11th NAPRECA Symposium Book of Proceedings*, Antananarivo: Madagascar, 2005, pp. 56–69.

Hanaki, H., Hososaka, Y., Yanagisawa, C., Otsuka, Y., Nagasawa, Z., Nakae, T., Sunakawa, K. Occurrence of vancomycin-intermediate-resistant *Staphylococcus aureus* in Japan. *J Infection Chemoth*. 2007, 13(2), 118–121.

Hancock, E. W., Strohl, W. R. Antimicrobials in the 21ᵗ century. *Curr Opin Microbiol.* 2001, 4, 491–545.

Haussler, S. Biofilm formation by the small colony variant phenotype of *Pseudomonas aeruginosa. Environ Microbiol.* 2004, 6(6), 546–551.

Hensel, R., Matussek, K., Michlke, K., Tacke, L., Tindall, B. J., Kohlhoff, M., Siebers, B., Dialenshnedder, J. *Sulfophobococcus zilligii* gen. nov., spec. nov., a novel hyperthermophilic archeum isolated from hot alkaline spring of Iceland. *Syst. Appl. Microbiol.* 1997, 20, 102–110.

Hiramatsu, K., Hanaki, H., Ino. T., Yabuta, K., Oguri, T., Tenover, F. C. Methicillin-resistant *Staphylococcus aureus* clinical strain with reduced vancomycin susceptibility. *J Antimicrob Chemoth.* 1997, 40(1), 135–136.

Hoa, N. Q., Larson, M., Kim C. N. T., Eriksson, B., Trung, N. V., Stalsby, C. L. Antibiotics and pediatric acute respiratory infections in rural Vietnam: health-care providers' knowledge, practical competence and reported practice. *Trop Med Int Health.* 2009, 14(5), 546–555.

Horikoshi, K. Discovering novel bacteria with an eye to biotechnological applications. *Curr Opin Biotech.* 1995, 6, 292–297.

Horikoshi, K. Alkaliphiles: some applications of their products for biotechnology. *Microbiol Mol Bio Rev.* 1999, 63(4), 735–750.

Hotta, K., Saito, N., Okami, Y. Studies on a new aminoglycoside antibiotic, istamycins, from an actinomycete isolated from a marine environment. *J Antibiot.* 1980, 33, 1502–1509.

Hsueh, P. R., Chen, W. H., Teng, L. J., Luh, K. T. Nosocomial infections due to methicillin-resistant *Staphylococcus aureus* and vancomycin-resistant enterococci at a University Hospital in Taiwan from 1991 to 2003: resistance trends, antibiotic usage and *in vitro* activities of newer antimicrobial agents. *Int J Antimicrob Ag.* 2005, 26(1), 43–49.

Kharat, K., Kharat, A., Hardikar, B. P. Antimicrobial and cytotoxic activity of *Streptomyces* sp. from Lonar Lake. *Afr J Biotechnol.* 2009, 8(23), 6645–6648.

Knowles, D. J. C. New strategies for antibacterial drug design. *Trend Microbiol.* 1997, 5, 379–383.

Kulkarni M. S., Kanekar P.P. Studies in the production of antibiotic like compounds from thermophilic actinomycetes isolated from niche habitats. *Proceedings of the World Biotechnology Congress.* OMICS Group, Sunnyvale CA, USA, 2011, pp. 1–14.

Kumar, S., Nussinov, R. How do thermophilic proteins deal with heat? A review. *Cell Mol Life Sci.* 2001, 58, 1216–1233.

Lacey, J. Actinomycetes in Compost. *Ann Agric Environ Med.* 1997, 4, 113–121.

Lam, K. S. Discovery of novel metabolites from marine actinomycetes. *Curr Opin Microbiol.* 2006, 9, 245–251.

Lange, L. Microbial metabolites-an infinite source of novel chemistry. *Pure Appl Chem.* 1996, 68(3), 745–748.

Langworthy, T. A. Microbial life in extreme pH values. In *Microbial Life in Extreme Environments*, Kushner, D. J. (Ed.), Academic Press: London, 1998, pp. 279–317.

Lautenbach, E., LaRosa, L. A., Marr, A. M., Nachamkin, I., Bilker, W. B., Fishman, N. O. Changes in the prevalence of vancomycin-resistant enterococci in response to antimicrobial formulary interventions: impact of progressive restrictions on use of vancomycin and third-generation cephalosporins. *Clin Infect Dis.* 2003, 36(4), 440–446.

Lawton, E. M., Cotter, P. D., Hill, C., Ross, R. P. Identification of a novel two peptide Lantibiotic, Haloduracin, produced by the alkaliphile *Bacillus halodurans* C–125. *FEMS Microbiol Lett.* 2007, 267, 64–71.

Lehrer, R. I., Ganz, T. Endogenous vertebrate antibiotics. Defensins, protegrins and other cysteine-rich antimicrobial peptides. *Ann NY Acad Sci*. 1996, 797, 228–239.

Lineweaver, C. H., Schwartzman, D. Cosmic thermobiology. Thermal constraints on the origin and evolution of life in the Universe. In *Genesis, Evolution and Diversity of Life*, Kluwer, J. S. (Ed.), The Netherlands, 2004, pp. 233–248.

Lopez, G. DNA supercoiling and temperature adaptation: a clue to early diversification of life. *J Mol Evol*. 1999, 46, 39–452.

Mann, J. Natural products as immunosuppressive agents. *Nat Prod Rep*. 2001, 18, 417–430.

Maplestone, R. A., Stone, M. J., Williams, D. H. The evolutionary role of secondary metabolites – a review. *Gene* 1992, 115(1–2), 151–157.

Marshall, S. H., Arenas, G. Antimicrobial peptides: a natural alternative to chemical antibiotics and a potential of applied biotechnology. *Elect. J Biotech*, 2003, 6(3), 1–17.

Martin, N. I., Hu, H., Moake, M. M., Churey, J. J., Whittal, R., Worobo, R. W., Verderas, J. C. Isolation, structural characterization and properties of Mattacin (Polymyxin M), a cyclic peptide antibiotic produced by *Paenibacillus kobensis* M. *J Biol Chem*. 2003, 278(15), 13124–13132.

Navon-Venezia, S., Feder, R., Gaidukov, L., Carmeli, Y., Mor, A. Antibacterial properties of dermaseptin S4 derivatives with *in vivo* activity. *Antimicrob Agents Ch*. 2002, 46(3), 689–694.

Omura, S. Trends in the search for bioactive microbial metabolites. *J Ind Microbiol*. 1992, 10,135–156.

Okami, Y., Okazaki, T., Kitahara, T., Umezawa, H. A new antibiotic, aplasmomycin, produced by a streptomycete isolated from shallow sea mud. *J Antibiot*. 1976, 29, 1019–1025.

Parekh, S., Vinci, V. A., Strobel, R. J. Improvement of microbial strains and fermentation processes. *Appl Microbiol Biot*. 2000, 54(3), 287–301.

Paterson, D. L., Bonomo, R. A. Extended-spectrum beta-lactamases: A clinical update. *Clin Microbiol Rev*. 2005, 18(4), 657–686.

Patke, D., Dey, S. Proteolytic activity from a thermophilic *Streptomyces megasporus* strain SDP4. *Lett Appl Microbiol*. 1998, 26, 171–174.

Perl, T. M. The threat of vancomycin resistance. *Am J Med*. 1999, 106(5A), 26S–37S.

Phoebe, C. H., Combie, J., Albert, F. G., Van Tran, K., Cabrera, J., Correira, H. J., Guo, Y., Lindermuth, J., Raufert, N., Galbraith, W., Seliternnikoff, C. P. Extremophilic organisms as an unexplored source for antifungal compounds. *J Antibiot*. 2001, 54, 56–65.

Prangishvili, D., Holz, I., Stieger, E., Nickell, S., Kristjansson, J. K., Zillig, W. Sulfolobicins, specific proteinaceous toxins produced by strains of the extremophilic archaeal genus *Sulfolobus*. *J Bacteriol*. 2000, 182, 2985–2988.

Roth, E., Schwenninger, S. M., Eugster-Meier, E., Lacroix, C. Facultative anaerobic halophilic and alkaliphilic bacteria isolated from a natural smear ecosystem inhibits *Listeria* growth in early ripening stages. *Int J Food Microbiol*. 2011, 147(1), 26–32.

Saiman, L., Chen, Y., Gabriel, P. S., Knirsch, C. Synergistic activities of macrolide antibiotics against *Pseudomonas aeruginosa*, *Burkholderia cepacia*, *Stenotrophomonas maltophilia*, and *Alcaligenes xylosoxidans* isolated from patients with cystic fibrosis. *Antimicrob Agents Ch*. 2002, 46(4), 1105–1107.

Sato, M., Beppu, T., Arima, K. Studies on antibiotics produced at high alkaline pH. *Agric Biol Chem*. 1983, 47(9), 2019–2027.

Stach, J. E. M., Bull, A. T. Estimating and comparing the diversity of marine actinobacteria. *Int J Gen Mol Microbiol* 2005, 87(1), 3–9.

Singh, L. S., Mazumder, S., Bora, T. C. Optimization of process parameters for growth and bioactive metabolite produced by a salt tolerant and alkaliphilic actinomycetes, *Streptomyces tanashiensis* strain A2D. *J Med Mycol.* 2009, 19(4), 225–233.

Spizek, J., Tichy, P. Some aspects of overproduction of secondary metabolites. *Folia Microbiol.* 1995, 40, 43–50.

Stone, M. J., Williams, D. H. On the evolution of functional secondary metabolites (natural products). *Mol Microbiol.* 1992, 6(1), 29–34.

Strohl, W. R. Antimicrobials. In *Microbial Diversity and Bioprospecting*, Bull, A. T. (Ed.), ASM Press: UK, 2004, pp. 336–355.

Tabarez, M. R. 7-*O*-malonyl macrolactin A, a new macrolactin antibiotic from *Bacillus subtilis* active against methicillin-resistant *Staphylococcus aureus*, vancomycin-resistant *Enterococci*, and a small-colony variant of *Burkholderia cepacia*. *Antimicrob Agents Ch.* 2005, 50(5), 1701–1709.

Tambekar, D., Dhundale, V. Isolation and characterization of antibacterial substance produced from Lonar Lake. *Int J Res Rev Pharma Appl Sci.* 2012, 2(1), 41–54.

Tenover, F. C., Weigel, L. M., Appelbaum, P. C., McDougal, L. K., Chaitram, J., McAllister, S., Clark, N., Killgore, G., O'Hara, C. M., Jevitt, L., Patel, J. B., Bozdogan, B. Vancomycin-resistant *Staphylococcus aureus* isolate from a patient in Pennsylvania. *Antimicrob Agents Ch.* 2004, 48(1), 275–280.

Thumar, J. T., Dhulia, K., Singh, S. P. Isolation and partial purification of an antimicrobial agent from halotolerant alkaliphilic *Streptomyces aburaviensis* strain Kut-8. *World J Microb Biot.* 2010, 26, 2081–2087.

Tsujibo, H., Sato, T., Inui, M., Yamamoto, H., Inamorai, Y. Intracellular accumulation of phenazine antibiotics produced by an alkaliphilic actinomycete. I. Taxonomy, isolation and identification of the phenazine antibiotics. *Agric Biol Chem.* 1988, 52(2), 301–306.

Ulukanli, Z., Digrak, M. Alkaliphilic microorganisms and habitats. *Turk J Biol.* 2002, 26, 181–190.

Vasavada, S. H., Thumar, J. T., Singh, S. P. Secretion of a potent antibiotic by salt-tolerant and alkaliphilic actinomycete *Streptomyces sannanensis* strain RJT–1. *Curr Sci.* 2006, 91(10), 1393–1397.

Waksman, S. A. The Actinomycetes. Watham Publishers, Massachusetts, USA, 1954.

Webster, N. S., Watts, J. E. M., Hill, R. T. Detection and phylogenetic analysis of novel crenarchaeote and Euryarchaeote 16S ribosomal RNA gene sequences from a Great Barrier Reef sponge. *Mar Biotechnol.* 2001, 3, 600–608.

Zahner, H., Fielder, H. P. The need for new antibiotics: possible ways forward. In *Fifty Years of Antimicrobials: Past Perspectives and Future Trends*, Hunter, P. A., Darby, G. K., Russell, N. J. (Eds.), Cambridge University Press: Cambridge, UK, 1995, pp. 67–83.

CHAPTER 3

INDUSTRIAL PRODUCTION AND APPLICATIONS OF YEAST AND YEAST PRODUCTS

REBECCA S. THOMBRE[1] and SONALI JOSHI[2]

[1]*Department of Biotechnology, Modern College of Arts, Science and Commerce, Pune, Maharashtra, 411005, India*

[2]*Department of Biotechnology, Fergusson College, Shivajinagar, Pune, Maharashtra, 411004, India*

CONTENTS

3.1   INTRODUCTION

Yeast is one of the most commonly used organisms with immense applications in food and dairy industry (Table 3.1). Yeast has been used in many ancient civilizations for the process of fermentation in the preparation of wine or bread. The ancient humans did not know that they 'leaven' that they were using are nothing but yeast. Yeasts are unicellular fungi that occur in the natural environment on fruits, leaves of plants, in soil and in aquatic and environments. Lodder (1970) defined yeasts as unicellular fungi that reproduced by budding or fission as their mode of asexual reproduction. However, this definition lacked the description of dimorphic yeast species that produce hyphae/pseudohyphae in addition to unicellular growth (Flegel, 1977; Kendrick, 1987).

3.2   CLASSIFICATION

Oberwinkler classified yeasts as unicellular organisms belonging to either ascomycetes or basidiomycetes group of fungi (Van der Walt, 1987). During sexual reproduction, Ascomycetous yeasts (i.e., yeasts of interest in this study) form asci that are sac like structures containing ascospores (Van der Walt, 2000). Whereas, basidiomycetous yeasts reproduce by budding or fission and during sexual reproduction, their spores are not enclosed in a fruiting body (Boekhout and Kurtzman, 1996; Kurtzman and Fell, 1998; Querol and Belloch, 2003). Yeast taxonomy includes identification, nomenclature and classification of yeast

TABLE 3.1 Common Genera of Yeast Occurring in Food

Genus	Features	Application	Examples
Brettanomyces	Terminally budding, asporogenous yeast	Production of Acetic acid from glucose	B. intermedius
Candida	Anamorphic yeast devoid of pigments, some species are pathogenic	Used for fermentation of cocoa beans and is used in Kefir grains	C. lipolytica C. albicans (pathogenic)
Cryptococcus	Asporogenous, budding yeast and have typical red-orange color	Commonly present on plants, in soil, on fruits and berries	
Debaryomyces	Pseudomycelia producing asporogenous yeast	Dairy fermentations	D. hansenii
Hanseniaspora	Apiculate yeast that show bipolar budding producing lemon shaped spores	Used in sugar fermentations	H. uvarum
Issatchenkia	Pseudomycelia producing yeast that multiply by lateral budding	Food fermentations	I. orientalis
Kluyveromyces	Ascospore producing yeast multiplies by binary fission	Production of β-galactosidase and laccase	K. fraglis K. lactis K. marxianus
Pichia	Largest genus of true yeast, produce pseudomycelia and arthrospcres	Forms membranes on food	P. membranaefaciens
Rhodotorula	Teliospore producing basidiomycetes, produce typical red-pink pigment	Cause spoilage of fish and dairy products	R. lactose

TABLE 3.1 Continued

Genus	Features	Application	Examples
Saccharomyces	Ascosporogenous yeast, multiply by multilateral budding and produce spherical ascospores	Prevalent in many food fermentations as a starter culture as Bakers, Brewers or Distillers yeast	S. cerevisiae S. bayanus S. boulardii S. dairenensis S. exiguous S. martiniae S. monacensis S. paradoxus S. uvarum
Schizosaccharomyces	Ascosporogenous yeast, multiply by lateral fission and produce true hyphae and ascospores	Degrades malic acid in wine causing wine spoilage	S. pombe
Torulaspora	Ascosporogenous yeast, multiply by multilateral budding	Sugar fermentors	T. delbrueckii
Trichosporon	Non ascospore forming yeast, multiply by budding and arthroconidia formation	Cocoa bean and idli fermentation	T. pullulans
Zygosaccharomyces	Ascosporogenous yeast, multiply by multilateral budding and produce bean shaped ascospores	Causes spoilage in food	Z. bailii

based on certain characteristics. Ascomycetes are classified as Hemiascomycetes and Euascomycetes on the basis of formation of asci enclosed within or out of the fruiting body. Barnett et al. (2000) have described 93 phenotypic characters for identification of yeast. The current classification of the ascomycetous yeasts is based on the biological species concept. The major yeast belonging to ascomycota are of the orders Schizosaccharomycetales (fission yeasts) and the Saccharomycetales. Yeast identification was based only conventional identification methods previously. But now advanced yeast identification techniques, fatty acids profiles, mean molar percentage guanine plus cytosine (mol% G+C) and the application of two universal and two species-specific primers derived from the D1/D2 region of the 26S rDNA is used for identification (Herzberg et al., 2002; Daniel and Meyer, 2003). For example, the common yeast *Saccharomyces cerevisiae* (Figure 3.1) can be classified as:

Domain: Eukarya
Kingdom: Fungi
Phylum: Ascomycota
Subphylum: Saccharomycotina
Class: Saccharomycetes
Order: Saccharomycetales
Family: Saccharomycetaceae
Genus: *Saccharomyces*
Species: *cerevisiae*

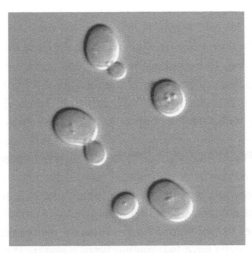

FIGURE 3.1 Typical budding yeast *Saccharomyces cerevisiae*.

3.3   PHYSIOLOGY AND BIOCHEMISTRY OF YEAST

Yeast produce larger cell size as compared to bacteria and demonstrate typical spherical, oval, elliptical or elongate cells ranging from 5–10 µm in size. Nutritionally yeasts are typical chemoheterotrophic organisms. They obtain energy by oxidation of organic compounds. They utilize carbohydrate sugars like hexoses and pentoses (Barnett, 1975). Yeasts are aerobic organisms and some species are facultative anaerobes. The yeast cell wall is composed of phosphorylated mannan, mannan, β-glucan, chitin and mannoprotein. The pH optimum for cultivation of yeast is around 5–7.5; however some species may be able to survive at a broader pH range. Similarly, most yeast grows optimally at room temperature, however yeast can tolerate high temperatures (Bakers yeast) and also low temperature (Watson, 1976). The common medium used for cultivation of yeast are Saborauds medium, Potato dextrose agar, Glucose yeast extract agar and Malt agar. The role of yeast in fermentation was described by Louis Pasteur. Since then, yeasts have been well known for the role as starter cultures in fermentations. Yeast can metabolize hexose sugar (glucose) via the glycolytic sequence (Embden Mayerhoff pathway). Some yeast like *Zymomonas* sp. utilizes the Entner-Duodorhoff pathway for breakdown of glucose to pyruvate. The key enzyme that yeast utilize in alcoholic fermentation is alcohol dehydrogenase.

3.4   GROWTH AND REPRODUCTION

Yeast reproduces by two methods, viz. asexual and sexual reproduction. In asexual reproduction, yeast divides by an asymmetric budding or by symmetric binary fission. In budding, the yeast cells (commonly referred to as mother cell) develop a protuberance or a bulge called as a bleb. The daughter cells buds from this bleb and forms a new cell. Sometimes chains of new cells are seen on a single mother cell (Figure 3.2). When there are plenty of nutrients available, *Saccharomyces cerevisiae*, the budding yeast reproduces by mitosis as diploid cells. But when cells are exposed to nutrient limitation, this yeast undergoes meiosis to form haploid spores. Binary fission is observed in *Schizosaccharomyces pombe* in which the mother cell divides by mitosis to form two identically sized daughter cells (Balasubramanian et al., 2004).

Yeast undergoes sexual reproduction or mating under nutrient or oxidative stress. Most of the cells have two mating types a and α. The two mating type

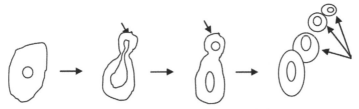

FIGURE 3.2 Budding and fission in yeast (asexual reproduction).

of cells, a and α fuse in a process called as plasmogamy after which the nuclei also fuse and this process is called as karyogamy. This gives rise to a diploid cell that undergoes meiosis in an ascus to produce four haploid spores (Figure 3.3). Alternatively, the diploid cell formed after karyogamy can also undergo budding (Bernstein and John, 1989; Davey, 1998; Neiman, 2005; Yeong, 2005).

3.5 YEAST PRODUCTS

Yeasts are one of the most versatile organisms having found applications in wide array of areas that include production of alcoholic beverages, baking industry, as probiotics, in bioremediation, nutritional supplements, in genetics, molecular biology and recombinant DNA technology to name a few. Some of these applications are discussed here.

3.5.1 ALCOHOLIC BEVERAGES

Beverages that contain ethanol are called alcoholic beverages. This ethanol is produced by the metabolism of carbohydrates, fermentation, by certain species of yeasts under anaerobic or microaerophillic conditions. The use of fermentation for production of alcoholic beverages like beer, wine dates back to many centuries. In fact production of alcohol is one of the earliest industries set up. Beverages such as wine and beer, mead, or distilled liquors all use yeast at some stage of their production. According to the intended use the yeasts are differently named as Brewing yeast, Distiller's yeast, Wine yeast and Sake yeast. Among the yeast species, selected and adapted strains of *Saccharomyces cerevisiae* are widely used for the production of beer, wine, distillates and ethanol (Bekatorou, 2006). The variety of yeast selected for brewery, distillery and winery operations function to effect maximum ethanol yield.

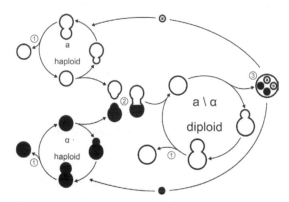

FIGURE 3.3 Life cycle of yeast: (1) Asexual reproduction (budding) with two different diploid mating types (a and α); (2) Sexual reproduction (conjugation); (3) Production of haploid spores by meiosis.

3.5.2 BREWING YEASTS

Pure cultures of Brewer's yeast are commercially manufactured and supplied to the brewing industry. Brewing yeasts are traditionally classed as "top-fermenting" and "bottom-fermenting" (Priest and Fergus, 2006). This distinction was introduced by the Dane Emil Christian Hansen (Hutkins, 2008). Top fermenting yeasts are those yeast that rise to the surface and are active the top of fermentation liquids. They are best used at temperatures ranging from 10 to 25°C. Top fermenting yeast, for example, *Saccharomyces cerevisiae* are used for producing ale, porter, stout and wheat beer. *S. uvarum* (earlier known as *S. carlsbergensis*), is a bottom fermenting yeast. They ferment well at low temperatures (7 to 15°C) and develop at the bottom of the vat. They are used for the production of lager beers like Pilsners, Bocks and American malt Liquor (Goldammer, 2000).

3.5.3 DISTILLER'S YEAST

Ethanol (as industrial solvent) and distilled liquors such as whiskey, tequila, rum and brandy are produced by using Distiller's yeast. They are usually isolated from beet or cane molasses and fermented fruits. Distiller's yeast must exhibit certain properties that will influence their selection for industrial use. Some of the prerequisite properties are rapid fermentation, high alcohol yield, high stress tolerance, high alcohol tolerance, low foam formation, lower amounts of higher alcohols, ethyl esters, aldehydes, fatty acids for fine quality products, and utilization of wide variety of substrates like wheat, barley, potato corn, etc. (Becze, 1964; Laube et al., 1987; Ibragimova, 1995).

3.5.4 WINE YEASTS

The strains of yeast are selected for the wine production for their alcoholic fermentation capability and organoleptic qualities, such as taste, flavor, aroma, color, tannin and glycerol contents, it imparts to the product. The selection of yeast strains is also influenced by geographical area, climate and type of grapes. A wide variety of pure yeast cultures, mainly *Saccharomyces* (*S. cerevisiae*, *S. bayanus*, *S. uvarum*, *S. oviformis*, *S. carsbergensis*, *S. chevalieri*, *S. diastaticus*, *S. fructum*, *S. pasteurianus*, *S. sake* and *S. vini*), are produced on large scale for wine production (Bekatorou, 2006). These pure cultures of yeasts, intended for use in wine manufacturing, have high tolerance to sulfur dioxide and can outgrow wild or indigenous yeasts present on the grapes. Some of the wine makers and researchers believe that, indigenous yeasts are better tailored to a particular must and have the potential of industrial starters. It is further argued that the utilization of selected autochthonous yeast strains would be a powerful tool to enhance organoleptic and sensory properties of typical regional wines. But compared to inoculated yeast, these ambient yeasts hold the risk of having a more unpredictable fermentation resulting in off-flavors/ aromas and higher volatile acidity and also the potential for a stuck fermentation if the indigenous yeast strains are not vigorous enough to fully convert all the sugars (Robinson, 2006). Pure yeast cultures are also used to conduct specific types of fermentations, like bottle fermentation of Champagne and sparkling wines, or to treat stuck and sluggish fermentations.

3.5.5 BAKER'S YEAST

Baker's yeast is the common name for the strains of yeast commonly used as a leavening agent in baking bread and bakery products, where it converts the fermentable sugars present in the dough into carbon dioxide and ethanol imparting soft and spongy character to the product. Baker's yeast is of the species *Saccharomyces cerevisiae* (Young et al., 2007). It is not known when yeast was first used to bake bread; the earliest definite records come from Ancient Egypt. Baker's yeast is available in a number of different forms, the main differences being the moisture contents (Young et al., 2007). They include Cream yeast, Compressed yeast, Active dry yeast (ADY) and Instant yeast. The choice primarily depends on the intended use. For industrial bakeries mainly cream and compressed yeasts are used which consists of *S. cerevisiae*. ADY are made by encapsulating yeast cells in the form of grains or beads, with leavening power and require rehydration before use. Instant yeasts are similar to active dry yeast, but have smaller granules (Reinhart, 2001), more perishable than active dry yeast but does not require rehydration. Inactive dry yeast is used for conditioning of dough properties in baking or for characteristic flavor development.

3.5.6 PROBIOTIC YEASTS

The microbial cell preparations or components of microbial cells that have beneficial effect on the health and well being of the host are called probiotics (Salminen et al., 1999). Yeasts have been reported to be resistant to antibiotics, safe, to colonize the intestine quickly and not to persist permanently in the intestine, which make them the most suitable candidate for use as probiotics. *S. cerevisiae* has been reported to survive in gastrointestinal (GI) tract and are antagonistic towards GI pathogens. The yeast, *S. boulardii*, a thermophilic non pathogen, has been widely used as livestock feed probiotic for many years. It is also used as therapeutic agent for treatment of diarrhoea. *Candida saitona, C. pintolopesii* are examples of other yeasts allowed and commonly used probiotic additives in animal feed (Bovill, 2001; Leuschner et al., 2004). Yeast and yeast derived preparations can provide inexpensive feed supplements that can have a major impact when used in poultry management system. They control the composition of microbial population in gastrointestinal tract, prevent colonization of pathogens, bind toxins and modulate immune system thereby increasing the efficiency of poultry production system (Dawson, 2001).

3.5.7 NUTRITIONAL SUPPLEMENTS

Yeasts are also used as nutritional supplements. It is often referred to as "nutritional yeast" when sold as a dietary supplement. Nutritional yeast, inactivated yeast, usually *S. cerevisiae*, is an excellent source of protein and vitamins, especially the B-complex vitamins, as well as other minerals and cofactors required for growth. It is also naturally low in fat and sodium. Nutritional yeast has a nutty, cheesy flavor that makes it popular as an ingredient in cheese substitutes. It is also used as a topping for popcorn, in mashed and fried potatoes, as well as in scrambled eggs.

3.5.8 TIBICOS

Tibicos are a culture of bacteria and yeasts held in a polysaccharide biofilm matrix created by the bacteria. The microbes present in tibicos act in symbiosis to maintain a stable culture. Tibicos can do this in many different sugary liquids, feeding off the sugar to produce lactic acid, alcohol (ethanol), and carbon dioxide to give carbonated drink. Tibicos is also known as tibi, water kefir grains, sugar kefir grains, Japanese water crystals and California bees, and in older literature as bébées, African bees, ale nuts, Australian bees, balm of Gilead, beer seeds, beer plant, bees, ginger bees, Japanese beer seeds and vinegar bees (Kebler, 1921). Tibicos are found around the world, with no two cultures being exactly the same. Typical tibicos have a mix of *Lactobacillus*, *Streptococcus*, *Pediococcus* and *Leuconostoc* bacteria with yeasts from *Saccharomyces*, *Candida*, *Kloeckera* and possibly others. *Lactobacillus brevis* has been identified as the species responsible for the production of the polysaccharide (dextran) that forms the grains (Horisberger, 1980).

3.5.9 YEAST EXTRACT

Various forms of processed yeast products made by extracting the cell contents (removing the cell walls) are called yeast extracts. Yeast extract is the product of enzymatic digestion of the yeast cell constituents by endogenous and exogenous yeast enzymes. It is rich in peptides, amino acids, nucleotides and vitamins; therefore it is good for use as supplement in culture media. It is also used in pharmaceuticals, as well as flavor and taste enhancer (replacing glutamates and nucleotides). Autolysed yeasts and hydrolyzed yeasts are also used as food additives or flavorings or as nutrients for bacterial culture media.

3.5.10 SOURDOUGH STARTER

Sourdough is a mixture of flour and water, containing yeasts and lactic acid bacteria, used as starter culture to leaven bread. Sourdoughs are produced at commercial level using various combinations of yeasts and bacteria, and are used for the conditioning of dough, improvement of preservation time and the development of breads and baking products with special organoleptic properties (Plessas, 2004).

3.6 YEASTS IN BIOREMEDIATION

Bioremediation is a process that involves the use of plants or microorganisms, viable or not, natural or genetically engineered to treat environments contaminated with organic molecules that are difficult to break down (xenobiotics) and to mitigate toxic heavy metals, by transforming them into substances with little or no toxicity, hence forming innocuous products (Dobson and Burgess, 2007; Li and Li, 2011). The use of microbial metabolic ability for degradation/removal of environmental pollutants provides an economic and safe alternative compared to other physicochemical methodologies. The potential applications of yeast cells in the field of bioremediation have been well investigated. Yeasts have been reported to degrade alkanes, fatty acids, TNT and other pollutants. Among the different contaminants, heavy metals have received special attention due to their strength and persistence in accumulating in ecosystems, where they cause damage by moving up the food chain to finally accrue in human beings, who are at the top of this chain (Volesky, 2001; Ahluwalia and Goyal, 2007; Machado et al., 2008). Several researchers have reported the application of yeasts in removal of heavy metals such as arsenic, chromium, etc. from industrial effluents. Live, dead or immobilized yeast cells can be used for bioremediation of heavy metals (Oswal, 2002; Jain et al., 2004; Fickers et al., 2005; Bankar et al., 2009; Soares and Soares, 2011, 2012).

3.7 YEASTS IN MEDICINE AND CANCER RESEARCH

Cancer is a devastating disease with a profound impact on society. In recent years, yeast has provided a valuable contribution with respect to uncovering the molecular mechanisms underlying this disease, allowing the identification of new targets and novel therapeutic opportunities. Indeed, several attributes make yeast an ideal model system for the study of human diseases. It combines a high level of conservation between its cellular processes and those of mammalian cells, with advantages such as a short generation time, ease of genetic manipulation and a wealth of experimental tools for genome- and proteome-wide analyzes. Additionally, the heterologous expression of disease-causing proteins in yeast has been successfully used to gain an understanding of the functions of these proteins and also to provide clues about the mechanisms of disease progression. Yeast research performed in recent years has demonstrated the tremendous potential of this model system, especially with the validation of findings obtained with yeast in more physiologically relevant models (Periera et al., 2012). One of the first scientists to discover some of these cancer-causing mutations was a biologist named Leland H. Hartwell. Hartwell eventually shared the 2001 Nobel Prize in Physiology or Medicine with Paul Nurse and R. Timothy Hunt, for their discoveries of protein molecules that control the division (duplication) of cells. Hartwell used the easy-to-manipulate, single-celled eukaryote, *Saccharomyces cerevisiae* as a model system for studying cancer and the cell cycle. The same genes that control the cell cycle in baker's yeast (and that malfunction in tumor cells) exist in more or less the same capacity in human cells. Thus, yeast cells have contributed in basic understanding of cell cycle regulation and cancer genetics (Pray, 2008).

Apoptosis in yeast is considered a new model system with applications in cell biology and medicine. Apoptosis in yeast promises to provide a better understanding of the genetics of apoptosis. During the past two years, scientists were successful in identifying new cell-death regulators of humans, plants and fungi using *Saccharomyces cerevisiae* (Madeo et al., 2002). Yeasts, such as *Saccharomyces cerevisiae*, have long served as useful models for the study of oxidative stress, an event associated with cell death and severe human pathologies. Farrugia and Balzan (2012) have reviewed oxidative stress in yeast, in terms of sources of reactive oxygen species (ROS), their role in ageing, their molecular targets, and the metabolic responses elicited by cellular ROS accumulation. Responses of yeast to accumulated ROS include upregulation of antioxidants mediated by complex transcriptional changes, activation of pro-survival pathways such as mitophagy, and programmed cell death (PCD) which, apart from apoptosis, includes pathways such as autophagy and necrosis, a form of cell death long considered accidental and uncoordinated.

A killer yeast is a yeast, such as *Saccharomyces cerevisiae*, which is able to secrete one of a number of toxic proteins which are lethal to receptive cells. The killer yeast system was first described in 1963 (Beaven and Makower, 1963). These yeast cells are immune to the toxic effects of the protein due to an intrinsic immunity (Breinig et al., 2006). The killer phenomenon in yeasts has been revealed to be a multicentric model for molecular biologists, virologists, phytopathologists, epidemiologists, industrial and medical microbiologists, mycologists, and pharmacologists. The best characterized toxin system is from yeast (*Saccharomyces cerevisiae*), which was found to spoil brewing of beer. In *S. cerevisiae*, these toxins are encoded by a double-stranded RNA virus, translated to a precursor protein, cleaved and secreted outside of the cells, where they may affect susceptible yeast. There are other killer systems in *S. cerevisiae*, such as KH (Gotto et al., 1990) and KHS (Gotto et al., 1991) genes encoded on chromosome. The susceptibility to toxins varies greatly between yeast species and strains. This property has been used to reliably identify different strains. Many researchers have used this basis to differentiate between strains of fungi, yeast and bacteria (Morace et al., 1984; Morace et al., 1989; Vaughan et al., 1996; Buzzini and Martini, 2001). Study of killer toxins helped to better understand the secretion pathway of yeast, which is similar to those of higher eukaryotes. It also can be used in treatment of some diseases, mainly that caused by fungi.

Others experimented with using killer yeasts to control undesirable yeasts. Palpacelli et al. (1991) found that *Kluyveromyces phaffii* was effective against *Kloeckera apiculata*, *Saccharomycodes ludwigii* and *Zygosaccharomyces rouxii* – all of which cause problems in the food industry (Palpacelli, 1991). Polonelli et al. (1994) used a killer yeast to vaccinate against *C. albicans* in rats. Lowes et al. (2000), created a synthetic gene for the toxin HMK normally produced by *Williopsis mrakii*, which they inserted into *Aspergillus niger* and showed that the engineered strain could control aerobic spoilage in maize silage and yoghurt. Ciani and Fatichenti (2001) used a toxin-producing strain of *Kluyveromyces phaffii* to control apiculate yeasts in wine-making. A toxin produced by *Candida nodaensis* effective at preventing spoilage of highly salted food by yeasts has been reported (Da Silvaa et al., 2007). Several experiments suggest that antibodies that mimic the biological activity of killer toxins have application as antifungal agents (Malgiani, 1997, 2004).

3.8   YEAST AS A MODEL ORGANISM FOR GENETIC STUDIES AND RECOMBINANT DNA TECHNOLOGY

The budding yeast *Saccharomyces cerevisiae* and *Schizosaccharomyces pombe* (fission yeast) have emerged as a versatile and robust model system of eukaryotic genetics. *S. cerevisiae* is an attractive model organism due to the fact that its genome has been sequenced, its genetics are easily manipulated, and it is very easy to maintain in the laboratory. Studies performed in yeast can help us to determine how a particular gene or protein functions in higher eukaryotes (including humans), because many yeast proteins are similar in sequence and function to those found in other organisms. Mutant screening and segregation analysis are simpler and easier to perform in yeast than in multicellular organisms; and fundamental eukaryotic biology, such as cell cycle control and alternation of generations, is well-conserved throughout eukaryotic taxa (Griffiths et al., 2000; Mell and Burgess, 2002). The yeast *Saccharomyces cerevisiae* has also become the most sophisticated eukaryotic model for recombinant DNA technology. One of the main reasons is that the transmission genetics of yeast is extremely well understood, and the stockpile of thousands of mutants affecting hundreds of different phenotypes is a valuable resource when using yeast as a molecular system. In yeast, another important advantage is the availability of a circular 6.3-kb natural yeast plasmid known as the "2-micron" plasmid forms the basis for several sophisticated cloning vectors (Griffiths et al., 2000).

3.8.1 YEAST VECTORS

A wide range of vectors are available to meet various requirements for insertion, deletion alteration and expression of genes in yeast. Most plasmids used for yeast studies are shuttle vectors, which contain sequences permitting them to be selected and propagated in *E. coli*, thus allowing for convenient amplification and subsequent alteration *in vitro*. The most common yeast vectors originated from pBR322 and contain an origin of replication (ori), promoting high copy-number maintenance in *E. coli*, and the selectable antibiotic markers, the β-lactamase gene, or AmpR), and sometime to tetracycline-resistance gene, tet or (TetR), conferring resistance to, respectively, ampicillin and tetracycline. Although there are numerous kinds of yeast shuttle vectors, those used currently can be broadly classified as integrative vectors, YIp; autonomously replicating high copy-number vectors, YEp; autonomously replicating low copy-number vectors, YCp or another type of vector YACs (Yeast Artificial Chromosomes), for cloning large fragments (Table 3.2).

TABLE 3.2 Types of Yeast Vectors

Type of vector	Description
YIp Vectors	The YpI integrative vectors do not replicate autonomously, but integrate into the genome at low frequencies by homologous recombination. Integration of circular plasmid DNA by homologous recombination leads to a copy of the vector sequence flanked by two direct copies of the yeast sequence.
YEp Vectors	The YEp yeast episomal plasmid vectors replicate autonomously because of the presence of a segment of the yeast 2 μm plasmid that serves as an origin of replication (2 μm ori). The 2 μm ori is responsible for the high copy-number and high frequency of transformation of YEp vectors.
YCp Vectors	The YCp yeast centromere plasmid vectors are autonomously replicating vectors containing centromere sequences, CEN, and autonomously replicating sequences, ARS.

3.8.2 YEAST ARTIFICIAL CHROMOSOMES

The initial step in the molecular characterization of eukaryotic genomes generally requires cloning of large chromosomal fragments, which is usually carried out by digestion with restriction endonucleases and ligation to specially developed cloning vectors. Usually 200 to 800 kb fragments are cloned as Yeast Artificial Chromosomes (YACs), and 100–200 kb fragments are cloned as Bacterial Artificial Chromosomes (BACs). The importance of YAC technology has been heightened by the recently developed methods for transferring YACs to cultured cells and to the germline of experimental animals (Figure 3.4). YAC cloning systems are based on yeast linear plasmids, denoted YLp, containing homologous or heterologous DNA sequences that function as telomeres (TEL) *in vivo*, as well as containing yeast ARS and CEN (centromeres) segments (Schneiter, 2004).

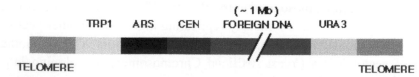

FIGURE 3.4 Yeast artificial chromosome (YAC) (TRP1, URA3: marker genes, ARS: Autonomously replicating sequence, CEN: centromere).

3.9 CONCLUSION

Yeasts are a diverse group of unicellular chemoorganotrophic eukaryotes. Their applications in brewing and fermentation are undoubtedly one of the most significant criteria of their importance. The unique heat and ethanol resistance makes it an ideal starter culture for food and beverage fermentations especially in bakeries and confectionaries. The increasing use of yeast as a model to study eukaryotic genetics signifies the relevance of this miniscule organism. Many yeast are known to be pathogenic. Though many new organisms are still being discovered, yeast like *Saccharomyces cerevisiae* still continue to dominate the fermentation, food and alcohol industry till date.

KEYWORDS

- Alcohol
- Bakers yeast
- Brewers yeast
- Fermentation
- Food industry
- Fungi
- Nutritional supplements
- Probiotics
- *Saccharomyces cerevisiae*
- Starter culture
- Wine
- Yeast
- Yeast extract

REFERENCES

Ahluwalia, S. S., Goyal, D. Microbial and plant derived biomass for removal of heavy metals from wastewater. *Bioresour Technol* 2007, 98(12), 2243–2257.

Arthur, H; Watson, K. Thermal adaptation in yeast: growth temperatures, membrane lipid, and composition of psychrophilic, mesophilic, and thermophilic yeast. *Journal of Bacteriology* 1976, 128(1), 56–68.

Balasubramanian, M. K., Bi, E., Glotzer, M. Comparative analysis of cytokinesis in budding yeast, fission yeast and animal cells. *Current Biology* 2004, 14(18), R806–818.

Bankar, A. V., Kumar, A. R., Zinjarde, S. S. Environmental and industrial applications of *Yarrowia lipolytica*. *Applied Microbiology and Biotechnology* 2009, 84(5), 847–865.

Bankar, A. V., Kumar, A. R., Zinjarde, S. S. Removal of chromium (VI) ions from aqueous solution by adsorption onto two marine isolates of *Yarrowia lipolytica*. *Journal of Hazardous Materials* 2009,170(1), 487–494.

Barne, J. A., Payne, R. W., Yarrow, D. How yeasts are classified. In *Yeasts: Characteristics and Identification*, Barnett, J. A., Payne, R. W., Yarrow, D., Ed., Cambridge University Press: Cambridge, 2000, pp. 15–22.

Barnett, J. A. The entry of D-ribose into some yeasts of the genus *Pichia*. *Journal of General Microbiology* 1975, 90(1), 1–12.

Bernstein, C., Johns, V. Sexual reproduction as a response to H_2O_2 damage in *Schizosaccharomyces pombe*. *J Bacteriol* 1989, 171(4), 1893–1897.

Bevan, E. A., Makower, M. The physiological basis of the killer character in yeast. *Proc XIth Int Congr Genet* 1963, 1, 202–203.

Boekhout, T., Kurtzman, C. P. Principles and methods used in yeast classification, and overview of currently accepted genera. In *Non conventional yeasts in Biotechnology, A Handbook*, Wolf, K., Ed., Springer-Verlag: Berlin, 1996, pp. 1–80.

Bovill, R., Bew, J., Robinson, S. Comparison of selective media for the recovery and enumeration of probiotic yeasts from animal feed. *Int J Food Microbiol* 2001, 67, 55–61.

Breinig, F., Sendzik, T., Eisfeld K., Schmitt, M. J. Dissecting toxin immunity in virus-infected killer yeast uncovers an intrinsic strategy of self-protection. *Proceedings of the National Academy of Sciences* 2006, 103(10), 3810–3815.

Buzzini, P., Martini, A. Discrimination between *Candida albicans* and other pathogenic species of the Genus *Candida* by their differential sensitivities to toxins of a panel of killer yeasts. *Journal of Clinical Microbiology* 2001, 39(9), 3362–3364.

Ciani, M., Fatichenti, F. Killer toxin of *Kluyveromyces phaffii* DBVPG 6076 as a biopreservative agent to control apiculate wine yeasts. *Applied and Environmental Microbiology* 2001, 67(7), 3058–3063.

Daniel, H. M., Meyer, W. Evaluation of ribosomal RNA and actin gene sequences for the identification of ascomycetous yeasts. *Int J Food Microbiol* 2003, 86, 61–78.

Dasilva, S., Calado, S., Lucas, C., Aguiar, C. Unusual properties of the halotolerant yeast *Candida nodaensis* killer toxin, CnKT. *Microbiological Research* 2008, 163(2), 243–251.

Davey, J. Fusion of a fission yeast. *Yeast* 1998, 14(16), 1529–1566.

Dawson, K. A. The application of yeast and yeast derivatives in the poultry industry. *Proc Aust Poult Sci Sym* 2001, 100–105.

De Becze, G. I. Reproduction of distillers' yeasts. *Biotechnol Bioeng* 1964, 6, 191–221.

Dobson, R. S., Burgess, J. E. Biological treatment of precious metal refinery wastewater: A review. *Miner Eng* 2007, 20, 519–532.

Duncan, J., Brady, D., Wilhelmi, B. Immobilization of yeast and algal cells for bioremediation of heavy metals. In *Bioremediation Protocols*, Sheehan, D., Ed., Humana Press Inc: New York, 1997, pp. 91–97.

Farrugia, G., Balzan, R. Oxidative stress and programmed cell death in yeast. *Front Oncol* 2012, 2, 64.

Fickers, P., Benetti, P. H., Wache, Y., Marty, A., Mauerberger, S., Smit, M. S., Nicaud, J. M. Hydrophobic substrate utilization by the yeast *Yarrowia lipolytica*, and its potential applications. *FEMS Yeast Research* 2005, 5(6–7), 527–543.

Flegel, T. W. Let's call yeast a yeast. *Can J Microbiol* 1977, 23, 945–946.

Goldammer, T. *The Brewers' Handbook: The Complete Book to Brewing Beer*, Apex Publishers: Clifton, Virginia, USA, 2000.

Goto, K., Iwatuki, Y., Kitano, K., Obata, T., Hara, S. Cloning and nucleotide sequence of the KHR killer gene of *Saccharomyces cerevisiae*. *Agricultural and Biological Chemistry* 1990, 54 (4), 979–984.

Goto, K., Fukuda, H., Kichise, K., Kitano, K., Hara, S. Cloning and nucleotide sequence of the KHS killer gene of *Saccharomyces cerevisiae*. *Agricultural and Biological Chemistry* 1991, 55(8), 1953–1958.

Griffiths, A. J. F., Miller, J. H., Suzuki, D. T., Lewontin, R. C., Gelbart, W. M. In *An Introduction to Genetic Analysis, 7th ed.*, W. H. Freeman: New York, 2000.

Herzberg, M., Fischer, R., Titze, A. Conflicting results obtained by RAPD-PCR and large subunit rDNA sequences in determining and comparing yeast isolates isolated from flowers: a comparison of two methods. *Int J Syst Evol Microbiol* 2002, 52, 1423–1433.

Horisberger, M., Bauer, H. The structural organization of the Tibi grain as revealed by light, scanning and transmission microscopy. *Archives of Microbiology* 1980, 128(2), 157–161.

Hutkins, R. W. In *Microbiology and Technology of Fermented Foods*. John Wiley and Sons: USA, 2006, 69.

Ibragimova, S. I., Kozlov, D. G., Kartasheva, N. N., Suntsov, N. I., Efremov, B. D., Benevolensky, S. V. A strategy for construction of industrial strains of distillers yeast. *Biotechnol Bioeng* 1995, 46, 285–290.

Jain, M. R., Zinjarde, S. S., Deobagkar, D. D., Deobagkar, D. N. 2,4,6-trinitrotoluene transformation by a tropical marine yeast *Yarrowia lipolytica* NCIM 3589. *Marine Pollution Bulletin* 2004, 49(9–10), 783–788.

Kebler, L. F. California bees. *J Pharm Sci* 1921, 10(12), 939–943.

Kendrick, B. Yeasts and yeast-like fungi – new concepts and new techniques. *Stud Mycol* 1987, 30, 479–486.

Kurtzman, C. P., Fell, J. W. Definition, classification and nomenclature of the yeasts. In Kurtzman, C. P., Fell, J. W. (Ed.), *The yeasts, a taxonomic study*, Elsevier Science BV: Amsterdam, The Netherlands, 1998, pp. 3–5.

Laube, K., Wesenberg, J., Lietz, P. Selection of distiller's yeasts with particular respect to non-*Saccharomyces* strains. *Acta Biotechnol* 1987, 7, 111–118.

Leuschner, R. G. K., Bew, J., Fourcassier, P., Bertin, G. Validation of the official control method based on polymerase chain reaction (PCR) for identification of authorized probiotic yeast in animal feed. *Syst Appl Microbiol* 2004, 27, 492–500.

Li, Y., Li D. Study on fungi-bacteria consortium bioremediation of petroleum contaminated mangrove sediments amended with mixed biosurfactants. *Adv Mat Res* 2011, 183, 1163–1167.

Lodder, J. General classification of the yeasts. In Lodder, J. (Ed.), *The yeasts, a taxonomic study*, North-Holland Publish. Co.: Amsterdam, 1970, pp. 1–33.

Lowes, K. F., Shearman, C. A., Payne, J., MacKenzie, D., Archer, D. B., Merry, R. J., Gasson, M. J. Prevention of yeast spoilage in feed and food by the yeast mycocin HMK. *Applied and Environmental Microbiology* 2000, 66(3), 1066–1076.

Machado, M. D., Santos, M. S. F., Gouveia, C., Soares, H. M. V. M., Soares, E. V. Removal of heavy metal using a brewer's yeast strain of *Saccharomyces cerevisiae*: The flocculation as a separation process. *Bioresour. Technol* 2008, 99, 2107–2115.

Madeo, F., Engelhardt, S., Herker, E., Lehmann, N., Maldener, C., Proksch, A., Wissing, S., Fröhlich, K. U. Apoptosis in yeast: a new model system with applications in cell biology and medicine. *Curr Genet* 2002, 41(4), 208–216.

Magliani, W., Cont, S., Gerloni, M., Bertolotti, D., Polonelli, L. Yeast killer systems. *Clin Microbiol Review* 1997, 10(3), 369–400.

Magliani, W., Conti, S., Salati, A., Vaccari, S., Ravanetti, L., Maffei, D., Polonelli, L. Therapeutic potential of yeast killer toxin-like antibodies and mimotopes. *FEMS Yeast Research* 2004, 5(1), 11–18.

Mell, J. C., Burgess, S. M. Yeast as a model genetic organism. In *Encyclopedia of Life Sciences*, MacMillan: NY, USA, 2002, 1–8.

Morace, G., Archibusacci, C., Sestito, M., Polonelli, L. Strain differentiation of pathogenic yeasts by the killer system. *Mycopathologia* 1984, 84(2–3), 81–85.

Morace, G., Manzara, S., Dettori, G., Fanti, F., Conti, S., Campani, L., Polonelli, L., Chezzi, C. Biotyping of bacterial isolates using the yeast killer system. *European Journal of Epidemiology* 1989, 5(3), 303–310.

Neiman, A. M. Ascospore formation in the yeast *Saccharomyces cerevisiae*. *Microbiology and Molecular Biology Reviews* 2005, 69(4), 565–584.

Oberwinkler, F. Heterobasidiomycetes with ontogenic yeast stages – systematic and phylogenetic aspects. *Stud Mycol* 1987, 30, 61–74.

Oswal, N., Sarma, P. M., Zinjarde, S. S., Pant, A. Palm oil mill effluent treatment by a tropical marine yeast. *Bioresource Technology* 2002, 85(1), 35–37.

Palpacelli, V., Ciani, M., Rosini, G. Activity of different 'killer' yeasts on strains of yeast species undesirable in the food industry. *FEMS Microbiology Letters* 1991, 84, 75.

Pereira, C., Coutinho, I., Soares, J., Bessa, C., Leão, M., Saraiva, L. New insights into cancer-related proteins provided by the yeast model. *FEBS J* 2012, 279(5), 697–712.

Plessas, S., Pherson, L., Bekatorou, A., Nigam, P., Koutinas, A. A. Bread making using kefir grains as baker's yeast. *Food Chem* 2005, 93, 585–589.

Polonelli, L., De Bernardis, F., Conti, S., Boccanera, M., Gerloni, M., Morace, G., Magliani, W., Chezzi, C., Cassone, A. Idiotypic intravaginal vaccination to protect against candidal vaginitis by secretory, yeast killer toxin-like anti-idiotypic antibodies. *The Journal of Immunology* 1994, 152(6), 3175–3182.

Pray, L. H. Hartwell's yeast: A model organism for studying somatic mutations and cancer. *Nature Education* 2008, 1(1), 183.

Priest, F. G. *Handbook of Brewing*. CRC Press: Florida, 2006, p. 84.

Reinhart, P. *The Bread Baker's Apprentice: Mastering the Art of Extraordinary Bread*. Ten Speed Press: Berkeley, California, 2001.

Robinson, J. *The Oxford Companion to Wine*. Oxford University Press: UK, 2006, pp. 778–780.

Salminen, S., Ouwehand, A., Benno, Y., Lee, Y. K. Probiotics: How should they be defined? *Trends Food Sci Technol* 1999, 10, 107–110.

Schneiter, R. *Genetics, Molecular and Cell Biology of Yeast*. Universite de Fribourg: Suisse, 2004, pp. 43–88.

Soares, E. V., Soares, H. M. Bioremediation of industrial effluents containing heavy metals using brewing cells of *Saccharomyces cerevisiae* as a green technology: a review. *Environmental Science and Pollution Research* 2012, 19(4), 1066–1083.

Van der Walt, J. P. The yeasts – a conspectus. *Stud Mycol* 1987, 30, 19–31.

Vaughan-Martini, A., Cardinali, G., Martini, A. Differential killer sensitivity as a tool for fingerprinting wine-yeast strains of *Saccharomyces cerevisiae*. *Journal of Industrial Microbiology* 1996, 17(2), 124–127.

Volesky, B. Detoxification of metal-bearing effluents: biosorption for the next century. *Hydrometallurgy* 2001, 59(2), 203–216.

Wickner, R. B. Double-stranded RNA replication in yeast: the killer system. *Annual Review of Biochemistry* 1986, 55, 373–395.

Yeong, F. M. Severing all ties between mother and daughter: cell separation in budding yeast. *Molecular Microbiology* 2005, 55(5), 1325–1331.

Young, L., Cauvain, S. P. In *Technology of Bread Making*, Springer: Berlin, 2007, pp. 77–79.

CHAPTER 4

INDUSTRIAL APPLICATIONS OF BIOSURFACTANTS

SHILPA MUJUMDAR,[1] SHRADHA BASHETTI,[1] SHEETAL PARDESHI,[1] and REBECCA S. THOMBRE[2]

[1]Department of Microbiology, Modern College of Arts, Science and Commerce, Shivajinagar, Pune, Maharashtra, India

[2]Department of Biotechnology, Modern College of Arts, Science and Commerce, Pune, Maharashtra, 411005, India

CONTENTS

4.1 INTRODUCTION

Surfactants are amphiphilic surface active compounds that possess both hydrophilic and hydrophobic groups in their structure. They are termed as surfactants because they reduce surface and interfacial tensions by accumulating at the interface between two immiscible fluids like oil and water a liquid or that between a liquid and a solid (Saharan et al., 2011). Emulsifiers are a subclass of surfactants which stabilize emulsions of two immiscible liquids (Ron and Rosenberg, 2001). Surfactants and emulsifiers can be of chemical or biological origin. Biologically originated surfactants are called as biosurfactants. Since microorganisms usually grow attached to surfaces and concentrate at interfaces, wide variety of diverse organisms are reported to produce biosurfactants (Ron and Rosenberg, 2001). Because of their function, surfactants have many potential industrial and environmental applications related to emulsification, dispersion, detergency, foaming and solubilisation of hydrophobic compounds and wetting (Banat et al., 2000; Dastgheib et al., 2008). Biosurfactants are less toxic, more biodegradable, and highly active at extreme conditions of pH, temperature, salinity; hence become an attractive alternative to chemical surfactants (Saharan et al., 2011). In this chapter, we have emphasized on different applications of biosurfactants in microbial enhanced oil recovery, bioremediation, medicine, food, pharmaceutical and agriculture industries.

Biosurfactants can be broadly grouped into two categories namely low and high molecular weight biosurfactants. Low and molecular weight biosurfactants consist of glycolipids and lipopeptides while high molecular weight biosurfactants consist of high molecular weight polymeric biosurfactants (Smyth et al., 2010). Structure based classification of biosurfactants is described in this chapter. Enormous amount of literature is now available on biosurfactants which describes its production in varied habitats and ecosystems. Different types of biosurfactants which also differ in their potency and function are produced in different ecosystems. This proves specific advantage of biosurfactants to microorganisms in specific environments (Ron and Rosenberg, 2001).

4.2 CLASSIFICATION OF BIOSURFACTANTS

Biosurfactants are classified based on their chemical composition and microbial origin. The chemical structure of biosurfactants contains hydrophilic and hydrophobic groups. Amino acids, peptides and polysaccharides can be present as hydrophilic moieties and saturated or unsaturated fatty acids can be present as hydrophobic moieties in the structure. Structure based classification has following major classes of biosurfactants (Guerra-Santos et al., 1987; Kooper and Goldenberg, 1987): (i) Glycolipids (Rhamnolipids, Trehalolipids and Sophorolipids); (ii) Lipopeptide and lipoprotein; (iii) Fatty acids, phospholipids and neutral lipids; (iv) Polymeric biosurfactants, and (v) Particulate biosurfactants.

Glycolipids, as the name indicates, are made of two structural components, carbohydrates and lipids. This type of biosurfactants is most abundantly found in microbial world. Lipids can be long chain aliphatic acids or hydroxyaliphatic acids (Gautam and Tyagi, 2006). Rhamnolipids have rhamnose in their structure which is connected to one or two molecules of β-hydroxydecanoic acid. The first report of production was in *Pseudomonas aeruginosa* (Edward and Hayashi, 1965; Hisatsuka et al., 1971). Like rhamnolipids, the carbohydrate moiety in trehalolipids is trehalose. Mycolic acid is found as the lipid counterpart in genera like *Mycobcterium*, *Corynebacterium* and *Nocardia*. The size and degree of mycolic acid moiety can vary in different organisms (Asselineau and Asselineau, 1978; Cooper et al., 1989). Sophorolipids have dimeric sophorose as in their structure along with long chain fatty acids. This type of glycolipid surfactants are mainly produced by yeast (Gautam and Tyagi, 2006).

Surfactin and iturin are well known lipopeptide biosurfactant produced by *Bacillus subtilis*. These are the most potent category of biosurfactants produced by microorganisms (Arima et al., 1968). Microorganisms are reported to produce fatty acids, phospholipids and neutral lipids when grown on n-alkane. *Acinetobucter* sp. and *Rhodobacter erythropolis* are reported to produce phosphatidylethanolamine (Kappeli and Finnerty, 1979; Ciriglino and Carman, 1985). Emulsan and liposan are well studied examples of polymeric biosurfactants. Emulsan is produced by *Acinetobacter calcoaceticus*, which is a very potent emulsifier and shows activity even at concentrations as low as 0.01–0.001% (Rosenberg et al., 1979). Liposan is reported to be produced by *Candida lipolytica*, and is extracellular water soluble emulsifier (Cirigliano and Carman, 1984). Particulate biosurfactants are extracellular membrane vesicles which form microemulsions and help microbial cells in uptake of alkane. This type is reported in *Acinetobacter* sp. strain HO1-N (Kappeli and Finnerty, 1979).

4.3 BIOSURFACTANTS IN MICROBIAL ENHANCED OIL RECOVERY

It has been well documented that only 40–50% of the oil present in the reservoir can be extracted using traditional oil recovery systems and these systems are costly for regular uses. Use of biosurfactants along with traditional systems has proved most cost effective (Sen, 2008; Banat et al., 2010). Microbial enhanced oil recovery (MEOR) employs biosurfactant producing microorganisms to extract remnants of oil from reservoirs which is trapped into the capillaries after primary (mechanical) and secondary (physical) recovery procedures (Banat et al., 2000; Pacwa-Plociniczak et al., 2011). MEOR was first proposed by Beckman in 1926, but it was given attention only after the work of Zobell and Russian investigators in the 1940s (Brown, 2010). Application of biosurfactants in bioremediation and oil spillage is reported since 1990s (Pattanathu et al., 2008).

There can be various mechanisms by which microorganisms can play an important role in oil recovery. They can produce certain products like acids from oil and other nutrients which ultimately dissolve carbonates and increases permeability. Some bacteria produce gases which dissolve in oil and reduce its viscosity. Production of biosurfactants is the most important mechanism followed by many bacteria. Biosurfactants act by reducing interfacial tension between oil/water and oil/rock. This causes reduction in the capillary forces which prevents oil from moving through rock pores. Biosurfactants also lead to formation of stable oil-water emulsion which allows removal of oil along with the injection water (Suthar et al., 2008). Apart from these mechanisms, biopolymers and enzymes produced by microorganisms also play role in oil recovery (Sen, 2008).

MEOR has also been reported to be used in areas where classical or modern enhanced oil recovery methods (EOR) have not been used or are non-profitable. Thus MEOR, since its discovery, has provided attractive alternative for the processes such as combustion, steam, miscible displacement, caustic surfactant-polymer flooding and others (Lazar et al., 2007).

There can be three different strategies of MEOR, first being production of biosurfactant separately and then its addition to the oil reservoir, second is inoculation of biosurfactant producing organisms in the oil reservoir and the third approach is to supply the oil reservoir with nutrients so that the native microflora will be enhanced and thus biosurfactant production will be stimulated (Singh et al., 2007). Among the three, the first approach is frequently adopted as the other two have major limitations with respect to tolerance of the organisms to the physiological conditions at oil reservoirs. But scientists are now trying to isolate biosurfactant producing organisms using extreme conditions of incubation which will select for strains which can survive at oil reservoirs (Agarwal and Sharma, 2009). Along with these research strategies, genetically modified *Pseudomonas* strains have been tried to increase production of biosurfactants which can be purified and then added to the oil reservoirs. But this approach was not successful due to limitations in stability of organism and regulation of biosurfactant production in *Pseudomonas* (Wang et al., 2007).

Rhamnolipid surfactants are mostly used for EOR. Whereas, other lipopeptides such as surfactin, emulsan and lichenysin have also proved useful (Sen, 2008). *Pseudomonas aeruginosa* rhamnolipid biosurfactants have been commercialized by Jeneil Biosurfactant, USA, mainly as a component to enhance bioremediation activity (Banat, 2010). Haddadin et al. (2009) have reported extraction of hydrocarbon from El-Lajjun oil shale in flask experiments using biosurfactants produced by *Rhodococcus ruber* and *R. erythropolis* (Jinfeng ct al., 2005). In various simulation studies using packed columns, *P. aeruginosa*, *B. subtilis* and *B. licheniformis* have been used for oil recovery; of which *Bacillus* species are found more effective (Bordoloi and Konwar, 2008; Pornsunthorntawee et al., 2008; Suthar et al., 2008).

In a study reported by Hossein et al. (2011), experiment was carried out for development of suitable technology for *ex situ* MEOR. *Pseudomonas aeruginosa* rhamnolipid biosurfactant was produced and was proposed to be a model for *ex situ* MEOR. Jinfeng et al. (2005) have given a detailed account of enhancing oil recovery using water flooding techniques in high temperature reservoirs. A report by Shafeeq et al. (1989) stated that *Pseudomonas aeruginosa* isolate S8, isolated from oil polluted sea water, degraded hydrocarbons by 45–60% when incubated for 28 days with hexadecane, octadecane and nonadecane. Microorganisms are also found to metabolize hydrocarbons in oil anaerobically (Kropp, 2000; Aiken, 2004; Brown, 2010).

Injection of microorganisms in oil reservoir can also be done for MEOR, but it has shown to cause plugging of the well. To overcome this, Hitzman

(1962) proposed use of spores as they have smaller size that vegetative cells. But this modification did not solve the problem. Lapin-Scott et al. (1988) then used ultramicrobacteria (UMB) that have a diameter of less than 0.3 mm. But UMB had less metabolic activity and thus less MEOR potential. Chang and Yen (1984) suggested use of lysogenic bacteria which can be triggered for metabolism by some inducer. Another problem in evaluation of MEOR is interpretation of field results as many parameters like penetration of microorganisms and oil reservoir conditions are variable during each recovery process (Ollivier and Magot, 2005). Also, the field effects have not been evaluated in many of the studies of MEOR.

Number of patents has been issued for MEOR. Zobell (1946) has described in his patent, use of the bacterium *Desulfovibrio hydrocarbonoclasticus* for MEOR method. In this study, he inoculated *D. hydrocarbonoclasticus* along with oxidized sulfur compounds and carbon source as lactose, but no field trials were performed. In his next patent, he introduced addition of oxygen free hydrogen produced by the action of a *Clostridium* species on a carbohydrate (Zobell, 1953).

4.4 APPLICATION OF BIOSURFACTANTS IN BIOREMEDIATION AND BIODEGRADATION

Biosurfactants have been explored for their bioremediation potential, and show promise for application to sites impacted by both organic and metal contaminants. Specific applications include enhancing contaminant biodegradation in sites that are contaminated with organics alone (Oberbremer et al., 1990; Jain et al., 1992; Herman et al., 1997; Bregnard et al., 1998) or that are co-contaminated with metals and organics (Maslin and Maier, 2000; Sandrin et al., 2000). Biosurfactants have also shown potential for use as additives to aid in cleaning or flushing organic (Van Dyke et al., 1993; Bai et al., 1998; Ivshina et al., 1998) or metal (Zosim et al., 1983; Torrens et al., 1998; Mulligan et al., 1999; Ochoa-Loza et al., 2001) contaminants out of tanks or soils.

In many cases, environmental contamination caused by industrial activity is due to accidental or deliberate release of organic and/or inorganic compounds into the environment. Such compounds pose problems for remediation, as they become easily bound to soil particles. The application of biosurfactants in the remediation of organic compounds, such as hydrocarbons, aims at increasing their bioavailability (biosurfactant-enhanced bioremediation) or mobilizing and removing the contaminants by pseudo-solubilisation and emulsification in a washing treatment.

The application of biosurfactants in the remediation of inorganic compounds such as heavy metals, on the other hand, is targeted at chelating and removal of such ions during a washing step facilitated by the chemical interactions between the amphiphiles and the metal ions. Amphiphiles are able to alter the physico-chemical conditions at the interfaces affecting the distribution of the chemicals among the phases (Tiehm, 1994). For instance, a hydrocarbon-contaminated soil contains at least six phases: bacteria, soil particles, water, air, immiscible liquid and solid hydrocarbon. The hydrocarbons can be partitioned among different states: solubilized in the water phase, ad/absorbed to soil particle, sorbed to cell surfaces and as a free/insoluble phase. Biosurfactants added to this system can interact with both the abiotic particles and the bacterial cells. This affects the mechanisms of interaction with environments with regard to the micellarisation and emulsification of organic contaminants, the interaction with sorbed contaminants and the sorption to soil particles which leads to the alteration of cell-envelope composition and hydrophobicity. The interactions between micelles and cells are among the main alterations to the bacterial component (Volkering et al., 1998). These phenomena on the one hand can be exploited to increase the bioavailability of poorly soluble contaminants, thus increasing biodegradation rate, or on the other hand, can result in an inhibition of biodegradation.

In spite of the publication bias which favors an over publication of successful applications, the main emerging feature of the large body of literature in this area is the contrasting result reported on efficiency. For instance, rhamnolipids can stimulate the degradation of n-hexadecane by the producer strain *P. aeruginosa*, but did not stimulate degradation by *Rhodococcus* strains showing strain specificity. In contrast, biosurfactants from *R. erythropolis* strain 3C-9 significantly increased the degradation rate of n-hexadecane by two phylogenetically distant species, *Alcanivorax dieselolei* and *Psychrobacter celer*, in flask tests (Noordman and Janssen, 2002; Peng et al., 2007). Therefore, with the current state of knowledge, the modeling of the effect of biosurfactant addition in bioremediation treatment is not predictable, and efficacy has to be evaluated experimentally (Franzetti et al., 2006; Franzetti et al., 2008b). To gain better insight into this problem, it is useful to review the current knowledge and recent advances regarding these interactions. Volkering et al. (1998) and Paria (2008) provides excellent reviews about interactions between surfactants and the environment.

The interactions between bacteria, contaminants and biosurfactant can be interpreted from a functional perspective, considering that the main natural role attributed to biosurfactants is their involvement in hydrocarbon uptake (Perfumo et al., 2010a). Microbial surfactants can promote the growth of bacteria on hydrocarbons by increasing the surface area between oil and water and through emulsification and increasing pseudosolubility of hydrocarbons through partitioning into micelles (Miller and Zhang, 1997; Volkering et al., 1998). High-molecular-weight biosurfactants (bioemulsifiers) have great potential for stabilizing emulsions between liquid hydrocarbons and water, thus increasing the surface area available for bacterial biodegradation. However, they have been rarely tested as enhancers of hydrocarbon biodegradation in bioremediation systems, and contrasting results are reported in the literature (Barkay et al., 1999; Franzetti et al., 2009a).

For low-molecular-weight biosurfactants, above the Critical Micelle Concentration (CMC), a significant fraction of the hydrophobic contaminant partitions in the surfactant micelle cores. In some cases, this results in a general increase in the bioavailability of contaminants for degrading microorganisms. Successful applications of rhamnolipids and surfactin in enhanced bioremediation have been recently reviewed (Mulligan, 2009). In addition, Wang and Mulligan (2009) studied the effect of ammonium ion concentration and pH on the potential application of rhamnolipid and surfactin for enhanced biodegradation of diesel. A lipopeptide and protein–starch–lipid produced by two strains of *P. aeruginosa* significantly enhanced the solubilisation of phenanthrene, pyrene and fluorene, increasing their metabolism and supporting sustained growth (Bordoloi and Konwar, 2009). Polycyclic Aromatic Hydrocarbons (PAH) biodegradation was also investigated by Das et al. (2008b); they used *Bacillus circulans* to increase the bioavailability of anthracene. Interestingly, the organism had better growth and biosurfactant production on glycerol containing mineral medium supplemented with anthracene, although it was unable to utilize anthracene as the sole carbon source. However, anthracene was used as a substrate for the production of the biosurfactant. The specific modes of hydrocarbon uptake, however, are not fully understood. Recently Cameotra and Singh (2009) elucidated the mechanism of n-hexadecane uptake mediated by rhamnolipids in *P. aeruginosa*. The rhamnolipids produced an emulsion with hexadecane, thus facilitating increased contact between the hydrocarbon substrate and the bacteria. It was also observed that uptake of the biosurfactant-coated hydrocarbon droplets occurred, suggesting a mechanism like pinocytosis taking place, a process not previously reported in bacterial hydrocarbon uptake systems.

In contrast, it is well known that the presence of a surfactant can detrimentally affect biodegradation. Micelle cores can trap organic contaminants, creating a barrier between microorganisms and organic molecules, resulting in the potential substrate becoming less rather than more available. For example, Witconol SN70, a non-ionic alcohol ethoxylate surfactant (Colores et al., 2000), reduced the biodegradation rate of hexadecane and phenanthrene, with biodegradation similarly inhibited by Tween 20, sodium dodecyl sulfonate, tetradecyl trimethyl ammonium bromide and Citrikleen at concentrations equal or greater than their CMCs (Billingsley et al., 1999).

Another proposed role of biosurfactants in hydrocarbon uptake is the regulation of cell attachment to hydrophobic and hydrophilic surfaces by exposing different parts of cell bound biosurfactants, thus changing cell-surface hydrophobicity (Rosenberg et al., 1987; Franzetti et al., 2008a). This natural role can be exploited by adding (bio) surfactants to increase the hydrophobicity of degrading microorganisms and to allow cells' easier access to hydrophobic substrates (Shreve et al., 1995). The release of LPS by *Pseudomonas* sp. induced by sub-CMC levels of rhamnolipids allowed a more efficient uptake of hexadecane by rendering the cell surface more hydrophobic (Al-Tahhan et al., 2000). Noordman and Janssen (2002) reported that rhamnolipid produced by *P. aeruginosa* UG2 facilitated the hydrocarbon uptake of the producer strain and increased the degradation of hexadecane, while the same product did not stimulate to the same extent the biodegradation of hexadecane by four unrelated species (*Acinetobacter lwoffii* RAG1, *R. erythropolis* ATCC 19558, *R. erythropolis* DSM 43066 and strain BCG112), nor was degradation of hexadecane stimulated by addition of the biosurfactants produced by these species themselves. Zhong et al. (2007) showed that the adsorption of dirhamnolipid biosurfactants on cells of *B. subtilis*, *P. aeruginosa* and *Candida lipolytica* depended on the physiological status of the cells and was specific to the microorganisms. Furthermore, the biosurfactant adsorption affected the cell-surface hydrophobicity depending on the rhamnolipid concentration and the physiological state of the cell. The effect of exogenous rhamnolipids on cell-surface composition of *P. aeruginosa* NBIMCC 1390 was recently studied by Sotirova et al. (2008). They showed that above the CMC, rhamnolipids caused a 22% reduction of total cellular LPS content, while at concentrations below the

CMC, they caused changes in the bacterial outer membrane protein composition yet did not affect the LPS component. Chang et al. (2009) demonstrated that the cell-surface hydrophobicity was enhanced by the accumulation at the cell surface of different fatty acids during growth on hydrocarbon in *R. erythropolis* NTU–1. A significant correlation between the modification of the cell surface by saponins and the degree of hydrocarbon biodegradation was reported by Kaczorek et al. (2008).

4.5 APPLICATIONS OF BIOSURFACTANT IN AGRICULTURE

Biosurfactants used in agriculture are also referred as green surfactants (Dhara and Cameotra, 2013). In agriculture industries biosurfactants can be used as mobilizing agents like chemical surfactants. A mobilizing agent can increase solubility of hydrophobic organic matters and also helps soil microorganisms in absorption and nutrient uptake (Makkar and Rockson, 2003; Fakruddin, 2012). Like synthetic surfactants, biosurfactants can also increase wettability of soil and can prevent cake formation of biofertilizers while storage (Makkar and Rockson, 2003). Following are some properties of biosurfactants because of which they gain importance in agriculture industry.

4.5.1 ANTIMICROBIAL ACTIVITY

Many biosurfactants showed antimicrobial activity against plant pathogens. Rhamnolipid type of biosurfactants produced by *Pseudomonas* sp. showed antimicrobial activity against many plant pathogenic bacteria and fungi, thus can be used in biocontrol agent against plant pathogens (Kachholz and Schlingmann, 1987). Rhamnolipids also inhibits zoospore formation of plant pathogen (Hutberg et al., 2008; Kim et al., 2011; Sha et al., 2011). It was noted that rhamnolipids also stimulates plant immunity and thus helps in reducing infection (Vasta et al., 2010). Lipopeptide type of biosurfactants produced by *Bacillus* sp. showed antagonistic activity against *Biopolaris sorokiniana*, *Aspergillus* sp. and *Fusarium* sp. (Velho et al., 2011). Another lipopeptide biosurfactant which is isoform of surfactin exhibited antifungal and antibacterial activity. Rhozobacterial biosurfactants also showed good antimicrobial activity (Nihorimbere et al., 2011). These biosurfactants facilitated biocontrol mechanisms of organisms such as competition, antibiosis, parasitism and hypovirulent activity (Singh et al., 2007). They also enhanced antagonistic properties of microbes (Jazzar and Hammad, 2003; Kim et al., 2004). Rhizosphere bacteria such as *Bacillus* and *Pseudomonas* exhibited biocontrol against *Pectobacterium* and *Dickeya* sp. (Krzyzarouska, 2012).

4.5.2 INSECTICIDE AND PESTICIDE ACTIVITY

Biosurfactants can also be used as biopesticide. One of the biosurfactant, lipopeptide, produced by various bacteria showed insecticide activity against *Drosophila melanogaster* (Mulligan, 2005). Rhamnolipid biosurfactant showed insecticidal activity against green aphid (*Myzus persicae*) (Kim et al., 2011). These green surfactants can be used as carbon source by soil micro-organisms (Scott and Jones, 2000; Takenaka et al., 2007; Lima et al., 2011). *Pseudomonas* sp. and *Burkholderia* sp. from paddy soil field has ability to degrade the toxic surfactants present in soil (Nishio et al., 2002). Camacho Chab et al. (2013) demonstrated bioemulsifier activity of micobactan, a non ionic glycolipoprotein produced by *Microbacterium* sp. MC3B–10 which may have applications in agriculture. Now a day's many agro based indus-tries made different combinations of biosurfactants with polymers and made excellent formulations for agriculture applications (Dhara and Cameotra, 2013).

4.5.3 BIOCONTROL ACTIVITY

Another lipopeptide biosurfactant which is an isoform of surfactin produced by *Brevibacillus brevis* HBO1 exhibited strong antifungal and antimicro-bial activity and can be developed as biocontrol agent (Haddad, 2008b). *Pseudomonas fluorescences* also showed good antibacterial and antifungal activity against many plant pathogens (Hultberg et al., 2008). Biosurfactant from *Bacillus subtilis* was reported to control *Colletotrichum gloeospori-oides*, a causative agent of anthracnose in papaya (Kim et al., 2010). Zang et al. (2011) found that these biosurfactants also has potential to accelerate microbial growth.

4.5.4 PLANT MICROBE INTERACTION

Rhamnolipid biosurfactant produced by *Pseudomonas* sp. regulate the quorum sensing, affects on mobility of microorganisms and biofilm formation

(Keans, 2003; Dusane et al., 2010; Ron and Rosenberg, 2011; Dhara and Cameotra, 2013). Thus these green surfactants can play an important role in beneficial association with microbes for environment and humans.

4.6 APPLICATIONS OF BIOEMULSIFIERS/BIOSURFACTANTS IN MEDICINE

From past few years several applications of biosurfactants as therapeutic agents have been reported. It was found that regardless of biological origin some biosurfactants can be used for biomedical applications and they are proved safe and effective therapeutic agents (Banat et al., 2000; Makkar and Cameotra, 2002; Maier, 2003; Benincasa et al., 2004; Singh and Cameotra, 2004). It is well known that biosurfactants are amphipathic molecules and they act on interfaces (Neu, 1996). Gottenbos et al. (2001) stated that bio-emulsifiers with positive charge biomaterial showed antimicrobial effect on adhering of Gram positive bacteria. There are few biosurfactants reported which showed great medicinal properties and hence used in medical as well as pharmaceutical industries. Following are those biosurfactants showed great potential as antimicrobial agents.

4.6.1 LIPOPEPTIDES

Lipopeptide biosurfactants are made up of lipids and peptides. They have high surface activities (Isoda et al., 1999). Lipopeptides such as Iturin A and Surfactin are mainly reported for their antimicrobial activities. Iturin produced by *Bacillus subtilis* showed antimicrobial, antifungal activity and also increased electrical conduction of bimolecular lipid membranes (Thiomon et al., 1995; Mitthenbuhler et al., 1997; Ahimou et al., 2001; Rodrigues and Teixeira, 2010). Like Iturin, *Bacillius subtilis* also produces another potential biosurfactant named Surfactin exhibited good antifungal activity as well as antimicrobial activity against multi drug resistant strains (Vellenbroich et al., 1997; Fernandes et al., 2007). In addition to this it also showed antitumor activity against Ehrlich's ascite carcinoma cells. It is also reported for antiviral activity against Human Immunodeficiency Virus-I (HIV-I), Herpes viruses and also has ability to induce apoptosis in human leukemia K562 (Itokawa et al., 1994; Vollenbroich et al., 1997a; Kracht et al., 1999; Wang et al., 2007). Surfactin showed hemolysis and inhibits fibrin clot formation, induces ion channels in lipid bi-layer membranes, inhibition to biofilm formation, suppression of inflammation by platelet cytosolic phospolipase, and anti-endotoxin property in animal models (Sen, 2010). Other lipopeptide biosurfactants reported were pulmiacidin and its variants, for example, pulmiacidin A, B, C, D, E, F and G for their antiviral activity against Herpes Simplex Virus I and inhibitory activity against H, K-ATPase and protection against gastric ulcers *in vivo* (Rodrigues et al., 2006; Rodrigues and Teixeira, 2010). *Bacillus licheniformis* produced biosurfactant named Lichenysin, showed antibacterial activity and also has chelating properties which helps in membrane disrupting effect of lipopeptides (Grangemard et al., 2001; Rodrigues and Teixeira, 2010).

4.6.2 GLYCOLIPIDS

Glycolipids are the biosurfactants with composition of carbohydrates and lipids. This type of biosurfactant was produced by *Pseudomonas aeruginosa* AT10 named Rhamnolipid. Near about seven different rhamnolipids were reported from *Pseudomonas aeruginosa* AT10. They showed antimicrobial activity against *Mycobacterium tuberculosis*, and also antifungal properties against a range of fungi, anti adhesive activity against several bacterial and yeasts isolated from voice prostheses. It also has ability to induce dose dependent hemolysis and coagulation of platelet poor plasma (Lang and Wullbrandt, 1999; Rodrigues et al., 2005; Rodrigues et al., 2006; Reis et al., 2012). Hossain et al. (2001) reported a glycolipid produced by *Borrelia burgdorferi* also showed antimicrobial activity. *Pseudozyma fusiformata*, an yeast reported for the production of low molecular weight glycolipid. It has fungicidal activity against 80 % of the 280 yeast and yeast like species under acidic conditions (pH 4.0) (Golubev et al., 2001). Glycolipid BS from *Streptococcus thermophilus* isolated from dairy showed good anti-adhesive activity against several bacterial and yeast strains isolated from voice prostheses (Rodrigues et al., 2004).

4.6.3 BIOSURFACTANTS WITH DIFFERENT COUNTERPARTS

Other biosurfactants which have lipid counterparts and have medicinal importance were reported by many workers. The major contribution is reported by fungi *Candida antartica* which produced mannosylerythritol lipids (MELS) which has great antimicrobial activity (particularly against Gram positive bacteria), neurological and immunological properties. It also induces cell differentiation in human promyelocytic leukemia cell line HL60, human leukemia cell line K562 and human basophilic leukemia cells line Ku812 (Isoda et al., 1997). It also induces PC12 cells in rat (Isoda et al., 1999). When studied in human it was found that it has ability to enhance gene transfection efficiency in human cervix and HeLa cells. *Rhodococcus erythropolis* produces succinoyl trehalose lipid biosurfactant which showed antiviral activity against HSV and influenza virus (Rodrigues et al., 2006). Protein A from this biosurfactant can be used in pulmonary disorders (Sano et al., 1999). Some biosurfactants gain importance in medicine as they have unique property of antiadhesive activity. These biosurfactant have ability to inhibit the adhesion of pathogenic organisms to solid surfaces or infection sites. Pre-coating of biosurfactants with surgical equipment's such as urethral catheters showed decrease in biofilm formation by *Salmonella typhimurium*, *Salmonella enterica*, *E. coli* and *Proteus mirabilis*. Biosurfactants from *S. mitis* showed antibacterial activity (Pratt et al., 1989). Biosurfactant named Surfactin, produced by *Lactobacillus* strains such as *Lactobacillus acidophilus*, *Lactobacillus fermentum* showed anti-adhesive property. They mostly inhibit uro-pathogenic bacteria including *Enterococcus faecalis* (Heinemann et al., 2000). Recent studies on *Lactobacillus* sp., probiotic bacteria, showed anti-adhesive activity against *E. coli* to intestinal epithelial cells. Bioproducts of strains such as biosurfactants from *Lactobacillus* GG and *L. rhamnosus* GRI are mainly effective in decolonization and protecting intestine as well as vaginal tract against microbial infection (Reid, 1999; Reid et al., 1999; Reid et al., 2001; Rastall et al., 2005). Biosurfactant also has great importance in defense against infection and inflammation in humans. Pulmonary surfactants are used which lowers the surface tension at air-liquid surface of lung in inflammatory lung disease (Sano et al., 1999).

4.7 APPLICATIONS OF BIOSURFACTANTS IN PHARMACEUTICAL INDUSTRY

Chemical surfactants are usually synthesized from petroleum derivatives and impose many environmental problems as well as can be toxic to human health. In contrast, naturally produced surfactants are ecofriendly and less toxic (Ashby et al., 2005; Mann and Bidwell, 2011). Biosurfactants are found to display excellent moisturizing properties along with good skin compatibility and low toxicity. Because of these characteristics of biosurfactants, they have also been used in cosmetic industry (Brown, 1991). Sophorolipids, rhamnolipids and mannosylerythritol lipids are the mostly used glycolipid biosurfactants in cosmetics (Lourith and Kanlayavattanakul, 2009). Sophorolipids possess antimicrobial properties (Pierce and Heilman, 1998; Kim et al., 2002; Yoo et al., 2005; Van Hamme et al., 2006) which make them potent bactericidal agent and thus can be used in acne treatment, dandruff treatment and for body odor problems (Mager, 1987). Sophorolipids also carry out varied functions like emulsification, solubilization, foaming and wetting agents, and detergents (Kleckner and Kosaric, 1993). Other properties, such as, stimulation of dermal fibroblast metabolism and collagen neosynthesis, desquamating and depigmenting agents in wound healing process (Hillion et al., 1998; Maingault, 1999; Borzeix, 2003), stimulation of leptin synthesis in adipocytes (Pellecier and Andre, 2004) and cellulite treatment (Lourith and Kanlayavattanakul, 2009) are also reported. They are used in pencil-shaped lip cream, lip rouge and eye shadow as well as in compressed powder cosmetics (Kawano et al., 1981). Rhamnolipids are also reported to have antimicrobial activity (Abalos et al., 2001; Benincasa et al., 2004). Like sophorolipids, rhamnolipids have been reported for their use in many products which include insect repellents, antacids, acne pads, anti-dandruff products, contact lens solutions, deodorants, nail care products and toothpastes (Kleckner and Kosaric, 1993; Maier and Soberon-Chavez, 2000). A patent has been issued for use of rhamnolipids in anti-wrinkle and anti-ageing products (Piljac and Piljac, 1999). Mannosyl erythritol lipids are found to be useful in products like detergents and dispersants (Fukuoka, 2007) because of their high hydrophilic properties and low critical aggregation concentration. Their low toxicity and skin compatibility make them potential candidates for future use in cosmetic industry (Eiko and Toshi, 2008).

Sophorolipids are commercially as humectants in Sofina and Soliance cosmetic make-up brands (Lourith and Kanlayavattanakul, 2009). A product containing sophorolipid:propylene glycol (1:12 mol) is shown to have specific compatibility to skin and has been used as moisturizer commercially. Patents have been issued stating use of rhamnolipids in cosmetics (Makkar and Cameotra, 2002). A mutant strain *Candida bomobicola* ATCC 22214 produces sophorolipids with properties like stimulation of fibroblast metabolism, anti-radical properties and hygroscopic properties to support healthy skin physiology (Bhardwaj et al., 2013). Number of other health care products uses surfactants in them like hair color and care products, deodorants, denture cleaners, baby products and antiseptics (Maier, 2003). Because of obvious advantages over chemical surfactants, biosurfactants find an attractive alternative in this industry.

Gene transfection is a great challenge in pharmaceutical industries (Fujita et al., 2009). To deliver foreign gene into target cells without side effects different methods are discovered. Among them lipofection is the most effective one and used frequently (Inoh et al., 2001). It was found by Kitamato et al. (2002) that liposomes based on biosurfactants showed increased gene transfection efficiency. Recently, Ueno et al. (2001) developed MEL-A containing liposome for gene transfection. Lipopeptides from bacteria were used as potent immunological adjuvants. They are non toxic and non pyrogenic when mixed with conventional antigens. Iturin AL, herbicolin A and microcystin showed marked humeral response rabbits and chickens (Rodrigues et al., 2006b; Eshrat et al., 2011).

Surfactants are used to permeabilise or lyse the cells to recover the product. *E. coli* cells can be made permeable to facilitate extraction by biosurfactants (Singh et al., 2007). Biosurfactants can be used as reagents for membrane permeabilisation for proteins (Desai and Banat, 1997). Recently Bhadoriya et al. (2013) described role of biosurfactants producing bacteria in development of pharmaceutical additives and its use in solubility enhancement of poorly soluble drugs. Antimicrobial activity and anti-adhesive properties of biosurfactants are also greatly used in pharmaceutical industries.

4.8   APPLICATIONS OF BIOSURFACTANTS IN FOOD INDUSTRY

The increasing demand of consumers for use of natural compounds in food in place of chemicals has drawn attention of many towards research in this field. Chemical emulsifiers are used in food industry, which stabilize the emulsions and thus the food products. Biosurfactants, being eco-friendly, are found to be an attractive replacement for chemical surfactants. Biosurfactants are often used as food additives where they act as emulsifiers. They form good stable emulsions, imparting good texture and creaminess to the food (Banat et al., 2010). Biosurfactants show very important properties such as surface activity, tolerance to temperature, pH, ionic strength, biodegradability, low toxicity, emulsifying and demulsifying ability and antimicrobial activity. These properties make them ideal candidates for use in food industries (Nitschke and Costa, 2007).

Many biosurfactants produced by different organisms have been shown be stable at conditions which are frequently encountered during food processing. To state some examples, lichenysin produced by *Bacillus licheniformis* JF-2 was stable at temperature up to 50°C, pH 4.5–9.0, and NaCl concentration up to 50 g/L and Ca concentrations up to 25 g/L, respectively (McInerney et al., 1990). *B. subtilis* LB5a biosurfactant could sustain autoclaving conditions, 121°C for 20 min and was also found to retain surface activity after 6 months of storage at −18°C. This lipopeptide was also stable at wide range of pH values (Nitschke and Pastore, 2006). Nitschke et al. (2007) have given a good review of low toxicity and better biodegradability of biosurfactants over chemical surfactants.

Many biosurfactants are known to show antimicrobial activity against bacteria, fungi, algae and also display antiviral properties (Nitschke, 2007). For example, *Bacillus subtilis* lipopeptide surfactant, iturin, sophorolipids and rhamnolipids showed antifungal activity (Besson et al., 1976), rhamnolipids are shown to inhibit some algal species (Wang et al., 2005) and inactivation of enveloped viruses, herpes virus and retrovirus by surfactin (Vollenbroich et al., 1997). *Pseudomonas aeruginosa* rhamnolipid and *Candida antarctica* glycolipid demonstrated antimicrobial activity (Nitschke, 2007). These properties of biosurfactants have definitive utility in food preservation. Along with the properties mentioned above, biosurfactants can have other functions in food as well, namely, stabilization of aerated systems, modification of rheological properties of wheat dough to control texture and shelf life improvement of starch containing products and fat containing products (Kachholz and Schlingmann, 1987). Kosaric (2001) has discussed the role of surfactants in ice cream formulations where they control consistency and used as stabilizers. Biosurfactants are also known for viscosity enhancement in certain products (Iyer et al., 2006). They can also be used as anti adhesive agents in food industry (Netschke, 2007). Because of their multipurpose function in food, biosurfactants have been given increasing attention for use in food industry. There is now great amount of literature available on production of biosurfactants using cheaper agro-industrial waste products which makes its economical application possible. The biosurfactants from *Candida lipolytica* and *Saccharomyces cerevisiae* are reported to be used in food and oil industries (Sarubbo et al., 2007). Liposan which is a bioemulsifier produced by *Candida lipolytica* was found to stabilize the emulsions of water and vegetative oils which included cottonseed oil, corn oil, soybean oil and peanut oil emulsions (Adamczak and Bednarski, 2000; Konishi et al., 2007; Bhardwaj et al., 2013).

4.9   APPLICATIONS OF BIOSURFACTANTS IN NANOTECHNOLOGY

Nanotechnology has many applications in chemical, industrial, mechanical and in varied applications including medicine. We can define nanoparticles as particles with size <100 nm. Various methods are used for the synthesis of metallic silver nanoparticles like chemical method and physical methods. Among all metallic nanoparticles, silver (Ag) and gold (Au) nanoparticles are used in diverse fields because of their peculiar physical, chemical as well as their property of cytotoxicity against cancer and microbial cells. Nanoparticles are produced by chemical methods including reductions of salts by different chemical reducing agents such as hydrazine hydrate or by physical methods including radiation or ultrasonication (Thombre et al., 2014). In recent years, the approach of production was shifted towards less harmful and ecofriendly biological methods which involve application of plant extracts or biological extracts for synthesis and stabilization of nanoparticles. This biological method of synthesis of nanoparticles is simple and a possible alternative to chemical synthesis procedures and physical methods (Thombre et al., 2012).

Taking into consideration the demand and need of greener biological methods and novel compounds for synthesis of nanoparticles, biosurfactants, and/or biosurfactant producing microbes are emerging as promising source for the rapid synthesis of nanoparticles. The common technique of preparation of microemulsion using an oil-water-surfactant mixture is a potential approach for nanoparticle synthesis. Biosurfactants are naturally present surfactants having microbial origin composed mostly of sugar and fatty acid moieties; they have higher biodegradability, less toxicity, and good biological activities. The biosurfactant mediated synthesis of nanoparticles is now considered a promising nontoxic and environmentally safe, "green chemistry" approach. The biosynthesis of nanoparticles using biosurfactants is considered to be superior to the methods of biological synthesis of nanoparticles since biosurfactants reduce the formation of aggregates due to the electrostatic forces of attraction and facilitate a uniform morphology of the nanoparticles (Kiran et al., 2011).

The bio-surfactant sophorolipid has been used as a capping and reducing agent in silver nanoparticle synthesis. Sophorolipids are molecules with a sophorose – a dimeric glucose – attached to ω or ω – 1 carbon of fatty acids (oleic acid, stearic acid, etc.). The final sophorolipid capped nanoparticles fall in the category of glyconanoparticles. These glyconanoparticles are being investigated for their potential application as cell mimicks (Kasture et al., 2008). Rhamnolipids from *Pseudomonas aeruginosa* strain BS–161R have been reported for synthesis of silver nanoparticles. The purified rhamnolipids in a pseudoternary system of n-heptane and water system along with n-butanol as a cosurfactant when added to the aqueous solutions of silver nitrate and sodium borohydride, form reverse micelles. When these micelles are mixed, they resulted in the rapid formation of silver nanoparticles (Kumar et al., 2013). Similarly functionalized iron oxide (Fe_3O_4) nanoparticles have been synthesized by using biosurfactant surfactin and rhamnolipid. The functionalization results in a dramatic alteration in the surface potential and hydrodynamic size due to the presence of coated moieties on the nanoparticle interface. Accordingly, surfactin and rhamnolipid coated nanoparticles were found to be cytotoxic and biocompatible (Sangeeta et al., 2013). A novel method for synthesis of NiO_2 nanoparticles using rhamnolipid biosurfactant for microemulsion synthesis has been demonstrated which offers an eco-friendly alternative to conventional microemulsion technique based oil organic surfactants (Palanisamy et al., 2009). Surfactin, a renewable, environmentally compatible, biodegradable surfactant has also been reported as a stabilizing agent for the synthesis of silver nanoparticles (Reddy et al., 2009).

4.10 CONCLUSION

Biosurfactants gain importance as they are biodegradable and lower toxic then chemical surfactants. They can be applied in various industries such as food, medicine, pharmaceutical, agricultural and can also be used in enhanced oil recovery, bioremediation and as biocontrol agent. Use of biosurfactants was limited to some industries because they are economically uncompetitive. Although in last few years there is noticeable increase in the use of biosurfactants in different industries as they have great potential. Defiantly increase in applications of biosurfactant will prove very helpful for environment and human health.

KEYWORDS

- **Biocontrol activity**
- **Biodegradation**
- **Bioremediation**
- **Biosurfactant**
- **Environment**
- **Glycolipids**
- **Green chemistry**
- **Human health**
- **Lipopeptides**
- **Microbial enhanced oil recovery**
- **Microemulsion**
- **Nanotechnology**
- **Rhamnolipids**
- **Sophorolipids**

REFERENCES

Abalos, A., Pinazo, A., Infante, M. R. Physicochemical and antimicrobial properties of new rhamnolipids produced by *Pseudomonas aeruginosa* AT10 from soy bean oil refinery wastes. *Langmuir* 2001, 17, 1367–1371.

Adamczak, M., Bednarski, W. Influence of medium composition and aeration on the synthesis of biosurfactants produced by *Candida antarctica*. *Biotechnol Lett.* 2000, 22, 313–316.

Agarwal, P., Sharma, D. K. Studies on the production of biosurfactant for the microbial enhanced oil recovery by using bacteria isolated from oil contaminated wet soil. *Pet Sci Technol.* 2009, 27, 1880–1893.

Ahimou, F., Jacques, P., Deleu, M. Surfactin and Iturin A effects on *Bacillus subtilis* surface hydrophobicity. *Enzyme Microb Technol.* 2001, 27, 749–754.

Aiken, C., Jones, D. M., Larter, S. R. Anaerobic hydrocarbon biodegradation in deep subsurface oil reservoirs. *Nature* 2004, 431, 291–294.

Amani, H., Müller, M. M., Syldatk, C., Hausmann, R. Production of microbial rhamnolipid by *Pseudomonas aeruginosa* MM1011 for *ex situ* enhanced oil recovery. *Appl Biochem Biotechnol.* 2013, 170, 1080–1093.

Arima, K., Kakinuma, A., Tamura, G. Surfactin, a crystalline peptide lipid surfactant produced by *Bacillus subtilis*: isolation, characterization and its inhibition of fibrin clot formation. *Biochem Biophys Res Commun.* 1968, 3, 488–494.

Asci, Y., Nurbas, M., Sagacikel, Y. A comparative study for the sorption of Cd(II) by soils with different clay contents and mineralogy and the recovery of Cd(II) using rhamnolipid biosurfactant. *J. Hazard Mat.* 2008, 154, 663–673.

Ashby, R. D., Nunez, A., Solaiman, D. K. Y., Foglia, T. A. Sophorolipid biosynthesis from a biodiesel co-product stream. *J. Am. Oil Chem. Soc.* 2005, 82, 625–630.

Asselineau, C., Asselineau, J. Trehalose containing glycolipids. *Prog Chem Ftas Lipids.* 1978, 16, 59–99.

Banat, I. M., Franzetti, A., Gandolfi, I., Bestett, G., Martinotti, M. G., Fracchia, L., Smyth, T. J., Marchant, R. Microbial biosurfactants production, applications and future potential. *Appl Microbiol Biotechnol.* 2010, 87, 427–444.

Banat, I. M., Makkar, R., Cameotra, S. Potential commercial applications of microbial surfactants. *Appl Microbiol Biotechnol.* 2000, 53, 495–508.

Benincasa, M., Abalos, A., Oliveria, I., Manresa, A. Chemical structure, surface properties and biological activities of the biosurfactant produced by *Pseudomonas aeruginosa* LBI from soapstock. *Antonie Van Leeuwenhoek.* 2004, 85, 1–8.

Bento, F. M., de Oliveira Camargo, F. A., Okeke, B. C., Frankenberger Jr, W. T. Diversity of biosurfactant producing microorganisms isolated from soils contaminated with diesel oil. *Microbiol Res.* 2005, 160, 249–255.

Besson, F., Peypoux, F., Michel, G., Delcambe, L. Characterization of iturin A in antibiotics from various strains of *Bacillus subtilis*. *J. Antibiotic.* 1976, 29(10), 1043–1049.

Bhadoriya, S. S., Madoriya, N., Shukla, K., Parihar, M. S. Biosurfactants: a new pharmaceutical additive for solubility enhancement and pharmaceutical development. *Biochem Pharmacol.* 2013, 2(2), 113.

Bhardwaj, G., Cameotra, S. S., Chopra, H. K. Biosurfactants from fungi: a review. *J. Pet. Environ. Biotechnol.* 2013, 4, 160.

Bodour, A. A., Drees, K. P., Maier, R. M. Distribution of biosurfactant producing bacteria in undisturbed and contaminated arid southwestern soils. *Appl Environ Microbiol.* 2003, 69(6), 3280–3287.

Bordoloi, N. K., Konwar, B. K. Microbial surfactant-enhanced mineral oil recovery under laboratory conditions. *Colloids Surf B: Biointerfaces* 2008, 63, 73–82.

Borzeix, C. F. Use of sophorolipids comprising diacetyl lactones as agent for stimulating skin fibroblast metabolism, US Patent 6596265, Institut Francais du Petrole, Malmaison, 2003.

Brown, L. R. Microbial enhanced oil recovery (MEOR). *Current Opinion in Microbiology* 2010, 13, 316–320.

Brown, M. J. Biosurfactants for cosmetic applications. *Int. J. Cosmet. Sci.* 1991, 13, 61–64.

Camacho Chab, C. J., Guez, J., Chan-Bacab, J. M., Rios-Leal, E., Sinquin, C. M., Salazar, R., Rosa Garcia, S. D., Estebanez, M. R., Ortega Morales, B. O. Emulsifying activity and stability of a non toxic bioemulsifier synthesized by *Microbacterium* sp. MC3B–10. *Int. J. Mol. Sci.* 2013, 14, 18959–18972.

Chang, P., Yen, T. F. Interaction of *Escherichia coli* B and B/4 and bacteriophage T4D with berea sandstone rock in relationship to enhanced oil recovery. *Appl Environ Microbiol.* 1984, 47, 544–550.

Cirigliano, M. C., Carman, G. M. Isolation of bioemulsifier from *Candida lipolytica*. *Appl. Environ. Microbiol.* 1984, 48, 747–750.

Ciriglino, M. C., Garman, G. M. Purification and characterization of liposan, a bioemulsifier from *Candida lipolytica*. *Appl Environ Microbiol.* 1985, 50, 846–850.

Cooper, D. G., Liss, S. N., Longay, R., Zajic, J. E. Surface activities of *Mycobacterium* and *Pseudomonas*. *J. Ferment. Technol.* 1989, 59, 97–101.

Dastgheib, S. M. M., Amoozegar, M. A., Elahi, E., Asad, A., Banat, I. M. Bioemulsifier production by a halothermophilic *Bacillus* strain with potential applications in microbially enhanced oil recovery. *Biotechnol Lett.* 2008, 30, 263–270.

Desai, J. D., Banat, I. M. Microbial production of biosurfactants and their commercial potential. *Microbiol. Mol. Biol. Rev.* 1997, 61, 47–64.

Dhara, P. S., Cameotra, S. S. Biosurfactants in agriculture. *Appl. Microbiol. Biotechnol.* 2013, 97, 1005–1016.

Dusane, D., Rehman, T., Zinzarde, S., Venugopalan, V., McLean, R., Weber, M. Quorum sensing, implication on rhamnolipid biosurfactant production. *Biotech. Genetic Eng. Rev.* 2010, 27, 159–184.

Eddouaouda, K., Mnif, S., Badis, A., Younes, S. B., Cherif, S., Ferhat, S., Mhiri, N., Chamaka, M., Sayadi, S. Characterization of a novel biosurfactant produced by *Streptococcus* sp. strain IE with potential application on hydrocarbon bioremediation. *J Basic Microbiol.* 2012, 52, 408–418.

Edward, J. R., Hayashi, J. A. Structure of rhamnolipid from *Pseudomonas aeruginosa*. *Arch. Biochem. Biophy.* 1965, 111, 415–421.

Eiko, K., Toshi, T. Dermatological anti-wrinkle agent, World Patent 2008/001921, Showa Denko, Tokyo, 2008.

Eshrat, G. F. Biosurfactants in pharmaceutical industry: A mini review. *American J Drug Discover Develop.* 2011, 1(1), 58–69.

Fakruddin, M. Biosurfactants: production and applications. *J. Pet. Environ. Biotechnol.* 2012, 3(4), 124–129.

Fernandes, P. A. V., De Arruda, I. R., Dos Santos, A. F. A. B., De Araujo, A. A., Souto Maior, A. M., Ximenes, E. Z. Antimicrobial activity of surfactants produced by *Bacillus subtilis* R14 against multidrug-resistant bacteria. *Braz. J. Microbiol.* 2007, 38, 704–709.

Fujita, T., Furuhata, M., Hattori, Y., Kawakami, H., Toma, K., Maitani, Y. Calcium enhanced delivery of tetraarginine-PEG-liquid coated DNA/Protamine complex. *Int. J. Pharm.* 2009, 368, 186–192.

Fukuoka, T., Morita, T., Konishi, H. Structural characterization and surface-active properties of a new glycolipid biosurfactant, mono-acylated mannosylerythritol lipid, produced from glucose by *Pseudozyma antarctica*. *Appl Microbiol Biotechnol.* 2007, 76, 801–810.

Gautam, K. K., Tyagi, V. K. Microbial surfactants: a review. *J. Oleo. Sci.* 2006, 55, 155–166.

Golubev, W. I., Kulakovskaya, T. V., Golubeva, W. The yeast *Pseudozyma fusiformata* VKM Y-2821 producing an antifungal glycolipid. *Microbiol.* 2001, 70, 553–556.

Gottenbos, B., Grijpma, D., Van der Mei, H. C., Feijen, J., Busscher, H. J. Antimicrobial effects of positively charged surfaces on adhering Gram positive and Gram negative bacteria. *J. Antimicrob. Chemo.* 2001, 48, 7–13.

Grangemard, I., Wallach, J., Maget Dana, R., Peypoux, F. Lichenysin: a more efficient cation chelator than surfactin. *Appl. Biochem. Biotechnol.* 2001, 90, 199–210.

Guerra-Santos, L. H., Kappeli, O., Fletcher, A. Dependence of *Pseudomonas aerugenosa* continuous culture biosurfactant production on nutritional and environmental factors. *Appl. Microbiol. Biotechnol.* 1986, 24, 443–448.

Haddad, N. I. Isolation and characterization of biosurfactant producing strain, *Brevibacillus brevis* HOB1. *J. Ind. Microbiol. Biotechnol.* 2008, 35, 1597–1604.

Haddadin, M. S. Y., Abou Arqoub, A. A., Abu Reesh, I., Haddadin, J. Kinetics of hydrocarbon extraction from oil shale using biosurfactant producing bacteria. *Energy Convers. Manag.* 2009, 50, 983–990.

Heinemann, C., Van Hylckama, V., Janssen, D. Purification and characterization of a surface-binding protein from *Lactobacillus fermentum* RC–14 that inhibits adhesion of *Enterococcus faecalis* 1131. *FEMS Microbiol Lett.* 2000, 190, 177–180.

Hillion, G., Marchal, R., Stoltz, C., Boreix, C. F. Use of a sophorolipid to provide free radical formation inhibiting activity or elastase inhibiting activity, US Patent 5756471, Institut Francais du Petrole, Malmaison, 1998.

Hisatsuka, K., Nakahara, T., Sano, N., Yamada, K. Formation of rhamnolipid by *Pseudomonas aerugenosa*: its function in hydrocarbon fermentations. *Agric Biol Chem.* 1971, 35, 686–692.

Hitzman, D. Microbiological secondary recovery, US Patent 3032472, 1962.

Hossain, H., Wellensick, H. J., Geyer, R., Lochnit, G. Structural analysis of glycolopids from *Borrelia burgdorferi. Biochimie* 2001, 83(7), 683–692.

Hossein, A., Muller, M. M., Syldatk, C., Hausmann, R. Production of microbial rhamnolipid by *Pseudomonas aeruginosa* MM1011 for *ex situ* enhanced oil recovery. *Appl. Biochem. Biotechnol.* 2013, 170 (5), 1080–1093.

Hultberg, M., Bergstrang, K. J., Khalil, S., Alsanius, B. Characterization of surfactant producing strains of florescent *Pseudomonas* in soil less cultivation system. *Ant. Von. Leeuwenhoek.* 2008, 94(2), 329–334.

Inoh, Y., Kitamoto, D., Hirashima, N., Nakanishi, M. Biosurfactant of MEL: an increase gene transfection mediated by cationic loposomes. *Biochem. Biophysic. Res. Commun.* 2001, 289, 57–61.

Isoda, H., Kitamoto, D., Shinmoto, H., Matsumura, M., Nakahara, T. Microbial extracellular glycolipid induction of differentiation and inhibition of protein kinase C activity of human promyelocytic leukemia cell line HL60. *Biosci Biotechnol Biochem.* 1997, 61, 609–614.

Isoda, H., Shinmoto, H., Matsumura, M., Nakahara, T. The Neurite-initiating effect of microbial extracellular glycolopid in PC12 cells. *Cytotechnology* 1999, 31, 165–172.

Itokawa, H., Miyashita, T., Morita, H., Takeya, K., Hirano, T., Homma, M., Oka, K. Structural and conformational studies of [Ile7] or [Leu7] surfactins from *Bacillus subtilis. Chem. Pharmacol. Bull.* 1994, 42, 604–607.

Iyer, A., Mody, K., Jha, B. Emulsifying properties of a marine bacterial exopolysaccharide. *Enzyme and Microbial Technology* 2006, 38, 220–222.

Jazzar, C., Hammad, E. A. The efficiency of enhanced aqueous extracts of *Melia azedarach* leaves and fruits integrated with the *Camptotylus reuteri* releases against the sweet potato whitefly nymphs. *Bull Insectol.* 2003, 56, 269–275.

Jinfeng, L., Lijun, M., Bozhong, M., Rulin, L., Fangtian, N., Jiaxi, Z. The field pilot of microbial enhanced oil recovery in a high temperature petroleum reservoir. *J. Petrol. Sci Eng.* 2005, 48, 265–271.

Juwarkar, A. A., Nair, A., Dubey, K. V., Singh, S. K., Devotta, S. Biosurfactant technology for remediation of cadmium and lead contaminated soils. *Chemosphere* 2007, 68, 1996–2002.

Kachholz, T., Schlingmann, M. Possible food and agricultural applications of microbial surfactants: an assessment, In *Biosurfactant and Biotechnology, Vol. 25*, Kosaric, N., Chains, W. L., Grey, N. C. C. (Eds.), Marcel Dekker Inc.: New York, 1987, pp. 183–208.

Kappeli, O., Finnerty, W. R. Partition of alkane by extracellular vesicle derived from hexadecane grown *Acinetobacter. J. Bact.* 1979, 140, 707–712.

Kasture, M. B., Patel, P., Prabhune, A. A., Ramana, C. V., Kulkarni, A. A., Prasad, B. L. V. Synthesis of silver nanoparticles by sophorolipids: Effect of temperature and sophoro-lipid structure on the size of particles. *J. Chem. Sci.* 2008, 120(6), 515–520.

Kasture, M., Singh, S., Patel, P., Joy, P. A., Prabhune, A. A., Ramana, C. V., Prasad, B. L. V. Synthesis of silver nanoparticles by sophorolipids: effect of temperature and sophoro-lipid structure on the size of particles. *Langmuir* 2007, 23, 1409–1412.

Kawano, J., Suzuki, T., Inoue, S., Hayashi, S. Powered compressed cosmetic material, US Patent 4305931, Kao Soap Co. Ltd., Tokyo, 1981.

Kawano, J., Suzuki, T., Inoue, S., Hayashi, S. Stick-shaped cosmetic material, US Patent 4305929, Kao Soap Co. Ltd., Tokyo, 1981.

Kearns, D. B., Losick, R. Swarming motility in undomesticated *Bacillus subtilis*. *Mol. Micro-biol.* 2003, 49, 581–590.

Kim, P. I., Kim, Y. H., Chi, Y. T. Production of biosurfactant lipopeptides, Iturin A, fengycin and surfactin A from *Bacillus subtilis* CMB32 for control of *Collectotrichum gloeospo-riodies*. *J. Microbiol. Biotechnol.* 2010, 20, 138–145.

Kim, P. I., Bai, H., Bai, D., Chae, H., Chung, S., Kim, Y., Park, R., Chi, Y. T. Purification and characterization of a lipopeptide produced by *Bacillus thuringiensis* CMB26. *J. Appl Microbiol.* 2004, 97, 942–949.

Kim, S. K., Kim, Y. C., Lee, S., Kim, J. C., Yun, M. Y., Kim, I. S. Insecticidal activity of rhamnolipid isolated from *Pseudomonas* spp. EP-3 against green peach aphid (*Myzus persicae*). *J. Agric. Food Chem.* 2011, 59, 934–938.

Kim, K., Yoo, D., Kim, Y. Characteristics of sophorolipid as an antimicrobial agent. *J. Microbiol. Biotechnol.* 2002, 12, 235–241.

Kiran, G. S., Selvin, J., Manilal, A., Sujith, S. Biosurfactants as green stabilizers for the bio-logical synthesis of nanoparticles. *Crit. Rev. Biotechnol.* 2011, 31, 354–364.

Kitamato D., Isoda, H., Hara, T. N. Functions and potential applications of glycolipid bio-surfactants from energy saving material to gene delivery carriers. *J. Bio. Sci. Bio. Eng.* 2002, 94, 187–201.

Kleckner, V., Kosaric, N. Biosurfactants for cosmetics. In *Biosurfactants: Production, Proper-ties, Applications*, Kosaric, N., Ed., Marcel Dekker: New York, 1993, pp. 329–389.

Konishi, M., Morita, T., Fukuoka, T., Imura, T., Kakugawa, K., Kitamoto, D. Production of different types of mannosylerythritol lipids as biosurfactants by the newly iso-lated yeast strains belonging to the genus *Pseudozyma*. *Appl. Microbiol. Biotechnol.* 2007, 75, 521–531.

Kooper, D. G., Goldenberg, B. G. Surface active agents from two *Bacillus* sp. *Appl. Environ. Microbiol.* 1987, 53, 224–229.

Kosaric, N. Biosurfactants and their application for soil bioremediation. *Food Technology and Biotechnology* 2001, 39(4), 295–304.

Kracht, M., Rokos, H., Ozel, M., Kowall, M., Pauli, G., Vater, J. Antiviral and hemolytic activities of surfactin isoforms and their methyl ester derivatives. *J. Antibiot.* 1999, 52, 613–619.

Kropp, K., Davidova, I. A., Suflita, J. M. Anaerobic oxidation of n-dodecane by an addition reaction in a sulfate-reducing bacterial enrichment culture. *Appl. Environ. Microbiol.* 2000, 66, 5393–5398.

Krujit, M., Tran, H., Raaijmakers, J. M. Functional, genetic and chemical characterization of biosurfactants produced by plant growth-promoting *Pseudomonas putida* 267. *J. Appl. Microbiol.* 2009, 107, 546–556.

Krzyzanowska, D. M., Potrykus, M., Golanowska, M., Polonia, K., Gwizdek-Wisniewska, A. E., Lojkowska, E., Jafra, S. Rhizosphere bacteria as potential biocontrol agents against soft rot caused by various *Pectobacterium* and *Dickeya* spp. starin. *J. Plant Pathol.* 2012, 94(2), 367–378.

Kumar, C. G., Mamidyala, S. K., Das, B., Sridhar, B., Devi, G. S., Karuna, M. S. Synthesis of biosurfactant-based silver nanoparticles with purified rhamnolipids isolated from *Pseudomonas aeruginosa* BS–161R. *J. Microbiol Biotechnol.* 2010, 20, 1061–1068.

Laha, S., Tansel, B., Ussawarujikulchai, A. Surfactant-soil interactions during surfactant-amended remediation of contaminated soils by hydrophobic organic compounds: a review. *J. Environ. Management.* 2009, 90, 95–100.

Lai, C. C., Huang, Y. C., Wei, Y. H., Chang, J. S. Biosurfactant-enhanced removal of total petroleum hydrocarbons from contaminated soil. *Journal of Hazardous Materials* 2009, 167, 609–614.

Lang, S., Wullbrandt, D. Rhamnose lipids – biosynthesis, microbial production and application potential. *Appl. Microbiol. Biotechnol.* 1999, 51, 22–32.

Lappin-Scott, H., Cusack, F., Costerton, J. W. Nutrient resuscitation and growth of starved cells in sandstone cores: a novel approach to enhanced oil recovery. *Appl Environ Microbiol.* 1988, 54, 1373–1382.

Lazar, I., Petrisor, I. G., Yen, T. F. Microbial Enhanced Oil Recovery (MEOR). *Pet. Sci. Technol.* 2007, 25, 1353–1366.

Lima, T. M., Procopio, L. C., Brandao, F. D., Leao, B. A., Totola, M. R., Borges, A. C. Evaluation of bacterial surfactant toxicity towards petroleum degrading microorganisms. *Bioresour Technol.* 2011, 102, 2957–2964.

Lourith, N., Kanlayavattanakul, M. Natural surfactants used in cosmetics: glycolipids. *Int J Cosmet Sci.* 2009, 31, 255–261.

Mager, H., Rothlisberger, R., Wagner, F. Use of sophorose-lipid lactone for the treatment of dandruffs and body odor, European Patent 0209783, Institut Francais du Petrole, Malmaison, 1987.

Maier, R. M. *Biosurfactants: evolution and diversity in bacteria*, Elsevier, USA, 2003.

Maier, R. M., Soberon-Chavez, G. *Pseudomonas aeruginosa* rhamnolipids: biosynthesis and potential applications. *Appl. Microbiol. Biotechnol.* 2000, 54, 625–633.

Maingault, M. Utilization of sophorolipids as therapeutically active substances or cosmetic products in particular for the treatment of the skin, US Patent 5981497, Institut Francais du Petrole, Malmaison, 1999.

Makkar, R. S., Rockne, K. J. Comparison of synthetic surfactants and biosurfactants in enhancing biodegradation of polycyclic aromatic hydrocarbons. *Environ Toxic Chem.* 2003, 22(10), 2280–2292.

Makkar, R. S., Cameotra, S. S. An update on the use of unconventional substrates for biosurfactant production and their new applications. *Appl. Microbiol. Biotechnol.* 2002, 58, 428–434.

Makkar, R. S., Rockne, K. J. Comparison of synthetic surfactants and biosurfactants in enhancing biodegradation of polycyclic aromatic hydrocarbons. *Environ. Toxic. Chem.* 2003, 22(10), 2280–2292.

Mann, R. M., Bidwell, J. R. The acute toxicity of agricultural surfactants to the tadpoles of four Australian and two exotic frogs. *Environ Pollut.* 2011, 114, 195–205.

Margesin, R., Schinner, F. Biodegradation and bioremediation of hydrocarbons in extreme environments. *Appl. Microbiol. Biotechnol.* 2001, 56, 650–663.

McInerney, M. J., Javaherim, M., Nagle, D. P. Properties of the biosurfactant produced by *Bacillus liqueniformis* strain JF-2. *Journal of Industrial Microbiology and Biotechnology* 1990, 5, 95–102.

Miller, R. M. Biosurfactant-facilitated remediation of metal contaminated soils. *Environ Health Perspect.* 1995, 103(S1), 59–62.

Millsap, K., Reid, G., Van der Mai, H. C. Adhesion of *Lactobacillus* species in urine and phosphate buffer to silicon rubber and glass under flow. *Biomaterials* 1996, 18, 87–91.

Mittenbuhler, K., Loleit, M., Baier, W., Fischer, B., Sedelmeier, E., Jung, G., Winkelmann, G., Jacobi, C., Weckesser, J., Erhard, M. H., Hofmann, A., Bessler, W., Hoffmann, P. Drug specific antibodies: T-cell epitope-lipopeptide conjugates are potent adjuvants for small antigens *in vivo* and *in vitro*. *Int. J. Immunopharmacol.* 1997, 19, 277–287.

Mnif, S., Chamkha, M., Labat, M., Sayadi, S. Simultaneous hydrocarbon biodegradation and biosurfactant production by oilfield-selected bacteria. *J. Appl. Microbiol.* 2011, 111, 525–536.

Mulligan, C. N. Environmental applications for biosurfactants. *Environ Pollut.* 2005, 133, 183–198.

Mullignan, C., Gibbs, B. Types, production and applications of biosurfactants. *Proc. Ind. Nat. Sci Acad.* 2004, B70, 31–55.

Neu, T. R. Significance of bacterial surface-active compounds in interaction of bacteria with interfaces. *Microbiol. Rev.* 1996, 60, 151–166.

Nihorimbere, V., Marc Ongena, M., Smargiassi, M., Thonart, P. Beneficial effect of the rhizosphere microbial community for plant growth and health. *Biotechnol. Agron. Soc. Environ.* 2011, 15, 327–337.

Nishio, E., Ichiki, Y., Tamura, H., Morita, S., Watanabe, K., Yoshikawa, H. Isolation of bacterial strain that produce the endocrine disruptor, octylphenol diethoxylates, in paddy fields. *Biosci Biotechnol Biochem.* 2002, 66, 1792–1798.

Nitschke, M., Costa, S. Biosurfactants in food industry. *Trends Food Sci. Technol.* 2007, 18, 252–259.

Nitschke, M., Pastore, G. M. Production and properties of a surfactant obtained from *Bacillus subtilis* grown on cassava wastewater. *Bioresource Technol.* 2006, 97, 336–341.

Ochoa-Loza, F. J., Noordman, W. H., Jannsen, D. B., Brusseau, M. L., Maier, R. M. Effect of clays, metal oxides and organic matter on rhamnolipid biosurfactant sorption by soil. *Chemosphere* 2007, 66, 1634–1642.

Ollivier, B., Magot, M. *Petroleum Microbiology*. ASM Press: Washington, DC, 2005.

Pacwa-Plociniczak, M., Plaza, G. A., Piotrowska-Seget, Z., Cameotra, S. S. Environmental applications of biosurfactants. *Recent Adv Int. J. Mol. Sci.* 2011, 12, 633–654.

Palanisamy, P., Raichur, A. Synthesis of spherical NiO nanoparticles through a novel biosurfactant mediated emulsion technique. *Materials Sci. Eng.* 2009, 29, 199–204.

Pattanathu, K. S. M., Gakpe, E. Production, characterization and applications of biosurfactants – review. *Biotechnol.* 2008, 7, 360–370.

Pellecier, F., Andre, P. Cosmetic use of sophorolipids as subcutaneous adipose cushion regulation agents and slimming application, World Patent 2004/108063, LVMH Recherche, Saint Jean de Braye, 2004.

Pierce, D., Heilman, T. J. Germicidal composition, World Patent 1998/16192, Alterna Inc., Los Angeles, CA, 1998.

Piljac, T., Piljac, G. Use of rhamnolipids in wound healing, treating burn shock, atherosclerosis, organ transplants, depression, schizophrenia and cosmetics, European Patent 1889623, Paradigm Biomedical Inc., New York, 1999.

Pirollo, M. P. S., Mariano, A. P., Lovaglio, R. B., Costa, S. G. V. A. O., Walter, V., Hausmann, R., Contiero, J. Biosurfactant synthesis by *Pseudomonas aeruginosa* LBI isolated from a hydrocarbon-contaminated site. *J. Appl Microbiol.* 2008, 105, 1484–1490.

Pornsunthorntawee, O., Arttaweeporn, N., Paisanjit, S., Somboonthanate, P., Abe, M., Rujiravanit, R., Chavadej, S. Isolation and comparison of biosurfactants produced by *Bacillus subtilis* PT2 and *Pseudomonas aeruginosa* SP4 for microbial surfactant enhanced oil recovery. *Biochem Eng J.* 2008, 42, 172–179.

Pratt-Terpstra, I. H., Weerkamp, A. H., Busscher, H. J. The effects of pellicle formation on streptococcal adhesion to human enamel and artificial substrata with various surface free-energies. *J. Dent. Res.* 1989, 68, 463–467.

Rastall, R. A., Gibson, G. R., Gill, H. S., Guarner, F., Klaenhammer, R. T., Pot, B., Reid, G., Rowland, I. R., Sanders, E. M. Modulation of the microbial ecology of the human colon by probiotics, prebiotics and synbiotics to enhance human health: an overview of enabling science and potential applications. *FEMS Microbiol Ecol.* 2005, 52(2), 145–152.

Reddy, S., Chien, Y. C., Baker, S. C., Chein, C. C., Jiin, S. J., Cheng, W. F., Hau, R. C. Synthesis of silver nanoparticles using surfactin: A biosurfactant as stabilizing agent. *Mat. Lett.* 2009, 63, 1227–1230.

Reid, G. Probiotic agents to protect the urogenital tract against infection. *Am. J. Clin. Nutr.* 2001, 73(2S), 437S–443S.

Reid, G., Heinemann, C., Velraeds, M., Henny, C., van der Mei, H. C., Henk, J. B. Biosurfactants produced by *Lactobacillus. Methods Enzymol.* 1999, 310, 426–433.

Reid, R. S., Pacheco, G. J., Pereira, A. G., Freire, D. M. G. Biosurfactants: Production and Applications. In *Biodegradation*, Chamy, R., Rosenkranz, F., Eds., Intech Publishing: Croatia, 2013, pp. 31–61.

Rodrigues, L., Teixeira, J. Biosurfactants: biomedical and therapeutic applications of biosurfactants. In *Biosurfactants*, Sen, R., Ed., Landes Bioscience: Austin, Texas, 2010.

Rodrigues, L., Moldes, A., Teixeira, J., Oliveira, R. Kinetic study of fermentative biosurfactant production by *Lactobacillus* strains. *Biochem. Eng. J.* 2006, 28(2), 109–116.

Rodrigues, L. R., Banat, I. M., Teixeira, J., Oliveira, R. Biosurfactants: potential applications in medicine. *J. Antimicrobiol Chemo.* 2006, 57, 609–618.

Rodrigues, L. R., Banat, I. M., Van der Mei, H. C., Teixeira, J., Oliveira, R. Interference in adhesion of bacteria and yeasts isolated from explanted voice prostheses to silicone rubber by rhamnolipid biosurfactants. *J. Appl. Microbiol.* 2005, 100, 470–480.

Rodrigues, L. R., Van der Mei, H. C., Banat, I. M. The influence of biosurfactants from probiotic bacteria on the formation of voice prosthetic biofilms. *Appl. Environ. Microbiol.* 2004, 70, 4408–4410.

Rodrigues, L. R., Van der Mei, H. C., Banat, I. M., Ibrahim, M., Teixeira, J. A., Oliveira, R. Inhibition of microbial adhesion to silicone rubber treated with biosurfactant from *Streptococcus thermophilus* A. *FEMS Immunol Med Microbiol.* 2006, 6, 107–125.

Rodrigues, L. R., Teixeira, J. A., Mei, H. C. V., Oliveira, R. Isolation and partial characterization of a biosurfactant produced by *Streptococcus thermophilus* A. *Coll. Surf B: Biointerfaces,* 2006, 53, 105–112.

Rodrigues, L. R., Teixeira, J. A., Oliveira, R. Low cost fermentative medium for biosurfactant production by probiotic bacteria. *Biochem. Eng. J.* 2006, 32, 135–142.

Ron, E. Z., Rosenberg, E. Biosurfactants and oil bioremediation. *Curr. Opn. Biotechnol.* 2002, 13, 249–252.

Ron, E. Z., Rosenberg, E. Natural roles in biosurfactants. *Environ Microbiol.* 2011, 13, 229–236.

Rosenberg, E., Zuckerberg, A., Rubinovitz, A., Gutnick, D. L. Emulsifier *Arthrobacter* Rag-1: isolation and emulsifying properties. *Appl. Environ. Microbiol.* 1979, 37, 402–408.

Saharan, B. S., Sahu, R. K., Sharma, D. A review on Biosurfactants: fermentation, current developments and perspectives. *Genetic Engg Biotechnol J.* 2011, 29, 1–14.

Sangeetha, J., Thomas, S., Arutchelvi, J., Doble, M., Philip, J. Functionalization of iron oxide nanoparticles with biosurfactants and biocompatibility studies. *J. Biomed. Nanotechnol.* 2013, 9, 751–764.

Sano, H., Sohma, H., Muta, T., Nomura, S., Voelker, D. R., Kuroki, Y. Pulmonary surfactant protein A modulates the cellular response to small rough polysaccharides by interaction with CD14. *J. Immunol.* 1999, 163(1), 387–395.

Sarubbo, L. A., Farias, C. B., Campos-Takaki, G. M. Co-utilization of canola oil and glucose on the production of a surfactant by *Candida lipolytica. Curr Microbiol.* 2007, 54, 68–73.

Satpute, S. K., Banat, I. M., Dhakephalkar, P. K., Banpurkar, A. G., Chopade, B. A. Biosurfactants, bioemulsifiers and exopolysaccharides from marine microorganisms. *Biotechnology Advances* 2010, 28, 436–450.

Scott, M. J., Jones, M. N. The biodegradation of surfactants in the environment. *Biochem Biophys Acta.* 2000, 1508, 235–251.

Sen, R. Biotechnology in petroleum recovery: the microbial EOR. *Prog Energ Combust.* 2008, 34, 714–724.

Sen, R. Surfactin: biosynthesis, genetics and potential applications. In *Biosurfactants*, Sen, R., Ed., Landes Bioscience: Austin, Texas, 2010, pp. 316–323.

Sha, R., Jiang, L., Meng, Q., Zhang, G., Song, Z. Producing cell free culture broth of rhamnolipids as a cost effective fungicide against plant pathogens. *J. Basic Microbiol.* 2011, 52, 458–466.

Shafeeq, M., Yokub, D., Khalid, Z. M., Khan, A., Malik, K. Degradation of different hydrocarbons and production of biosurfactant by *Pseudomonas aerugenosa* isolated from coastal waters. *J. Appl. Microbiol. Biotech.* 1989, 5, 505–510.

Singh, A., van Hamme, J. D., Ward, O. P. Surfactants in microbiology and biotechnology: Part 2. Application aspects. *Biotechnol Adv.* 2007, 25, 99–121.

Singh, P., Cameotra, S. Potential applications of microbial surfactants in biomedical sciences. *Trends Biotechnol.* 2004, 22, 142–146.

Smyth, T. J. P., Perfumo, A., Marchant, R., Banat, I. M. Isolation and analysis of lipopeptides and high molecular weight biosurfactants. In *Hydrocarbon and Lipid Microbiology*, Timmis, K. T. (Ed.), Springer-Verlag: Berlin, Heidelberg, 2010, pp. 436–450.

Satpute, S. K., Bhuyan, S. S., Pardesi, K. R., Mujumdar, S. S., Dhakephalkar, P. K., Shete, A. M., Chopade, B. A. Molecular genetics of biosurfactant synthesis in microorganisms. *Adv Exp Med Biol* 2010, 672, 14–41.

Suthar, H., Hingurao, K., Desai, A., Nerurkar, A. Evaluation of bioemulsifier mediated microbial enhanced oil recovery using sand pack column. *J. Microbiol. Methods.* 2008, 75, 225–230.

Takenaka, S., Tonoki, T., Taira, K., Murakami, S., Aoki, K. Adaptation of *Pseudomonas* sp. strain 7–6 to quaternary ammonium compounds and their degradation via dual pathways. *Appl. Environ. Microbiol.* 2007, 73, 1797–1802.

Thimon, L., Peypoux, F., Wallach, J., Michel, G. Effect of lipopeptide antibiotic iturin on morphology and membrane ultrastructure of yeast cells. *FEMS Microbiol Lett.* 1995, 128, 101–106.

Thombre, R. S., Parekh, F., Patil, N, M A facile method for green synthesis of stabilized silver nanoparticles using *Argyeria nervosa. Int. J. Pharm. Biosci.* 2014, 5, 114–119.

Thombre, R. S., Parekh, F., Lekshminarayanan, P., Francis, G. Studies on antibacterial and antifungal activity of silver nanoparticles synthesized using *Artocarpus heterophyllus* leaf extract. *Biotechnol. Bioinf. Bioeng.* 2012, 2, 632–637.

Ueno, Y., Hirashima, N., Inoh, Y., Furuno, T., Nakanishi, M. Characterization of biosurfactant-containing liposomes and their efficiency for gene transfection. *Biol Pharm Bull.* 2007, 30(1), 169–172.

Updegraff, D., Wren, G. B. Secondary recovery of petroleum oil by *Desulfovibrio*. US Patent 2660550, 1953.

Van Hamme, J. D., Singh, A., Ward, O. P. Physiological aspects: Part 1, surfactants in microbiology and biotechnology. *Biotechnol. Adv.* 2006, 24, 604–620.

Vasta, P., Sanchez, L., Clement, C., Baillieul, F., Dorey, S. Rhamnolipid biosurfactants as new players in animal and plant defense against microbes. *Int. J. Mol. Sci.* 2010, 11, 5095–5108.

Velho, R. V., Meina, L. F., Segalin, J., Brandelli, A. Production of lipopeptides among *Bacillus* strain showing growth inhibition of phytopathogenic fungi. *Folia Microbiol (Praha).* 2011, 56, 297–303.

Vollenbroich, D., Ozel, M., Vater, J., Kamp, R., Pauli, G. Mechanism of inactivation of enveloped viruses by the biosurfactant surfactin from *Bacillus subtilis. Biologicals* 1997, 25(3), 289–297.

Vollenbroich, D., Pauli, G., Ozel, M., Vater, J. Antimycoplasma properties and applications in cell culture of surfactin, a lipopeptide antibiotic from *Bacillus subtilis. Appl Environ Microbiol.* 1997, 63, 44–49.

Wang, C. L., Ng, T. B., Yuan, F., Liu, Z. K., Liu, F. Induction of apoptosis in human leukemia K562 cells by cyclic lipo peptide from *Bacillus subtilis* natto T-2. *Peptides* 2007, 28, 1344–1350.

Wang, Q., Fang, X., Bai, B., Liang, X., Shuler, P. J., Goddard, W. A. III., Tang, Y. Engineering bacteria for production of rhamnolipid as an agent for enhanced oil recovery. *Biotechnol Bioeng.* 2007, 98, 842–853.

Wang, X., Gong, L., Liang, S., Han, X., Zhu, C., Li, Y. Algicidal activity of rhamnolipid biosurfactants produced by *Pseudomonas aeruginosa. Harmful Algae* 2005, 4, 433–443.

Yoo, D. S., Lee, B. S., Kim, E. K. Characteristics of microbial biosurfactant as an antifungal agent against plant pathogenic fungus. *J. Microbiol. Biotechnol.* 2005, 15, 1164–1169.

Zang, C., Wang, S., Yan, Y. Isomerization and biodegradation of beta-cypermethrin by *Pseudomonas aeruginosa* CH-7 with biosurfactant production. *Bioresour Technol.* 2011, 102, 7139–7146.

Zobell, C. Action of microorganisms on hydrocarbons. *Bacteriol Rev.* 1946, 10, 1–49.

Zobell, C. Recovery of hydrocarbons, US Patent 2641566, 1953.

CHAPTER 5

HALOPHILES: PHARMACEUTICAL POTENTIAL AND BIOTECHNOLOGICAL APPLICATIONS

REBECCA S. THOMBRE, VAISHNAVI S. JOSHI, and RADHIKA S. OKE

Department of Biotechnology, Modern College of Arts, Science and Commerce, Shivajinagar, Pune, 411005, Maharashtra, India

CONTENTS

5.1   INTRODUCTION

'Halophiles' are salt loving organisms requiring 3–5 M salt concentration for their growth and survival. These organisms are known to dwell in hypersaline marshes, solar salterns – manmade and naturally occurring, saline ponds, hot springs, etc. Halophiles are exposed to very strict conditions of growth like the UV radiations, high salinity, high temperature, alkaline pH, hence they have adapted a mechanism to survive the environment by the means of 'osmoadaptation.'

Extremophiles have been proved to be of great interest as the source of different enzymes, biomolecules and compounds of pharmaceutical relevance since they can thrive well under unique set of conditions. Among these organisms, enzymes produced from halophiles are of more interest as these organisms thrive in an environment of low water activity. They are the group of organisms that carry out all their activities in saline conditions. The enzymes produced from halophiles have unique set of properties and structures which enables them to function well in presence of high salt (Karan and Khare, 2011). Halophiles have predominant negatively charged residues on the solvent-exposed surfaces of the protein. These negatively charged residues attract water molecules thus keeping the proteins hydrated (Balasubramanian et al., 2002). It has been found that halophilic enzymes show high activity in low level of water activity. Studies have shown that these enzymes show substantial activity in organic solvents, and this makes it an ideal choice as biocatalysts (Lanyi and Stevenson, 1969). Halophiles in this context have been found to be a potential source of enzymes and they have known to produce many different enzymes since past (Haddar et al., 2009; Karan and Khare, 2010).

5.2   PHARMACEUTICAL POTENTIAL OF COMPATIBLE SOLUTES

Compatible solutes are synthesized by the cell for counter reaction or they are transported from the medium into the cell. To combat with the high ionic strength, halophilies particularly halophilic bacteria accumulate solutes inside the cell. These solutes are called "compatible solutes" as they show compatibility with the vital functioning of the cellular functions. Compatible solutes are soluble, uncharged, having low molecular weight organic molecules. These solutes do not have any interference with the routine functioning/metabolic activities of the cell of an organism. Compatible solutes are of different kinds produced by various kinds of bacteria and eukaryotes like the polyols, sugars, aminoacids, peptides, ectoine, betaine and their derivatives (Galinski, 1995). Microorganisms belonging to the archaeal domain and extremely halophilic bacteria use another strategy to survive the high salinity environment, "salt-in strategy" wherein they accumulate the K^+ ions in their cytoplasm. Bacteria, any eukaryotes generally accumulate neutral solutes whereas Archaea accumulate negatively charged solutes by modifying the neutral solutes accumulated by bacteria and eukaryotes. These solutes help in maintaining the turgor pressure, volume and size of the cell, help in efficient cell growth and proliferation and also aid in thermoprotection of the enzymatic machinery and other molecules of the cell (Martin et al., 1999; Roberts, 2004).

5.2.1   ECTOINE

Ectoine is a cyclic tetrahydropyrimidine (1,4,5,6-tetrahydro-2-methyl-4-pyrimidinecarboxylic acid) and increases in the cell with the increase in the extracellular NaCl concentration (Figure 5.1). A variant solute of ectoine known as the hydroxyectoine is also known to accumulate in some halotolerant species growing in high salt concentration. Ectoine only accumulates in the exponential phase of growth (Regev et al., 1990; Galinski et al., 1995; Kuhlmann and Bremer, 2002). The biosynthetic pathway was first studied in *H. elongate* DSM 2581 (Peters et al., 1990). Phosphorylation of L-aspartate to L-aspartate phosphate semialdehyde takes place in the presence of L-aspartate phosphate. The reaction is catalyzed by L-aspartate-beta-semialdehyde dehydrogenase. In the first step L-diaminobutyric acid is formed from L-aspartate-beta-semialdehyde dehydrogenase by the enzyme L-diaminobutyric acid transaminase. This is then acetylated to N (4)-acetyl-L-2, 4-diaminobutyric acid by the enzyme L-diaminobutyric acid acetyl transferase. Lastly, the acetylated N (4)-acetyl-L-2,4-diaminobutyric acid is converted to ectoine by cyclic condensation catalyzed by the enzyme ectoine synthase (Grammel, 2000; Gracia-Estepa et al., 2006). Genes encoding the enzymes for the biosynthesis of ectoines have been identified in fifty bacterial

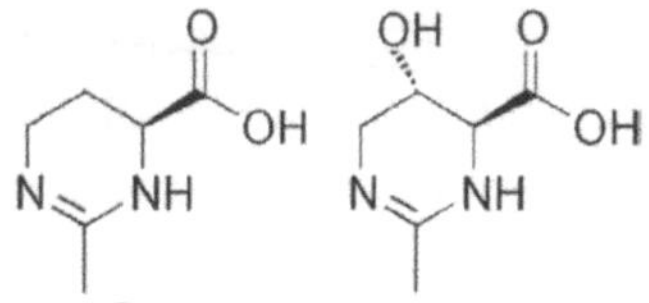

Ectoine   Hydroxyectoine

FIGURE 5.1   Structure of ectoine and its derivative hydroxyectoine.

and one archaeal species (Lo et al., 2009). The genes are clustered in one single operon. It is important to optimize knowledge of ectoine synthesis as new and genetically engineered strains of ectoine producers are essential. Ectoines offer macromolecule protection against freezing, drying, high salinity, oxygen radicals, radiation, etc. (da Costa et al., 1998). Ectoines bring about conformational change in the structure of DNA which aids them is camouflaging from getting cleaved by the restriction endonucleases. Ectoines also display a protective role in diseases like Ischemia (Grant et al., 1990), emphysema, chronic obstructive pulmonary disease, cancer and fibroses, cardiovascular and immune system (Peter et al., 2004; Macnee, 2007).

5.2.2   BETAINE

Another compatible solute produced by halophilic bacteria to maintain osmoadaptation is betaine (Figure 5.2). This solute is formed by methylation of primary amine glycine to form a quaternary structure (Imhoff and Rodriguez-Valera, 1984). Concentration of betaine accumulated in the cell varies with the varying external NaCl concentration (Robertson et al., 1990). It is generally transported from the medium into the cell except a few bacteria and one methanogen has the potential to synthesize it (Roberts et al., 1992; Canovas et al., 1998; Nyyssola et al., 2000). They are normally produced by halophilic phototrophic bacteria, chemotrophic bacteria and archeaebacteria. There are two different pathways for betaine synthesis: (i) by oxidation, using a single soluble enzyme choline oxidase, or (ii) by membrane associated system coded by four genes in an operon having (bet A), (bet B), (bet T) and (bet I) coding for enzymes choline dehydrogenase, betaine-aldehyde dehydrogenase, choline transporter and putative regulation. The choline dehydrogenase helps in catalyzing the oxidation reaction of choline to betaine aldehyde which is further oxidized to betaine by betaine-aldehyde dehydrogenase. This process consumes an O_2 molecule with the release of H_2O_2. In the final oxidation step

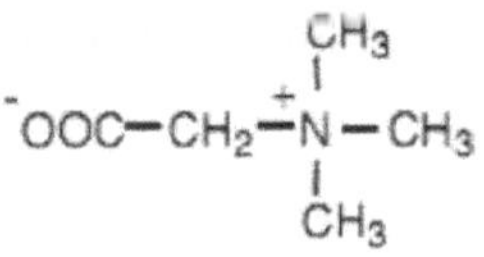

$$\overset{\overset{\displaystyle CH_3}{\displaystyle |}}{^-OOC-CH_2-\overset{+}{N}-CH_3}$$
$$\underset{\displaystyle CH_3}{\displaystyle |}$$

FIGURE 5.2 Chemical structure of betaine.

NADP$^+$ is reduced. Betaine synthesis from glycine in some halotolerant organisms is carried out by GSMT (glycine sarcosine methyl transferase) and SDMT (sarcosine dimethylglycine methyl transferase). It is suggested that the accumulation of betaine in the cell is by accumulation of internal K$^+$ concentration. Betaine is transported across the cells by betaine transporters which use the proton motive force or sodium motive force. There is an ATP binding cassette which couples ATP hydrolysis to the betaine uptake. They are used in therapeutics to treat the patients with liver prophylaxis (Detkova and Boltyanskaya, 2007). They are used as anticoagulants to decrease the thrombus formation, thus reducing the chances of heart attacks and strokes (Messadek, 2005). They also aid to amplify the GC rich DNA templates which can help is amplifying the product yield. They are used as an effective cryo-protectant for long term storage (Cleland et al., 2004).

5.3 PHARMACEUTICAL IMPORTANCE OF BACTERIORUBERIN AND PIGMENTS

Microorganisms have proved a good source of diverse natural products which are used in different fields like food industry, pharmaceutical industry and in other industrial products (Naziri et al., 2014). Microorganisms in hypersaline environments have gained importance due to their ability to produce many useful products like bioactive compounds (pigments, antimicrobial compounds), compatible solutes, biosurfactants, etc. (Abbes et al., 2013). These halophilic archaea can be extreme, moderate and slight halophilic. One of the best examples studied in the context of production of bioactive molecules is of carotenoids. These are accessory, photosynthetic, fat soluble pigments found in all photosynthetic organisms. It consists of two, six-membered rings as its structure. The yellow, orange, red and purple color of plants, microorganisms and animals is due to presence of these pigments (Naziri et al., 2014). All the archaea produces carotenoids, so the colonies they produce are red and orange in color. Carotenoids have wide application as a precursor, a colorant, as a feed additive, as anti-oxidant, in medicine, pharmaceutical and cosmetic industry. It also has proved its role in immunology as enhancers of *in vitro* antibody production. The role of this pigment depends upon its structural properties like its molecular weight, number of side chains, number of functional groups attached and number of bonds present (Furubayashi et al., 2012).

Bacterioruberin is an important pigment with wide applications among the many carotenoids produced, especially in haloarchaea (Figure 5.3). Raman spectroscopy studies in haloarchaea have shown that major carotenoid produced in them is bacterioruberin. It is because of this pigment that the archaea appears in different shades of red or orange in their natural forms. It is believed that bacterioruberin protects these microorganisms from strong sunlight in their natural habitat. Bacterioruberin ($C_{50}H_{76}O_4$) is a 50-carbon open chain carotenoid with 13 pairs of conjugated double carbon bonds and with an approximate molecular weight of 740 Daltons (Rodriguez-Amay, 2001). This pigment has wide variety of applications, pharmaceutical application being of prime importance. Due to presence of carbon double bonds, this pigment shows hydroxyl free-radical scavenging activity which protects the organism producing it from DNA damage and H_2O_2 exposure (Shahmohammadi et al., 1998). Bacterioruberin also acts as a water barrier which is responsible for oxygen and other molecules permeability that enhances the survival rate of the microorganisms living in hypersaline environments and thus helps in membrane fluidity (Strand et al., 1997). Other light pigments produced by archaea includes β carotene, phytoene, lycopene and salinixanthin also exhibit good pharmaceutical activity.

Carotenoids serve as pro-vitamin A in human beings. Consumption of carotenoid rich food increases Vitamin A levels in the body (Zeb and Mehmood, 2004). Vitamin A is the major source of energy in human body and its deficiency leads to several diseases. Presence of carotenoid pigment in the photoreceptor of eyes protects eyes from harmful effects of light as it absorbs light of short wavelength. Lycopene is a light pigment isolated from

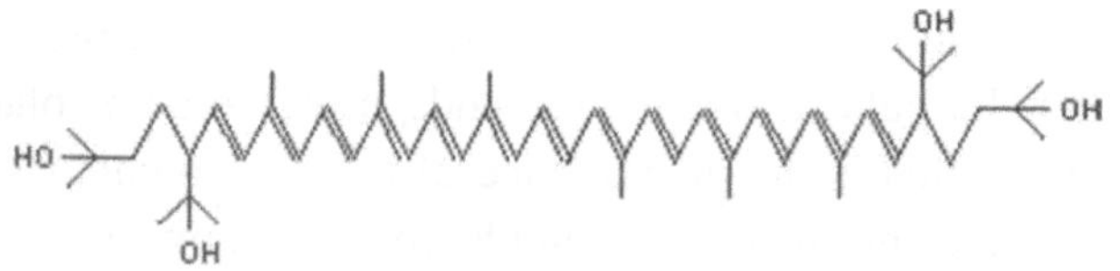

FIGURE 5.3   Structure of bacterioruberin.

different archaea has also got pharmaceutical applications. It is a hydrocarbon with antioxidant properties It helps in reduction of oxidation reactions thus preventing from many chronic diseases. Oxidation of cholesterol (LDL – Low Density Lipoprotein), reduction in risk of development of atherosclerosis and other heart related disorders can also be contributed to lycopene pigment (Zeb and Mehmood, 2004).

5.4 PHARMACEUTICAL POTENTIALS OF HALOCINS

Haloclines are peptidal bacteriocins, proteinaceous in nature produced by the members of extremely halophilic members of the archaeal family Halobacteriaceae and released in the external environment and can inhibit the growth of other archaeal species with whom they share the same growth requirements using the strategy of "Survival of the Fittest." These peptides were first discovered by Francisco Rodriguez-Valera and his co-workers in 1982 (Rodriguez and Valera, 1982). He isolated about 40 different species of extreme halophiles and screened them for halocin production. The mechanism of action of the bactericidal activity of halocins includes inhibition of transcription, translation, DNA and RNA nuclease activity, disruption of cell membrane, modifying the cell permeability or inhibiting the Na^+/H^+ antiporter and proton flux.

Some halocins are salt dependent and hence lose their activity when the salt concentration decreases below a minimum level (Rodriguez-Valera et al., 1982; Price and Shand, 2000). Halocin H4 was the first halocin to be studied which was produced by *Haloferax mediterranei* R4 (Messeguer and Rodriguez-Valera, 1985; Messeguer and Rodriguez-Valera, 1986). Later on they discovered a second kind of halocin which was characteristically similar to H4 and produced by *Haloferax gibonsii* nomenclatured as H6. Until recently, many haloclines have been successfully discovered from different archaeal species. Purification of halocins is comparatively simpler as they are secreted in the environment by the organism in the stationary phase (Cheung et al., 1997; Price and Shand, 2000; Li et al., 2003).

Although, there are many species of Archaea reported till date but only a few of them are known to produce halocins and are studied at molecular and gene levels. Halocin A4, G1, R1, H1, H2 (O'Connor and Shand, 2002); H3, H5 (Rodriguez-Valera et al., 1982; O'Connor and Shand, 2002); H4 (O'Connor and Shand, 2002; Sun et al., 2005); H6/H7 (O'Connor and Shand, 2002; Li et al., 2003); S8 (O'Connor and Shand, 2002); C8 (Li et al., 2003; Sun et al., 2005), Secha (Pasic et al., 2008) and KPS1 (Kavitha et al., 2011).

On the basis of their molecular weight the halocins are divided into two classes halocins (30–40kDa) and microhalocins (>10 kDa) (Meseguer and Rodríguez-Valera, 1985; Shand and Leyva, 2007). Large sized haloclines are heat labile and dependent on salt. Most of the halocline genes are produced between the exponential and the stationary phase and these genes are located on the mega plasmids of about 100 kb in size. They consist of TATA boxes and TFB recognition elements (BRE). The transcriptional site is either coincident or few base pairs upstream of the ATG initiation codon. Halocins can inhibit a wide range of organisms belonging to different genera such as Euryarchaea and Crenarchaea (O'Connor and Shand, 2002). The production of halocins plays a vital role in maintaining the species diversity in extremely halophilic environments as the competition for the growth is quiet competitive. Halocin producers (inhibitor strain) are known to lyse the cells of the sensitive strain and use the same for its nutritional benefits (Torreblanca et al., 1994; Shand and Leyva, 2007). Secretion of the halocins is known to occur using the Tat pathway as the precursors exhibit the signal sequences at their amino terminus of twin-arginine translocation pathway (Rose et al., 2002). Halocin production was determined based on the two-fold end point dilution to extinction method and is expressed as arbitrary units (AU). The zone of inhibition which depicts the activity of halocin production is calculated in centimeters (Meseguer et al., 1986; Torreblanca et al., 1989). Halocins are used in leather industry due to their antimicrobial nature to prevent the growth of halophilic organisms when applied to the hides (Birbir et al., 2004).

5.5   PHARMACEUTICAL POTENTIAL OF ARCHEOSOMES AND LIPOSOMES

Liposomes are small spherical enclosed structures composed of lipid molecules. The liposomes are mainly composed of phospholipids containing water soluble hydrophilic head region and lipid soluble hydrophobic tail section (Bangham, 1961). Liposomes are classified on the basis of the carrier system, viz. (1) virosomes – they contain the reconstituted viral proteins and carry artificial viruses (Lasic, 1993); (2) stealth liposomes – they are used for long life circulation of the biological particle in the body without it undergoing phagocytosis (Gref et al., 1994); (3) transferosomes – synthesized for transdermal (across the skin) delivery of the bioactive molecule (Mozafari and Mortazavi, 2005); (4) archeosomes – synthesized from polar lipids of Archaea, more stable, distinctly absorbed by the body and are superior to liposomes in all aspects (Krieg, 2001); (5) cochleates – they are used for oral and systemic deliver of antioxidants and other moieties (Mozafari et al., 2006); and (6) nanoliposomes – as the name suggests they are a minute version of liposomes, deliver vitamin E and ascorbic acid at the site of oxidation in the digestion system (Mozafari et al., 2006).

Liposomes have varied applications in the pharmaceutical, food and cosmetic industry. Archeosomes are similar to liposomes aiding in drug delivery. This term was coined by Sprott and his co-workers to describe the liposomes prepared from ether linked Archaeal lipids thus conferring the stability to liposomes (Sprott et al., 1997). They are small nano-shaped bodies prepared from the lipids extracted from specific species of Archaea, hence the name. Archaeal lipids consist of ether linkages containing phytanyl, bysphytanylediyl archaeols and cladarchaeols in sn-2, 3-enantomeric confirmation which are different to bacterial and eukaryal species. Ethers are resistant to acid hydrolysis. They are resistant to oxidation and high temperatures than ester linked lipids (Van de Vossenberg, 1998). Also to bile salts pH, activity of phospholipases and serum proteins (Jacquamet et al., 2009). Archeosomes are stable and do not form colloidal masses during storage for at least 4 months (De Rosa, 1996). One of the exciting of archeosomes is its antileak property (Gambarcota, 1994). In recent days due to changes in various aspects of lifestyle, the viruses and certain bacteria have become susceptible to the available drugs, hence there is a need to coat these available drugs with specific soluble protein antigens that can give a deliver a targeted immune response. The use of adjuvants in drug delivery has been in use since early days but its mechanism was poorly understood then.

The methods of production of archeosomes are by lipid hydration method, detergent dialysis method, reverse phase evaporation method, sonication, membrane extrusion, freeze and thaw method (Weiner and Cannon, 1989; Relini et al., 1994; Watwe and Bellare, 1995). As carrier vehicles in vaccine formulations, as delivery systems for drugs, genes, or cancer screening agents (Allen, 1997). Primary work on mouse has been carried out which consists of intravenous, oral and subcutaneous administration of these archaeosomes and it has reported to be non-toxic and safe (Patel et al., 2004).

5.6 BIOTECHNOLOGICAL POTENTIAL OF HALOPHILIC BACTERIA AND ARCHAEA

With the renaissance of biotechnology, science has demanded the search of new enzymes by employing the techniques of genetic engineering or by searching a new source of enzymes (Oren, 2002). Scientists have therefore diverted their focus on the new domain called Archaea – the extremophiles. Archaea can withstand harsh and strict conditions hence their use in industries like environmental remediation, food processing, detergents industry, need such biocatalysts. Until recently the source of these enzymes have been plants, animals, fungi and microbes from bacterial domain, but as the enzymes of these organisms work efficiently at high pH, salinity, temperature, they have gained a lot of importance (Synowiecki et al., 2006; Ozcan et al., 2009). These enzymes are efficient in catalyzing the polymerase chain reaction. Taq polymerase is isolated from thermophilic bacteria is used extensively in PCR reactions. The demand for such enzymes is growing manifold and there has arisen the need to clone the genes encoding these enzymes in some other organisms (Alqueres et al., 2007).

In starch processing industry, thermostable enzymes are used to convert starch into its monosaccharides and disaccharides. The complete process of converting starch requires high temperatures to liquefy starch so that it becomes available to the enzymes for conversion. Most of the archaeal amylases are reported to be stable at high temperatures ranging from 80°C–100°C (Egorova and Antranikian, 2005). Lipases are produced in the nature by animals, plants, fungi and some bacteria. It has been reported that halophilic archaea *Haloarcula marismortui* produce esterase intracellularly and lipase extracellularly (Camcho et al., 2009).

Biofuel industry is one of the upcoming industries who are in search of new and different strategies for the conversion of biomass into biofuel products. Since Archaea are exposed to strict sunlight, they have adapted a mechanism to convert solar energy by the transversion of protons and chloride ions for phototaxis. They harbor light sensitive proteins called the bacterial rhodopsin, a 25 kDa protein that carries the retinal group linked to lysine-216 by the Schiff base action. The bacteriorhodopsin protein is used in holography, spatial modulators, artificial retina and volumetric and associative optical memories (Alqueres et al., 2007).

Archaeal lipids have been exploited for its use in drug delivery system in therapeutics. Liposomes easily assimilate with the cells and introduce the substance directly into it. They can also deliver non biological agents into the human body without any degradation. These ether linked lipids are resistant to high temperature and oxidation events than ester-linked core lipids, hence it contributes to its thermostability (Van de Vossenberg, 1998). Thermostability is essential as they have to go through the steps of sterility at high temperature of autoclaving at 121°C. With this nature it is essential to have an engulfed envelope stable to protect the inner encapsulated substance from the loss of its molecular weight and being stable to heat. Ether based lipids are also resistant to phospholipids and do not aggregate when stored at 4°C (De Rosa, 1996). The substances entrapped in the archeosomes are also leak free (Gambacorta, 1994). Archeosomes are taken up by the cells more vigorously than normal liposome which increases the efficiency of the immune response (Sprott et al., 1997).

5.6.1 AMYLASE

Scrutinizing and screening new and novel enzymes which are industrially important is the key area of research in recent times. Enzymes are proteins in nature and they help in catalyzing all the biochemical reaction in optimum conditions. There are a number of sources of enzymes, viz. animal, plant and microorganisms. Microbial sources are preferred sources of enzymes as their biomass is easy to cultivate and enzyme production is particularly easy. Enzymes like amylases, lipases, proteases and galactosidase have gained importance in recent times as the demand for enzymatic use in various industries is increasing and α amylase (1,4-α-D- glucanhydrolase, EC.3.2.1.1) hydrolyze the glycoside bonds of the residues (Arikan, 2008; Gupta et al., 2008). Kirchoff was the first scientist to discover α-amylase in 1811. Ohlsson is the father of classification of amylase as α- and β-amylase based on the anomeric type of sugars produced in the enzymatic reaction (Gupta et al., 2003). Starch is the main source of carbohydrates consisting of amylase and amylopectin. The endo-acting enzyme breaks down α, 1,4-glucosidic linkage of starch polymer to produce oligosaccharides like maltose and glucose (Parka and Son, 2007). Amylases are one of the industrially important enzymes having a wide spectrum of applications in starch processing, alcohol production, food, textile, pharmaceutical, detergent industry. Microorganisms from extreme environments such as the halophilic archaea have gained a lot of importance lately due to their stable nature and high yield. Lately, genetic modifications are applied on the microbes to increase the yield of the productivity of the enzymes. Media optimization is also one of the essential factors for increased productivity (Li et al., 2007).

Amylase activity is measured using the DNSA method as described by Fischer and Stein (Fischer and Stein, 1961) which involves mixing 1 ml of enzyme solution and 1 ml of Starch followed by incubation at 37°C for 3 minutes. Amylases are widely used in various industries. α-amylase is used for the production of glucose. This enzyme hydrolyzes the starch and converts it into the smaller units. It breaks down the glucosidic α–1,4 linkages in the starch to produce glucose and maltose. In baking industry α amylase can be used to improve the taste, aroma of the bakery products. It is widely used in European countries. Anti-salting in baking is also controlled by α-amylase (Shiau and Hung, 2003; Demirkan et al., 2005; Saxena et al., 2007; Gupta et al., 2008). Conversion of starch into fermentable sugars is also carried out by α-amylase in alcohol industry. Ethyl alcohol production is carried out by fermentation of starch like potatoes and certain grains (Juge et al., 2006). Stabilizing the bleaching effect in the detergent industry is one of the major uses of the amylase enzyme in detergent industry (Haq et al., 2010). For the castle feed industry the addition of α-amylase improves the body weight gain and feed conversion ratio. The digestion of the starch in grains is improved by hydrolyzing the starch polymers (Iji et al., 2003; Silva et al., 2006; Sidkey et al., 2011). α-amylase is known to improve the paper quality, its stiffness and also helps in sizing and coating of the paper.

5.6.2   PROTEASES

Proteases are known for their wide applications in different industries like food, pharmaceuticals, leather, detergent industry and in bioremediation (Haki and Rakshit, 2003). They are also used in production of soy products. Out of these, proteases are primarily used in detergent industry to enhance soap formulations and thus to increase washing efficiency (Maurer, 2004). Depending upon activity of the microbial proteases in alkaline, acidic or neutral conditions or on basis of their different active sites like metallo, cysteine and serine type, they are classified into various types (Gupta et al., 2002). Protease enzymes from different organisms like *Haloferax mediteranei* (Kamekura et al., 1996), *Natronococcus occultus* (Studdert et al., 1997; Elsztein et al., 2001), *Natrialba asiatica* (Kamekura and Seno, 1990), *Natrialba magadi* (Gimenez et al., 2000). Lama et al. (2005) reported a 38 kDa serine metalloprotease from *Salinivibrio* sp. strain AG18 by a synergistic effect of calcium and sodium chloride concentrations. *Salinivibrio proteolyticus* produced a 38 kDa (zinc-metalloprotease) (Karbalaei-Heidari et al., 2007) and 29 kDa (serine-metalloprotease) protease enzyme which was studied using Sodium Dodecyl Sulphate-Polyacrylamide Gel Electrophoresis

(SDS-PAGE). The industrial potential of this enzyme has been studied extensively and more work is underway. An extracellular protease, halo-protease CP1 has been isolated from the moderately halophilic bacterium *Pseudoalteromonas ruthenica* (Chand and Mishra, 2003). *Virgibacillus* is the unique genus among halophiles for production for multiple proteases of different molecular masses (Sinsuwan et al., 2008). *Virgibacillus* sp. SK33 produced six different proteases, but purification of these enzymes was difficult. Only two proteases of molecular weight 19 kDa and 32 kDa were purified successfully (Sinsuwan et al., 2010). The activity of the enzyme isolated from different halophiles was studied over various ranges of pH and temperature and the pH and temperature range in which the enzyme shows highest activity (optimum conditions) was found out. It has been found that the production of protease starts in the early exponential phase and is maximum in the mid-stationary phase; this pattern however differs in different halophilic bacteria. Enzyme activity can be studied using standard spectrophotometric procedure with absorbance at 280 nm.

Protease is used in commercial detergents to enhance their washing efficiency. Such enzymes should be able to work properly in harsh conditions of pH, temperature, salinity and presence of other detergents. The protease enzyme isolated from halophilic organisms can withstands high saline conditions and they have shown to be more effective in removing stains from clothes as compared to other detergents found in the market. Fish sauces used as condiments in certain parts of world are prepared from raw fishes themselves or from saturated brines. The protease enzyme produced from haloarchaea helps in degradation of the fish proteins into its constituent amino acids (Thongthai et al., 1992). The tolerances shown by halophilic bacteria to different organic solvents are mainly because of the presence of protease enzymes in them.

5.6.3 LIPASE

Lipases isolated from microbes have been considered to be important due to their applications in food processing, synthetic chemistry, environmental monitoring including biodegradation of waste material (Jaeger et al., 1994; Jaeger and Reetz, 1998). Industrial processes are usually carried out at specific physicochemical conditions, but these conditions may not be optimal for activity of the enzyme in question. Therefore, it is preferable to have an enzyme that has optimal activity at various ranges of salt concentrations, pH, temperatures and salinity (Rohban et al., 2009). Due to this, halophiles have proved to be the best source of the enzymes as they being salt loving, also can withstand different ranges of temperature and pH (Gomez and Steiner, 2004). There are many reports about synthesis of lipases enzyme from halophiles (Boutaiba et al., 2006; Amoozegar et al., 2008; Ozcan et al., 2009; Rohban et al., 2009), but success rate in terms of its purification and characterization is very less. Moderate halophiles have been proved to an effective source of the industrially important enzymes including lipase. Lipase research mainly includes its kinetic studies, determination of its 3D structure and genetic modification in lipase producing genes (Alberghina et al., 1991). Lipases are produced by plants, animals and microorganisms but those produced from microorganisms have gained more importance due to its special selectivity and broad substrate specificity (Dutra et al., 2008; Griebeler et al., 2009).

Lipase (triacylglycerol acylhydrolase) is a water soluble enzyme that catalyses hydrolysis of ester bonds of triacylglycerol and converts it into monoglycerides and free fatty acids (Sharma et al., 2001). The lipase activity can be studied using spectrophotometric analysis where absorbance is taken at 410 nm. One unit is defined as the amount of enzyme liberating 1 μmol of fatty acid per minute. The specific activity is expressed in the units of enzyme activity per milligram of protein. Purification of the enzyme is done using chromatographic techniques and dialysis and protein estimation can be done using standard biochemical test like Lowry et al. (1951) and its molecular weight can be determined using SDS-PAGE. The optimum conditions of temperature and pH can be determined for production of lipase.

Lipases have been used in development of aroma and flavor in cheese ripening, bakery products, beverages and yoghurt (Jaeger et al., 1994). It is widely used in paper and cosmetic industry, in chemical and pharmaceutical industry and in processing of oils and fats (Rubin and Dennis, 1997a,b). Lipase enzyme has shown to be effective in fish sauce production. The volatile fatty acids produced from lipase enzyme can give a better aroma and flavor to fish sauce. A halophilic lipase isolated from halotolerant *Staphylococcus warneri* PB233, helps in increasing the aroma and flavor of fish sauce as it has been shown that optimum values for production of this enzyme is similar to actual fermentation conditions of fish sauce production (Kanlayakrit and Boonpan, 2007).

5.6.4   POLYMERS

After the industrial revolution took place, there is an increasing use of petroleum-based plastics. This in turn has caused many problems of oil sources depletion, rising of oil prices and accumulation of harmful petroleum based plastics in nature. This problem has become extremely severe in past few years and scientists have been busy finding alternative methods for production of eco-friendly plastics which can be degraded by biological means (biodegradable plastics) under normal conditions and which can be available from cheap sources like waste products or microorganisms. Such bioplastics produced are from renewable sources and are environmental friendly (Jain et al., 2010).

Microorganisms living in hypersaline environments, for example, halophiles are known to be potential sources of many different industrially important products ranging from enzymes to plastics. Halophiles are considered to be a good and economic source of polymer production. As these organisms are grown in high salt content, there is no chance of contamination by normal microbes making the cultivation process easy and much sterility conditions need not be maintained. Such synthesis processes can also be done in large ponds to obtain large amount of polymers (Lillo and Rodriguez-Valera, 1990). Microbiologically produced most common form of plastics is polyhydroxyalkonates (PHA) which is the most versatile class of polymers (Jain et al., 2010; Ojumu et al., 2004). They are a group of biodegradable polymers of biological origin (Figure 5.4). They have proved to be

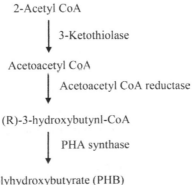

FIGURE 5.4   Pathway for PHB synthesis from halophilic archaea (Ojumu et al., 2004).

a good substitute for conventional petroleum plastics since they have similar properties to these plastics and are completely degradable by biological means (Sudesh et al., 2000).

The most common form of PHA produced from microorganisms is poly-hydroxybutyrate (PHB). It is the most common polymer stored in prokaryotes (Steinbüchel and Füchtenbush, 1998). It is one of the best characterized forms of PHA which has properties similar to polypropylene (in terms of its mechanical properties like the tensile strength and young's modulus, etc.). It has been successfully used in production of compost bags, bottles and films. Microorganisms convert sugars and fatty acids into PHA's through three different mechanisms (Philip et al., 2007). PHB is produced by microorganisms when they are grown in stress conditions. These stress conditions usually include growth in excess of carbon and absence of essential nutrients like proteins, oxygen, nitrogen, sulfur or absence of trace elements such as magnesium, phosphorus or iron (Lee, 1996; Rehm, 2003). PHB produced from 3-hydroxybutyric acid polymers by polymerization reaction of three enzymes: 3-ketothiolase, acetoacetyl CoA reductase and PHA synthase.

Primary molecular studies of these enzymes in the pathway present in the polymer producing archaea have been done (Han et al., 2007; Lu et al., 2008). The components present in PHB dictates its properties like the elasticity and tensile strength. PHB has good moisture resistance, water insolubility, good oxygen impermeability and optical purity. Its degradation takes place when microorganisms thrives on the surface of the polymer and secretes enzyme that degrades the polymer, though they do not degrade under normal conditions and they are stable in air (Ojumu et al., 2004). Certain haloarcheal strains like *Haloferax*, *Haloarcula*, *Haloquadratum*, *Halorubrum*, *Halobiforma*, *Halorhabdus*, *Halalkalicoccus* and *Halobacterium* have shown to produce PHB. Out of these, at present only one haloarchaea, *Haloferax mediterranei* have shown to produce good amount of PHB as compared to other organisms. It has shown to produce 65 wt % PHA (with 22% salt concentration in the cultivation media, Chen et al., 2006) with respect to its dry cell weight using starch or glucose as the substrate under phosphorus limiting conditions in batch cultures (Lillo and Rodriguez-Valera, 1990). This archaea accumulates a biopolymer (a co-polymer) poly (3-hydroxybutyrate-co-3-hydroxyvalerate) in huge amounts by using proper substrates like glucose and starch. Chemical composition and molecular weight of PHA produced from this organism can be modified depending upon the substrate utilized by the organism as a precursor. Recent studies have shown that *Cobetia marina* accumulates 81 wt% of PHA in a two-step cultivation system with glucose and valerate as carbon sources (Biwas et al., 2009). Organisms of genus *Halomonadaceae* requires almost similar nutrients for polymer production so, its co-culture can also be done to produce good amount of polymer.

Exopolysaccharides (EPS) have also been isolated from some halophilic archaea like *Halomonas eurihalina* (Quesada et al., 1993; Bejar et al., 1998). EPS polymers have been found useful in biotechnology, mostly in food, pharmaceutical and industrial sector (Sutherland, 1998; Tombs and Harding, 1998). To better study the EPS production of microorganisms, the DNA studies of these organisms is necessary that will tell whether the organism under study is capable of producing the polymer or not. Apart from this natural production of polymers from halophilic organisms, it has also been seen that the amount of polymers can be significantly increased by mutagenesis studies and genetic engineering. PHA has been proved to be a useful material for synthesis of certain packing films to some artifacts used in medicine like artificial heart valves, bone implants, sutures and drug delivery systems.

5.7 BIODEGRADATION OF TOXIC COMPOUNDS

Compounds that cause hazardous effects to the environment or an organism are termed as toxic compounds. Different kinds of metals, plastics, oils and other substances cause damage to the environment in which they are released. Different microorganisms are capable of degrading these toxic compounds found in nature. One of the groups among them is of halophiles. These are the organisms that are usually found in salt water, brines, salt pans, salted products, cold saline habitats and unusual saline habitats (Ventosa et al., 1998). Reports of biodegradation of heavy oils, phenolic and other organic compounds and aromatic compounds have been published. Different halophilic strains have been isolated and subjected to their bioremediation activity. Usually, saline and hyper-saline environments get contaminated by organic compounds, pesticides and phenols as a result of human activities and in the process of manufacturing pharmaceuticals, pesticides, etc. (Oren et al., 1992; Margesin and Schinner, 2001). Brines which have some amount of organic compounds and minimum of 3.5% w/v of total dissolved solids are termed as hypersaline wastewaters (Saha et al., 1999). Contamination of these hypersaline environments has caused a major problem mainly due to high amount of toxicity of organic compounds. Different biodegradation procedure adopted are restricted to primary toxicity removal and these being traditional, are less effective in saline conditions especially when salinity increases above the sea level (Oren et al., 1992). As a result, it becomes difficult to treat saline wastewaters which are contaminated by various mining, industrial and other activities. But, one of the alternative method for this is use of the habitant organisms of saline waters, for example, halophiles which are well adapted for such environments for treatment of wastewaters.

Phenols are the main source of contamination in saline wastewaters. They are used in different processes like pharmaceutical production, oil refining and resin manufacturing units. Moderately halophilic organisms have shown to degrade these phenolic compounds, but very few reports on it have been published. Salt tolerant phenol degrading organisms were iso-lated from the Amazonian soil (Bastos et al., 2000). Among the halophilic organisms, *Halomonas organivorans* has shown to degrade low molecular weight phenolic compounds at a salt concentration of 10% (Garcia et al., 2005). Attempts have also been made to enrich the consortium of halophilic microorganisms that can degrade phenols (Gayathri and Vasudevan, 2010).

Heavy oil biodegradation is also possible with the help of halophilic organ-isms, physical methods used being expensive and have other associated prob-lems. Certain microbial metabolites like polysaccharides play a major role in oil recovery due to their bioemulsifying and surfactant activity (Banat, 1995). A halophilic bacterium, TM–1, an extreme halophile was shown to degrade heavy oil by changing the chemical properties of crude oils. This change was brought by decreasing the concentration of high-molecular alkanes in the oil. Moderate halophiles belonging to the family Halomonadaceae, isolated from highly saline sites, have been shown to utilize chloroaromatic compounds as sources of carbon and energy (Maltseva et al., 1996).

5.8   APPLICATIONS OF BACTERIORHODOPSIN

In contrast to complex photosynthesis process brought about by green plants and some microorganisms, where sunlight is captured to make basic sugars which are then used as precursor for building more complex molecules, haloarchaea have got an easy and rapid method of light trapping and its utilization with the help of a membrane bound protein, bacteriorhodopsin (bR). This protein with 26 kDa molecular weight is a molecular pump that transports protons across membrane in outward direction making inside 10,000 folds alkaline than outside and this reaction is powered by sunlight. In short, it converts light energy into a proton gradient. Bacteriorhodopsin is homologous to G-protein coupled protein (Figure 5.5). It has three α helical protein chains which traverse the cytoplasmic membranes seven times. The bacteriorhodopsin (bR) molecule is a flexible molecule which goes into a bent form from straight form that powers the proton pumping action. Bacteriorhodopsin being a proton pump, it converts light energy into protons; it has got several applications in industries where prolonged energy supply is needed. In these places, this pump can be used in electronic circuits, to drive light batteries or for photochromic data storage and information processing. The protein also has got applications in artificial retinal implants (Saeedi et al., 2012).

Due to the ability of bR to convert light energy from sun into electrical energy, it has got immense applications in optical appliances, for therapeutic/medical use and in research. The application of this protein in different fields stems from the fact that it is stable towards thermal, chemical and photochemical degradation. Moreover, it has got good photoelectric and photochromic properties.

Bacteriorhodopsin is a kind of device that collects sunlight and which can then be converted into other sources of energy like light energy, heat or mechanical energy. The energy produced can be used or stored in batteries as per the need. Upon illumination of bR with xenon light, a photo voltage is induced which can be used for different applications (Hellingwerf et al., 1979). Bacteriorhodopsin is similar to rhodopsin protein (a retinal protein) that converts light energy into chemical (or electrical) energy during its cycle (Brauchle et al., 1991; Frydrych et al., 2000). The most important property of bacteriorhodopsin is that it maintain its photocycle even if it is separated from the purple membrane and integrated in an artificial

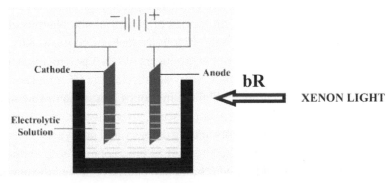

FIGURE 5.5   Schematic diagram of battery using bacteriorhodopsin.

membrane (Tukiainen et al., 2007). Hence, this protein can be used as artificial retina, for example, artificial vision apparatus. There is a need of more advanced and less surgically invasive retinal implants in the treatment of vision disorders of blind people (Chen et al., 2010). A vision sensor made up of genetically modified bR against wild type can be a good option in this sense. Such sensors are capable of detecting slight light intensity changes and moving images (Borman, 1992; Saeedi et al., 2011).

Bacteriorhodopsin has strong affinity for microwaves that are in the range of 3–40 GHz. Due to this, the materials coated with microwaves can be made undetectable by IR detectors. The military uniforms and concealments can be made of these biomaterials (bR prototypes) that will be camouflaged effectively and hence can be used in biodefense (Armstrong and Warner, 2003). Since bR is a photoactive protein, it can be used as a photo sensor and information processor for optical computing in defense. It will also help in reducing damages caused by bio-weapons (Da Silva, 1999). When there is a limitation of nutrients, bR can act as light-converting enzyme. It is the power house of the protein that in times of extreme conditions goes back and forth between purple and yellow colors. If this is controllable, it could be a valuable source to built battery conserving and long lasting computer display panels (Katoch et al., 2012).

5.9   CONCLUSION

In the very early age of the archaeal discovery, it was prejudiced that the entire topic would remain a mythical story, but this was proved wrong by all the scientists working towards exploring the potentials of the microbial community with a length and breadth of interest. Their efforts have hence led to the beginning of new era in the field of microbiology and biotechnology. This chapter can be concluded by collating the procurables of the halophilic organisms as probable source of variety of enzymes and molecules of industrial and pharmaceutical importance. They are very strong and obvious yielders of commercially relevant enzymes. Besides that, they are the producers of DNA polymerases, proof reading enzymes aiding in PCR products by amplifying the GC rich reactions. The compatible solutes accumulated by the microbes play a major role in correcting the cancer and pulmonary heart disease, liver disease, etc. The ether linked lipids can be used as adjuvants in drug delivery and new vaccine production. They can be used as biosurfactants and as biocolors in food, textile and cosmetic industry. Carotenoid pigments and halocins have served the mankind with a pharmaceutical boon. Production of bioplastics has been made possible which could help is environment conservation. By exploring new source of energy to combat the depletion of the available resources of energy is of importance and archaea have successfully served the need by being able to convert the solar energy into other forms of energy by possessing the rhodopsin pump. The enzymes like proteases and lipases produced by archaea can also be synergistically used in numerous economically important reactions.

KEYWORDS

- Archaebacteria
- Archeosomes
- Bacteriorhodopsin
- Bacterioruberin
- Betaine
- Carotenoids
- Compatible solutes
- Ecotine
- Exopolysaccharides
- Extremophiles
- Halobacteriaceae
- Halocins
- Halophiles
- Liposomes
- Osmoadaptation
- Polyhydroxyalkonate
- Polyhydroxybutyrate
- Raman Spectroscopy

REFERENCES

Abbes, M., Baati, H., Guermazi, S., Messina, C., Santulli, A., Gharsallah, N., Ammar, E. Biological properties of carotenoids extracted from *Halobacterium halobium* isolated from a Tunisia solar saltern. *BMC Complementary and Alternative Medicine* 2013, 13, 255.

Adwitiya, P., Ashwini, P., Avinash, A. K., Badri, R., Kajal, D., Vomsi, P., Srividya, S. Mutagenesis of *Bacillus thuringiensis* IAM 12077 for increasing poly opic(-β-) hydroxybutyrate (PHB) production. *Turk J Biol.* 2009, 33, 225–230.

Alberghina, L., Schmid, R. D., Verger, R. Lipases: structure, mechanism and genetic engineering. VCH: Weinheim, 1991, pp. 1–16.

Allen, T. M. Liposomes. Opportunities in drug delivery. *Drugs* 1997, 54, 8–14.

Alqueres, S. M. C., Almeida, R. V., Clementino, M. M., Vieira, R. P., Almeida, W. I., Cardoso, A. M., Martins, O. B. Exploring the biotechnological applications in the archaeal domain. *Braz J Microbiol.* 2007, 38, 398–405.

Amoozegar, M. A., Salehghamari, E., Khajeh, K., Kabiri, M., Naddaf, S. Production of an extracellular thermohalophilic lipase from a moderately halophilic bacterium, *Salinivibrio* sp. strain SA-2. *J. Basic. Microbiol.* 2008, 48, 3, 160–167.

Amstrong, R. E., Warner, J. B. Biology and the Battlefield. *Defense Horizons* 2003, 25, 1–8.

Arikan, B. Highly thermostable, thermophilic, alkaline, SDS and chelator resistant amylase from a thermophilic *Bacillus* sp. isolate A3–15. *Bioresour. Technol.* 2008, 99, 3071–3076.

Balasubramanian, S., Pal, S., Bagchi, B. Hydrogen bond dynamics near a micellar surface: origin of the universal slow relaxation at complex aqueous interface. *Phys Rev Lett.* 2002, 89, 115505.

Banat, I. M. Biosurfactants production and possible uses in microbial enhanced oil recovery and oil pollution remediation: a review. *Biores Tech.* 1995, 51, 1–12.

Bangham, A.D. A correlation between surface charge and coagulant action of phospholipids. *Nature* 1961, 192(4808), 1197–1198.

Bastos, A. E., Moon, D. H., Rossi, A., Trevors, J. T., Tsai, S. M. Salt-tolerant phenol-degrading microorganisms isolated from Amazonian soil samples. *Arch Microbiol.* 2000, 174, 346–352.

Bejar, V., Llamas, I., Calvo, C., Quesada, E. Characterization of exopolysaccharides produced by 19 halophilic strains included in the species *Halomonas eurihalina*. *J Biotechnol.* 1998, 61, 135–141.

Birbir, M., Eryilmaz, S., Ogan, A. Prevention of halophilic bacterial damage on brine cured hide with halocins. *J. Soc. Leather. Technol. Chem.* 2004, 88(3), 99–104.

Biwas, A., Patra, A., Paul, A. K. Production of poly-3- hydroxyalkanoic acids by a moderate halophilic bacterium *Halomonas marina* HMA 103 isolated from solar saltern of Orissa, India. *Acta Microbiol Inmunol. Hung* 2009, 56, 125–143.

Borman, S. Photoreceptor based on Bacteriorhodopsin. *Chem Eng News.* 1992, 70, 5–6.

Boutaiba, S., Bhatnagar, T., Hacene, H., Mitchell, D. A., Baratti, J. C. Preliminary characterization of a lipolytic activity from an extremely halophilic archaeon, *Natronococcus* sp. *J. Mol. Catal. B Enzym.* 2006, 41, 21–26.

Brauchle, C., Hampp, N., Oesterhelt D. Optical applications of bacteriorhodopsin and its mutated variants. *Adv Mater.* 1991, 3, 420–428.

Bringardner, B. D., Baran, C. P., Eubank, T. D., Marsh, C. B. The role of inflammation in the pathogenesis of idiopathic pulmonary fibrosis. *Antioxid Redox Signal.* 2008, 10, 287–301.

Camacho, R. M., Mateos, J. C., Gonzalez-Reynoso, O., Prado, L. A., Cordova, A. Production and characterization of esterase and lipase from *Haloarcula marismortui. J. Ind. Microbiol. Biotechnol.* 2009, 36, 901–909.

Canovas, D., Vargas, C., Csonka, L. N., Ventosa, A., Nieto, J. J. Synthesis of glycine betaine from exogenous choline in the moderately halophilic bacterium *Halomonas elongata. Appl Environ Microbiol.* 1998, 64, 4095–4097.

Chand, S., Mishra, P. Research and application of microbial enzymes, India's contribution. *Adv. Biochem. Eng. Biotechnol.* 2003, 85, 95–124.

Chen, Z., Birge, R. R. Protein-based artificial retinas, US Patent 2010/0226957A1.

Cheung, J., Danna, K. J., O'Connor, F. M., Pilee, L. B., Shand, R. F. Isolation, sequence, and expression of the gene encoding halocin H4, a bacteriocin from the halophilic archaeon *Haloferax mediterranei* R4. *J Bacteriol.* 1997, 179, 548–551.

Cleland, D., Krader, P., McCree, C., Tang, J., Emerson, D. Glycine betaine as a cryoprotectant for prokaryotes. *J. Microbiol. Methods* 2004, 58, 31–38.

Da Costa, M. S., Santos, H., Galinski, E. A., Antranikian, G. An overview of the role and diversity of compatible solutes in Bacteria and Archaea. *Adv Biochem Eng Biotechnol.* 1998, 61, 117–153.

Da Silva, E. J. Biological warfare, bioterrorism, biodefense and the biological and toxin weapons convention. *Elect J Biotechnol.* 1999, 2, 109–139.

De Rosa, M. Archaeal lipids: structural features and supramolecular organization. *Thin Solid Films* 1996, 284–285, 13–17.

Demirkan, B. M., Adachi, A., Higasa, T., Utsumi, S. α-amylase from *Bacillus amyloliquefaciens*, purification, characterization, raw starch degradation and expression in *E. coli. Process Biochem.* 2005, 40, 2629–2636.

Detkova, E. N., Boltyanskaya, Y. V. Osmoadaptation of haloalkaliphilic bacteria: role of osmoregulators and their possible practical application. *Microbiology* 2007, 76, 511–522.

Dutra, J. C. V., Terzi, S. C., Bevilaqua, J. V., Damaso, M. C. T., Couri, S., Langone, M. A. P., Senna, L. F. Lipase production in solid state fermentation monitoring biomass growth of *Aspergillus niger* using digital image processing. *Appl. Biochem. Biotechnol.* 2008, 147, 63–75.

Egorova, K., Antranikian, G. Industrial relevance of thermophilic Archaea. *Current Opinion in Microbiology* 2005, 8, 649–655.

Elsztein, C., Herrera Seitz, M. K., Sanchez, J. J., de Castro, R. E. *J Basic Microbiol.* 2001, 41, 319–327.

Fischer, E. H., Stein, E. A. Use of dinitrosalicylic acid. *Biochemical Preparations* 1961, 8, 27.

Frydrych, M., Silfsten, P., Parkkinen, S., Parkkinen, J., Jaaskelainen, T. Color sensitive retina based on bacteriorhodopsin. *Biosystems* 2000, 54, 131–140.

Furubayashi, M., Umeno, D. Directed evolution of carotenoid synthases for the production of unnatural carotenoids. In *Microbial Carotenoids from Bacteria and Microalgae: Methods and Protocols*, Barredo, J. (Ed.), Springer: USA, 2012, pp. 245–253.

Galinski, E. A., Pfeiffer, H. P., Trüper, H. G. 1,4,5,6-Tetrahydro-2-methyl-4-pyrimidinecarboxylic acid. A novel cyclic amino acid from halophilic phototrophic bacteria of the genus *Ectothiorhodospira*. *Eur J Biochem.* 1985, 149, 135–139.

Gambacorta, A., Trincone, A., Nicolaus, B., Lama, L., de Rosa M. Unique features of lipids of archaea. *Systematic and Applied Microbiology* 1994, 16, 518–527.

Garcia, M. T., Ventosa, A., Mellado, E. Catabolic versatility of aromatic compound degrading halophilic bacteria. *FEMS Microbiol Ecol.* 2005, 54, 97–109.

Garcia-Estepa, R., Argandoña, M., Reina-Bueno, M., Capote, N., Iglesias-Guerra, F., Nieto, J. J., Vargas, C. The ectD gene, which is involved in the synthesis of the compatible solute hydroxyectoine, is essential for thermoprotection of the halophilic bacterium *Chromohalobacter salexigens*. *J Bacteriol.* 2006, 188, 3774–3784.

Gayathri, K. V., Vasudevan, N. Enrichment of phenol degrading moderately halophilic bacterial consortium from saline environment. *J Bioremed Biodegrad.* 2010, 1, 104.

Gimenez, M. I., Studdert, C. A., Sanchez, J. J., De Castro R. E. Extracellular protease of *Natrialba magadii*: purification and biochemical characterization. *Extremophiles* 2000, 4, 181–188.

Grammel, N. Molekulargenetische und biochemische analyze der biosynthese von 2-methyl-4-carboxy-3,4,5,6-tetrahydropyrimidin und seinem 5-hydroxyderivat, zwei salzstreβinduzierbaren osmolyten in *Streptomyces chrysomallus*. Technische Universität Berlin: Berlin, Germany, 2000.

Grant, D., Wall, W., Mimeault, R., Zhong, R., Ghent, C., Garcia, B., Stiller, C., Duff, J. Successful small-bowel/liver transplantation. *Lancet* 1990, 335, 181–184.

Gref, R., Minamitake, Y., Peracchia, M. T., Trubetskoy, V., Torchilin, V., Langer, R. Biodegradable long-circulating polymeric nanospheres. *Sci. Cult.* 1994, 263, 1600–1603.

Griebeler, N. A. E., Polloni, D., Remonatto, F., Arbter, R., Vardanega, J. L., Cechet, M. D., Luccio, D., de Oliveira, H., Treichel, R. L., Cansian, E., Rigo, A., Ninow, J. L. Isolation and screening of lipase producing fungi with hydrolytic activity. *Food Bioprocess Tech.* 2009, 4, 578–586.

Gupta, A., Gupta, V. K., Modi D. R., Yadava, L. P. Production and characterization of α-amylase from *Aspergillus niger*. *Biotechnology* 2008, 1, 1–6.

Gupta, R., Beg, Q. K., Lorenz, P. Bacterial alkaline proteases: molecular approaches and industrial applications. *Appl Microbiol Biotechnol*. 2002, 59, 15–32.

Haddar, A., Agrebi, R., Bougatef, A., Hmidet, N., Kamoun, S., Nasri, M. Two detergent stable alkaline serine-proteases from *Bacillus mojavensis* A21: purification, characterization and potential application as an laundry detergent additive. *Bioresour Technol*. 2009, 100, 3366–3373.

Haki, G. D., Rakshit, S. K. Developments in industrially stable thermo stable enzymes: a review. *Bioresour Technol*. 2003, 89, 17–34.

Han, J., Lu, Q., Zhou, L., Zhou, J., Xiang, H. Molecular characterization of the $phaEC_{Hm}$ genes, required for biosynthesis of poly(3-hydroxybutyrate) in the extremely halophilic archaeon *Haloarcula marismortui*. *Appl Env Microbiol*. 2007, 73, 6058–6065.

Haq, I. U., Muhammad, M. J., Uzma, H., Fazal, A. Kinetics and thermodynamic studies of alpha amylase from *Bacillus licheniformis* mutant. *Pak. J. Bot*. 2010, 42, 3507–3516.

Hellingwerf, K. J., Arents, J. C., Scholte, B. J., Westerhoff, H. V. Bacteriorhodopsin in liposomes. II. Experimental evidence in support of a theoretical model. *Biochim Biophys Acta*. 1979, 547, 561–582.

Hezayen, F. F., Gutierrez, M. C., Steinbuchel, A., Tindall, B. J., Rehm, B. H. A. *Halopiger aswanensis* sp. nov., a polymer-producing and extremely halophilic archaeon isolated from hypersaline soil. *Int J Syst Evol Micr.* 2010, 60, 633–637.

Hezayen, F. F., Tindallm B. J., Steinbuchel, A., Rehm, B. H. Characterization of a novel halophilic archaeon, *Halobiforma haloterrestris* gen. nov., sp. nov., and transfer of *Natronobacterium nitratireducens* to *Halobiforma nitratireducens* comb. nov. *Int J Syst Evol Micr.* 2002, 52, 2271–2280.

Iji, P. A., Khumalo, K., Slippers, S., Gous, R. M. Intestinal function and body growth of broiler chickens fed on diets based on maize dried at different temperatures and supplemented with a microbial enzyme. *Reprod. Nutr. Dev*. 2003, 43, 77–90.

Imhoff, J. F., Rodriguez-Valera, F. Betaine is the main compatible solute of halophilic eubacteria. *J Bacteriol*. 1984, 160, 478–479.

Jacquemet, A., Barbeau, J., Lemiegre, L., Benvegnu, T. Archaeal tetra ether bipolar lipids: structure, function and applications. *Biochimie* 2009, 91, 711–717.

Jaeger, K. E., Ransac, S., Dijkstra, B. W., Colson, C., Van Heuvel, M., Misset, O. Bacterial lipases. *FEMS Microbiol Rev.* 1994, 15, 29–63.

Jaeger, K. E., Reetz, M. T. Microbial lipases form versatile tools for biotechnology. *Trends Biotechnol.* 1998, 16, 396–403.

Jain, R., Costa, S., Tiwari, A. Polyhydroxyalkanoates: a way to sustainable development of bioplastics. *Chron Joung Sci.* 2010, 1, 10–15.

Juge, N., Nohr, J., Coeffet, M. F. L. G., Kramhoft, B., Furniss, C. S. M., Planchot, V., Archer, D. B., Williamson, G., Svensson, B. The activity of barley α-amylase on starch granules is enhanced by fusion of a starch binding domain from *Aspergillus niger* glucoamylase. *Biochem. Biophys. Acta.* 2006, 8, 275–284.

Kamekura, M., Seno, Y., Dyall-Smith, M. Halolysin R4, a serine proteinase from the halophilic archaeon *Haloferax mediterranei*: gene cloning, expression and structural studies. *Biochim Biophys Acta.* 1996, 1294, 159–167.

Kamekura M., Seno Y. A halophilic extracellular protease from a halophilic archaebacterium strain 172p1. *Biochem. Cell Biol.* 1990, 68, 352–359.

Kamekura, M., Seno, Y., Holmes, M. L., Dyall-Smith, M. L. Molecular cloning and sequencing of the gene for a halophilic alkaline serine protease (halolysin) from an unidentified halophilic archaea strain (172P1) and expression of the gene in *Haloferax volcanii*. *Biochem Cell Biol.* 1992, 174, 736–742.

Karan, R., Khare, S. K. Purification and characterization of a solvent stable protease from *Geomicrobium* sp. EMB2. *Environ Technol.* 2010, 10, 1061–1072.

Karan, R., Khare, S. K. Stability of haloalkaliphilic *Geomicrobium* sp. modulated by salt. *Biochemistry* 2011, 76, 6867–693.

Karbalaei-Heidari, H. R., Ziaee, A. A., Amoozegar, M. A. Purification and biochemical characterization of a protease secreted by the *Salinivibrio* sp. strain AF-2004 and its behavior in organic solvents. *Extremophiles* 2007, 11, 237–243.

Kavitha, P., Lipton, A. P., Sarika, A. K., Aishwarya, M. S. Growth characteristics and halocin production by a new isolate *Haloferax volcanii* KPS1 from Kovalam Solar Saltern (India). *Res. J. Biol. Sci.* 2011, 6, 257–262.

Krieg, A. M. From bugs to drugs: therapeutic immunomodulation with oligodeoxynucleotides containing CpG sequences from bacterial DNA. *Antisense Nucleic Acid Drug Dev.* 2001, 11, 181–188.

Kuhlmann, A. U., Bremer. E. Osmotically regulated synthesis of the compatible solute ectoine in *Bacillus pasteurii* and related *Bacillus* spp. *Appl Environ Microbiol.* 2002, 68, 772–783.

Laemmli, U. K. Cleavage of structural proteins during the assembly of the head of bacteriophage T4. *Nature* 1970, 227, 680–685.

Lama, L., Romano, I., Calandrelli, V., Nicolaus, B., Gambacorta, A. Purification and characterization of a protease produced by an aerobic haloalkaliphilic species belonging to the *Salinivibrio* genus. *Res Microbiol.* 2005, 156, 478–484.

Lanyi, J. K., Stevenson, J. Effect of salts and organic solvents on the activity of *Halobacterium cutirubrum* catalase. *J Bacteriol.* 1969, 98, 611–616.

Lasic, D. D. Kinetic and thermodynamic effects in the formation of amphiphilic colloidal particles. *J. Liposome Res.* 1993, 3, 257–273.

Lee, S. Y. Bacterial polyhydroxyalkanoates. *Biotechnol Bioeng.* 1996, 49, 1–14.

Legat, A., Gruber, C., Zangger, K., Wanner, G., Stan-Lotter, H. Identification of polyhydroxy alkanoates in *Halococcus* and other haloarchaeal species. *Appl Microbiol Biotechnol.* 2010, 87, 1119–1127.

Li, Y., Xiang, H., Liu, J., Zhou, M., Tan, H. Purification and biological characterization of halocin C8, a novel peptide antibiotic from *Halobacterium* strain AS7092. *Extremophiles* 2003, 7, 401–407.

Lillo, J. G., Rodriguez-Valera, F. Effects of culture conditions on poly(β-hydroxybutyric) acid production by *Haloferax mediterranei*. *Appl Environ Microbiol.* 1990, 56, 2517–2521.

Lo, C. C., Bonner, C. A., Xie, G., D'Souza, M., Jensen, R. A. Cohesion group approach for evolutionary analysis of aspartokinase, an enzyme that feeds a branched network of many biochemical pathways. *Microbiol Mol Biol Rev.* 2009, 73, 594–551.

Lu, Q., Han, J., Zhou, L., Zhou, J., Xiang, H. Genetic and biochemical characterization of the poly(3-hydroxybutyrate-co- 3-hydroxyvalerate) synthase in *Haloferax mediterranei. J. Bacteriol.* 2008, 190, 4173–4180.

Macnee, W. Pathogenesis of chronic obstructive pulmonary disease. *Clin Chest Med.* 2007, 28, 479–513.

Maltseva, O., McGowan, C., Fulthorpe, R., Oriel, P. Degradation of 2,4-dichlorophenoxyacetic acid by haloalkaliphilic bacteria. *Microbiology* 1996, 142, 1115–1122.

Margesin, R., Schinner, F. Biodegradation and bioremediation of hydrocarbons in extreme environments. *Appl Microbiol Biotechnol.* 2001, 56, 650–663.

Martin, D. D., Ciulla, R. A., Roberts, M. F. Osmoadaptation in archaea. *Appl Environ Microbiol.* 1999, 65, 1815–1825.

Maurer, K. Detergent proteases. *Curr Opin Biotechnol.* 2004, 15, 330–334.

Meseguer, I., Rodríguez-Valera, F. Production and purification of halocin H4. *FEMS Microbiol Lett.* 1985, 28, 177–182.

Messadek, J. Glycine betaine and its use. US Patent 6855734, 2005.

Messeguer, I., Rodriguez-Valera, F. Effect of halocin H4 on cells of *Halobacterium halobium. Journal of General Microbiology* 1986, 132, 3061–3068.

Mozafari, M. R., Flanagan, J., Matia-Merino, L., Awati, A., Omri, A., Suntres, Z. E., Singh, H. Recent trends in the lipid-based nanoencapsulation of antioxidants and their role in foods. *J. Sci. Food Agric.* 2006, 86, 2038–2045.

Mozafari, M. R., Mortazavi M. S. *Nanoliposomes: from fundamentals to recent developments*. Trafford Publishing Ltd. UK, 2005.

Naziri, D., Hamidi, M., Hassanzadeh, S., Tarhriz, V., Zanjani, B. M., Nazemyieh, H., Hejazi, M. A., Hejazi, M. S. Analysis of carotenoid production by *Halorubrum* sp. TBZ126; an extremely halophilic archaeon from Urmia lake. *Advanced Pharmaceutical Bulletin* 2014, 4(1), 61–67.

Nyyssola, A., Kerovuo, J., Kaukinen, P., von Weymarn, N., Reinikaiuem, T. Extreme halophiles synthesize betaine from glycine by methylation. *J Biol Chem.* 2000, 275, 22196–22201.

O'Connor, E. M., Shand, R. F. Halocins and sulfolobicins: the emerging story of archaeal protein and peptide antibiotics. *J. Ind. Microbiol. Biotechnol.* 2002, 28, 23–31.

Ojumu, T. V., Yu, J., Solomon, B. O. Production of polyhydroxyalkanoates, a bacterial biodegradable polymer. *Afr. J. Biotechnol.* 2004, 3, 18–24.

Oren, A. Diversity of halophilic microorganisms: environments, phylogeny, physiology, and applications. *J Indust. Microbiol. Biotechnol.* 2002, 1, 56–63.

Oren, A., Peter, G., Azachi, M., Henis, Y. Microbial degradation of pollutants at high salt concentrations. *Biodegradation* 1992, 3, 387–398.

Ozcan, B., Ozyilmaz, G., Cokmus, C., Caliskan, M. Characterization of extracellular esterase and lipase activities from five halophilic archaeal strains. *J. Ind. Microbiol. Biotech.* 2009, 36, 105–110.

Parka, G. T., Son, H. J. Keratinolytic activity of *Bacillus megaterium* F7-1, a feather-degrading mesophilic bacterium. *Microbiol. Res.* 2007, 164, 478–485.

Pasic, L., Velikonja, B. H., Ulrih, N. P. Optimization of the culture conditions for the production of a bacteriocin from halophilic archaeon Sech7a. *Prep. Biochem. Biotechnol.* 2008, 38, 229–245.

Patel, G. B., Zhou, H., Kuo Lee, R., Chen, W. Archaeosomes as adjuvants for combination vaccines. *J. Liposome Res.* 2004, 14, 191–202.

Peters, A., von Klot, S., Heier, M., Trentinaglia, I., Hormann, A., Wichmann, H. E., Löwel, H. Exposure to traffic and the onset of myocardial infarction. *N Engl J Med.* 2004, 351, 1721–1730.

Peters, P., Galinski, E. A., Trüper, H. G. The biosynthesis of ectoine. *FEMS Microbiol Lett.* 1990, 71, 157–162.

Philip, S., Keshavarz, T., Roy, I. Polyhydroxyalkanoates: biodegradable polymers with a range of applications. *J Chem Technol Biotechnol.* 2007, 82, 233–247.

Price, L. B., Shand, R. F. Halocin S8: a 36-aminoacid microhalocin from the haloarchaeal strain S8a. *J Bacteriol.* 2000, 182, 4951–4958.

Quesada, E., Bejar, V., Calvo, C. Exopolysaccharide production by *Volcaniella eurihalina*. *Experientia* 1993, 49, 1037–1041.

Regev, R., Peri, I., Gilboa, H., Avi-Dor, Y. ^{13}C NMR study of the interrelation between synthesis and uptake of compatible solutes in two moderately halophilic eubacteria, *Bacterium* Ba1 and *Vibro costicola*. *Arch Biochem Biophys.* 1990, 278, 106–112.

Rehm, B. H. Polyester synthases: natural catalysts for plastics. *Biochem J.* 2003, 376, 15–33.

Relini, A., Cassinadri, D., Mirghani, Z., Brandt, O., Gambacorta, A. Calcium-induced interaction and fusion of archaeobacterial lipid vesicles: a fluorescence study. *Biochem Biophys Acta.* 1994, 1194, 17–24.

Roberts, M. F. Osmoadaptation and osmoregulation in archaea: update 2004. *Front Biosci.* 2004, 9, 1999–2019.

Roberts, M. F., Lai, M. C., Gunsalus, R. P. Biosynthetic pathway of the osmolytes N-ε-acetyl-β-lysine, β-glutamine, and betaine in *Methanohalophilus* strain FDF1 suggested by nuclear magnetic resonance analyzes. *J Bacteriol.* 1992, 174, 6688–6693.

Robertson, D. E., Noll, D., Roberts, M. F., Menaia, J. A., Boone, D. R. Detection of the osmoregulator betaine in methanogens. *Appl Environ Microbiol.* 1990, 56, 563–565.

Rodriguez-Amaya, D. B. *A Guide to Carotenoid Analysis in Foods*. ILSI Press: Washington, 2001, pp. 14–22.

Rodriguez-Valera, F., Juez, G., Kushner, D. J. Halocins: salt-dependent bacteriocins produced by extremely halophilic rods. *Can J Microbiol.* 1982, 28, 151–154.

Rohban, R., Amoozegar, M., Ventosa, A. Screening and isolation of halophilic bacteria producing extracellular hydrolysis from Howz Soltan Lake, Iran. *J Ind Microbiol Biotechnol.* 2009, 36, 333–340.

Rose, R. W., Bruser, T., Kissinger, J. C., Pohlschroder, M. Adaptation of protein secretion to extremely high-salt conditions by extensive use of the twin-arginine translocation pathway. *Mol Microbiol.* 2002, 45, 943–950.

Rubin. B., Dennis, E. A. Lipases: Part A. Biotechnology. Academic Press: New York, 1997a, pp. 1–408.

Rubin, B., Dennis, E. A. Lipases: Part B. Enzyme characterization and utilization. Academic Press: New York, 1997b, pp. 1–563.

Saeedi, P., Sebtahmadi, S. S., Meharbadi, J. F., Behmanesh, M., Mekhilef, S. Potential applications of bacteriorhodopsin mutants. *Bioengineered.* 2012, 3(6), 326–328.

Saeedi, P., Moosaabadi, M. J., Meharbadi, F. J., Behmanesh, M. Bacteriorhodopsin and its mutants allude a breakthrough impending to artificial retina construction and strategies for curing blindness. *Journal of Paramedical Sciences* 2011, 2, 33–40.

Saha, N. C., Bhunia, F., Kaviraj, A. Toxicity of phenol to fish and aquatic ecosystems. *Bull Environ Contam Toxicol.* 1999, 63, 195–202.

Saxena, K. R., Dutt, K., Agarwal, L., Nayyar, P. A highly and thermostable alkaline amylase from a *Bacillus* sp. PN5. *Bioresour. Technol.* 2007, 98, 260–265.

Shahmohammadi, H. R., Asgarani, E., Terato, H., Saito, T., Ohyama, Y., Gekko, K., Yamamoto, O., Ide, H. Protective roles of Bacterioruberin and intracellular KCl in the resistance of *Halobacterium salinarium* against DNA-damaging agents. *J. Radiat Res.* 1998, 39, 251–262.

Katoch, S., Chaturvedi, A., Pareek, S., Deshmukh, A. Potential applications of proteins in IT. *Int. J. Comp. Appln* 2012, 3, 17–21.

Shand, R. F., Leyva, K. J. Peptide and protein antibiotics from the domain *Archaea*: halocins and sulfolobicins. In *Bacteriocins: Ecology and Evolution*, Riley, M. A., Chavan, M.A. (Eds.), Springer: New York, 2007, pp. 93–109.

Sharma, R., Chisti, Y., Banerjee, U. C. Production, purification, characterization and applications of lipases. *Biotechnology Advances* 2001, 19, 627–662.

Shiau, R., Hung, J. H. Improving the thermostability of raw-starch-digesting amylase from a *Cytophaga* sp. by site-directed mutagenesis. *Appl. Environ. Microbiol.* 2003, 69, 2383–2385.

Sidkey, N. M., Maha, A., Reham, B., Reham, S., Ghadeer, B. Purification and characterization of α-amylase from a newly isolated *Aspergillus flavus* F2Mbb. *Int. Res. J. Microbiol.* 2011, 2, 96–103.

Silva, M. T. S. L., Santo, F. E., Pereira P. T., Poseiro, C. P. Phenotypic characterization of food waste degrading *Bacillus* strains isolated from aerobic bioreactors. *J. Basic. Microb.* 2006, 46, 34–46.

Sinsuwan, S., Rodtong, R., Yongsawatdigul, J. A NaCl stable serine protease *Virgibacillus* sp. SK33 isolated from Thai fish sauce. *Food Chem.* 2010, 119, 573–579.

Sinsuwan, S., Rodtong, R., Yongsawatdigul, J. Production and characterization of NaCl activated of proteinases from *Virgibacillus* sp. SK33 from fish sauce fermentation. *Process Biochem.* 2008, 43, 185–192.

Sprott, G., Tolson, D., Patel, G. Archaeosomes as novel antigen delivery systems. *FEMS Microbiology Letters* 1997, 154, 17–22.

Steinbüchel, A., Füchtenbush, B. Bacterial and other biological systems for polyester production. *Trends Biotechnol.* 1998, 16, 419–427.

Strand, A., Shivaji, S., Liaaen, J. S. Bacterial carotenoids, C-50-carotenoids, revised structures of carotenoids associated with membranes in psychrotrophic *Micrococcus roseus*. *Biochem Syst Ecol.* 1997, 25, 547–552.

Studdert, C. A., De Castro, R. E., Seitz, K. H., Sanchez, J. J. Detection and preliminary characterization of extracellular proteolytic activities of the haloalkaliphilic archaeon *Natronococcus occultus*. *Arch Microbiol.* 1997, 168, 6, 532–535.

Sudesh, K., Abe, H., Doi, Y. Synthesis, structure and properties of polyhydroxyalkanoates: biological polyesters. *Prog Polym Sci.* 2000, 25, 1503–1555.

Sun, C., Li, Y., Mei, S., Lu, Q., Zhou, L., Xiang, H. A single gene directs both production and immunity of halocin C8 in a haloarchaeal strain AS7092. *Mol. Microbiol.* 2005, 57, 537–549.

Sutherland, I. W. Novel and established applications of microbial polysaccharides. *Trends Biotechnol.* 1998, 16, 41–46.

Synowiecki, J., Grzybowska, B., Zdzieblo, A. Sources, properties and suitability of new thermostable enzymes in food processing. *Crit. Rev. Food Sci. Nutrit.* 2006, 46, 197–205.

Thongthai, C., McGenity, T. J., Suntinanalert, P., Grant, W. D. Isolation and characterization of an extremely halophilic archaebacterium from traditionally fermented Thai fish sauce (nam-pla). *Lett. Appl. Microbiol.* 1992, 14, 111–114.

Tombs, M., Harding, S. E. In *An Introduction to Polysaccharide Biotechnology.* Taylor and Francis: London, 1998, pp. 144–147.

Torreblanca, M., Meseguer, I., Rodriguez, F. V. Halocin H6, a bacteriocin from *Haloferax gibbonsii. J. Gen Microbiol.* 1989, 135, 2655–2661.

Torreblanca, M., Meseguer, I., Ventosa, A. Production of halocin is a practically universal feature of archaeal halophilic rods. *Lett. Appl. Microbiol.* 1994, 19, 201–205.

Tukiainen, T., Lensu, L., Parkkinen, J. Temporal characteristics of artificial retina based on bacteriorhodopsin and its variants. In *Advances in Brain, Vision, and Artificial Intelligence,* Mele, F., Ramella, G., Santillo, S., Ventriglia, F. (Eds.), Springer: Berlin, Heidelberg, 2007, pp. 94–103.

Van de Vossenberg, J. The essence of being extremophilic: the role of the unique archaeal membrane lipids. *Extremophiles* 1998, 2, 163–170.

Van de Vossenberg, J. L., Driessen, A. J., Zillig, W., Konings W. N. Bioenergetics and cytoplasmic membrane stability of the extremely acidophilic, thermophilic archaeon *Picrophilus oshimae. Extremophiles* 1998, 2, 67–74.

Ventosa, A., Nieto, J. J., Oren, A. Biology of moderately halophilic aerobic bacteria. *Microbiol Mol Biol Rev.* 1998, 62, 504–544.

Watwe, R. M., Bellare, J. R. Manufacture of liposomes: A review. *Curr Sci.* 1995, 68, 715–724.

Weiner, A. L., Cannon, J. B. Commercial approaches to the delivery of macromolecular drugs with liposomes. In *Controlled Release of Drugs: Polymers and Aggregate Systems,* Rosoff, M. (Ed.), VCH Publishers: New York, 1989, pp. 217–252.

Kanlayakrit, W., Boonpan, B. Screening of halophilic lipase-producing bacteria and characterization of enzyme for fish sauce quality improvement. *Kasetsart J. Nat. Sci.* 2007, 41, 576–585.

Zeb, A., Mehmood, S. Carotenoids content from various sources and their potential health applications. *Pak J Nutr.* 2004, 3, 199–204.

CHAPTER 6

CYCLODEXTRIN GLYCOSYL TRANSFERASE (CGTase): AN OVERVIEW OF THEIR PRODUCTION AND BIOTECHNOLOGICAL APPLICATIONS

REBECCA S. THOMBRE and PRADNYA P. KANEKAR

Department of Biotechnology, Modern College, Pune 411005, Maharashtra, India

CONTENTS

6.1   INTRODUCTION

The increasing industrialization poses a great threat to the environment because of use of harsh chemicals. Biological enzymes provide a solution to develop an eco-friendly and sustainable process for industrial applications. Enzymes are proteins that are used in food, detergent, paper, leather and other industries. The common enzymes used in industry are amylase, protease, lipase, nuclease, ligase, phytase, cellulase, and xylanase. Horikoshi (1999) expected that production of enzymes used in detergent industry may increase to 60%. However, the major share in the market is of enzymes used in food industry, viz. amylase, pullulanase, pectinase and invertase. Amylases account for 30% of the world's enzyme production (Sivaramakrishnan et al., 2006). Cyclodextrin glycosyl transferase (CGTase; EC 2.4.1.19) is the enzyme that belongs to the amylase family and converts starch into cyclodextrins (CD's) which are closed-ring structures having six or more glucose units joined by means of α-1, 4 glucosidic bonds. Cyclodextrin glycosyltransferase (CGTase) is a common starch degrading enzyme produced by bacteria.

CGTases are catalyzing four different transferase reactions: cyclisation, coupling, disproportionation, and hydrolysis. Three major types of Cyclodextrins are produced by CGTases depending on number of glucose units, α-CD, β-CD and γ-CD. Biwer et al. (2002) have reported that cyclodextrin glycosyl transferase produced by *Bacillus macerans* has the highest market share. This enzyme is important because of its application in the industrial production of cyclodextrin that is used in pharmaceutical, food and cosmetic industry.

6.2   DISCOVERY OF CYCLODEXTRIN GLYCOSYL TRANSFERASE (CGTase) AND CYCLODEXTRIN

Villiers (1891) reported the production of a precipitate by *Bacillus amylobacter* and called it cellulosine. It was later discovered by Franz Schardinger that the precipitate was a crystalline dextrin produced from starch by the action of *Bacillus macerans*. It was Tilden and Hudson (1939) who reported that a cell free enzyme of *B. macerans* converts starch to Schradingers dextrins, for example, Cyclodextrin (CD). Hence the enzyme was known as *B. macerans* amylase (BMA). Schwimmer (1953) reported the first evidence on purity of CGTase which he then called as Schardingers dextrinogenase.

It was during this period that the focus of all researchers was on CD's but the key enzyme CGTase still remained overshadowed. French (1957) published the first review on CDs. In 1969, Corn Products International started producing CD's using CGTase from *Bacillus macerans* (Horikoshi, 1999). The first International Symposium on cyclodextrins was organized in 1981, but the structural and genetic studies on CGTase were reported only in the late 1980's. Kaneko et al. (1988) elucidated the nucleotide sequence and studied the cloning of genes coding for CGTase. Maekalea et al. (1988) purified and studied the enzyme in the same year.

Nakamura et al. (1993) investigated the active residues in the substrate binding sites of CGTase and identified them as three histidine residues. Haga et al. (1994) crystallized CGTase from alkaliphilic *Bacillus* and performed some preliminary experiments on X-Ray Diffraction studies. New cyclodextrin-based technologies are constantly being developed and, thus, even 100 years after their discovery cyclodextrins are still regarded as novel excipients of unexplored potential (Loftsson and Duchêne et al., 2007).

6.3 SOURCE OF CGTase

Bacteria are the only known sources of CGTases. Besides mesophiles, extremophiles, for example, psychrophiles, alkaliphiles and thermophiles also secrete CGTase. *Bacillus macerans* (Takano et al., 1986; Fujiwara et al., 1992), *Thermoanaerobacterium thermosulfurigenes* EM1 (Wind et al., 1998) and *Bacillus stearothermophilus* (Fujiwara et al., 1992) are known to be producers of α-CGTase while *Bacillus circulans* strain 251 (Lawson et al., 1994), *Bacillus ohbensis* (Sin et al., 1991), alkalophilic *Bacillus* sp. 38–2 (Horikoshi, 1999) and *Bacillus* sp. 1011 (Kimura et al., 1987) are known to be producers of β-CGTase. Bacteria that produce γ-CGTase from starch are *Bacillus* sp. AL-6 (Fujita et al., 1990) and *Brevibacillus brevis* CD162 (Myung et al., 1998). Most of the CGTase producing bacteria belong to the genus *Bacillus* and most *Bacillus* sp. produce extracellular CGTase (Yong et al., 1996).

Some of the *Bacillus* sp. producing CGTase are listed in Table 6.1. A CGTase produced by *Thermoanaerobacter* sp. has been reported by Norman and Jorgensen (1992). CGTases from thermophilic actinomycetes have also been reported (Abelain et al., 2002). Gawande and Patkar (1998) have described production of cyclodextrin glycosyltransferase from *Klebsiella pneumoniae* AS-22 and *Bacillus firmus*.

Thatai et al. (1999) have studied CGTase enzyme produced by *Bacillus* isolated from soil. The enzyme has been reported from *Paenibacillus pabuli*, *P. graminis* and *Thermoanaerobicum* spp. Antranikian et al. (2009) have isolated and identified gene for CGTase production from extremophile *Anaerobranca gottschalkii* an extreme thermo alkaliphile. CGTase production had also been reported by halophilic archeon *Haloferax medditerrani* (Bautista et al., 2012) and alkaliphilic *Amphibacillus* spp. (Ibrahim et al., 2012). Thus it is evident that there are still many unidentified CGTase producing organisms, and they need to be studied as the panorama of applications of CD's keeps extending daily.

6.4   ALKALIPHILES AS SOURCE OF CGTase

Corn Products International Co. (United States) started the industrial production of *B. macerans* CGTase in 1969. Similarly, Teijin Ltd. (Japan) initiated the manufacture of β-CD using CGTase produced by *B. macerans* in a pilot plant. Two major technical issues were encountered during the

TABLE 6.1   Some of the *Bacillus* sp. Reported to Produce CGTase

Organism	CGTase type	References
Alkalophilic *Bacillus* sp. 1011	α	Kimura et al. (1987)
Alkalophilic *Bacillus* sp. 38–2	β	Horikoshi (1999)
Bacillus ohbensis	β/γ	Sin et al. (1991)
Bacillus circulans 251	β	Lawson et al. (1994)
Bacillus sp. E–1	γ	Yong et al. (1996)
Brevibacillus brevis CD162	γ	Myung et al. (1998)
Bacillus macerans	α	Horikoshi (1999)
Bacillus agaradhaerens strain LS-3C	β	Martin and Hoseney (2003)
Paenibacillus sp. L 55	β	Thombre et al. (2013)
Bacillus licheniformis	β	Thombre et al. (2013)

production of CGTase. The yield of CD was low as a mixture of all the three CDs was produced. Thus the cost of the product was high. Secondly, the recovery of product was difficult from the mixture of CD's and various harmful solvents were used for precipitation. The recovery of product using hazardous and toxic chemicals was a threat to the environment. The CGTase of alkaliphilic *Bacillus* sp. strain 38–2 produced 85 to 90% CD from amylose due to high conversion rate. The CDs could be directly crystallized from the fermentation broth without the addition of toxic precipitating agents. The industrial production of CD by this alkaliphile decreased the production cost drastically and lead to mass production of CD. The application of CD's increased as their cost decreased. The application of CD's increased extensively in foodstuffs, chemicals, and pharmaceuticals. Since then it is known that alkaliphiles are candidate organisms for CGTase production especially since they have greater specificity and more yield (Horikoshi, 1999).

6.5   PHYSIOLOGICAL FUNCTION OF CGTase

Bacillus species secrete a variety of important industrial enzymes. One of these enzymes is cyclodextrin glucanotransferase. The function of CGTase is to generate CD's as illustrated in Figure 6.1. CGTase cleaves α-(1, 4) bond in starch molecule, linking the reducing and non-reducing end to produce a cyclic molecule known as cyclodextrin (CD). The bacterium builds up an external storage of glucose by producing CD which is not accessible for other organisms because they are not able to metabolize CD (Figure 6.1). The CD's protect the bacteria from toxic compounds by forming inclusion complex with it (Aeckersberg et al., 1991). The CD's produced by CGTase outside the cell are transported by transporter proteins inside the cell. The CD's are hydrolyzed by cyclodextrinase (CDase) during starvation to release glucose molecules which are metabolized by the EMP pathway.

6.6   CHARACTERISTICS AND CATALYTIC MECHANISM OF CGTase

CGTases have a molecular weight varying from 60–110 kDa and consist of about 700 amino acids. Most require calcium as a protective agent against heat denaturation (Wind et al., 1998) and slightly acidic range at pH values 4.5–7.0 but alkalophilic CGTases display an optimum pH

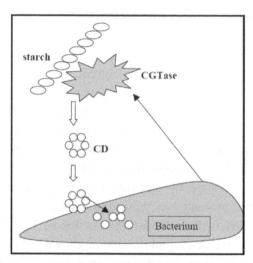

FIGURE 6.1   Schematic representation of function of CGTase (Aeckersberg et al., 1991).

of 9–10. Maximal temperature for most bacterial CGTases range from 40°C to 85°C. Most of the CGTases are strongly inhibited by Zn^{2+}, Cu^{2+} and Fe^{2+} (Tonkova, 1998). CGTases produce a mixture of three types of CDs, α-, β- and γ-CDs, with specific equilibrium distributions. CGTases are classified in the α-amylase family and are known to catalyze four different transferase reactions: cyclization (A), coupling (B), dispro-portionation (C), and hydrolysis (D) (Wind et al., 1998) as illustrated in Figure 6.2. The disproportionation reaction is also performed by several other enzymes of the amylase family. Cyclization, for example, formation of CD, is the characteristic property of CGTase in which the acceptor is a part of the cleaved donor. The reverse reaction is called as the coupling reaction. The next reaction of CGTase is hydrolysis. The most common substrate of CGTase is starch.

Starch is one of the largest molecules in nature found in potato, sweet potato, corn, wheat, barley, potato, rice, oat, tapioca and sago. It is car-bohydrate polymer composed of two high molecular weight compounds amylopectin (75–85%) and amylose (15–25%) comprising of glucose units linked by α-(1, 4) or α-(1, 6) glucosidic bonds. Starch has a highly branched structure, which consists of short amylose chains. Amylose is a linear polymer containing up to 6000 glucose units and is connected by α-(1,4) linkages. Amylopectin chains are linked to each other by α-(1,6)

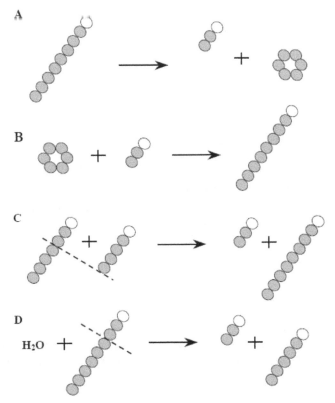

FIGURE 6.2 The reactions catalyzed by CGTase (van der Veen et al., 2000).

bonds. The major application of starch hydrolyzing enzymes like cyclo-dextrin glucanotransferase (CGTase), glucoamylase, α-amylase, pullulanase, and α-glucosidase is in the production of low molecular weight oligosaccharides by complete breakdown of starch (Leemhuis et al., 2003). The action of various starch hydrolyzing enzymes is outlined in Figure 6.3.

It can be concluded that starch hydrolyzing enzymes release different types of oligosaccharides according to their difference in mode of action. The α-amylase family is a large enzyme family that consists about 20 different reaction and product specificities, including exo/endo specificity. Takata et al. (1992) defined the α-amylase family as enzymes that hydrolyze or transfer α-glycosidic bonds and have four conserved sequence motifs which have the catalytic sites containing Asp, Glu and Asp residues.

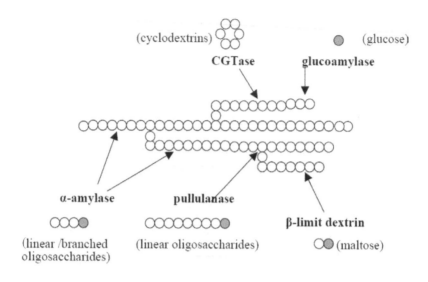

○ glucose molecule without reducing end

◐ glucose molecule with reducing end

FIGURE 6.3   Action of α-amylase family enzymes involved in degradation of starch. Arrows indicate points in the starch molecule where the enzyme attacks (Takata et al., 1992).

6.7   THREE DIMENSIONAL STRUCTURE OF THE α-AMYLASE FAMILY

α-amylase from *Aspergillus oryzae* was the first α-amylase with the three dimensional structure that has been solved (Matsuura et al., 1984). α-amylase family contains mainly three domains which are known as domain A, B and C. Domain A is the catalytic $(\alpha/\beta)_8$ domain, present in entire α-amylase family. This domain contains 300–400 amino acid residues and the catalytic residues are located at the C-terminal ends of the β strands in domain A. Four short conserved regions typical for the α-amylase family in this domain. Domain B is an extended loop region inserted after the third β-strand of domain A. This domain consists of 44–133 amino acid residues and contributes to substrate binding. Domain C has an antiparallel β-sandwich fold. It is known to be a maltose binding site observed in the structure derived from maltose dependent crystals (Lawson et al., 1994). This maltose binding site is involved in raw starch binding (Penninga et al., 1996), suggesting a role of the C-domain in substrate binding. There are a few other domains such as domain D and domain E for some of the enzymes in α-amylase family. The overall sequence similarity within the α-amylase family enzymes is relatively low, less than 30% (Tao, 1991). However there are four highly conserved regions that have been identified in this family (Nakajima et al., 1986). All four regions are directly involved in catalysis, either through substrate binding, bond cleavage or transition stabilization or as ligands of a calcium binding site present near the active site. Three carboxylic acid groups, one glutamic acid and two aspartic acid residues, were found to be essential for catalytic activity in α-amylase family. The amino acids are equivalent to Asp 229, Glu 257 and Asp 328 for CGTase from *Bacillus circulans* strain 251 (Klein et al., 1992). Two conserved histidine residues; His 140 and His 327 are involved in substrate binding and transition state stabilization (Nakamura et al., 1993; Uitdehaag et al., 1999). Lastly, the third histidine His 233 is present only in some α-amylase and CGTase, is involved in substrate binding and acts as calcium ligand with its carbonyl oxygen (Lawson et al., 1994).

6.8 PRODUCTION OF CGTase

Studies on scale up production of bioactive molecules are carried out to evaluate efficiency of the microbial strain and efficacy of the process for industrial production and eventually enhance the yield by optimizing process parameters. Production of enzyme in bioreactor is greatly influenced by aeration, agitation and geometry of the bioreactor. CGTase is an industrially important extracellular enzyme which is used for the production of CD. Large scale production of enzymes like proteases, amylases and xylanases has been studied extensively. However there are only a few reports on scale up production of CGTase. Studies on CGTase production in 3–5 L fermentor are very limited (Gawande and Patkar, 1998; Savergeve et al., 2008). Most of the optimization procedures are reported for shake flask (Rosso et al., 2002; Mahat et al., 2004). There are many methods for the industrial production of CGTase (Biwer et al., 2002). The production of CGTase is very similar to other enzyme-manufacturing processes (Gawande and Patkar, 2001). Mahat et al. (2004) have reported the production of CGTase from *Bacillus* sp. TS 1–1 in shake flask. Major reports on production of CGTase are in shake flask with a medium volume of 50 to 250 ml volume (Gawande and Patkar, 1998; Rosso et al., 2002; Higuti et al., 2004; Noi et al., 2008). There are a very few reports on optimization of CGTase production at fermentor level. Frietas et al. (2004) have reported the production of CGTase from alkaliphilic *Bacillus* CG strain II isolated from waste water of a flour industry in a 5 L fermentor.

Most bacteria have been reported to produce CGTase during late stationary phase. Rosso et al. (2002) have also reported a typical late – log production/secretion profile not associated with growth for *Bacillus circulans* in shakes flask fermentation. However some of the reports have indicated growth associated and/or biphasic profiles. Bacilleae show prolonged lag phase due to which the onset of CGTase production is only after 20–24 h. It is believed that the enzyme is attached to the cell wall and then is released in the broth. Kabaivanova (1999) have reported that sporulation in *Bacillus* spp. triggers CGTase production. The literature suggests that CGTase production reaches peak within 16 to 20 hours of incubation period (Jamuna et al., 1993). Makela et al. (1988) observed CGTase activity peak after a 40 hours lag period. Thatai et al. (1999) have produced 7.5 U/ml CGTase after 24 h.

To enhance the production of active CGTase, several methods have been applied. The first method comprises an optimization of the cultivation conditions of the CGTase-producing bacterial strain, including optimization of the growth medium (Jamuna et al., 1993; Gawande and Patkar, 1998; Rosso et al., 2002). Heterologous expression of the CGTase gene in another method used to enhance CGTase production. The production of the enzyme can also be increased by the over expression of CGTase gene. It is estimated that half of the total commercial CGTase production processes use heterologous production. A CGTase production which was three times higher than the extracellular production of the enzyme by the wild-type *Bacillus* sp. 1011 was observed in a combination of the trp promoter and the CGTase gene starting from the 48 nucleotide position in the presence of the inducer IAA (Kimura et al., 1990).

Two types of CD production processes are generally used at an industrial scale (Biwer et al., 2002). The solvent process, which is mainly used, requires an organic complexing agent to precipitate a specific CD selectively and to obtain one main product. A non-solvent process does not require complexing agents and produces a mixture of CDs. The total yield of CD produced by CGTase depends on the enzyme employed and the chosen reaction conditions. Lima et al. (1998) proposed a CD production process where the CGTase synthesis reaction is combined with yeast fermentation. The yeast consumes compounds inhibitory for CD synthesis by CGTase, which are produced by the enzymatic conversion, for example, glucose or maltose, while the ethanol produced by the yeast further increases the yield of CD.

CGTases can also be employed to produce large-ring CDs in high yields, by adjusting the reaction conditions to prevent their conversion to small CDs (Terada et al., 1997; Qi et al., 2004).

6.9 THE PRODUCT OF CGTase: CYCLODEXTRIN

Cyclodextrins (CD's also known as cellulosines or Schardingers dextrins) are produced from starch by means of enzymatic conversion by cyclodextrin glycosyl transferase enzyme. The CD's produced by CGTase are also called as cycloamyloses and are cyclic oligosaccharides composed of 5 or more α-D-glucopyranoside units. The interior of the cyclodextrin is not hydrophobic, but considerably less hydrophilic than the aqueous environment and thus able to host other hydrophobic molecules. In contrast, the exterior is sufficiently hydrophilic to impart water solubility to cyclodextrins or their complexes (Li et al., 2007). Typical cyclodextrins contain a number of glucose monomers ranging from six to eight units in a ring, thus denoting α-cyclodextrin: six sugar ring molecule, β-cyclodextrin: seven sugar ring molecule and γ-cyclodextrin: eight sugar ring molecule. Now a days, a cyclodextrin containing 9 glucose monomer is termed as theta- cyclodextrin. Structure containing more than 10 glucose monomers is called as Large Ring Cyclodextrins (LR- CD's). Cyclodextrin containing upto 60 glucose monomers are also reported. The cyclodextrins are soluble in water and have an inner cavity suitable for complexation.

6.10 MECHANISM OF CD FORMATION BY CGTase

CGTase hydrolyses any α-1,4-linkage within amylose molecule and then transfers the newly formed reducing end of the oligosaccharide to its own non-reducing end (Terada et al., 1997). During longer reaction times, the larger CDs will be subsequently converted to small CDs (CD_6–CD_8), due to their susceptibility to the coupling and hydrolytic reactions of the enzyme. This mechanism of CD formation can be divided into five steps as described by van der Maarel et al. (2002). In the first step of the reaction (bond cleavage) a covalently linked oligosaccharide intermediate is formed. In the second reaction step this oligosaccharide is transferred to an acceptor molecule.

The five steps in mechanism are as given below: (1) The substrate binds to the active site and glutamic acid transfers a proton to the glycosidic bond oxygen, while the nucleophilic aspartate residue attacks the C1 of glucose at subsite −1; (2) There is formation of oxocarbonium ion-like transition state after which a covalent intermediate formed; (3) The protonated molecule of glucose at subsite +1 leaves the active site and another acceptor glucose molecule attacks the covalent bond between the glucose molecule at subsite −1 and the aspartate residue; (4) There is reformation of oxocarbonium ion-like transition state (Uitdehaag et al., 1999), and (5) The glutamate accepts a hydrogen ion from glucose molecule at subsite +1 and the oxygen of the incoming glucose molecule at subsite +1 replaces the oxocarbonium bond between the glucose molecule at subsite −1 and the aspartate residue leading to formation of a new hydroxyl group at the C1 position of the new glycosidic bond between the glucose at subsite −1 and +1.

The role of the three important carboxylic amino acids in this mechanism was clarified with acarbose, a potent pseudotetraose inhibitor, bound at the active site. Glu257 is the general acid catalyst, acting as proton donor, Asp229 serves as the nucleophile, stabilizing the intermediate, and Asp328 has an important role in substrate-binding (Klein et al., 1992; Nakamura et al., 1993).

6.11   APPLICATIONS OF CGTase AND CD IN BIOTECHNOLOGY

The most important application of CGTases is the synthesis of CDs (Mori et al., 1994; Goel and Nene, 1995; Gawande and Patkar, 1998; Wong et al., 2008). Cyclodextrins are used for the stabilization of active compounds, reduction in volatility of drug molecules, and masking of odors and bitter tastes. Normally drug substance has to have a certain level of water solubility to be readily delivered to the cellular membrane, but it needs to be hydrophobic enough to cross the membrane. The majority of pharmaceutical active agents do not have sufficient solubility in water and these insoluble drugs can be complexed with CD's to increase their solubility (Tonkova, 1998).

CDs have many industrial applications, but their use is commercially restricted due to high price of CD's. Immobilization of the enzyme cyclodextrin glycosyl transferase (CGTase) has been pursued as a means of reducing the production cost of cyclodextrins (CDs) from starch. The immobilization of bacteria in polymers can facilitate the recovery of the free enzyme activity, maintain an appropriate pH, affinity for the substrate, and enzyme stability because in this case immobilization may cause benign changes in enzyme microenvironment, or in the of mobility of the protein. The extension of these changes depends upon the enzyme, support and reaction conditions used for immobilization. Among these factors, the choice of support is the most important factor. An ideal support for enzyme immobilization must be selected considering some essential properties, such as, chemical stability, hydrophilic behavior, rigidity, mechanical stability, large surface area, and resistance to microbial attack. There are not many publications describing immobilization of CGTase producers and CGTase. *Bacillus* spp. producing CGTase have been entrapped in different matrices, but with limited success (Ishmail et al., 1996; Park et al., 2000). Immobilization of the CGTase producer and exploring the enzyme production using agro- waste based media or medium components can minimize the cost of CD production.

CGTase can also be used to synthesize linear oligosaccharides and their derivatives by its coupling and disproportionating reactions. It has been shown that CGTase can use a variety of carbohydrates and other compounds as acceptors in transglcosylation reactions. CGTase is used to for the enzymatic transglycosylation and to improve solubility of compounds like rutin, curcumin, stevioside and naringin. Lee et al. (2002) have reported the use of CGTase produced by *B. stearothermophilus* as an antistaling enzyme. Gujral et al. (2003) have improved the quality of rice bread using CGTase.

CDs form inclusion complexes with a wide variety of hydrophobic guest molecules due to the unique molecular structure. The properties like solubility and stability of the guest molecule can be modified after encapsulation in the CDs. Therefore, there are immense applications of CDs in cosmetic, pharmaceutical, food industry and agriculture (Szejtli, 1998). Cyclodextrins can improve the stability of active pharmaceutical ingredients and increase the shelf life of drugs (Szejtli, 2004). They can improve the cord strength of polyester fibers used for reinforcement of rubbers (Szejtli, 2004). Fava et al. (1998) found that γ-Cyclodextrin has the potential of being successfully used in the bioremediation of chronically polychlorinated biphenyl-contaminated soils. It has been proposed that CDs can be used in all kinds of food and nutraceutical applications as a food ingredient and additive. Cyclodextrin can also stabilize emulsions of fats and oils. This property is useful for the preparation of bread spreads, dairy ice creams and breads (Munro et al., 2004).

Many undesirable changes take place in bread during storage which is called staling. Staling is defined as increase in firmness and loss in freshness due to gradual changes in the structure of starch (Min et al., 1998). The changes in starch take place due to inter or intramolecular association of starch molecules via hydrogen bonds that changes the amorphous state of starch to highly crystalline state (Hebeda et al., 1991). This change in bread firmness during staling is called as retrogradation. Enzymes with Intermediate Temperature Stability (ITS) like CGTase could be used as antistaling agents in bread and other products because they have optimum activity at gelatinizing temperature of starch and are inactivated at baking temperatures (Lee et al., 2002). Enzymes like amylase can decrease retrogradation by hydrolyzing starch, amylose and amylopectin to smaller molecules leading to lesser crystallization (Boyle and Hebeda, 1990; Martin and Hoseney, 1991). Lee et al. (2002) have reported the use of CGTase produced by *B. stearothermophilus* as an antistaling enzyme. They reported that maltose and water soluble dextrin were most effective in preserving crumb softness. However, the use of α-amylase has become limited despite its antistaling effect because too much of the enzyme can cause stickiness in bread. De Stefanis and Turner (1981) explained that it is the production of branched maltooligosaccharides by α-amylase that leads to gumminess in bread.

6.12   CONCLUSION AND FUTURE PROSPECTS

CGTase is an important enzyme due to its application in manufacture of cyclodextrins. CDs form inclusion complexes with a wide variety of hydrophobic guest molecules due to the unique molecular structure. The molecular structure and form of CD's confers the ability to act as molecular containers by entrapping guest molecules in their interior cavity (Szejtli, 2004). CD molecules have hydrophobic interior cavity and hydrophilic exterior surface and thus can form complexes with hydrophobic guest molecules. These inclusion complexes are formed with drugs, dyes, insoluble compounds, volatile chemicals and pharmaceutical compounds. There are various applications of CD's in encapsulation of pharmaceutical drugs and nanotubes. The property of CD's of encapsulating guest molecules inside their cavity is called "molecular encapsulation." Nowadays, nanoparticles are also encapsulated in the cavity of CD's leading to the development of a new field of nanoencapsulation. CGTase and CDs have immense applications in cosmetic, pharmaceutical and food industry. Studies reveals that extremophiles especially alkaliphiles are candidate organisms for the industrial production of CGTase. The significance of CGTase is still increasing day by day as the panaroma of applications of cyclodextrins keeps on widening.

KEYWORDS

- Alkaliphilic bacteria
- Antistaling enzyme
- Bread making
- Cyclodextrin
- Cyclodextrin glycosyl transferase
- Drug carrier
- Intermediate thermostable enzyme
- Pharmaceutical applications
- Retrogradation
- Starch
- Transglycosylation
- α-Amylase family

REFERENCES

Abelian, V. A., Balaian, A. M., Manukian, L. S., Afian, K. B., Meliksetian, V. S., Andreasian, N. A., Markocian, A. A. Characteristics of cyclodextrin glucosyltransferases of various groups of microorganisms. *Prikl Biokhim Mikrobiol* 2002, 38, 616–624.

Aeckersberg, F., Bak, F., Widdel, F. Anaerobic oxidation of saturated hydrocarbons to CO_2 by a new type of sulfate-reducing bacterium. *Arch Microbiol* 1991, 156, 5–14.

Antranikian, G., Ruepp, A., Gordon, P. M. K., Ballschmiter, M., Zibat, A., Stark, M., Sensen, C. W., Frishman, D., Liebl, W., Klenk, H. P. Rapid access to genes of biotechnologically useful enzymes by partial genome sequencing: the thermoalkaliphile *Anaerobranca gottschalkii*. *J Mol Microbiol Biotechnol* 2009, 16, 81–90.

Bautista, V., Esclapez, J., Pérez-Pomares, F., Martínez-Espinosa, R. M., Camacho, M., Bonete, M. J. Cyclodextrin glycosyltransferase: a key enzyme in the assimilation of starch by the halophilic archaeon *Haloferax mediterranei*. *Extremophiles* 2012, 16(1), 147–159.

Biwer, A., Antranikian, G., Heinzle, E. Enzymatic production of CDs. *Appl Microbiol Biotechnol* 2002, 59, 609–617.

Boyle, P. J., Hebeda, R. E. Antistaling enzyme for baked goods. *Food Technology* 1990, 44, 129.

De Stefanis, V. A., Turner, E. W. US Patent 1981, 4, 299–848.

Fava, F., Gioia, D., Marchetti, L. Cyclodextrin effects on the *ex situ* bioremediation of a chronically polychlorobiphenyl-contaminated soil. *Biotechnol Bioeng* 1998, 58, 345–355.

Freitas, T. L., Monti, R., Contiero, J., Production of CGTase by a Bacillus alkalophilic CGII strain isolated from waste water of a manioc flour industry. *Braz J Microbiol* 2004, 35, 255–260.

French, D. The Schardinger dextrins. *Adv Carbohydr Chem* 1957, 12, 189–260.

Fujiwara, S., Kakihara, H., Woo, K. B., Lejeune, A., Kanemoto, M., Sakaguchi, K., Imanaka, T. Cyclization characteristics of cyclodextrin glucanotransferase are conferred by the NH2-terminal region of the enzyme. *Appl Environ Microbiol* 1992, 58, 4016–4025.

Fujita, Y., Tsubouchi, H., Inagi, Y., Tomita, K., Ozaki, A., Nakanishi, K. Purification and properties of cyclodextrin glycosyltransferase from *Bacillus* sp. AL-6. *J Ferment Bioeng* 1990, 70, 150–154.

Gawande, B. N., Singh, R. K., Chauhan, A. K., Goel, A., Patkar, A. Y. Optimization of cyclomaltodextrin glucanotransferase production from *Bacillus firmus*. *Enzyme and Microbial Technology* 1998, 22, 288–291.

Gawande, B. N., Patkar, A. Alpha-cyclodextrin production using cyclodextrin glycosyltransferase from *Klebsiella pneumoniae* AS-22. *Starch* 2001, 53, 75–83.

Goel, A., Nene, S. A novel cyclomaltodextrin glucanotransferase from *Bacillus firmus* that degrades raw starch. *Biotechnology Letters* 1995, 17(4), 411–416.

Gujral, H. S., Guardiola, I., Carbonell, J. V., Rosell, C. M. Effect of cyclodextrin glycosyl transferase on dough rheology and bread quality from rice flour. *J Agric Food Chem* 2003, 18, 51(13), 3814–3818.

Haga, K., Harata, K., Nakamura, A., Yamane, K. Crystallization and preliminary X-ray studies of cyclodextrin glucanotransferase from alkaliphilic *Bacillus* sp. 1101. *J Mol Biol* 1994, 237, 163–164.

Hebeda, R. E., Bowles, L. K., Teague, W. M. Use of intermediate temperature stability enzyme for retarding staling in baked goods. *Cereal Foods World* 1991, 36, 619.

Higuti, I. H., Silva, P. A. D., Papp, J., Okiyama, V. M. D. E., Andrade, E. A. D., Marcondes, A. D. A., Nascimento, A. J. D. Colorimetric determination of α and β-cyclodextrins and studies on optimization of CGTase production from *B. firmus* using factorial designs. *Braz Arch Biol Technol* 2004, 47, 837–841.

Horokoshi, K. Alkaliphiles: some applications of their products in biotechnology. *Microbiol Mol Biol Rev* 1999, 63, 735–750.

Ibrahim, A., Al-Salamah, A., El-Tayeb, M. A., El-Badawi, Y. B., Antranikian, G. A novel cyclodextrin glycosyltransferase from alkaliphilic *Amphibacillus* sp. NPST–10: purification and properties. *Int J Mol Sci* 2012, 13(8), 10505–10522.

Ishmail, A. S., Sobieh, U. I., Abdel-Fattah, A. F. Biosynthesis of cyclodextrin glycosyl transferase and cyclodextrin *Bacillus macerans* 314 and properties of the crude enzyme. *Chem Eng* J 1996, 61, 247–253.

Jamuna, R., Saswathi, N., Sheela, R., Ramakrishna, S. V. Synthesis of cyclodextrin glucosyl transferase by *Bacillus cereus* for the production of cyclodextrins. *Appl Biochem Biotechnol* 1993, 43, 163–176.

Kabaivanova, L., Dobreva, E., Miteva, V. Production of cyclodextrin glucosyltransferase by *Bacillus stearothermophilus* R2 strain isolated from a Bulgarian hot spring. *Journal of Applied Microbiology* 1999, 86, 1017–1023.

Kaneko, T., Hamamoto, T., Horikoshi, K. Molecular cloning and nucleotide sequence of the cyclomaltodextrin glucanotransferase gene from the alkalophilic *Bacillus* sp. strain no. 38–2. *J Gen Microbiol* 1988, 134, 97–105.

Kimura, K., Kataoka, S., Ishii, Y., Takano, T., Yamane, K. Nucleotide sequence of the β-cyclodextrin glucanotransferase gene of alkalophilic *Bacillus* sp. strain 1011 and similarity of its amino acid sequence to those of α-amylases. *J Bacteriol* 1987, 169, 4399–4402.

Klein, C., Hollender, J., Bender, H., Schulz, G. E. Catalytic center for cyclodextrin glycosyltransferase derived from X-ray structure analysis combined with site-directed mutagenesis. *Journal of Biochemistry* 1992, 31, 8740–8746.

Lawson, C. L., van Montfort, R., Strpkoptov, B., Rozeboom, H. J., Kalk, K. H., de Vries, G., Penninga, D., Dijkhuizen, L., Dijkstra, B. W. Nucleotide sequence and X-ray structure of cyclodextrin glycosyltransferase from *Bacillus circulans* strain 251 in a maltose-dependent crystal form. *Journal of Molecular Biology* 1994, 236, 590–600.

Lee, S. H., Kim, Y. W., Lee, S., Auh, J. H., Yoo, S. S., Kim, T. J., Kim, J. W., Kim, S. T., Rho, H. J., Choi, J. H., Kim, Y. B., Park, K. H. Modulation of cyclizing activity and thermostability of cyclodextrin glucanotransferase and its application as an antistaling enzyme. *J Agric Food Chem* 2002, 50, 1411–1415.

Leemhuis, H., Rozeboom, H. J., Dijkstra, B. W., Dijkhuizen, L. The fully conserved Asp residue in conserved sequence region I of the α-amylase family is crucial for the catalytic site architecture and activity. *FEBS Letter* 2003, 541, 47–51.

Li, Z. F., Wang, M., Wang, F., Gu, Z. B., Du, G. C., Wu, J., Chen, J. γ-Cyclodextrin: a review on enzymatic production and applications. *Appl Microbiol Biotechnol* 2007, 77, 245–255.

Lima, H. O., De Moraes, F. F., Zanin, G. M. Beta-cyclodextrin production by simultaneous fermentation and cyclization. *Appl Biochem Biotechnol* 1998, 70–72, 789–804.

Loftsson, T., Duchêne, D. Cyclodextrins and their pharmaceutical applications. *Int J Pharm* 2007, 329, 1–11.

Maekelae, M., Mattsson, P., Schinina, M. E., Korpela. Purification and properties of cyclomaltodextrin glucanotransferase from an alkalophilic *Bacillus*. *Biotechnol Appl Biochem* 1988, 10, 414–427.

Mahat, M. K., Illias, R. M., Rahman, R. A., Rashid, N. A., Mahmood, N. A. N., Hassan, O., Suraini, A. A., Kamaruddin, K. Production of cyclodextrin glucanotransferase (CGTase) from alkalophilic *Bacillus* sp. TS1–1: media optimization using experimental design. *Enzyme and Microbial Technology* 2004, 35, 467–473.

Martin, M., Hoseney, R. C. A mechanism of bread firming. II. Role of starch hydrolyzing enzymes. *Cereal Chem* 1991, 68, 503–507.

Matsuura, Y., Kusunoki, M., Harada, W., Kakudo, M. Structure and possible catalytic residues of Taka-amylase A. *J Biochem* 1984, 95, 697–702.

Min, B. C., Yoon, S. H., Kim, J. W., Lee, Y. W., Kim, Y. B., Park, K. H. Cloning of novel maltooligosaccharide-producing amylases as antistaling agents for bread. *Journal of Agricultural and Food Chemistry* 1998, 46(2), 779–782.

Mori, S., Hirose, S., Oya, T., Kitahata, S. Purification and properties of cyclodextrin glucanotrasferase from *Brevibacterium* sp. no. 9605. *J Biosci Biotechnol Biochem* 1994, 58, 1968–1972.

Munro, I. C., Newberne, P. M., Young, V. R., Bär, A. Safety assessment of γ-cyclodextrin. *Regul Toxicol Pharm* 2004, 39, S3–S13.

Myung, H. K., Cheon, B. S., Tae, K. O. Cloning and sequencing of a cyclodextrin glycosltransferase gene from *Brevibacillus brevis* CD 162 and its expression in *Escherichia coli*. *FEMS Microbiology Letters* 1998, 164, 411–418.

Nakamura, A., Haga, K., Yamane, K. Three histidine residues in the active center of cyclodextrin glucanotransferase from alkaliphilic *Bacillus* sp. 1011: effects of replacement on pH dependence and transition–state stabilization. *Biochemistry* 1993, 32, 6624–31.

Nakajima, R., Imanaka, T., Aiba, S. Comparison of amino acid sequences of eleven different α-amylases. *Appl Microbiol Biotechnol* 1986, 23, 355–360.

Norman, B. E., Jorgensen, S. T. *Thermoanaerobacter* sp. CGTase its properties and application. *Denpun Kagaku* 1992, 39, 101–108.

Park, C. S., Park, K. H., Kim, S. H. A rapid screening method for alkaline β cyclodextrin-methyl orange containing solid medium. *Agric Biol Chem* 1989, 53, 1167–1169.

Park, K. H., Kim, T. J., Cheong, T. K., Kim, J. W. Structure, specificity and function of cyclomaltodextrinase, a multispecific enzyme of the α-amylase family. *Biochimica et Biophysica Acta* 2000, 1478, 165–185.

Penninga, D., van der Veen, B., Knegtel, R. M. A., van Hijum, S. A. F. T., Rozeboom, H. J., Kalk, K. H., Dijkstra, B. W., Dijkhuizen, L. The raw starch binding domain of cyclodextrin glycosyltransferase from *Bacillus circulans* strain 251. *The Journal of Biological Chemistry* 1996, 271(51), 32777–32784.

Qi, Q., Zimmermann, W. Cyclodextrin glucanotransferase: from gene to applications. *Appl Microbiol Biotechnol* 2005, 66, 475–485.

Rosso, A. M., Ferrarotti, S. A., Krymkiewicz, N., Nudel, B. C. Optimization of batch culture conditions for cyclodextrin glycosyl transferase production from *Bacillus circulans* DF 9F. *Microbial Cell factories* 2002, 1, 3.

Sauvaphap, A. N., Suraini, A. A., Norjahan, A., Osman, H., Mohamed, A. K. Optimization of cyclodextrin glycosyltransferase production by response surface methodology approach. *Biotechnology* 2008, 7(1), 10–18.

Savergave, L. S., Dhule, S. S., Jogdand, V. V., Nene, S. N., Gadre, R. V. Production and single step purification of cyclodextrin glycosyltransferase from alkalophilic *Bacillus firmus* by ion exchange chromatography. *Biochem Eng J* 2008, 39, 510–515.

Schwimmer, S. Evidence for the purity of schardinger dextrinogenase. *Arch Biochem Biophys* 1953, 43, 108–117.

Sin, K., Nakamura, A., Kobayashi, K., Masaki, H., Uozumi, T. Cloning and sequencing of a cyclodextrin glucanotransferase gene from *Bacillus obhensis* and its expression in *Escherichia coli*. *Applied Microbiology and Biotechnology* 1991, 35, 600–605.

Szejtli, J. Utilization of cyclodextrins in industrial products and processes. *Journal of Material Chemistry* 1997, 7(4), 575–587.

Szejtli, J. Introduction and general overview of cyclodextrin chemistry. *Chemical Reviews* 1998, 98, 1743–1753.

Szejtli, J. Past, present, and future of cyclodextrin research. *Pure Appl Chem* 2004, 6, 1825–1845.

Sivaramakrishnan, S., Gangadharan, D., Madhavan, K., Soccol, C. R., Pandey, A. α-Amylases from microbial sources. *Food Technol Biotechnol* 2006, 44 (2), 173–184.

Takano, T., Fukuda, M., Monma, M., Kobayashi, S., Kainuma, K., Yamane, K. Molecular cloning, DNA nucleotide sequencing and expression in *Bacillus subtilis* cells of *Bacillus macerans* cyclodextrin glucanotransferase gene. *Journal of Bacteriology* 1986, 166(3), 1118–1122.

Takata, H., Kuriki, T., Okada, S., Takesada, Y., Iizuka, M., Minamiura, N., Imanaka, T., Action of neopullulanase. *J Biol Chem* 1992, 267(26), 18447–18452.

Tao, B. Y. Cyclodextrin glucanotransferases: technology and biocatalyst design. In *Enzyme in Biomass Conversion*, Himmel, M. E., Leatham, G. F. (Eds.), American Chemical Society: Washington, DC, 1991, pp. 372–383.

Terada, Y., Yanase, M., Takata, H., Takaha, T., Okada, S. Cyclodextrins are not the major cyclic α-1, 4-glucans produced by the initial action of cyclodextrin glucanotransferase on amylose. *J Biol Chem* 1997, 272, 15729–15733.

Thatai, A., Kumar, M., Mukherjee, M. Single step purification process for cyclodextrin glucanotransferase from a *Bacillus* sp. isolated from soil. *Preparative Biochemistry and Biotechnology* 1999, 29, 35–47.

Thombre, R. S., Kanekar, P. P., Rajwade, J. M. Production of cyclodextrin glycosyl transferase from alkaliphilic *Paenibacillus* sp. L55 MCM B-1034 isolated from alkaline Lonar lake, India. *International Journal of Pharma and Biosciences* 2013, 4(1), B1–9.

Tilden, E. B., Hudson, C. S. The conversion of starch to crystalline dextrins by the action of a new type of amylase separated from cultures of *Aerobacillus macerans*. *J Am Chem Soc* 1939, 61, 2900–2902.

Tonkova, A. Bacterial cyclodextrin glucanotransferase. *Enzyme and Microbial Technology* 1998, 22, 678–686.

Uitdehaag, J. C. M., Kalk, K. H., van der Veen, B. A., Dijkuizen, L., Dijkstra, B. W. The cyclization mechanism of cyclodextrin glycosyltransferase (CGTase) as revealed by a γ-cyclodextrin-CGTase complex at 1.8-Å resolution. *The Journal of Biological Chemistry* 1999, 274(49), 34868–34876.

van der Maarel, M. J. E. C., van der Veen, B. A., Uitdehaag, J. C. M., Leemhuis, H., Dijkhuizen, L. Properties and applications of starch-converting enzymes of the α-amylase family. *J Biotechnol* 2002, 94, 137–155.

van der Veen, B. A., van Aleebek, G. J. W. M., Uitdehaag, J. C. M., Dijkhuizen, L. The three transglycosylation reactions catalyzed by cyclodextrin glycosyltransferase from *Bacillus circulans* (strain 251) proceed via different kinetic mechanisms. *European Journal of Biochemistry* 2000, 267, 658–665.

Villiers, A. Sur la transformation de la fécule en dextrine par le ferment butyrique. *Compt Rend Fr Acad Sci* 1891, 435, 8.

Wind, R. D., Uitdehaag, C. M., Buitelaar, R. M., Dijkstra, B. W., Dijkhuizen, L. Engineering of cyclodextrin product specificity and pH optima of the thermostable cyclodextrin glycosyltransferase from *Thermoanaerobacterium thermosulfurigenes* EM1. *The Journal of Biological Chemistry* 1998, 273(10), 5771–5779.

Wong, T. E., Rahman, R. A., Ismail, A. F., Salleh, M. M., Hassan, O., Kamaruddin, K. Initial screening of fermentation variables for the production of cyclomaltodextrin glucanotransferase (CGTase) from local isolated *Bacillus stearothermophillus* HR1. *Developments in Chemical Engineering and Mineral Processing* 2008, 13(5–6), 541–547.

Yong, J., Choi, J. N., Park, S. S., Park, C. S., Park, K. H., Choi, Y. D. Secretion of heterologous cyclodextrin glycosyltransferase of *Bacillus* sp. E1 from *Escherichia coli*. *Biotechnology Letters* 1996, 18, 1223–1228.

CHAPTER 7

INDUSTRIAL APPLICATIONS OF FUNGAL CHITINASES: AN UPDATE

PADMANABH MISHRA,[1] S. K. SINGH,[2] and
SMITA S. NILEGAONKAR[3]

[1]*Molecular Biophysics Unit, Indian Institute of Science, Bangalore, Karnataka, 560012, India*

[2]*Biodiversity and Palaeobiology Group, Agharkar Research Institute,, Pune, Maharashtra, 411004, India*

[3]*Microbial Sciences Division, Agharkar Research Institute, Pune, Maharashtra, 411004, India*

CONTENTS

7.1   INTRODUCTION

Chitin, a tough and pliable homopolymer of β 1–4 linked 2-acetamido-2-deoxy-D-glucose (N-acetyl-D-glucosamine or GlcNAc), is the most abundant polysaccharide after cellulose, existing in nature (Khoushab and Yamabhai, 2010). It constitutes a major structural component of many biological systems particularly, insects, crustaceans, molluscs, fungi, algae and marine invertebrates (Shaikh and Deshpande, 1993) (Figure 7.1).

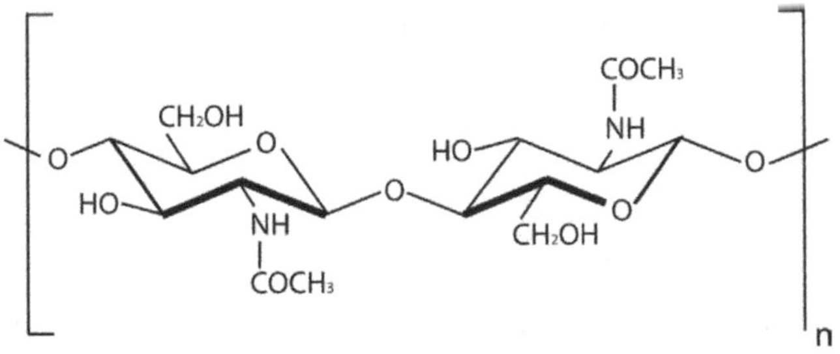

FIGURE 7.1   Structure of chitin showing two repeating N-acetylglucosamine monomers that form long-chain polymers through covalent β-1,4 linkages.

7.1.1   OCCURRENCE

Chitin occurs widely in nature as a structural polymer in the integument of insects and crustaceans and the cell wall of true fungi (Jeuniaux, 1971). The prominence in the mantles of the insects led to its name which comes from the Greek 'chiton' means a 'coat of mail.' The cell wall of fungi is reported to contain 22–44% but in green algae it is only 3–5% (Teng and Whistler, 1973). In arthropods, coelenterates, and nematodes, it accounts for 25–50% of the cuticles (Jeuniaux, 1971).

7.1.2   STRUCTURE AND FUNCTION OF CHITIN

Chemically, chitin may be termed as a derivative of cellulose, in which hydroxyl groups have been replaced by acetamido residues. In case of chitosan, the acetylated amino group of chitin is deacetylated.

7.1.3   PHYSICAL AND CHEMICAL PROPERTIES OF CHITIN

Due to the insoluble nature of chitin, its isolation frequently requires use of drastic chemical treatments to remove the contaminants. Inorganic contaminants are generally removed by digestion with dilute mineral acids, whereas treatment with hot, dilute alkali helps to remove proteinaceous and other organic impurities. The most preferable and common source of chitin for experimental purposes is crustaceans shell and typical isolation procedure for the same has been described by Hackman (1954) using 2N HCl, which was further modified by Horowitz et al. (1957). The acid hydrolysis (conc. HCl) of chitin offers D-glucosamine and acetic acid in 1:1 proportions.

Acid hydrolysis under mild condition lead to 2 acetamido-2-deoxy-D-glucose, while partial hydrolysis, gives chitobiose, chitotriose, chitotetraose and higher oligosaccharides. Chitin is insoluble in water due to its inter-molecular hydrogen bonds (Minke and Blackwell, 1978). But water-soluble chitin-based derivatives such as chitosan or carboxymethyl chitin (CMC) can be obtained. One of their most important features is their flexibility to be shaped into different forms such as fibers, hydrogels, beads, sponges, and membranes (Mano et al., 2007). The source of chitin affects its physical and chemical properties such as crystallinity, purity, polymer chain arrangement, and dictates its properties (Rinaudo, 2006; Khoushab and Yamabhai, 2010).

7.2   CHITOSAN

Chitosan, a non-crystalline fully deacetylated derivative of chitin, occurs in the walls of a limited but medically important group of fungi, the Zygomycetes (Kreger, 1954; Bartnicki, 1968). These fungi are opportunistic invaders to human and can be major pathogens in burn wounds (Bruck et al., 1970). When chitin is treated with concentrated alkali at high temperatures, it undergoes various degrees of deacetylation and degradation to give rise to chitosan. However, chitosan is a family name for a group of partially deacetylated products rather than a single substance. In Japan, huge quantities of chitin (million kg) from crab shells are processed annually to obtain chitosan for use as a flocculating agent in the clarification of sewage water (Imeri and Knorr, 1998). Chitosan is also being produced on commercial scale by fermentation of a Mucoralean fungus *Absidia coerulea*, which possess chitosan as its cell wall component.

7.3   INDUSTRIAL AND BIOMEDICAL APPLICATIONS OF CHITIN, CHITOSAN AND CHITOOLIGOSACCHARIDES

Chitin and chitosan have various characteristics that are not found in other natural polymers and biopolymers. However, for a long time, chitin and chitosan were unutilized bioresources. In the investigation of the importance of natural polymers, chitin and chitosan, as well as cellulose, are being considered functional polysaccharides, and are actively being studied and applied in various fields, such as medical treatment, medicine, food, chemical industries, fibers, and others (Uragami et al., 2001; Sashiwa and Aiba, 2004;

Muzzarelli, 2010). Chitin and chitosan have high organic solvent resistance, which is advantageous for preparation of separation membranes typically used with organic solvents, such as gas permeation membranes, dialysis membrane, reverse osmosis membranes, ultrafiltration membranes, water/organic selective membranes, and career transport membranes (Nakatsuka and Andrady, 1992; Qurashi et al., 1992; Uragami et al., 2002). Chitin and its derivatives have also been used in food and beverages industry, as wine clarification and stabilizing agent, de-acidifying and de-hazing agent in fruit juices, antimicrobial agent in food preservation and in agriculture like horticulture development and as plant growth promoters in orchid cultivation (Lefebvre et al., 2000). Chitin and chitosan have also been explored for their unique biomedical applications as in chitosan-DNA nanoparticles delivery system for gene therapy, in tissue engineering and wound healing, antioxidant activity, use of D-glucosamine in dentistry and bone regeneration, in veterinary medicine, chitosan based materials as carriers for anticancer drugs, ant diabetic effect, cholesterol lowering effect, and in cosmetics (Panos et al., 2008; Tian and Sun, 2008; Vyas et al., 2011). These polysaccharides are also being used as complexing agents for heavy metals and in waste water treatment (Onsoyen et al., 1990).

7.4   CHITINASE

The enzymatic hydrolysis of chitin to free GlcNAc units is performed by a chitinolytic system, which is found in a variety of organisms such as actinomycetes, bacteria, fungi, yeasts, plants, protozoans, coelenterates, nematodes, mollusks, arthropods and also in human beings (Kuranda and Robbins, 1991; Hawtin et al., 1995; Renkema, 1995; Patil et al., 2000; Kasprzewska, 2003). Chitinase is the basic protein responsible to catalyze the hydrolysis of chitin. Relationship between mycolytic enzymes, chitinases and β-1, 3 glucanases produced by mycoparasitic fungi and their significance in fungal cell wall lysis and degradation has been well established (Elad et al., 1980; Elad et al., 1999). The complete enzymatic hydrolysis of chitin to free GlcNAc is performed by a chitinolytic system, the action of which is known to be synergistic and consecutive (Deshpande, 1986; Shaikh and Deshpande, 1993). Thus assay of chitinase could be used as a basis for screening potential biocontrol agents (Elad et al., 1982).

In recent years, chitinases have received greater attention due to their wider range of biotechnological applications especially in the biocontrol of fungal phytopathogens (Mathivanan et al., 1998) and harmful insect pests (Mendosa et al., 1996; Pinto et al., 1997). Chitinases have also been used in the preparation of sphaeroplasts and protoplasts from yeast and fungal species (Peberdy, 1985; Mizuno et al., 1997). Some other significant applications of chitinases include bioconversions of chitin waste to single cell protein (SCP) and ethanol (Vyas and Deshpande, 1991) and fertilizers (Sakai et al., 1986).

7.4.1   SOURCES OF CHITINASES

7.4.1.1   Plant Chitinases

Chitinases are constitutively present in plants, seeds, stems, tubers and flow-ers, which are tissue-specific as well as developmentally regulated. Plant chitinases are induced by the attack of phytopathogens as pathogenesis-related proteins in plant self-defense or by contact with elicitors such as chitooligosaccharides or growth regulators such as ethylene (Gooday, 1996; Koga et al., 1996; Kim and Chung, 2002). The agricultural biotechnology offers new tools for the protection of plants from pathogens by using disease resistant transgenic plants. For instance, isolation of genes of chitinases from *Trichoderma* spp. were reported and transferred to plants in order to increase the resistance against phytopathogens (Lorito et al., 1996; Carsolio et al., 1999; Yedidia et al., 1999).

7.4.1.2   Insect Chitinases

The chitinases present in the insects have been described from *Bombyx mori* and *Manduca sexta*. These enzymes play important roles as degradative enzymes during ecdysis, where endochitinases break randomly the cuticle to chitooli-gosaccharides that afterwards are hydrolyzed by exoenzymes to GlcNAc. The monomer is reused to synthesize a new cuticle. Insect chitinases also have a defensive role against their own parasites. The production of enzymes in insects is regulated by hormones during the transformation of the larvae (Kramer and Muthukrishnan, 1997; Arakane and Muthukrishnan, 2010).

7.4.1.3   Microbial Chitinases

The microorganisms able to degrade chitin are widely distributed in nature (Deshpande, 1986; Felse and Panda, 2000). Due to its insolubility, size, molecular complexity and heterogeneous composition, the chitin is not degraded inside the cell, but the microorganisms secrete enzymes with different specificity, to transform or hydrolyse chitin (Cottrell et al., 1999). The microorganisms produce chitinases in higher amounts than animals and plants, generally as inducible extracelullar that are of the two types, endochitinases and exochitinases. Among the bacteria producers are found the *Serratia, Chromobacterium, Klebsiella, Bacillus, Streptomyces* and fungi like *Trichoderma, Penicillium, Lecanicillium, Neurospora, Mucor, Metarhizium, Beauveria, Lycoperdon, Aspergillus* and *Basidiobolus* (Leger et al., 1998; Rojas-Avelizapa et al., 1999; Van Aalten et al., 2000; Matsumoto et al., 2006; Mishra et al., 2011). The microbial chitinases in particular fungal chitinases have grabbed wide attention in the bioconversion process to recycle crab and shellfish chitin waste (Wang and Chio, 1998; He et al., 2006). As in the bioconversion of cellulose to SCP, the production of chitinase is thought to be one of the primary economic variables, estimated to account for 12% of the total production cost (Cosio et al., 1982). Though chitinases are present in a variety of living organisms such as snails, bean seeds, insects and microbes, last one being the most preferable source for industrial scale production (Cottrell, 1999). Microbial chitinases have been produced by liquid batch fermentation, continuous fermentation, and fed-batch fermentation. In addition to these, solid-state fermentation and biphasic cell systems have also been used for the production of chitinase. Generally, chitinases produced from microorganisms are inducible in nature (Dahiya et al., 2006).

7.4.2 CLASSIFICATION OF CHITINASE

Chitinase are classified in two different ways, on their two innate properties, first, their mode of action on chitin and second, their amino acid sequence. Chitinolytic system, responsible for the enzymatic hydrolysis of chitin can be classified into two major categories. Endochitinases (EC 3.2.1.14) which cleave chitin randomly at internal sites, generating low molecular mass multimers of GlcNAc, such as chitotetraose, chitotriose, and diacetylchitobiose. Exochitinases can be divided into two subcategories: chitobiosidases (EC 3.2.1.29), which catalyse the progressive release of diacetylchitobiose starting at the nonreducing end of chitin microfibril, and β-(1, 4) N-acetyl glucosaminidases (EC 3.2.1.30), which cleave the oligomeric products of endochitinases and chitobiosidases, generating monomers of GlcNAc (Sahai and Manocha, 1993). An alternative pathway involves the deacetylation of chitin to chitosan, which is finally converted to glucosamine residues by the action of chitosanase (EC 3.2.1.132) (Sahai and Manocha, 1993).

Secondly, based on amino acid sequence similarity, chitinolytic enzymes are grouped into families 18, 19, and 20 of glycosyl hydrolases (Henrissat and Bairoch, 1993). Family 18 is diverse in evolutionary terms and contains chitinases from bacteria, fungi, viruses, animals, and some plant chitinases. Family 19 consists of plant chitinases (classes I, II, and IV) and some *Streptomyces* chitinases (Hart et al., 1995). The chitinases of the two families, that is, 18 and 19, do not share amino acid sequence similarity. They have completely different 3-D structures and molecular mechanisms and are therefore likely to have evolved from different ancestors (Suzuki et al., 1999). Family 20 includes β-N-acetylhexosaminidases from bacteria, streptomycetes, and humans (Dahiya et al., 2006).

In fungi, chitinases are considered to play autolytic, nutritional, and morphogenetic roles (Adams, 2004). Chitinases in mycoparasitic fungi are most commonly suggested to be involved in mycoparasitism (Haran et al., 1996). Fungal and bacterial chitinases are shown to play a role in the digestion of chitin for utilization as a carbon and energy source and recycling chitin in nature (Tsujibo et al., 1993; Park et al., 1997; Svitil et al., 1997). In insects, chitinases are associated with post embryonic development and degradation of old cuticle (Merzendorfer and Zimoch, 2003). Plant chitinases are mainly involved in defense and development (Graham and Sticklen, 1994). Chitinases encoded by viruses have roles in pathogenesis (Patil et al., 2000). Human chitinases are suggested to play a role in defense against chitinous human pathogens (Boot et al., 2001; Boot et al., 2005; Van Eijk et al., 2005). On the other hand, chitinases have shown immense potential applications in agricultural, biological and environmental fields. Due to important biophysiological functions and applications of chitinase, a considerable amount of research on fungal chitinases has been carried out in recent years (Chuan 2006; Hartl et al., 2012).

7.4.3 BINDING MODE OF CHITINASE

The chitooligosacchrides and their reduced or methylated derivatives have been used to elucidate the mode of binding of chitinase and their action pattern, for instance, from the analysis of the hydrolysis products of the oligosaccharides, it has been suggested that the chitinase from *Pycnoporus cinnabarinus* had an exo-type action, predominantly hydrolysing the second β-N-acetylglucosaminide linkage from the non-reducing end (Ohtakara and Mitsutomi, 1988). Most of the carbohydrases show the similar protein-saccharide interactions, although they show widely diverse three-dimensional structures and binding site topologies.

7.4.4 MOLECULAR GENETICS FOR ENHANCED PRODUCTION OF CHITINASE

A number of naturally occurring organisms serve as a major source of chitinolytic enzymes but genetic improvement plays a significant role in their biotechnological applications. The conventional procedures for the strain improvement are chemical, physical mutation and protoplast fusion. In addition to mutation and protoplast fusion, molecular cloning is being effectively used to achieve overproduction of chitinases lately, to change in their localization, such as periplasmic or extracellular and to understand the organism itself. Molecular cloning of chitinase genes has been reviewed earlier by different workers world over (Flach et al., 1992; Shaikh and Deshpande, 1993; Sahai and Manocha, 1993). Few years back, for the constitutive, extracellular activity, the co-transformation of *Trichoderma reesei* protoplasts with *Aphanocladium album* chitinase was reported (Kunz et al., 1992). The 6.5 fold higher activity expressed in transformant was found to be useful in bioremediation and biocontrol activities (Deane et al., 1999). Number of reports on molecular cloning for chitinases, either to increase biocontrol efficiency of *Bacillus thuringiensis*, to prepare highly active chitinase preparation or to develop transgenic plants for the increased resistance have been described (Wiwat et al., 1996; Tantimavanich et al., 1997). The *B. thuringiensis* strains are known to produce d-toxin that kills number of insect pests however a combination of d-toxin and chitinase has been reported to be more effective. Studies related to molecular aspects of chitinolytic enzymes are becoming indispensible for designing a more efficient chitinase producer and production of transgenic plants that can be used for the control of fungal and insect pathogens. Furthermore, biochemical and molecular studies could lead to a better understanding of the chitinase secretory process and the development of cloning strategies suitable for secretion of desired products.

7.4.5 INDUSTRIAL APPLICATIONS OF CHITINASES

During recent times, chitinases have gained interest in different biotechnological applications due to their ability to degrade chitin in the fungal cell wall and insect exoskeleton, leading to their use as antimicrobial or insecticidal agents (Karasuda et al., 2003; Mostafa et al., 2009). Another interesting application of chitinase is for bioconversion of chitin, a cheap biomaterial, into pharmacological active products, namely GlcNAc and chito-oligosaccharides (Bhattacharya et al., 2007). Production of chitin derivatives with suitable enzymes is more appropriate for sustaining the environment than using chemical reactions (Songsiriritthigul et al., 2009). Other interesting applications include the preparation of protoplasts from filamentous fungi, bio-control of insects and mosquitoes as well as the production of SCP (Dahiya et al., 2006; Hayes et al., 2006). Thus, there have been many reports on cloning, expression and characterization of chitinases from various organisms, including bacteria, fungi, plants and animals (Deshpande, 1986; Fukamizo, 2000). The major applications of chitinases are discussed and summarized in Figure 7.2.

7.4.5.1   Cytochemical Localization of Chitin/Chitosan Using Chitinase/Chitosanase Gold Complexes

Chitin and chitosan are the most ubiquitous polymers of fungal cell walls. Wheat germ agglutinin-gold complex and chitinase gold complex have been used as probes for the detection of GlcNAc residues in the secondary cell walls of plants and in pathogenic fungi (Benhamou and Asselin, 1989). Grenier et al. (1991) reported the tagging of a barley chitosanase with colloidal gold particles for the localisation of chitosan in spore and hyphal cell walls of fungi. This technique was used for the detection of chitosan in the cell walls of *Ophiostoma ulmi* and *Aspergillus niger*. Chitinase gold-labeled complexes have also been used for the immunocytochemical and cytochemical localization of chitin and GlcNAc residues in a biotrophic mycoparasite, *Piptocephalis virginiana* (Manocha and Zhonghua, 1997).

7.4.5.2   Production of Single-Cell Protein

The solid waste from shellfish processing is mainly composed of chitin, $CaCO_3$, and protein. Revah-Moiseev and Carrod (1981) suggested the use of shellfish

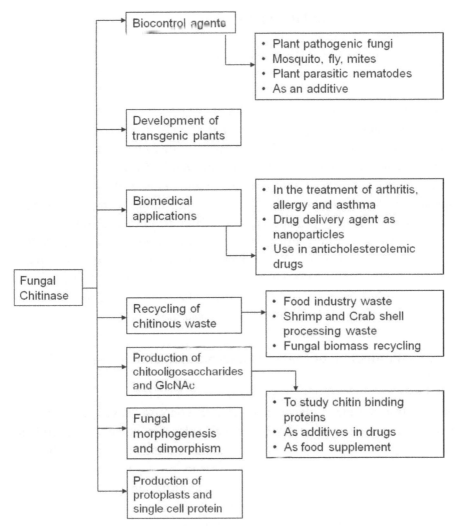

FIGURE 7.2 Applications of fungal chitinase: a summarized view.

waste for the bioconversion of chitin to yeast single-cell protein (SCP) using chitinolytic enzymes. They used the *Serratia marcescens* chitinase system to hydrolyze the chitin and *Pichia kudriavazevii* to yield SCP (with 45% protein and 8–11% nucleic acids). The commonly used fungi as the source of SCP are *Hansenula polymorpha*, *Candida tropicalis*, *Saccharomyces cerevisiae*, and *Myrothecium verrucaria* (Wang and Hwang, 2001). Vyas and Deshpande (1991) utilized the chitinolytic enzymes of *M. verrucaria* and *S. cerevisiae* for the production of SCP from chitinous waste.

7.4.5.3   Production of Protoplasts

Fungal protoplasts have been used as an effective experimental tool in studying cell wall synthesis, enzyme synthesis, and secretion, as well as in strain improvement for biotechnological applications. Since fungi have chitin in their cell walls, the chitinolytic enzyme seems to be essential along with other wall-degrading enzymes for protoplast formation from fungi. Dahiya et al. (2005) reported the effectiveness of *Enterobacter* sp. NRG4 chitinase in the generation of protoplasts from *Trichoderma reesei, Pleurotus florida, Agaricus bisporus* and *A. niger*. Mizuno et al. (1997) isolated protoplast from *Schizophyllum commune* using the culture filtrate of *Bacillus circulans* KA-304. An enzyme complex from *B. circulans* WL–12 with high chitinase activity was effective in generating protoplasts from *Phaffia rhodozyme* (Johnson et al., 1979). Gautam et al. (1996) reported protoplast formation from *Malbranchea sulfurea* using chitinase from *Paecilomyces variotii*. Similarly, Kitomoto (1988) showed the protoplast generation from various fungi using chitinase from *Trichoderma harzianum*.

7.4.5.4   Production of Chitooligosaccharides, Glucosamine and GLcNAc

Chitooligosaccharides, glucosamines, and GlcNAc have an immense pharmaceutical potential. Chitooligosaccharides are potentially useful in medicines for mankind. For example, chitohexose and chitoheptose showed antitumor activity. Chitosan and chito-oligosaccharides have also proved to be of immense usefulness in numerous applications in agriculture, cosmetics, water treatment and medicine (Aam et al., 2010; Chen et al., 2010). A chitinase from *Vibrio alginolyticus* was used to prepare chitopentaose and chitotriose from colloidal chitin (Murao et al., 1992). The chitobiose produced was subjected to chemical modifications to give novel disaccharide derivatives of 2-acetamido 2-deoxy D-allopyranose moieties that are potential intermediates for the synthesis of an enzyme inhibitor, that is, N, N'-diacetyl-β-chitobiosyl allosamizoline (Terayama et al., 1993). Specific combinations of chitinolytic enzymes would be necessary to obtain the desired chain length of the oligomer. For example, the production of chitooligosaccharides requires high levels of endochitinase and low levels of N-acetylglucosaminidase and exochitinase, whereas the production of GlcNAc requires higher proportion of exochitinase and N-acetylglucosaminidase (Aloise et al., 1996).

7.4.5.5 Detection of Fungal Biomass

A variety of methods have been described to quantify fungi in soil. The techniques include direct microscopic observation and extraction of fungus-specific indicator molecules such as glucosamine ergosterol. A strong correlation has been reported between chitinase activity and fungal population in soils. Such correlation was not found for bacteria and actinomycetes. Thus, chitinase activity appears to be a suitable indicator of actively growing fungi in soil. Miller et al. (1998) reported the correlation of chitinase activity with the content of fungus-specific indicator molecules 18:2ωb phospholipid fatty acid and ergosterol using specific methylumbelliferyl substrates. Similarly, chitinase and chitin-binding proteins can be used for the detection of fungal infections in humans (Laine and Lo, 1996).

7.4.5.6 Control of Mosquito

The worldwide socioeconomic aspects of diseases spread by mosquitoes made them potential targets for various pest control agents. In case of mosquitoes, entomopathogenic fungus such as *Beauveria bassiana* could not infect the eggs of *Aedes aegypti*, a vector of yellow fever and dengue, and other related species due to the aquatic environment. The scarabaeid eggs laid in the soil were found to be susceptible to *B. bassiana* (Ferron, 1985). *Myrothecium verrucaria*, a saprophytic fungus, produces a total complex of an insect cuticle-degrading enzyme (Shaikh and Desphande, 1993). It has been seen that both first and fourth instar larvae of mosquito *A. aegypti* can be killed within 48 h with the help of the crude preparation from *M. verrucaria* (Mendosa et al., 1996). Though 100% mortality was observed within 48 h, purified endochitinase lethal times (LT50) were 48 and 120 h for first and fourth instar larvae, respectively. However, the time period was found to be decreased, corresponding to 24 h and 48 h, when the purified chitinase was supplemented with lipolytic activity.

7.4.5.7 Fungal and Insect Morphogenesis

Chitinases play an important role in filamentous fungi, yeast and insect morphogenesis. They also play a major role in dimorphism, a phenomenon shown by a few filamentous fungi. Kuranda and Robbins (1991) reported the role of chitinases in cell separation during growth in *S. cerevisiae*. Similarly, functional expression of chitinase and chitosanase and their effects on morphogenesis in the yeast *Schizosaccharomyces pombe* were studied by Shimono et al. (2002). Chitinases of filamentous fungi are also required during cell separation, sporulation, spore germination, hyphal elongation and hyphal autolysis (Kuranda and Robbins, 1991; Dünkler et al., 2005).

7.4.5.8   Control of Phytopathogenic Fungi

A biological control agent of fungal root pathogen should exhibit a sufficient amount of antagonistic activity. The chitinase produced by *Enterobacter* sp. was highly active toward *Fusarium moniliforme*, *A. niger*, *Mucor rouxi* and *Rhizopus nigricans* (Dahiya et al., 2005). The chitinase from *Alcaligenes xylosoxydans* inhibited the growth of *Fusarium udum* and *Rhizoctonia bataticola* (Vaidya et al., 2001; Palma-Guerrero et al., 2008). Mahadevan and Crawford (1997) reported the antagonistic action of *Streptomyces lydicus* WXEC108 against *Pythium ultimum* and *Rhizoctonia solani*, which cause disease in cotton and pea respectively. In a first of its kind of study, antifungal chitinase is reported from *Basidiobolus ranarum* isolated from frog excrement which is normally known as insect pathogenic fungi and studied first time for chitinase production. Mycelia degradation ability of extra cellular chitinase from *Basidiobolus* was shown successfully against all the test fungal pathogens, such as, *Alternaria alternata*, *Fusarium solani*, *Rhizoctonia solani*, and *Aspergillus flavus*. The results obtained indicate possible use of chitinase from *B. ranarum* in the biocontrol of phytopathogenic fungi using chitin as a target molecule (Mishra et al., 2012). Chitinases can be added as a supplement to the commonly used fungicides and insecticides not only to make them more potent, but also to minimize the concentration of chemically synthesized active ingredients of the fungicides and insecticides that are otherwise harmful to the environment and human health.

7.4.5.9   Development of Disease Resistant Transgenic Plant Varieties

It has been an established fact that plant chitinases play a significant role in disease resistance (Schlumbaum et al., 1986; Karasuda et al., 2003; Cletus et al., 2013). Molecular plant pathologists world over have cloned and expressed number of genes or cDNAs responsible for the plant defense mechanism in a variety of plants (Van der Biezen, 2001; Chen et al., 2006; Ignacimuthu and Ceasar, 2012; Kovacs et al., 2013). A study showed the development of a successful model of transgenic tobacco plant which expressed a bean endochitinase gene. The developed transgenic tobacco plants exhibited nil or negligible susceptibility to infection by *Rhizoctonia solani* (Broglie et al., 1991). Similarly, an endochitinase from *Trichoderma harzianum* has been expressed in Broccoli to show increased resistance against *Alternaria* sp. (Mora and Earle, 2001). With the help of modern molecular tools, whole genomes sequencing of many model fungi (*Saccharomyces cerevisiae, Candida albicans, Coccidioides inmitis, Neurospora crassa, Gibberella zeae, Magnaporthe grisea, Aspergillus nidulans, A. fumigates* and *T. reesei*) has been completed thus contributed significantly the whole range and variety of fungal chitinases (Chuan, 2006). Such studies suggest that expression of antifungal genes could be used as an effective measure to safeguard the crops from fungal plant pathogens.

7.4.5.10   Antagonistic Agent Against Nematode Eggs

During the recent decades, a large chunk of agricultural crops has suffered serious economic losses caused by nematodes (Gortari and Hours, 2008). Mainly chemical compounds are used to control root-knot nematodes in the soil. This situation has further given rise to resistance to conventional nematicides and serious environmental concerns like imbalance in the eco-biodiversity, and contamination of the soil, raising the necessity to seek out for more efficient and environmental biocontrol strategies (Thamsborg et al., 1999; Akhtar and Malik, 2000). The nematode egg shell is composed of chitin-protein complex which is required for the structural strength of the eggs, and is susceptible to enzymatic degradation which has been the target molecule for biocontrol studies conducted so far (Spindler et al., 1990, Veronico et al., 2001; Khan et al., 2004; Morton et al., 2004; De Jin et al., 2005). *Pochonia chlamydosporia* and *Paecilomyces lilacinus* are examples of soil-borne fungi that have been studied as antagonistic agents against nematode eggs. However, *P. chlamydosporia* has also been described as a pathogen of insects (Stirling et al., 1998; Monfort et al., 2005). These results have paved the way for further exploration leading to the development of an effective biocontrol agent against plant parasitic nematodes.

7.4.5.11   Recycling of Chitinous Bio-Waste

Worldwide production of chitin has been estimated to be approximately 10^{11} tons per annum making it one of the most abundant natural compounds on earth (Kurita, 2006). Shrimps, crabs, lobsters, krill and squid wastes from the marine processing industry have become the major resource used today (Teng et al., 2001). During last two decades there has been a lot of focus on the production of GlcNAc by enzymatic hydrolysis of chitin (Songsiriritthigul et al., 2010). Tom and Carroad (1981) have described a process for the bio-conversion of shell-fish waste to GlcNAc and other value added products. In another such study, chitinolytic potential of *Basidiobolus haptosporus*, a member of saprophytic entomophthorales group, isolated from frog excrement, has been elaborated for the first time. During this study, fungal mycelia from different fungi such as *Aspergillus niger*, *Aspergillus flavus*, *Penicillium* sp., *Paecilomyces variotii*, and *Acremonium* sp. was used as an alternate carbon source for the chitinase production using *Basidiobolus haptosporus*. Significant hydrolysis of fungal mycelia was observed with mycelium of *Aspergillus niger*, after 72 h incubation. The comparative chitinase production was calculated against chitin-containing medium. The study provides presumptive information about one of the less explored fungal genera, namely *Basidiobolus*, which may have possible applications in the biocontrol of insect pests and phytopathogenic fungi and in the bioconversion of chitin to free N-acetyl glucosamine units (Mishra et al., 2011). Utilization of fungal mycelia for the chitinase production from fungi, actinomycetes, and bacteria has already been reported and there is plenty of scope for new and unconventional fungal isolates to be utilised for this very purpose (El-Katatny et al., 2000; Anitha and Rebeeth, 2009; Anitha and Rebeeth, 2010).

7.4.5.12   Energy Production: An Emerging Application

Insects are most dynamic widely distributed on earth, comprising 80% of all known species (Zhang et al., 2007). Robots with the ability to hunt, digest insects and obtain energy from them can serve various purposes by performing missions in dangerous situations. A robot that contains a microbial fuel cell was created to digest and metabolize chitin by microbial chitinase. This process produces electrons that act as horsepower of the system (Ieropoulos et al., 2004). The system can take advantage of the wide distribution of arthropods and mollusks because chitin is abundant in both phyla. Since Arthropoda and Mollusca rank first and second in species diversity in all animal phyla (Giribet et al., 2006) and are a major source of native chitin, this strategy can be highly applicable in both land and marine environments. The advantage of using chitin in this way is that most chitin sources are waste materials, such as shrimp shells; therefore, there is no need to be concerned about pressure on food supplies.

7.4.5.13   Biomedical Applications

Chitinases can be employed in human health care, such as making ophthalmic preparations with chitinases and microbiocides. A direct medical use has been suggested for chitinases in the therapy for fungal diseases in potentiating the activity of antifungal drugs (Pope and Davis, 1979; Orunsi and Trinci, 1985). They can also be used as potential additives in antifungal creams and ointments due to their topical applications. Although some of the studied chitinase like proteins (CLPs) lack enzymatic activity, most of them possess a chitin-binding domain, which is similar to chitinases. Though, very few studies have been conducted to test the chitin binding ability of CLPs from microbial in particular from fungal sources.

7.4.5.14   Drug Delivery Carriers

In order to develop an efficient drug delivery carrier it is important for it to possess the following qualities. It should be immediately and completely removed after delivering drugs, it should not be accumulative or toxic in nature. Recent studies have proved the applicability of some chitin derivatives such as carboxymethyl chitin (CMC), chitosan hydrogel, and hydroxyethyl chitin have been shown to possess such characteristics (Ishihara et al., 2006; Dev et al., 2010).

7.4.5.15   Targeting Chitinases in the Treatment of Allergy and Asthma

Chitinase has also been a target molecule in the development of allergy and asthma therapeutics. Chitinase inhibitors are relatively a different class of drugs which have been studied extensively in agricultural and food industry for many decades. Although the use of chitinase inhibitors as mammalian therapeutic agent is more recent, taking the lead from the studies implicating chitinases in infection and allergy. One natural inhibitor of chitinase enzymatic activity is allosamidin, reported from a strain of *Streptomyces* which is being developed as a therapeutic agent (Andersen, 2005). Allosamidin, and some other chitinase inhibitors from natural sources, were initially proposed as potential bio-pesticides, as they exhibited the inhibitory activity against of mites and housefly larva (Sakuda, 1987). Allosamidin again has been on the forefront for study as a potential treatment for allergy and asthma with promising initial studies which may eventually give rise to the development of novel chitinase inhibitors (Schuttelkopf et al., 2006).

7.5 CONCLUDING REMARKS AND FUTURE TRENDS

From the day of discovery till date, research on the second most abundant biopolymer, has been a major source of applications in various fields ranging from agriculture to therapeutics, as described by many researchers over the world. However, there is still scope for further research especially fundamental aspects on chitin and its modulating enzyme chitinase. The exploration of fungal diversity for novel molecule has been a topic of intense research in recent years. Globally, over the years emphasis has been on screening the rapidly growing and easily cultured species of fungi. Recent trends however have shifted towards realizing that the lesser investigated taxa belonging to diverse taxonomic groups which may also be a potential source of hitherto unrecognised bioactive potential; chitinase contributing the significant part of the bio prospecting studies of fungi which can play an important role in generating bio-economy by exploring the genetic and biochemical diversity captured in natural resources. It is the need of hour to review and updates our knowledge on fungal chitinase and their industrial applications.

For decades, fungi have been a valuable and preferred source of enzymes and secondary metabolites because of their ease of isolation, readily manipulative cultivation conditions, greater biomass production, and downstream processing. Fungal chitinases, in particular, strains of *Trichoderma* have been studied extensively to identify, isolate, clone and express a number of genes responsible for a variety of chitinases production especially in the development of transgenic plant varieties. Other than the biocontrol activities against the plant pathogenic fungi, insects-pests, plant parasitic nematodes, fungal chitinases have also received considerable attention to be used in the production of chitooligosaccharides including monomers of GlcNAc which further finds applications to be used as therapeutic agents in the treatment of osteoarthritis, allergy, asthma, and inflammatory bowel disease. Moreover, they have also been on the center stage of studies related to the localisation and detection of fungal biomass, production of single cell proteins and protoplasts, recycling of chitinous waste. In addition to chitinase, the more recent studies are focused on the chitin binding proteins and chitinase like proteins from fungi which have the potential to be the biomacromolecules for next generation therapeutics. Hence, one is expected to be ready for more scientific breakthroughs and pleasant surprises from the family of fungal chitinases as the studies progress in these directions and become more and more sophisticated.

KEYWORDS

- **Biocontrol**
- **Bioconversion**
- **Bioremediation**
- **Chitin**
- **Chitinase**
- **Fungi**
- **Industrial application**
- **N-acetyl-D-glucosamine**
- **Phytopathogens**

REFERENCES

Aam, B. B., Heggset, E. B., Norberg, A. L., Sorlie, M., Varum, K. M., Eijsink, V. G. Production of chitooligosaccharides and their potential applications in medicine. *Mar Drugs* 2010, 5, 1482–1517.

Adams, D. J. Fungal cell wall chitinases and glucanases. *Microbiology* 2004, 150, 2029–2035.

Akhtar, M., Malik, A. Roles of organic soil amendments and soil organisms in the biological control of plant-parasitic nematodes: a review. *Biores Technol* 2000, 74, 35–47.

Aloise, P. A., Lumme, M., Haynes, C. A. N-acetyl D-glucosamine production from chitin waste using chitinase from *Serratia marcescens*. In *Chitin enzymology*, Muzzarelli, R. A. A. (Ed.), European Chitin Society: Grottammare, 1996, pp. 581–594.

Andersen, O. A., Dixon, M. J., Eggleston, I. M., Van Aalten, D. M. Natural product family 18 chitinase inhibitors. *Nat Prod Rep* 2005, 22, 563–579.

Anitha, A., Rebeeth, M. Degradation of fungal cell walls of phytopathogenic fungi by lytic enzyme of *Streptomyces griseus*. *Afr J Plant Sci* 2010, 4(3), 61–66.

Anitha, A., Rebeeth, M. *In vitro* antifungal activity of *Streptomyces griseus* against phytopathogenic fungi of tomato field. *Acad J Plant Sci* 2009, 2(2), 119–123.

Arakane, Y., Muthukrishnan, S. Insect chitinase and chitinase-like proteins. *Cell Mol Life Sci* 2010, 67(2), 201–216.

Bartnicki, G. S. Cell wall chemistry, morphogenesis, and taxonomy of fungi. *Annu Rev Microbiol* 1968, 22, 87–108.

Benhamou, N., Asselin, A. Attempted localization of a substrate for chitinase in plant cells reveals abundant N-acetyl D-glucosamine residues in secondary walls. *Biol Cell* 1989, 67, 341–350.

Bhattacharya, D., Nagpure, A., Gupta, R. K. Bacterial chitinases: properties and potential. *Crit Rev Biotechnol* 2007, 27, 21–28.

Boot, R. G., Bussink, A. P., Verhoek, M., de Boer, P. A., Moorman, A. F., Aerts, J. M. Marked differences in tissue-specific expression of chitinases in mouse and man. *J Histochem Cytochem* 2005, 53, 1283–1292.

Boot, R. G., Renkema, G. H., Strijland, A., Van Zonneveld, A. J., Aerts, J. M. Identification of a novel acidic mammalian chitinase distinct from chitotriosidase. *J Biol Chem* 2001, 276, 6770–6778.

Broglie, K., Chet, I., Holliday, M., Cressman, R., Biddle, K. S., Mauvais, J., Broglie, R. M. Transgenic plants with enhanced resistance to the fungal pathogen, *Rhizoctonia solani*. *Science* 1991, 254, 1194–1197.

Bruck, H. M., Nash, G., Foley, F. O. Opportunistic fungal infection of the burn wound with *Phycomycetes*, *Aspergillus*. *Arch Surg* 1970, 102, 476–482.

Carsolio, C., Nicole, B., Shoshan, H., Carlos, C., Ana, G., Chet, I., Alfredo, H. E. Role of *Trichoderma harzianum* endochitinase gene, ech42, in mycoparasitism. *Curr Genet* 1999, 65(3), 929–935.

Chen, J. K., Shen, C. R., Liu, C. L. N-acetylglucosamine: production and applications. *Mar Drugs* 2010, 8(9), 2493–2516.

Chen, S. C., Liu, A. R., Zou, Z. R. Overexpression of glucanase gene and defensin gene in transgenic tomato enhances resistance to *Ralstonia solanacearum*. *Plant Sci* 2006, 5, 2134–2140.

Chuan, L. D. Review of fungal chitinases. *Mycopathologia* 2006, 161, 345–360.

Cletus, J., Balasubramanian, V., Vashisht, D., Sakthivel, N. Transgenic expression of plant chitinases to enhance disease resistance. *Biotechnol Lett* 2013, 35, 1719–1732.

Cosio, I. G., Fisher, R. A., Carroad, P. A. Bioconversion of shellfish chitin waste: waste pretreatment, enzyme production, process design, and economic analysis. *J Food Sci* 1982, 47, 901–905.

Cottrell, M. T., Moore, J. A., Kirchman, D. L. Chitinase from uncultured marine microorganisms. *Appl Environ Microbiol* 1999, 65, 2553–2557.

Dahiya, N., Tewari, R., Hoondal, G. S. Biotechnological aspects of chitinolytic enzymes: a review. *Appl Microbiol Biotechnol* 2006, 71, 773–782.

Dahiya, N., Tewari, R., Tiwari, R. P., Hoondal, G. S. Production of an antifungal chitinase from *Enterobacter* sp. NRG4 and its application in protoplast production. *World J Microbiol Biotechnol* 2005, 21, 1611–1616.

De Jin, R., Suh, J. W., Park, R. D., Kim, Y. W., Krishnan, H. B., Kim, K. Y. Effect of chitin compost and broth on biological control of *Meloidogyne incognita* on tomato (*Lycopersicon esculentum* Mill.). *Nematology* 2005, 7(1), 125–132.

Deane, E. E., Whipps, J. M., Lynch, J. M., Peberdy, J. F. Transformation of *Trichoderma reesei* with a constitutively expressed heterologous fungal chitinase gene. *Enzyme Microb Technol* 1999, 24, 419–424.

Deshpande, M. V. Enzymatic degradation of chitin and its biological applications. *J Sci Ind Res* 1986, 45, 273–281.

Dev, A., Mohan, J. C., Sreeja, V., Tamura, H., Patzke, G. R., Hussain, F., Weyeneth, S., Nair, S. V., Jayakumar, R. Novel carboxymethyl chitin nanoparticles for cancer drug delivery applications. *Carbohydr Polym* 2010, 79, 1073–1079.

Dünkler, A., Walther, A., Specht, C. A., Wendland, J. *Candida albicans* CHT3 encodes the functional homolog of the Cts1 chitinase of *Saccharomyces cerevisiae*. *Fungal Genet Biol* 2005, 42, 935–947.

Elad, Y., Chet, I., Katan, P. *Trichoderma harzianum*: a biocontrol agent effective against *Sclerotium rolfsii* and *Rhizoctonia solani*. *Phytopathology* 1980, 70, 119–121.

Elad, Y., David, D. R., Levi, T., Kapat, A., Kirshner, B., Gavrin, E., Levine, A. *Trichoderma harzianum* T39 mechanisms of biocontrol of foliar pathogens. In *Modern fungicides and antifungal compounds II*, Lyr, H. (Ed.), Intercept Ltd: Andover, Hampshire, UK, 1999, pp. 459–467.

Elad, Y., Hader, Y., Chet, I., Heni, Y. Prevention with *Trichoderma harzianum* Rifai aggr. of reinfestation by *Sclerotium rolfsii* Sacc. and *Rhizoctonia solani* Kühn of soil fumigated with methyl bromide, and improvement of disease control in tomatoes and peanuts. *Crop Prot* 1982, 1, 199–211.

El-Katatny, M. H., Somitsch, W., Robra, K. H., El-Katatny, M. S., Gübitz, G. M. Production of chitinase and β-1, 3-glucanase by *Trichoderma harzianum* for control of the phytopathogenic fungus *Sclerotium rolfsii*. *Food Technol Biotechnol* 2000, 38(3), 173–180.

Felse, P. A., Panda, T. Production of microbial chitinase: a revisit. *Bioproc Eng* 2000, 23, 127–134.

Ferron, P. Fungal control. In *Comprehensive Insect Biochemistry and Pharmacology, Vol. 12*, Kerkut, G. A., Gilbert, L. I. (Eds.), Pergamon: Oxford, 1985, pp. 313–346.

Flach, J., Pilet, P. E., Jolles, P. What's new in chitinase research? *Experientia* 1992, 48, 701–716.

Fukamizo, T. Chitinolytic enzymes: catalysis, substrate binding, and their application. *Curr Protein Pept Sci* 2000, 1(1), 105–124.

Gautam, S. P., Gupta, A. K., Shrivastava, R., Awasthi, M. Protoplast formation from the thermophilic fungus *Malbranchea sulfurea*, using the thermostable chitinase and lamarinase of *Paecilomyces variotii*. *World J Microb Biotechnol* 1996, 12, 99–100.

Giribet, G., Okusu, A., Lindgren, A. R., Huff, S. W., Schrodl, M., Nishiguchi, M. K. Evidence for a clade composed of molluscs with serially repeated structures: monoplacophorans are related to chitons. *Proc Natl Acad Sci* 2006, 103, 7723–7728.

Gooday, G. W., Humphreys, A. M., McIntosh, W. H. Roles of chitinases in fungal growth. In *Chitin in Nature and Technology*, Muzzarelli, R. A. A., Jeuniaux, C., Gooday, G. W. (Eds.), Plenum Press: New York, 1986, pp. 83–91.

Gortari, M. C., Hours, R. A. Fungal chitinases and their biological role in the antagonism onto nematode eggs: a review. *Mycol Progress* 2008, 7, 221–238.

Graham, L. S., Sticklen, M. B. Plant chitinases. *Can J Bot* 1994, 72, 1057–1083.

Grenier, J., Benhamou, N., Asselin, A. Colloidal gold-complexed chitosanase: a new probe for ultrastructural localization of chitosan in fungi. *J Gen Microbiol* 1991, 137, 2007–2015.

Hackman, R. H. Studies on chitin. I. Enzymic degradation of chitin and chitin esters. *Aust J Biol Sci* 1954, 7(2), 168–178.

Haran, S., Scitlckler, H., Chet, I. Molecular mechanisms of lytic enzymes involved in the biocontrol activity of *Trichoderma harzianum*. *Microbiology* 1996, 142, 2321–2331.

Hart, P. J., Pfluger, H. D., Monzingo, A. F., Hoihi, T., Robertus, J. D. The refined crystal structure of an endochitinsae from *Hordeum vulgare* L. seeds at 1.8Å resolution. *J Mol Biol* 1995, 248, 402–413.

Hartl, L., Zach, S., Seidl-Seiboth, V. Fungal chitinases: diversity, mechanistic properties and biotechnological potential. *Appl Microbiol Biotechnol* 2012, 93, 533–543.

Hawtin, R. E., Arnold, K., Ayres, M. D., Zanotto, P. M., Howard, S. C., Gooday, G. W., Chappell, L. H., Kitts, P. A., King, L. A., Possee, R. D. Identification and preliminary characterization of a chitinase gene in the *Autographa californica* nuclear polyhedrosis virus genome. *Virology* 1995, 212, 673–685.

Hayes, M., Ross, R. P., Fitzgerald, G. F., Hill, C., Stanton, C. Casein-derived antimicrobial peptides generated by *Lactobacillus acidophilus* DPC6026. *Appl Environ Microbiol* 2006, 72, 2260–2264.

He, H., Chen, X., Sun, C., Zhang, Y., Gao, P. Preparation and functional evaluation of oligo-peptide-enriched hydrolysate from shrimp (*Acetes chinensis*) treated with crude protease from *Bacillus* sp. SM98011. *Bioresour Technol* 2006, 97, 385–390.

Henrissat, B., Bairoch, A. New families in the classification of glycosyl hydrolases based on amino acid sequence similarities. *Biochem J* 1993, 293, 781–788.

Horowitz, S. T., Roseman, S., Blumenthal, H. J. Preparation of glucosamine oligosaccharides. 1. Separation. *J Am Chem Soc* 1957, 79, 5046–5049.

Ieropoulos, I., Melhuish, C., Greenman, J. Energetically autonomous robots. In *Intelligent Autonomous Systems*, Groen, F. (Ed.), IOS Press: Amsterdam, 2004, pp. 128–135.

Ignacimuthu, S., Ceasar, S. A. Development of transgenic fingermillet (*Eleusine coracana* (L.) Gaertn.) resistant to leaf blast disease. *J Biosci* 2012, 37(1), 135–147.

Imeri, A. G., Knorr, D. Effects of chitosan on yield and compositional data of carrot and apple juice. *J Food Sci* 1998, 53, 1707–1709.

Ishihara, M., Obara, K., Nakamura, S., Fujita, M., Masuoka, K., Kanatani, Y., Takase, B., Hattori, H., Morimoto, Y., Ishihara, M., Maehara, T., Kikuchi, M. Chitosan hydrogel as a drug delivery carrier to control angiogenesis. *J Artif Organs* 2006, 9, 8–16.

Jeuniaux, C. *Comprehensive Biochemistry*. Elsevier Publishing Co: Amsterdam, 1971, pp. 595.

Johnson, E. A., Villa, T. G., Lewis, M. J., Phaff, H. J. Lysis of cell wall of yeast *Phaffia rhodozyme* by a lytic complex from *Bacillus circulans* WL–12. *J Appl Biochem* 1979, 1, 272–282.

Karasuda, S., Tanaka, S., Kajihara, H., Yamamoto, Y., Koga, D. Plant chitinase as a possible biocontrol agent for use instead of chemical fungicides. *Biosci Biotechnol Biochem* 2003, 67, 221–224.

Kasprzewska, A. Plant chitinases - regulation and function. *Cell Mol Biol Lett* 2003, 8, 809–824.

Khan, A., Williams, K. L., Nevalainen, H. K. M. Effects of *Paecilomyces lilacinus* protease and chitinase on the eggshell structures and hatching of *Meloidogyne javanica* juveniles. *Biol Cont* 2004, 31, 346–352.

Khoushab, F., Yamabhai, M. Chitin research revisited. *Mar Drugs* 2010, 8, 1988–2012.

Kim, H. B., Chung, S. A. Differential expression patterns of an acidic chitinase and a basic chitinase in the root nodule of *Elaeagnus umbellate*. *Mol Plant Microbe Interact* 2002, 15, 209–215.

Kitomoto, Y., Mori, N., Yamamoto, M., Ohiwa, T., Ichiwaka, Y. A simple method of protoplast formation and protoplast regeneration from various fungi using an enzyme from *Trichoderma harzianum*. *Appl Microbiol Biotechnol* 1988, 28, 445–450.

Koga, D., Hoshika, H., Matsushita, M., Tanaka, A., Ide, A., Kono, M. Purification and characterization of β-N-acetylhexosaminidase from the liver of a prawn, *Penaeus japonicus*. *Biosci Biotech Bioch* 1996, 60(2), 194–199.

Kovacs, G., Sagi, L., Jacon, G., Arinaitwe, G., Busogoro, J. P., Thiry, E., Strosse. H., Swennen, R., Remy, S. Expression of a rice chitinase gene in transgenic banana ('Gros Michel,' AAA genome group) confers resistance to black leaf streak disease. *Transgen Res* 2013, 22, 117–130.

Kramer, K. J., Muthukrishnan, S. Insect chitinases: molecular biology and potential use as biopesticides. *Insect Biochem Mol* 1997, 27, 887–900.

Kreger, D. R. Observations on cell walls of yeasts and some other fungi by X-ray diffraction and solubility tests. *Biochim Biophys Acta* 1954, 13, 1–9.

Kunz, C., Sellam, O., Bertheau, Y. Purification and characterization of a chitinase from the hyperparasitic fungus *Aphanocladium album*. *Physiol Mol Plant Pathol* 1992, 40, 117–131.

Kuranda, M. J., Robbins, P. W. Chitinase is required for cell separation during growth of *Saccharomyces cerevisiae*. *J Biol Chem* 1991, 266, 19758–19767.

Kurita, K. Chitin and chitosan: functional biopolymers from marine crustaceans. *Mar Biotechnol* 2006, 8, 203–226.

Laine, L. A., Lo, C. J. Diagnosis of fungal infections with a chitinase. PCT Int. Appl. Wo 9802742 A122 (CA 128, 86184), 1996.

Lefebvre, S., Maury, M., Gazzalo, M. Nouvelles colles vegetales: Origine, proprietes et performances. *Revue Française D'oenologie* 2000, 184, 28–32.

Leger, R. J., Joshi, L., Roberts, D. Ambient pH is a major determinant in the expression of cuticle-degrading enzymes and hydrophobin by *Metarhizium anisopliae*. *Appl Environ Microbiol* 1998, 64(2), 709–713.

Lorito, M., Sheridan, I., Gary, E. H., Sposato, P., Muccifora, S., Scala, F. Genes encoding for chitinolytic enzymes from biocontrol fungi: applications for plant disease control. In *Chitin Enzymology*, Muzzarelli, R. A. A., Ed., Atec Edizioni: Italia, 1996, pp. 95–101.

Mahadevan, B., Crawford, D. L. Properties of the antifungal biocontrol agent *Streptomyces lydicus* WYFC 108. *Enzyme Microb Technol* 1997, 20, 489–493.

Mano, J. F., Silva, G. A., Azevedo, H. S., Malafaya, P. B., Sousa, R. A., Silva, S. S., Boesel, L. F., Oliveira, J. M., Santos, T. C., Marques, A. P., Neves, N. M., Reis, R. L. Natural origin biodegradable systems in tissue engineering and regenerative medicine: present status and some moving trends. *J R Soc Interface* 2007, 4, 999–1030.

Manocha, M. S., Zhonghua, Z. Immunochemical and cytochemical localization of chitinase and chitin in infected hosts of a biotropic mycoparasite *Piptocephalis virginiana*. *Mycologia* 1997, 86, 185–194.

Mathivanan, N., Kabilan, V., Murugesan, K. Purification, characterization and anti-fungal activity from *Fusarium chlamydosporum*, a mycoparasite to groundnut rust, *Puccinia arachidis*. *Can J Microbiol* 1998, 44, 646–651.

Matsumoto, K. S. Fungal chitinases. In *Advances in Agricultural and Food Biotechnology*, Guevara-González, R. G., Torres-Pacheco, I. (Eds.), Research Signpost: Trivandrum, India, 2006, pp. 289–304.

Mendosa, E. S., Vartak, P. H., Rao, J. U., Deshpande, M. V. An enzyme from *Myrothecium verrucaria* that degrades insect cuticle for biocontrol of *Aedes aegypti* mosquito. *Biotechnol Lett* 1996, 18, 373–376.

Merzendorfer, H., Zimoch, L. Chitin metabolism in insects: structure, function and regulation of chitin synthases and chitinases. *J Exp Biol* 2003, 206, 4393–4412.

Miller, M., Palofarvi, A., Rangger, A., Reeslev, M., Kjoller, A. The use of fluorogenic substrates to measure fungal presence and activity in soil. *Appl Environ Microbiol* 1998, 64, 613–617.

Minke, R., Blackwell, J. The structure of [alpha]-chitin. *J Mol Biol* 1978, 120, 167–181.

Mishra, P., Kshirsagar, P., Nilegaonkar, S. S., Singh, S. K. Statistical optimization of medium components for production of extracellular chitinase by *Basidiobolus ranarum*: a novel biocontrol agent against plant pathogenic fungi. *J Basic Microbiol* 2012, 52(5), 539–548.

Mishra, P., Singh, S. K., Nilegaonkar, S. S. Extracellular chitinase production by some members of the saprophytic Entomophthorales group. *Mycoscience* 2011, 52, 271–277.

Mizuno, K., Kimura, O., Tachiki, T. Protoplast formation from *Schizophyllum commune* by a culture filtrate of *Bacillus circulans* KA-304 grown on a cell-wall preparation of *S. commune* as a carbon source. *Biosci Biotechnol Biochem* 1997, 61, 852–857.

Monfort, E., López Llorca, L. V., Jansson, H. B., Salinas, J., Park, J. O., Sivasithamparan, K. Colonization of seminal roots of wheat and barley by egg-parasitic nematophagous fungi and their effects on *Gaeumannomyces graminis* var. *tritici* and development of root-rot. *Soil Biol Biochem* 2005, 37, 1229–1235.

Mora, A., Earle, E. D. Combination of *Trichoderma harzianum* endochitinase and a membrane-affecting fungicide on control of *Alternaria* leaf spot in transgenic broccoli plants. *Appl Microbiol Biotechnol* 2001, 55, 306–310.

Morton, C. O., Hirsch, P. R., Kerry, B. R. Infection of plant-parasitic nematodes by nematophagous fungi – a review of the application of molecular biology to understand infection processes and to improve biological control. *Nematology* 2004, 6(2), 161–170.

Mostafa, S. A., Mahmoud, M. S., Mohamed, Z. K., Enan, M. R. Cloning and molecular characterization of chitinase from *Bacillus licheniformis* MS-3. *J Gen Appl Microbiol* 2009, 55, 241–246.

Murao, S., Kuwada, T., Itoh, H., Oyama, H., Shin, T. Purification and characterization of a novel type of chitinase from *Vibrio alginolyticus* TK-22. *Biosci Biotechnol Biochem* 1992, 56, 368–369.

Muzzarelli, R. A. A. Chitins and chitosans as immunoadjuvants and non-allergenic drug carriers. *Mar Drugs* 2010, 8, 292–312.

Nakatsuka, S., Andrady, A. C. Permeability of vitamin B12 in chitosan membranes, effect of crosslinking and blending with poly (vinyl alcohol) on permeability. *J Appl Polym Sci* 1992, 44, 17–28.

Ohtakara, A., Mitsutomi, M. Analysis of chitooligosaccharides and reduced chitooligosaccharides by high-performance liquid chromatography. *Methods Enzymol* 1988, 161, 453–457.

Onsoyen, E., Skaugrud, O. Metal recovery using chitosan. *J Chem Technol Biotechnol* 1990, 49(4), 395–404.

Orunsi, N. A., Trinci, A. P. J. Growth of bacteria on chitin, fungal cell walls and fungal biomass, and the effect of extracellular enzymes produced by these cultures on the antifungal activity of amphotericin B. *Microbiol* 1985, 43, 17–30.

Palma-Guerrero, J., Jansson, H. B., Salinas, J., Lopez-Llorca, L. V. Effect of chitosan on hyphal growth and spore germination of plant pathogenic and biocontrol fungi. *J Appl Microbiol* 2008, 104(2), 541–553.

Panos, I., Acosta, N., Heras, A. New drug delivery systems based on chitosan. *Curr Drug Discov Technol* 2008, 5, 333–341.

Park, J. K., Morita, K., Fukumoto, I., Yamasaki, Y., Nakagawa, T., Kamamukai, M., Matsuda, H. Purification and characterization of the chitinase (ChiA) from *Enterobacter* sp. G–1. *Biosci Biotechnol Biochem* 1997, 61, 684–689.

Patil, S. R., Ghormade, V., Deshpande, M. V. Chitinolytic enzymes: an exploration. *Enzyme Microb Technol* 2000, 26, 473–483.

Peberdy, J. F. Mycolytic enzymes. In *Fungal protoplasts - applications in biochemistry and genetics*, Peberdy, J. F., Ferenczy, L., Ed., Marcel Dekker: New York, 1985, pp. 31–44.

Pinto, A. S., Barreto, C. C., Schrank, A., Ulhao, C. J., Vainstein, M. H. Purification and characterization of an extracellular chitinase from the entomopathogen *Metarhizium anisopliae. Can J Microbiol* 1997, 43, 322–327.

Pope, A. M. S., Davis, D. A. L. The influence of carbohydrates on the growth of fungal pathogens *in vitro* and *in vivo. Postgrad Med J* 1979, 55, 674–676.

Qurashi, M. T., Blair, H. S., Allen, S. Studies on modified chitosan membranes. II. Dialysis of low molecular weight metabolites. *J Appl Polym Sci* 1992, 46, 263–269.

Renkema, G. H., Boot, R. G., Muijsers, A. O., Donker-Koopman, W. E., Aerts, J. M. F. G. Purification and characterization of human chitotriosidase, a novel member of the chitinase family of proteins. *J Biol Chem* 1995, 270, 2198–2202.

Revah-Moiseev, S., Carrod, P. A. Conversion of the enzymatic hydrolysate of shellfish waste chitin to single cell protein. *Biotechnol Bioeng* 1981, 23, 1067–1078.

Rinaudo, M. Chitin and chitosan: properties and applications. *Prog Polym Sci* 2006, 31, 603–632.

Rojas-Avelizapa, L. I., Cruz-Camarillo, R., Guerrero, M. I., Rodriguez-Vazquez, R., Ibarra, J. E. Selection and characterization of a proteo-chitinolytic strain of *Bacillus thuringiensis*, able to grow in shrimp waste media. *World J Microbiol Biotechnol* 1999, 15, 299–308.

Sahai, A. S., Manocha, M. S. Chitinases of fungi and plants: their involvement in morphogenesis and host parasite interaction. *FEMS Microbiol Rev* 1993, 11, 317–338.

Sakai, T., Koo, K., Saitoh, K. Use of the protoplast fusion for the development of rapid starch-fermenting strains of *Saccharomyces diastaticus. Agric Biol Chem* 1986, 50(2), 297–306.

Sakuda, S., Isogai, A., Matsumoto, S., Suzuki, A. Search for microbial insect growth regulators II. Allosamidin, a novel insect chitinase inhibitor. *J Antibiot* 1987, 40, 296–300.

Sashiwa, H., Aiba, S. Chemically modified chitin and chitosan as biomaterials. *Prog Polym Sci* 2004, 29, 887–908.

Schlumbaum, A., Mauch, F., Vogeli, U., Boller, T. Plant chitinases are potent inhibitors of fungal growth. *Nature* 1986, 324, 365–367.

Schuttelkopf, A. W., Andersen, O. A., Rao, F. V. Screening-based discovery and structural dissection of a novel family 18 chitinase inhibitor. *J Biol Chem* 2006, 281, 27278–27285.

Shaikh, S. A., Deshpande, M. V. Chitinolytic enzymes: their contribution to basic and applied research. *World J Microbiol Biotechnol* 1993, 9, 468–475.

Shimono, K., Matsuda, H., Kawamukai, M. Functional expression of chitinase and chitosanase, and their effects on morphologies in the yeast *Schizosaccharomyces pombe*. *Biosci Biotechnol Biochem* 2002, 66, 1143–1147.

Songsiriritthigul, C., Pesatcha, P., Eijsink, V. G., Yamabhai, M. Directed evolution of a *Bacillus* chitinase. *Biotechnol J* 2009, 4, 501–509.

Songsiriritthigul, C., Lapboonrueng, S., Pechsrichuang, P., Pesatcha, P., Yamabhai, M. Expression and characterization of *Bacillus licheniformis* chitinase (ChiA), suitable for bioconversion of chitin waste. *Bioresour Technol* 2010, 101, 4096–4103.

Spindler, K. D., Splinder-Barth, M., Londershausen, M. Chitin metabolism: a target for drugs against parasites. *Parasitol Res* 1990, 76, 283–288.

Stirling, G. R., Licastro, K. A., West, L. M., Smith, L. J. Development of commercially acceptable formulations of the nematophagous fungus *Verticillium chlamydosporium*. *Biol Control* 1998, 11, 217–223.

Suzuki, K., Taiyoji, M., Sugawara, N., Nikaidou, N., Henrissat, B., Watanabe, T. The third chitinase gene (chiC) of *Serratia marcescens* 2170 and the relationship of its product to other bacterial chitinases. *Biochem J* 1999, 343, 587–596.

Svitil, A. L., Chadhain, S. M., Moore, J. A., Kirchman, D. L. Chitin degradation proteins produced by the marine bacterium *Vibrio harveyi* growing on different forms of chitin. *Appl Environ Microbiol* 1997, 63, 408–413.

Tantimavanich, S., Pantuwatana, S., Bhumiratana, S., Panbangred, W. Cloning of a chitinase gene in to *Bacillus thuringiensis* spp. *aizawai* for enhanced insecticidal activity. *J Gen Appl Microbiol* 1997, 43, 341–347.

Teng, J., Whistler, R. L. Cellulose and chitin. In *Phytochemistry*, Miller, L. P., Ed., Van Nostrand Reinhold: New York, 1973, pp. 249–269.

Teng, W. L., Khor, E., Tan, T. K., Lim, L. Y., Tan, S. C. Concurrent production of chitin from shrimp shells and fungi. *Carbohydr Res* 2001, 332, 305–316.

Terayama, H., Takahashi, S., Kuzuhara, H. Large scale preparation on N, N'-diacetylchitobiose by enzymatic degradation of chitin and its chemical modification. *J Carbohydr Chem* 1993, 12, 81–93.

Thamsborg, S. M., Roepstorff, A., Larsen, M. Integrated and biological control of parasites in organic and conventional production systems. *Vet Parasitol* 1999, 84, 169–186.

Tian, J., Yu, J., Sun, X. Chitosan microspheres as candidate plasmid vaccine carrier for oral immunization of Japanese flounder (*Paralichthys olivaceus*). *Vet Immunol Immunopathol* 2008, 126, 220–229.

Tom, R. A., Carroad, P. A. Effect of reaction conditions on hydrolysis of chitin by *Serratia marcescens* QM B 1466 chitinase. *J Food Sci* 1981, 46, 646–647.

Tsujibo, H., Orikoshi, H., Tanno, H., Fujimoto, K., Miyamoto, K., Imada, C., Okami, Y., Inamori, Y. Cloning, sequence, and expression of a chitinase gene from a marine bacterium, *Alteromonas* sp. strain O-7. *J Bacteriol* 1993, 175, 176–181.

Uragami, T., Kurita, K., Fukamizo, T. Chitin and Chitosan in Life Science. Kodansha Scientific Ltd: Tokyo, Japan, 2001, pp. 51–54.

Uragami, T., Takuno, M., Miyata, T. Evapomeation characteristics of cross-linked quaternized chitosan membranes for the separation of an ethanol/water azeotrope. *Macromol Chem Phys* 2002, 203, 1162–1170.

Vaidya, R. J., Shah, I. M., Vyas, P. R., Chhatpar, H. S. Production of chitinase and its optimization from a novel isolate *Alcaligenes xylosoxydans*: potential antifungal biocontrol. *World J Microbiol Biotechnol* 2001, 1, 62–69.

Van Aalten, D. M. F., Synstad, B., Brurberg, M. B., Hough, E., Riise, B. W., Eijsink, V. G. H., Wierenga, R. K. Structure of a two-domain chitotriosidase from *Serratia marcescens* at 1.9-angstrom resolution. *Proc Natl Acad Sci* 2000, 97, 5842–5847.

Van der Biezen, E. A. Quest for antimicrobial genes to engineer disease-resistant crops. *Trends Plant Sci* 2001, 6, 89–91.

Van Eijk, M., van Roomen, C. P., Renkema, G. H., Characterization of human phagocyte-derived chitotriosidase, a component of innate immunity. *Int Immunol* 2005, 17(11), 1505–1512.

Veronico, P., Gray, L. J., Jones, J. T., Bazzicalupo, P., Arbucci, S., Cortese, M. R., Di Vito M., De Giorgi C. Nematode chitin synthases: gene structure, expression and function in *Caenorhabditis elegans* and the plant parasitic nematode *Meloidogyne artiellia*. *Mol Genet Genomics* 2001, 266, 28–34.

Vyas, P. R., Deshpande, M. V. Enzymatic hydrolysis of chitin by *Myrothecium verrucaria* chitinase complex and its utilization to produce SCP. *J Gen Appl Microbiol* 1991, 37, 267–275.

Vyas, S. P., Paliwal, R., Paliwal, S. R. Chitosan/Chitosan derivatives as carriers and immunoadjuvants in vaccine delivery. In *Chitin, Chitosan, Oligosaccharides and Their Derivatives, Biological Activities and Applications*, Kim, S. –K., Ed., Taylor and Francis: New York, 2011, pp. 339–356.

Wang, S. L., Chio, S. H. Deproteinization of shrimp and crabshell with the protease of *Pseudomonas aeruginosa* K–187. *Enzyme Microb Technol* 1998, 22, 629–633.

Wang, S., Hwang, J. Microbial reclamation of shellfish wastes for the production of chitinases. *Enzyme Microb Technol* 2001, 28, 376–82.

Wiwat, C., Lertcanawanichakul, M., Siwayapram, P., Pantu-watana, S., Bhumiratana, A. Expression of chitinase encoding genes from *Aeromonas hydrophila* and *Pseudomonas maftophila* in *Bacillus thuringiensis* subsp. *israefensis*. *Gene* 1996, 179, 119–126.

Yedidia, I., Benhamou, N., Chet, I. Induction of defense responses in cucumber plants (*Cucumis sativus* L.) by the biocontrol agent *Trichoderma harzianum*. *Appl Environ Microbiol* 1999, 65(3), 1061–1070.

Zhang, F. Y., Feng, B., Li, W. Induction of tobacco genes in response to oligochitosan. *Mol Biol Rep* 2007, 34(1), 35–40.

CHAPTER 8

PROTEASES FROM THERMOPHILES AND THEIR INDUSTRIAL IMPORTANCE

D. R. MAJUMDER[1] and PRADNYA P. KANEKAR[2]

[1]*Department of Microbiology, Abeda Inamdar Senior College, 2390-KB Hidayatullah Road, Azam Campus, Camp, Pune 411001, Maharashtra, India*

[2]*Department of Biotechnology, Modern College, Pune 411005, Maharashtra, India*

CONTENTS

8.1   INTRODUCTION

The term 'thermophile' refers to any microbe capable of growth at 55°C. The rapidity of growth and senescence in thermophilic cultures at elevated temperatures made it desirable to use shorter time intervals for incubation than those recommended for the mesophilic organisms. A mechanism characterized by entropic stabilization could be responsible for the higher thermostability of the thermophilic enzymes. The cellular proteins of thermophiles are of immense research interest in basic and applied areas like evolution, industrial, medical, biotechnology and molecular biology. A number of enzymes and bioactive molecules from several thermophilic microorganisms have been identified in recent years of which thermophilic proteases have the largest market share. Proteases have ecofriendly application with a profound market value. Therefore, concerted effort is focused on isolating and identifying new protease producing microorganism to meet the market (industry) demand.

8.2   THE HISTORY OF EVOLUTION OF THERMOPHILES

Microorganisms constitute a very unique group of living organisms, which appeared on the Earth's surface almost 3000 million years ago. The study of the origin and evolution of microorganisms are extremely complex. Primitive organisms capable of metabolism (ability to accumulate and modify nutrients and energy) and reproduction first appeared ~ 3.6 Bya. They were most likely thermophilic anaerobes, and may have depended on RNA for both enzymatic and genetic activities. Oxygenic photosynthesizers evolved ~1.2 Bya after the first organisms. Aerobic thermophiles evolved after adequate oxygen concentration on earth was available around ~ 0.5 Bya. Biogenesis and emergence of the Universal Ancestor common to all later forms of life known to currently exist on earth was established. This is easily understood from the rRNA phylogenetic tree. Thus evolutionary studies have shown that the last universal common ancestor was most likely a thermophilic organism living in a hydrothermal ecosystem (Reysenbach et al., 2001; Lineweaver et al., 2004). Plate tectonics play an important role in astrobiology. It is involved in the evolution of global planetary climate and life. The recent focus has been on the present day Earth where plate tectonics is already occurring (Solomatov, 2004). Natural geothermal habitats have a worldwide distribution and are primarily associated with tectonically active zones, where major movements of the earth's crust occur (Brock, 1986). There is strong molecular and physiological evidence from present day microorganisms that the 'Universal ancestor' was capable of growing at high temperatures. Thus, the importance of thermophiles in the evolutionary process is well established.

8.3 THE HABITATS OF THERMOPHILES

Thermophilic microorganisms are ubiquitous. The temperature conditions were obviously favorable in hot springs for the growth of thermophiles. Therefore, the thermophilic bacteria, since early nineteenth century was known as a normal flora of hot springs. Gradually, they became a subject of a lively field of investigation, and many of the fundamental facts of their existence were discovered. Presence of thermophilic spore forming bacteria is found in meteoritic crater too (Patke et al., 1998). The thermophilic spore forming bacteria are abundant in various geothermally heated region of the earth such as hot springs and deep sea hydrothermal vents. But true to the old belief, the temperature conditions are obviously more favorable in hot springs and their sediments (Allen, 1953; Majumder and Kanekar, 2012; Majumder et al., 2013). Sampling from natural lake, soil, etc. and extreme environment like hot springs, saline lakes, meteoritic craters, acid soils, etc., was conducted for isolation of thermophilic spore forming bacteria (Patke et al., 1998; Chitte et al., 1999; Chitte et al., 2000; Yallop et al., 1997; Rath and Subramanyam, 1998). Thermotolerant enzyme activities of thermophiles isolated from hot springs have immense biotechnological significance (Rath and Subramanyam, 1996).

The thermophilic bacteria were isolated from soil, mud, water, air, desert sand, ocean bottom, sea water, sewage, animal and human faces, cultivated and garden soils, masses of decaying plant materials, moldy fodders and other vegetable matter, including cotton, straw, cereal grain, hay, composts, manures, mushroom compost, heated sugarcane bagasse, soil peat, sewage, dung and grass compost, deep sea hydrothermal vent, marine solfatare, decomposed plant samples from a lake, compost of fermenting citrus peels, coffee and tea extract residues, Korean salt, fermented anchovy, garbage dump, compost treated with artichoke juice, Lonar lake silt and water sample, deep sea, thermophilic digester, grassy marshland, Thai fish sauce, hot acidic soil, waste activated sludge, kefir, sewage sludge (Bergey, 1919; Allen, 1953; Goodfellow et al., 1984; Lacey, 1989).

A great majority of bacteria isolated at high temperature were proved to be aerobic spore forming bacteria. The term 'thermophilic bacteria' refers to any microbe capable of growth at 55°C (Bergey, 1919; Waksman, 1959). Thermophilic actinomycetes are most abundant in moldy fodders and other vegetable matter including cotton, straw and cereal grain, hay, composts, manures, mushroom compost, heated sugarcane bagasse, soil peat, sewage, dung and grass compost Bulgarian soil, forest soil, poultry compost (Goodfellow et al., 1984; Lacey, 1989; De Azeredo et al., 2006; Petrova, 2006; Lazim et al., 2009; Chitte et al., 2011; Habbeche et al., 2014).

Thermophilic fungi are isolated from organic compost, industrial wastepile, dust, Legume field soil, compost, municipal waste, coal spoil tips, air surrounding coal spoil tips, Compost, Hot Spring of Barquizin valley in North Baikal (Bazarzhapov et al., 2006; Macchione et al., 2008; O'Donoghue et al., 2008; Li et al., 2009; Zanphorlin et al., 2010; Li et al., 2011; Liao et al., 2012; Silva et al., 2014).

8.4   THE PHYSIOLOGY OF THERMOPHILES

The rapidity of growth and senescence in thermophilic cultures at elevated temperatures made it desirable to use shorter time intervals for incubation than those recommended for the mesophilic spore forming bacteria. Vegetative cells are formed from spores within 8–10 hours post incubation at elevated temperature. The principal factor is the extraordinary requirement of oxygen by the thermophiles. The solubility of oxygen in water at 55°C is one half its values at 30°C. Therefore, this is a big hurdle to overcome for the growth of thermophiles in laboratories. The limitations of growth are due to exhaustion of the dissolved oxygen. This was proved by an experiment designed by Hansen (Allen, 1953). The most interesting problem posed by the thermophilic microorganism is a physiological one. It is fascinating to know how these microbes are able to live and grow at high temperatures at which many proteins are coagulated and thus the existence of life appears as a biochemical anomaly. The obvious answer to this anomaly would be that thermopilic bacteria, living at high temperature may gain more energy metabolizing available nutrients than life forms that live in more temperate environments (Allen, 1953).

Currently, a number of publications have extensively discussed developments in these areas of production of heat stable enzymes from thermophiles

(Neihaus et al., 1999) and structure and function relationships of thermozymes. The external environment of the thermophilic organisms, their structural features including structural modifications at molecular level and dynamic effects of metabolism in preventing or repairing thermal damage, leads to the thermal resistance of the organisms. Thermophiles are reported to contain specialized proteins known as 'Chaperonins,' which are thermostable and resist denaturation by refolding the proteins to their native forms and restoring their functions (Kumar et al., 2001). Thus thermophiles can successfully resist proteolysis at high temperature. The cell membrane of thermophile is made up of saturated fatty acids, which provide a hydrophobic environment for the cell and keep the cell rigid enough to live at elevated temperatures. The DNA of thermophiles contains a reverse DNA 'gyrase,' which produces positive super coils in the DNA (Lopez, 1999). This raises the melting point of the DNA (the temperature at which the strands of the double helix separate) to at least as high as the 'organisms' maximum temperature for growth. Thermophiles also tolerate high temperature by using increased interactions than non-thermotolerant organisms use, viz. electrostatic, disulphide bridge and hydrophobic interaction. While determining the conformational entropy of both enzymes, the folded state showed a higher structural flexibility for the thermophilic protein than the mesophilic homologue. Thus, it has been assumed that a mechanism characterized by entropic stabilization could be responsible for the higher thermostability of the thermophilic enzymes.

8.5 PROTEASES FROM THERMOPHILES AND THEIR BIOTECHNOLOGICAL POTENTIAL

Several species of thermophilic microorganisms produce thermostable proteases (Table 8.1) which are of immense research interest in basic and applied areas like evolution, industrial, medical, biotechnology and molecular biology (Zierenberg et al., 2000). Thermophiles were known to mankind as far back as the last century. But the search for them has intensified recently, as scientists have recognized that 'survival kits' possessed by the thermophiles can potentially serve an array of applications (Peek et al., 1992). A number of proteases from several thermophilic microorganisms have been identified in recent years. Cost savings, associated with longer storage stability and higher activity at higher temperatures of thermophilic enzymes, provide good reason to select and develop thermophilic enzymes

TABLE 8.1   Thermophilic Organisms Reported to Produce Thermostable Protease

Microorganisms		References
Bacteria	*Thermus* sp.	Rakshit et al. (2003)
	Bacillus sp. W.N. 11	
	Staphilothermus marimus	
	Thermococcus litoralis	
	Clostridium absonum CFR-702	
	Bacillus thermoleovocans ID–1	
	Bacillus strain MH–1	
	Bacillus stearothermophilus CH-4	
	Bacillus sp. KYJ963	
	Pyrococcus abyssi	
	Bacillus sp. 3183	
	Bacillus circulans	
	Bacillus sp.	
	Bacillus thermoproteolyticus	
	Bacillus sp. JB-99	
	Bacillus brevis	
	Bacillus licheniformis	
	Bacillus sp. No. AH–101	
	Bacillus thermoruber	
	Pyrococcus sp. KODI	
	Staphylothermus marninus	
	Thermoacidophiles (archeal and bacterial origin)	
	Thermococcus aggreganes	
	Thermococcus celer	
	Thermococcus litoralis	
	Thermococcus maritema	
	Pyrococcus horikoshii	Zhan et al. (2014)
	Coprothermobacter proteolyticus	Majeed et al. (2013)
	Bacillus sp. WF146	Zhu et al. (2013)
	Bacillus sp. (phylum of *Firmicutes*)	Jang et al. (2013)
	Coprothermobacter proteolytics	Toplak et al. (2013)
	Meiothermus ruber H328	Kataoka et al. (2013)

TABLE 8.1 Continued

Microorganisms	References
Keratinibaculum paraultunense	Huang et al. (2013)
Bacillus amyloliquefaciens	Panda et al. (2013)
Bacillus cereus	Saleem et al. (2012)
Bacillus thermophilic bacteria AT07–1	Tang et al. (2012)
Geobacillius thermodenitrificans	Charbonneau et al. (2012)
Bacillus caldolyticus	Bader et al. (2012)
Bacillus stearothermophilus F1	Rahman et al. (2012)
Thermococcus onnurineus NA1	Yun et al. (2011)
Thermus aquaticus	Arnórsdóttir et al. (2011)
Virgibacillus sp. SK37	Phrommao et al. (2011)
Bacillus subtilis JB1	Sung et al. (2010)
Poenibacillus tezpurensis sp. nov. AS-S24-II	Rai et al. (2010)
Brevibacillus thermoruber T1E	Bihari et al. (2010)
Thermoplasma volcanium GSS1	Kocabiyik et al. (2009)
Brevibacillus sp. KH3	Li et al. (2009)
Bacillus stereothermophilus	Ugwuanyi et al. (2008)
Lactobacillus helveticus	Valasaki et al. (2008)
Thermus aquaticus YT–1	Sakaguchi et al. (2008)
Thermus thermophilus HB27	Maehara et al. (2008)
Bacillus subtilis DB 104	Zhang et al. (2008)
Sulfolobus solfactaricus	De Felice et al. (2007)
Geobacillus collagenovorans MO–1	Itoi et al. (2006)
Pyrococcus furiosus	Juhász et al. (2006)
Mucor bacilliformis	Machalinski et al. (2006)

TABLE 8.1   Continued

Microorganisms		References
	Ferridobacterium islandicum	Gödde et al. (2005)
	Geobacillus tepidamans GS5–97T	Kählig et al. (2005)
	Borrelia burgetoferi	Bertin et al. (2005)
	Bacillus thermoproteolyticus	Adekoya et al. (2005)
	Filobacillus sp. RF2–5	Hiraga et al. (2005)
	Geobacillus caldoproteolyticus	Chen et al. (2004)
	Bacillus stearothermophilus	Matsubara (1970)
	Bacillus stearothermophilus TP26	
Actinomycetes	*Thermoactinomyces thalpophilus* MCMB-380	Majumder and Kanekar (2012); Majumder et al. (2013)
	Streptomyces megasporus strain SDP4	Patke et al. (1998)
	Streptomyces thermoviolaceus strain SD8	Chitte et al. (1999)
	Streptomyces megasporus strain SD5	Patke et al. (1998)
	Themoactinomyces sp. 27a	Zabolotskaya et al. (2004)
	Actinomadura keratinilytica Cpt 29	Habbeche et al. (2013)
	Actinomadura keratinilytica T16–1	
	Streptomyces sp. 594	De Azeredo et al. (2006)
	Thermoactinomyces sp. 21E	Petrova et al. (2006)
	Thermophilic streptomyces sp. MCMB 379	Chitte et al. (2011)
	Streptomyces sp. CN902	Lazim et al. (2009)
Fungi	*Thermomucor indicae seudaticae* N31	Silva et al. (2013)
	Thermophilic aspergillus terreus	Liao et al. (2012)
	Thermoascus aurantiacus var. *levisporus*	Li et al. (2011)
	Myceliophthora sp.	Zanphorlin et al. (2010)
	Chaetomium thermophilus	Li et al. (2009)

TABLE 8.1 Continued

Microorganisms	References
Talaromyces emersonii	O'Donoghue et al. (2008)
Thermoascus aurantiacus Miehe	Macchione et al. (2008)
Thermomyces lanuginosus	
Thermomyces lanuginosus (TO.03)	
Aspergillus flavus 1.2	
Aspergillus sp. 13.33, 13.34, 13.35	
Rhizomucor pusillus 13.36	
Rhizomucos sp. 13.37	
Paecelomyces variotii	Bazarzhapov et al. (2006)
Aspergillus carneus	

for industrial and biotechnological applications. Normally, biocatalysts are highly specific and highly active under mild environmental conditions. Despite their obvious advantages, biocatalysts, generally are liable proteins. Therefore, stable and robust biocatalysts which can withstand harsh operational condition, is the need of the hour (Illanes, 1999).

The general advantages of enzymes from thermophiles are as follows (Rakshit et al., 2003): (1) Lesser stringency in the maintenance of sterile conditions as contamination is lower due to the increased temperature; (2) Fermentation allows a higher operational temperature which has a significant influence on the bioavailability and solubility of organic compounds; (3) Elevated process temperatures leads to higher reaction rates due to a decrease in viscosity and an increase in diffusion coefficient of substrates and higher process yield due to increased solubility of substrates and products and favorable equilibrium displacement in endothermic reactions, and (4) Volatile products can be easily recovered. Stability of bioactive molecules against temperature, denaturing agents like detergents and organic solvents, extreme pH conditions enhances their industrial and biotechnological applications. The fermentation is cost effective as elaborate cooling conditions are not required. Thermostable enzymes can also be used as models for the understanding of thermostability and thermoactivity, which is useful for protein engineering.

Microbiologists use 'screening techniques,' the most important approach towards achieving the goal of detecting the novel microbial metabolites. Successful screening methods are normally desired by a permutation and combination of knowledge obtained from microbiology, chemistry, biochemistry, engineering and literature survey. 'Screening' is knowledge based and helps to detect and identify new metabolite of commercial interest. Enzymes are the excellent biological catalysts, which catalyse myriad of chemical reactions in living organisms at different pH and temperature conditions. General criteria for screening the microbes for enzyme production are as follows (O'Donnel et al., 1994):

1. Organism should produce the enzyme in short cultivation period with good yield preferably in submerged condition.
2. Organism should be easily separable from the medium after fermentation.
3. Organism should be genetically stable and non-pathogenic.

Screening naturally occurring microorganisms is still one of the promising ways to achieve the above mentioned criteria. Since the discovery of commercially viable enzymes from extremophiles, microbiologists have earnestly diverted their attention to fast growing microorganisms from pristine habitats. Today, with recent advances in basic screening and relevant technology, cultivating these extremophiles has also become somewhat easy. Screening for novel biocatalysts can be conducted for the following characteristics (Stanbury et al., 1995): resistance to high temperature and organic solvents, type of enzyme – extracellular, nature of enzyme – constitutive, and easy to extract and purify.

Thermophilic bacteria are known to produce novel bioactive molecules. Proteases are a unique class of enzymes since they are of immense physiological as well as commercial importance. They possess both degradative and synthetic properties. Thermophilic enzyme Biolysin from *Bacillus stearothermophilus* (Durrschmidt et al., 2001) was reported for the first time. The authors studied differentiation between conformational and autoproteolytic stability of this neutral protease. Thermolysin from *Bacillus thermoproteolyticus* Rokko (Matsubara, 1970), thermolysin and biolysin are thermostable proteases.

Thermophilic actinomycetes have been found to be predominant in production of thermophilic proteases. Some proteases of actinomycete origin are highly resistant to heat and denaturing agents. The high thermostability of proteases isolated from thermophilic actinomycete is well known and all are used on an industrial scale. Thermophilic protease Aqualysin-1 from *Thermus aquaticus* (Arnórsdóttir et al., 2011). There are reports of other proteases too from thermophilic actinomycetes. Among thermophilic actinomycetes, *Thermoactinomyces* sp. are known producers of the enzyme protease – viz. thermitase and the serine protease was investigated by Lacey (1989) from *Thermoactinomyces vulgaris*. Goodfellow et al. (1988) have done pioneering work on thermophilic actinomycetes. The thermophilic enzyme carboxypeptidase T was reported from *Thermoactinomyces* sp.

Several other proteases from *Thermoactinomyces fusca* (Desai et al., 1969), *Thermoactinomyces albus* (Goodfellow et al., 1988), *Thermoactinomyces thalpophilus* (Odibo et al., 1988), *Thermoactinomyces sacchari* (Georgieva et al., 2000), *Thermoactinomyces* sp. HS628 (Georgieva et al., 2000), *Thermoactinomyces* sp. 27a (Zabolotskaya et al., 2004), *Thermoactinomyces* sp. (Yallop et al., 1997), have been isolated. Thermolysin – the enzyme was first studied from *Bacillus thermoproteolyticus* Rokko by Matsubara (1970). Thermophilic actinomycete, *Thermoactinomyces thalpophilus* isolated from thermal spring in Maharashtra, India was found to produce thermolysin like protease (TLP). The enzyme has unique property of condensation of two amino acids to form Aspartame, an artificial sweetner. The enzyme is a thermostable metalloprotease (Majumder and Kanekar, 2012; Majumder et al., 2013). *Thermoactinomyces thalpophilus* was reported to produce TLP for the first time which had commercial importance with respect to production of aspartame. Thermolysin is a thermostable metalloprotease, which has a role in biotransformation of amino acids into an important condensation product like Aspartame (Rao et al., 1998). The enzyme actinokinase is produced by thermophilic actinomycete, *Streptomyces megasporus* isolated from thermal spring in Maharashtra, India. It dissolves blood clot but does not affect other plasma proteins. Thus the enzyme has potential application in treatment of myocardial infarction (Chitte and Dey, 2000; Chitte and Dey, 2002; Chitte et al., 2011).

Thermophilic protease from fungi have contributed to food industry in a big way. Besides that they have broad substrate specificity and contribute to processing of biomass into fuel (Bazarzhapov et al., 2006; O'Donoghue et al., 2008; Macchione et al., 2008; Li et al., 2009; Zanphorlin et al., 2010; Li et al., 2011; Liao et al., 2012; Silva et al., 2014). Broadly, protease enzyme has wide applications in dairy, detergents, pharmaceuticals and food industries. In food industries thermoresistant protease enzymes are widely used as a meat tenderizer, production of amino acid concentrates, etc.

Specific applications of thermophilic protease are as follows: unearthing new allosteric sites of the C56 family of peptidases could unravel opportunities in understanding the biological processes involving this family specifically in the fields of pharmaceutical development (Charbonneau et al., 2012; Panda et al., 2013; Zhan et al., 2014), potential as food processing aid and biotechnological applications in halophilic conditions. Fungal rennet can replace the traditional one in near future (Machalinski et al., 2006; Sung et al., 2010; Phrommao et al., 2011; Silva et al., 2013). Efficient aerobic and anaerobic sludge dissolution and digestion of organic particulates from high strength food waste water, waste activated sludge (WAS), fishery waste degradation would bring a new dimension to microbial waste treatment (Chen et al., 2004; Chenel et al., 2008; Ugwuanyi et al., 2008; Li et al., 2009; Kim et al., 2012; Tang et al., 2012; Jang et al., 2013). Ecofriendly valorization of keratin containing waste by aerobic and anaerobic processes is also reported earlier (De Azeredo et al., 2006; Liang et al., 2010; Huang et al., 2013; Kataoka et al., 2014). Proteolysis, alkaline proteases are stable in presence of harsh conditions, thereby compatible with several detergents (Gödde et al., 2005; Rai et al., 2010; Chitte et al., 2011; Saleem et al., 2012; Majeed et al., 2013; Toplak et al., 2013). Ion pair effect on thermostability of F1 protease was studied through computational approach to give a better understanding between protein structure and biological function, thereby identifying opportunities for protein engineering (Rahman et al., 2012). The principle of corresponding states allows successful thermostability optimization and also improves enzyme activity in protein engineering (Berezovsky et al., 2005; Bian et al., 2006; Ditursi et al., 2006; Radestock et al., 2011).

8.6 INDUSTRIAL SIGNIFICANCE OF THERMOSTABLE PROTEASES

Thermophiles have a lower biomass yield but an increased product to substrate ratio. They require less incubation time for product formation, higher incubation temperature and early sporulation, which enhances industrial feasibility. They are known producers of heterologous extracellular enzymes. Among the extracellular enzymes, 'proteases' have high specific activity compared to their mesophilic counterparts. In the global market, extracellular enzymes account for two third of the total world market because they are easily extractable and purifiable (Illanes, 1999; Rakshit et al., 2003). 'Proteases' are a unique class of enzymes, which occupy a pivotal position with respect to their application in both physiological and commercial fields. They possess both degradative and synthetic properties. Proteolytic enzymes catalyse the cleavage of peptide bonds in other protein. Since proteases are physiologically necessary, they occur ubiquitously in animals, plants and microbes. However, microbes become a goldmine of proteases and represent the preferred sources of enzymes in view of their rapid growth and ready accessibility to genetic manipulation. Microbial proteases have been extensively used in the food, barking, dairy, pharmaceutical, detergent, leather and textile industries (Rao et al., 1998).

Protease is also used in basic research where selective peptide bond cleavage is used in the elucidation of structure – function relationship, in the sequencing of proteins etc. In addition to the multitude of activities that are already assigned to proteases, many more new functions are likely to emerge in the near future (Rao et al., 1998).

8.7 GLOBAL MARKET FOR INDUSTRIAL ENZYMES

The present focus is specially on exploring thermophilic proteases for industrial application with special reference to enzyme as part of everyday life. The best studied and most widely used enzymes in the industry are most effective over a temperature range from 40°C to ~50°C. At these temperatures, substrate conversion requires long reaction times. One way to overcome these obstacles is to raise the reaction temperature. However, implementing higher reaction temperatures requires the deployment of enzymes that are more thermostable. Therefore, thermostable enzymes and thermophilic cell factories may afford economic advantages in the production of many chemicals and biomass-based fuels (Berka et al., 2011). The present review focuses specially on exploring thermophilic proteases for industrial application with special reference to TLP production and exploitation for condensation reaction from thermophilic actinomycetes.

The global market for industrial enzymes was valued at $3.1 billion in 2009 and reached about $3.6 billion in 2010. The estimated market for 2011 is about $3.9 billion. BCC Research projects this market to grow at a compounded annual growth rate (CAGR) of 9.1% to reach $6 billion by 2016. Food and beverage enzymes comprise the largest segment of the industrial enzymes industry with revenues of nearly $1.2 billion in 2010. This market is expected to reach $1.3 billion by 2011 and further it will grow to $2.1 billion by 2016, a CAGR of 10.4%. The second-largest category is technical enzymes with revenues of about $1.1 billion in 2010 and nearly $1.2 billion in 2011. This global enzyme market based on application sectors is further expected to grow to $1.7 billion by 2016, a CAGR of 8.2%.

During this growth, demand for specialty enzymes would outpace industrial enzymes. Developing country like India will offer some of the best growth opportunities in future. Proteases represent one of the largest groups of industrial enzymes and account for more than 65% of the world market. Microbial proteases account for approx. 40% of the total worldwide enzymes market (Godfrey and West, 1996).

8.8   CONCLUSION AND FUTURE OUTLOOK

Commercially market driven demand is to isolate novel wild type thermophilic protease producing micro-organisms which will have good activity and stability in the presence of harsh conditions like organic solvents, detergents, surfactants and oxidants – 'robust protease.' Waste degradation by thermophilic proteases from aerobic and anaerobic organisms is a viable area for active research. New structural insights related to catalytic activity of the protease could be studied theoretically by quantum mechanical calculations and molecular dynamic stimulation. Understanding the principle of 'corresponding states' could enhance successful thermostability optimization and designing of experiments in order to improve enzyme activity in protein engineering. Novel entropic mechanisms of protein thermostability because of residual dynamics of rotamer isomerization in native state could be shown with its immediate proteomic implications. Last but not the least, thermophilic fungi producing thermostable proteases should be exploited for strong biotechnological potential.

KEYWORDS

- **Biotechnological applications**
- **Effluent treatment**
- **Evolution**
- **Global market**

- Habitats
- Industrial significance
- Physiology
- Robust protease
- Thermophiles
- Thermostable proteases

REFERENCES

Adekoya, O., Willassen, N. P., Sylte, I. The protein-protein interactions between SMPI and thermolysin studied by molecular dynamics and MM/PBSA calculations. *J Biomol Struct Dyn* 2005, 22(5), 521–531.

Allen, M. B. The thermophilic aerobic spore-forming bacteria. *Bacteriological Reviews* 1953, 17(2), 125–173.

Arnórsdóttir, J., Magnúsdóttir, M., Friðjónsson, O. H., Kristjánsson, M. M. The effect of deleting a putative salt bridge on the properties of the thermostable subtilisin-like proteinase, aqualysin I. *Protein Pept Lett* 2011, 18(6), 545–551.

Bader, J., Skelac, L., Wewetzer, S., Senz, M., Popović, M. K., Bajpai, R. Effect of partial pressure of CO_2 on the production of thermostable alpha-amylase and neutral protease by *Bacillus caldolyticus*. *Prikl Biokhim Mikrobiol* 2012, 48(2), 206–211.

Bazarzhapov, B. B., Lavrentev, E. V., Dunaevskiĭ, I. E., Bilanenko, E. N., Namsaraev, B. B. Extracellular proteolytic enzymes of microscopic fungi from thermal springs of the Barguzin Valley (Northern Baikal region). *Prikl Biokhim Mikrobiol* 2006, 42(2), 209–212.

Berezovsky, I. N., Chen, W. W., Choi, P. J., Shakhnovich, E. I. Entropic stabilization of proteins and its proteomic consequences. *PLoS Comput Biol* 2005, 1(4), e47.

Bergey, D. H. Thermophilic bacteria. *Journal of Bacteriology* 1919, 4(4), 301–306.

Berka, R. M. Comparative genomic analysis of the thermophilic biomass-degrading fungi *Myceliophthora thermophila* and *Thielavia terrestris*. *Nature Biotechnology* 2011, 29, 922–927.

Bertin, P. B., Lozzi, S. P., Howell, J. K., Restrepo-Cadavid, G., Neves, D., Teixeira, A. R., de Sousa M. V., Norris, S. J., Santana, J. M. The thermophilic, homohexameric aminopeptidase of *Borrelia burgdorferi* is a member of the M29 family of metallopeptidases. *Infect Immun* 2005, 73(4), 2253–2261.

Bian, Y., Liang, X., Fang, N., Tang, X. F., Tang, B., Shen, P., Peng Z. The roles of surface loop insertions and disulfide bond in the stabilization of thermophilic WF146 protease. *FEBS Lett* 2006, 580(25), 6007–6014.

Bihari, Z., Vidéki, D., Mihalik, E., Szvetnik, A., Szabó, Z., Balázs, M., Kesseru, P., Kiss, I. Degradation of native feathers by a novel keratinase-producing, thermophilic isolate, *Brevibacillus thermoruber* T1E. *Z Naturforsch* 2010, 65(1–2), 134–140.

Brock, T. D. The value of basic research: discovery of *Thermus aquaticus* and other extreme thermophiles. *Genetics* 1997, 146, 1207–1210.

Brock, T. D. Introduction: An overview of the thermophiles. In *Thermophiles: General, Molecular and Applied Microbiology*. Brock, T. D. (Ed.), John Wiley and Sons: New York, 1986, pp. 1–16.

Charbonneau, D. M., Meddeb-Mouelhi, F., Boissinot, M., Sirois, M., Beauregard, M. Identification of thermophilic bacterial strains producing thermotolerant hydrolytic enzymes from manure compost. *Indian J Microbiol* 2012, 52(1), 41–47.

Chen, X. G., Stabnikova, O., Tay, J. H., Wang, J. Y., Tay, S. T. Thermoactive extracellular proteases of *Geobacillus caldoproteolyticus* sp. nov., from sewage sludge. *Extremophiles* 2004, 8(6), 489–498.

Chenel, J. P., Tyagi, R. D., Surampalli, R. Y. Production of thermostable protease enzyme in wastewater sludge using thermophilic bacterial strains isolated from sludge. *Water Sci Technol* 2008, 57(5), 639–645.

Chitte R. R., Dey S. Potent fibrinolytic enzyme from a thermophilic *Streptomyces megasporus* strain SD5. *Letters in Applied Microbiology* 2000, 31, 405–410.

Chitte, R. R., Nalawade V. K., Dey, S. Keratinolytic activity from the broth of a feather degrading thermophilic *Streptomyces thermoviolaceus* strain SD8. *Letters in Applied Microbiology* 1999, 28, 131–136.

Chitte, R. R., Deshmukh, S. V., Kanekar, P. P. Production, purification, and biochemical characterization of a fibrinolytic enzyme from thermophilic *Streptomyces* sp. MCMB-379. *Appl Biochem Biotechnol* 2011, 165(5–6), 1406–1413.

Chitte, R. R., Dey, S. Production of a fibrinolytic enzyme by thermophilic *Streptomyces* species. *World Journal of Microbiology and Biotechnology* 2002, 18, 289–284.

De Azeredo, L. A., De Lima, M. B., Coelho, R. R., Freire, D. M. A low-cost fermentation medium for thermophilic protease production by *Streptomyces* sp. 594 using feather meal and corn steep liquor. *Curr Microbiol* 2006, 53(4), 335–339.

De Azeredo, L. A., De Lima, M. B., Coelho, R. R., Freire, D. M. Thermophilic protease production by *Streptomyces* sp. 594 in submerged and solid-state fermentations using feather meal. *J Appl Microbiol* 2006, 100(4), 641–647.

De Felice, M., Medagli, B., Esposito, L., De Falco, M., Pucci, B., Rossi, M., Grùz, P., Nohmi, T., Pisani, F. M. Biochemical evidence of a physical interaction between *Sulfolobus solfataricus* B-family and Y-family DNA polymerases. *Extremophiles* 2007, 11(2), 277–282.

Desai, A. J., Dhala, S. A. Purification and properties of proteolytic enzymes from thermophilic actinomycetes. *Journal of Bacteriology* 1969, 100(1), 149–155.

Ditursi, M. K., Kwon, S. J., Reeder, P. J., Dordick, J. S. Bioinformatics-driven, rational engineering of protein thermostability. *Protein Eng Des Sel.* 2006, 19(11), 517–524.

O'Donnell, A. G., Goodfellow, M., Hawksworth, D. L. Theoretical and practical aspects of the quantification of biodiversity among microorganisms. *Philosophical Transactions: Biological Sciences* 1994, 345(1311), 65–73.

Durrschmidt, P. Differentiation between conformational and autoproteolytic stability of the neutral protease from *Bacillus stearothermophilus* containing an engineered disulfide bond. *Eur J Biochem* 2001, 268, 3612–3618.

Georgieva, D. N. Specificity of a neutral Zn-dependent proteinase from *Thermoactinomyces sacchari* toward the oxidized insulin B chain. *Current Microbiology* 2000, 41(1), 70–72.

Gödde, C., Sahm, K., Brouns, S. J., Kluskens, L. D., van der Oost, J., de Vos, W. M., Antranikian, G. Cloning and expression of islandisin, a new thermostable subtilisin from *Fervidobacterium islandicum*, in *Escherichia coli. Appl Environ Microbiol* 2005, 71(7), 3951–3958.

Godfrey, T., West, S. In *Industrial Enzymology*, 2nd ed., Macmillan Publishers Inc.: New York, 1996, pp. 1–4.

Goodfellow, M., Mordarski, M., Williams, S. T. *Actinomycetes in Biotechnology*, Academic Press: London, UK, 1988, pp. 219–480.

Goodfellow, M., Mordarski, M., Williams, S. T. *The Biology of the Actinomycetes*. Academic Press: New York, 1984, pp. 481–513.

Habbeche, A., Saoudi, B., Jaouadi, B., Haberra, S., Kerouaz, B., Boudelaa, M., Badis, A., Ladjama, A. Purification and biochemical characterization of a detergent-stable keratinase from a newly thermophilic actinomycete *Actinomadura keratinilytica* strain Cpt29 isolated from poultry compost. *J Biosci Bioeng* 2014, 117(4), 413–421.

Hiraga, K., Nishikata, Y., Namwong, S., Tanasupawat, S., Takada, K., Oda, K. Purification and characterization of serine proteinase from a halophilic bacterium, *Filobacillus* sp. RF2–5. *Biosci Biotechnol Biochem* 2005, 69(1), 38–44.

Huang, Y., Sun, Y., Ma, S., Chen, L., Zhang, H., Deng, Y. Isolation and characterization of *Keratinibaculum paraultunense* gen. nov., sp. nov., a novel thermophilic, anaerobic bacterium with keratinolytic activity. *FEMS Microbiol Lett* 2013, 345(1), 56–63.

Illanes A. Stability of biocatalysts. *Electronic Journal of Biotechnology* 1999, 2(1), 8–15.

Itoi, Y., Horinaka, M., Tsujimoto, Y., Matsui, H., Watanabe, K. Characteristic features in the structure and collagen-binding ability of a thermophilic collagenolytic protease from the thermophile *Geobacillus collagenovorans* MO–1. *J Bacteriol* 2006, 188(18), 6572–6579.

Jang, H. M., Lee, J. W., Ha, J. H., Park, J. M. Effects of organic loading rates on reactor performance and microbial community changes during thermophilic aerobic digestion process of high-strength food wastewater. *Bioresour Technol* 2013, 148, 261–269.

Juhász, T., Szeltner, Z., Polgár, L. Properties of the prolyl oligopeptidase homologue from *Pyrococcus furiosus*. *FEBS Lett* 2006, 580(14), 3493–3497.

Kählig, H., Kolarich, D., Zayni, S., Scheberl, A., Kosma, P., Schäffer, C., Messner, P. N-acetylmuramic acid as capping element of alpha-D-fucose-containing S-layer glycoprotein glycans from *Geobacillus tepidamans* GS5–97T. *J Biol Chem*. 2005, 280(21), 20292–20299.

Kataoka, M., Yamaoka, A., Kawasaki, K., Shigeri Y., Watanabe, K. Extraordinary denaturant tolerance of keratinolytic protease complex assemblies produced by *Meiothermus ruber* H328. *Appl Microbiol Biotechnol* 2014, 98(7), 2973–2980.

Kim, H. W., Nam, J. Y., Kang, S. T., Kim, D. H., Jung, K. W., Shin, H. S. Hydrolytic activities of extracellular enzymes in thermophilic and mesophilic anaerobic sequencing-batch reactors treating organic fractions of municipal solid wastes. *Bioresour Technol* 2012, 110, 130–134.

Kocabiyik, S., Demirok, B. Cloning and overexpression of a thermostable signal peptide peptidase (SppA) from *Thermoplasma volcanium* GSS1 in *E. coli*. *Biotechnol J* 2009, 4(7), 1055–1065.

Kumar, S., Nussinov, R. How do thermophilic proteins deal with heat? A review. *Cell Mol Life Sci* 2001, 58, 1216–1233.

Lacey, J. Actinomycetes. *Bergey's Manual of Systematic Bacteriology* 1989, 4, 2573–2585.

Lazim, H., Mankai, H., Slama, N., Barkallah, I., Limam, F. Production and optimization of thermophilic alkaline protease in solid-state fermentation by *Streptomyces* sp. CN902. *J Ind Microbiol Biotechnol* 2009, 36(4), 531–537.

Li, A. N., Li, D. C. Cloning, expression and characterization of the serine protease gene from *Chaetomium thermophilum*. *J Appl Microbiol* 2009, 106(2), 369–380.

Li, A. N., Xie, C., Zhang, J., Zhang, J., Li, D. C. Cloning, expression, and characterization of serine protease from thermophilic fungus *Thermoascus aurantiacus* var. *levisporus*. *J Microbiol*. 2011, 49(1), 121–129.

Li, X., Ma, H., Wang, Q., Matsumoto, S., Maeda, T., Ogawa, H. I. Isolation, identification of sludge-lysing strain and its utilization in thermophilic aerobic digestion for waste activated sludge. *Bioresour Technol* 2009, 100(9), 2475–2481.

Liang, X., Bian, Y., Tang, X. F., Xiao, G., Tang, B. Enhancement of keratinolytic activity of a thermophilic subtilase by improving its autolysis resistance and thermostability under reducing conditions. *Appl Microbiol Biotechnol* 2010, 87(3), 999–1006.

Liao, W. Y., Shen, C. N., Lin, L. H., Yang, Y. L., Han, H. Y., Chen, J. W., Kuo, S. C., Wu, S. H., Liaw, C. C. Asperjinone, a nor-neolignan, and terrein, a suppressor of ABCG2-expressing breast cancer cells, from thermophilic *Aspergillus terreus*. *J Nat Prod* 2012, 75(4), 630–635.

Lineweaver, C. H., Schwartzman, D. Cosmec thermobiology, thermal constraints on the origin and evolution of life in the Universe. In *Genesis, Evolution and Diversity of Life*, Seekbach, J. (Ed.), Kluwer: Dordrecht, 2004, pp. 233–248.

Lopez, G. DNA super coiling and temperature adaptation: a clue to early diversification of life. *J. Mol. Evol* 1999, 46, 439–452.

Macchione, M. M., Merheb, C. W., Gomes, E., da Silva, R. Protease production by different thermophilic fungi. *Appl Biochem Biotechnol* 2008, 146(1–3), 223–230.

Machalinski, C., Pirpignani, M. L., Marino, C., Mantegazza, A., de Jiménez Bonino, M. B. Structural aspects of the *Mucor bacilliformis* proteinase, a new member of the aspartyl-proteinase family. *J Biotechnol* 2006, 123(4), 443–452.

Maehara, T., Hoshino, T., Nakamura, A. Characterization of three putative Lon proteases of *Thermus thermophilus* HB27 and use of their defective mutants as hosts for production of heterologous proteins. *Extremophiles* 2008, 12(2), 285–296.

Majeed, T., Tabassum, R., Orts, W. J., Lee, C. C. Expression and characterization of *Coprothermobacter proteolyticus* alkaline serine protease. *Scientific World Journal* 2013, 25, 396156.

Majumder D. R., Kanekar P. P. Different aspects of production of thermolysin-like protease from *Thermoactinomyces thalpophilus*. *International Journal of Pharma and Bio Sciences* 2012, 3(1B), 610–627.

Majumder, D. R., Gaikwad, S. M., Kanekar, P. P. Purification and characterization of a thermolysin like protease from *Thermoactinomyces thalpophilus* MCMB-38. *Protein and Peptide Letters* 2013, 20(8), 918–925.

Matsubara, H. Purification and assay of thermolysin. *Methods in Enzymology* 1970, 19, 642–651.

Neihaus, F. Extremophiles as a source of novel enzymes for industrial application. *Applied Microbiology and Biotechnology* 1999, 51(6), 711–729.

Odibo, F. J. C., Obi, S. K. C. Purification and some properties of a thermostable protease of *Thermoactinomyces thalpophilus*. *World Journal of Microbiology and Biotechnology* 1988, 4(3), 327–332.

O'Donoghue, A. J., Mahon, C. S., Goetz, D. H., O'Malley, J. M., Gallagher, D. M., Zhou, M., Murray, P. G., Craik, C. S., Tuohy, M. G. Inhibition of a secreted glutamic peptidase prevents growth of the fungus *Talaromyces emersonii*. *J Biol Chem* 2008, 283(43), 29186–29195.

Pace, N. R. A molecular view of microbial diversity and the biosphere. *Science* 1997, 276 (5313), 734–740.

Panda, M. K., Sahu, M. K., Tayung, K. Isolation and characterization of a thermophilic *Bacillus* sp. with protease activity isolated from hot spring of Tarabalo, Odisha, India. *Iran J Microbiol* 2013, 5(2), 159–165.

Patke, D., Dey, S. Proteolytic activity from a thermophilic *Streptomyces megasporus*, strain SDP4. *Letters in Applied Microbiology* 1998, 26, 171–174.

Peek, K., Daniel, R. M., Monk, C., Parker, L., Coolbear, T. Purification and characterization of a thermostable proteinase isolated from *Thermos* sp. Rt 41A. *Eur. J. Biochem.* 1992, 207, 1035–1044.

Petrova, D. H., Shishkov, S. A., Vlahov, S. S. Novel thermostable serine collagenase from *Thermoactinomyces* sp. 21E: purification and some properties. *J Basic Microbiol* 2006, 46(4), 275–285.

Phrommao, E., Rodtong, S., Yongsawatdigul, J. Identification of novel halotolerant bacillopeptidase F-like proteinases from a moderately halophilic bacterium, *Virgibacillus* sp. SK37. *J Appl Microbiol* 2011, 110(1), 191–201.

Radestock, S., Gohlke, H. Protein rigidity and thermophilic adaptation. *Proteins* 2011, 79(4), 1089–1108.

Rahman R. N., Muhd Noor, N. D., Ibrahim, N. A., Salleh, A. B., Basri, M. Effect of ion pair on thermostability of F1 protease: integration of computational and experimental approaches. *J Microbiol Biotechnol* 2012, 22(1), 34–45.

Rai, S. K., Roy, J. K., Mukherjee, A. K. Characterization of a detergent-stable alkaline protease from a novel thermophilic strain *Paenibacillus tezpurensis* sp. nov. AS-S24-II. *Appl Microbiol Biotechnol* 2010, 85(5), 1437–1450.

Rakshit, S. K., Haki, G. D. Developments in industrially important thermostable enzymes: a review. *Biosource Technology* 2003, 89, 17–34.

Rao, M. B. Molecular and biotechnological aspects of microbial proteases. *Microbiology and Molecular Biology Reviews* 1998, 62(3), 597–635.

Rath, C. C., Subramanyam, V. R. Isolation of thermophilic bacteria from hot springs of Orissa, India. *Geobiol* 1998, 25, 113–119.

Rath, C. C., Subramanyam, V. R. Thermotolerant enzyme activities of *Bacillus* sp. isolated from the hot springs of Orissa. *Microbios* 1996, 86, 157–161.

Reysenbach, A. L., Cady, S. L. Microbiology of ancient and modern hydrothermal systems. *Trends in Microbiology* 2001, 9(2), 79–86.

Sakaguchi, M., Niimiya, K., Takezawa, M., Toki, T., Sugahara, Y., Kawakita, M. Construction of an expression system for aqualysin I in *Escherichia coli* that gives a markedly improved yield of the enzyme protein. *Biosci Biotechnol Biochem* 2008, 72(8), 2012–2018.

Saleem, M., Rehman, A., Yasmin, R., Munir, B. Biochemical analysis and investigation on the prospective applications of alkaline protease from a *Bacillus cereus* strain. *Mol Biol Rep* 2012, 39(6), 6399–6408.

Silva, B. L., Geraldes, F. M., Murari, C. S., Gomes, E., Da-Silva R. Production and characterization of a milk-clotting protease produced in submerged fermentation by the thermophilic fungus *Thermomucor indicae-seudaticae* N31. *Appl Biochem Biotechnol* 2014, 172(4), 1999–2011.

Solomatov, V. S. The role of plate tectonics. *Journal of Geophysical Research* 2004, 108, 1029.

Stanbury, P. F., Whitaker, A., Hall, S. J. *Principles of Fermentation Technology, 2nd ed.*, Elsevier Science and Technology Books: Exeter, Great Britain, 1995, pp. 34–91.

Sung, J. H., Ahn, S. J., Kim, N. Y., Jeong, S. K., Kim, J. K., Chung, J. K., Lee, H. H. Purification, molecular cloning, and biochemical characterization of subtilisin JB1 from a newly isolated *Bacillus subtilis* JB1. *Appl Biochem Biotechnol* 2010, 162(3), 900–911.

Tang, Y., Yang, Y. L., Li, X. M., Yang, Q., Wang, D. B., Zeng, G. M. The isolation, identification of sludge-lysing thermophilic bacteria and its utilization in solubilization for excess sludge. *Environ Technol* 2012, 33(7–9), 961–966.

Toplak, A., Wu, B., Fusetti, F., Quaedflieg, P. J., Janssen, D. B. Proteolysin, a novel highly thermostable and cosolvent-compatible protease from the thermophilic bacterium *Coprothermobacter proteolyticus*. *Appl Environ Microbiol* 2013, 79(18), 5625–5632.

Ugwuanyi, J. O., Harvey, L. M., McNeil, B. Protein enrichment of corn cob heteroxylan waste slurry by thermophilic aerobic digestion using *Bacillus stearothermophilus*. *Bioresour Technol*. 2008, 99(15), 6974–6985.

Valasaki, K., Staikou, A., Theodorou, L. G., Charamopoulou, V., Zacharaki, P., Papamichael, E. M. Purification and kinetics of two novel thermophilic extracellular proteases from *Lactobacillus helveticus*, from kefir with possible biotechnological interest. *Bioresour Technol* 2008, 99(13), 5804–5813.

Yallop, C. A. Isolation and growth physiology of novel thermoactinomycetes. *Journal of Applied Microbiology* 1997, 83, 685–692.

Yun, S. H., Choi, C. W., Kwon, S. O., Lee, Y. G., Chung, Y. H., Jung, H. J., Kim, Y. J., Lee, J. H., Choi, J. S., Kim, S., Kim, S. I. Enrichment and proteome analysis of a hyperthermostable protein set of archaeon *Thermococcus onnurineus* NA1. *Extremophiles* 2011, 15(4), 451–461.

Zabolotskaya, M. V. A novel neutral protease from thermoactinomyces species 27a: sequencing of the gene, purification, and characterization of the enzyme. *The Protein Journal* 2004, 23(7), 483–492.

Zanphorlin, L. M., Facchini, F. D., Vasconcelos, F., Bonugli-Santos, R. C., Rodrigues, A., Sette, L. D., Gomes, E., Bonilla-Rodriguez, G. O. Production, partial characterization, and immobilization in alginate beads of an alkaline protease from a new thermophilic fungus *Myceliophthora* sp. *J Microbiol* 2010, 48(3), 331–336.

Zhan, D., Sun, J., Feng, Y., Han, W. Theoretical study on the allosteric regulation of an oligomeric protease from *Pyrococcus horikoshii* by Cl- ion. *Molecules* 2014, 19(2), 1828–1842.

Zhang, M., Zhao, C., Du, L., Lu, F., Gao, C. Expression, purification, and characterization of a thermophilic neutral protease from *Bacillus stearothermophilus* in *Bacillus subtilis*. *Sci China C Life Sci* 2008, 51(1), 52–59.

Zhu, H., Xu, B. L., Liang, X., Yang, Y. R., Tang, X. F., Tang, B. Molecular basis for auto- and hetero-catalytic maturation of a thermostable subtilase from thermophilic *Bacillus* sp. WF146. *J Biol Chem*. 2013, 288(48), 34826–34838.

Zierenberg, R. A. Life in extreme environments: hydrothermal vents. *Proc. Natl. Acad. Sci. USA* 2000, 97(24), 12961–12962.

CHAPTER 9

PROTEASES IN LEATHER PROCESSING

VASUDEO P. ZAMBARE[1] and SMITA S. NILEGAONKAR[2]

[1]Center for Bioprocessing Research and Development, South Dakota School of Mines and Technology, E. Saint Joseph Street, Rapid City, SD, USA

[2]Microbial Sciences Division, Agharkar Research Institute, Pune, Maharashtra, 411004, India

CONTENTS

9.1   INTRODUCTION

Enzymes, the buzzword in vogue since centuries, are the point where biological sciences meet the physical sciences. Enzymes are crux molecules for sustaining life which depends on a complex network of chemical reactions brought about by specific enzymes, any modification of which having far reaching consequences on the living organism. Proteases have a wide range of functions in nature. Proteases are degradative enzymes that catalyze the hydrolysis of proteins. Extracellular proteases contribute to the nutritional well being of the organism by hydrolyzing large polypeptide substrates into smaller molecules that the cell can absorb. Proteases are also involved in the regulation of biological processes such as spore formation, spore germination, protein mutation in viral assembly, and activation of certain viruses with importance for pathogenicity, various stages of the mammalian fertilization processes, blood coagulation, fibrinolysis, complement activation, phagocytosis and blood pressure control (Rao et al., 1998).

In all cell systems the protein turnover is delicately balanced between its synthesis and degradation by different intracellular proteases. For example, in a growing bacterium, 1–3% of the cell protein is degraded into amino acids every hour. Apart from this, protein turnover is an essential factor in the environmental adaptation of a cell, particularly in an environment lacking in protein or amino acids (Tanaka et al., 1998). There is a plethora of evidence to indicate that proteases are involved in modulation of gene expression, modification of the enzyme and its secretion processes and in the control of translation by modification of enzyme associated with ribozyme (Atsumi et al., 1989; Guzzo et al., 1991; Traidej et al., 2003; Peng et al., 2004).

The estimated value of the worldwide sales of industrial enzymes was $1 billion in 1998 (Anwar and Saleemuddin, 1998), $1.5 billion in 2000 and $3.3 billion in 2010 (Sarrouh et al., 2012). Proteases represent one of the three largest groups of industrial enzymes and account for about 60% of the total worldwide sale (Figure 9.1). They differ in the properties such as substrate specificity, active sites, catalytic mechanism, pH, temperature and activity profiles. Figure 9.2 presents a breakdown of the market shares for industrial proteases. These data clearly show that bacterial proteases are the significant segment representing 60% of the total industrial protease turnover.

Commercial proteases have application in a range of processes, which take advantage of the unique physical and catalytic properties of

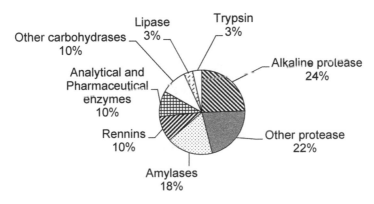

FIGURE 9.1 Distribution of worldwide sale of enzymes.

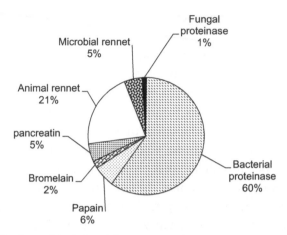

FIGURE 9.2   Pie chart of the industrial protease market.

individual proteolytic enzyme (Gupta et al., 2002). Proteases have a long history of application in the food, pharmaceutical and detergent industries (Zambare et al., 2011b; Zambare et al., 2014). Their application in the leather industry for soaking, dehairing and bating of hides, to substitute currently used toxic chemicals, is a relatively new development and has added biotechnological importance (Rao et al., 1998). The vast diversity of proteases, in contrast to the specificity of their action, has attracted worldwide attention in attempts to exploit their physiological and biotechnological applications (Anwar and Saleemuddin, 1998; Gupta et al., 2002). Proteases have numerous research applications like peptide synthesis, cell culturing. The major producers of proteases worldwide are listed in Table 9.1.

TABLE 9.1   Worldwide Major Protease Producers

Company	Country	Market share (%)
Novo Nordics Industries	Denmark	40
Gist-Brocades	Netherlands	20
Genencor International	Unites States	10
Miles Laboratories	United States	10
Other		20

9.2 SOURCES OF PROTEASES

Since proteases are physiologically essential for living organisms, they are ubiquitous, being found in a wide diversity of sources such as plants, animals, and microorganisms. The choice of a particular source depends on the efficiency of production, degree of control, steady production, composition of enzyme and degree of purity of the preparation. The isolation of enzymes from plant or animal sources involves a high cost factor and low yield as compared to that from microbial sources. Table 9.2 shows different sources of protease for various industrial applications. Thus, although proteases are widespread in nature, microbes serve as a preferred source of these enzymes because of their rapid growth, high enzyme yield, short duration for production, easy recovery of the product, the limited space required for their cultivation, and the ease with which they can be genetically manipulated to generate new enzymes with altered properties that are desirable for their various applications.

9.3 CLASSIFICATION OF PROTEASES

According to the Nomenclature Committee of the International Union of Biochemistry and Molecular Biology, proteases are classified into the 4th subgroup of group 3, which classifies the hydrolases (International Union of Biochemistry, 1992). However, proteases do not comply easily with the general system of enzyme nomenclature due to their huge diversity of action and structure. Currently, proteases are classified on the basis of three major criteria, type of reaction catalyzed, chemical nature of the catalytic site and evolutionary relationship with reference to structure (Barett, 1994).

Proteases are grossly subdivided into two major groups, such as exopeptidases and endopeptidases, depending on their site of action. Based on the functional group present at the active site, proteases are further classified into four prominent groups, such as serine, aspartic, cysteine, and metalloproteases (Hartley, 1996). There are a few miscellaneous proteases that do not precisely fit into the standard classification and one of them is ATP-dependent proteases (Menon et al., 1987). The flow sheet for classification of peptide hydrolases is given in Figure 9.3.

TABLE 9.2 Different Sources of Proteases

Category	Proteases	Sources	Industrial applications	References
Plants	Papain	*Carica papaya* fruits	Food	Khalil et al. (2006)
	Bromelain	Stem and juice of pineapples	Food	Priem et al. (1993)
Animals	Trypsin	Animal intestine	Food	Gupta et al. (2002)
	Chymotrypsin	Animal pancreas	Food	Sarah et al. (1996)
	Pepsin	Vertebrate's stomach	Pharmaceuticals	Kim et al. (2000)
	Rennin	Nursing mammal stomach	Dairy	Rao et al. (1998)
Microorganisms	Bacteria			
	Neutral	*Bacillus* sp.	Brewing, detergent, Silver recovery	Anwar and Saleemuddin (1998); Zambare et al. (2011b)
	Alkaline			
	Fungi			
	Acid	*Aspergillus oryzae*	Detergent and leather	Poza et al. (2001)
	Neutral			
	Alkaline			
	Viruses			
	Serine	Potyviruses,		Patick and Potts (1998)
	Aspartic	Retrovirus	—	
	Cysteine	Tobacco etch virus		

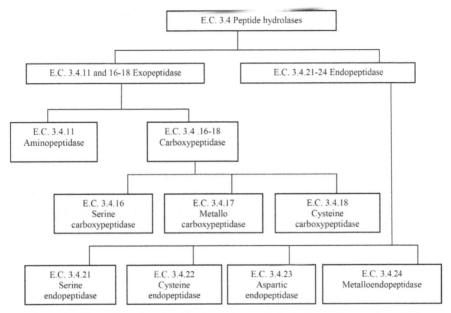

FIGURE 9.3 Flow sheet for classification of peptide hydrolases (International Union of Biochemistry, 1992).

9.3.1 EXOPEPTIDASES

The exopeptidases act only at the ends of polypeptide chains. Based on their site of action at the N or C terminus, they are classified as amino- and carboxypeptidases respectively.

9.3.1.1 Aminopeptidases

Aminopeptidases act at a free N terminal of the polypeptide chain and liberate a single amino acid residue, a dipeptide or a tripeptide. Aminopeptidases occur in a wide variety of microbial species including bacteria and fungi (Kumar and Takagi, 1999). The substrate specificities of the enzymes from bacteria and fungi are distinctly different in that the organisms can be differentiated on the basis of the profiles of the products of hydrolysis (Cerny, 1978).

9.3.1.2 Carboxypeptidases

The carboxypeptidases act at C terminal of the polypeptide chain and liberate a single amino acid or a dipeptide. Carboxypeptidases can be divided into three major groups, serine carboxypeptidases, metallocarboxypepti-dases, and cysteine carboxypeptidases, based on the nature of the amino acid residues at the active site of the enzymes. The enzymes can also hydrolyze the peptides in which the peptidyl group is replaced by a pteroyl moiety or by acyl groups.

9.3.2 ENDOPEPTIDASES

Endopeptidases are characterized by their preferential action at the pep-tide bonds in the inner regions of the polypeptide chain, away from the N and C termini. The endopeptidases are divided into four subgroups based on their catalytic mechanism as, (a) serine, (b) aspartic, (c) cysteine, and (d) metalloproteases.

9.3.2.1 Serine Endopeptidases

Serine proteases are characterized by the presence of a serine group in their active site. They are numerous and widespread among viruses, bacteria, and eukaryotes, suggesting that they are vital to the organisms. Based on their structural similarities, serine proteases have been grouped into 20 families, which have been further subdivided into six clans with common ancestors (Barett, 1994). The primary structures of the members of four clans, chy-motrypsin (SA), subtilisin (SB), carboxypeptidase C (SC), and Escherichia D-Ala-D-Ala peptidase A (SE) are totally unrelated, suggesting that there are at least four separate evolutionary origins for serine proteases. Clans SA, SB, and SC have a common reaction mechanism consisting of a com-mon catalytic triad of the three amino acids – serine (nucleophile), aspartate (electrophile), and histidine (base). Although the geometric orientations of these residues are similar, the protein folds are quite different, forming a typical example of a convergent evolution.

Serine proteases are recognized by their irreversible inhibition by 3,4-dichloroisocoumarin (3,4-DCI), L-3-carboxytrans 2,3-epoxypropyl-leucylamido (4-guanidine) butane (E.64), diisopropylfluorophosphate (DFP), phenylmethanelsulfonyl fluoride (PMSF) and tosyl-L-lysine chloromethyl ketone (TLCK). Some of the serine proteases are inhibited by thiol reagents such as p-chloromercuribenzoate (PCMB) due to the presence of a cysteine residue near the active site. Serine proteases are generally active at neutral and alkaline pH, with an optimum between pH 7 and 11. They have broad substrate specificities including esterase and amidase activity. Serine alkaline proteases that are active at highly alkaline pH represent the largest subgroup of serine proteases. Serine endopeptidases are of two types. A novel subtilisin like keratinase was isolated from *B. subtilis* that has application in leather industry as a dehairing agent (Alexander et al., 2005).

9.3.2.2   Cysteine/Thiol Endopeptidases

Cysteine proteases occur in both prokaryotes and eukaryotes. About 20 families of cysteine proteases have been recognized. The activity of all cysteine proteases depends on a catalytic dyad consisting of cysteine and histidine. The order of Cys and His (Cys-His or His-Cys) residues differ among the families (Barett, 1994). Generally, cysteine proteases are active only in the presence of reducing agents such as HCN or cysteine. Based on their side chain specificity, they are broadly divided into four groups, (i) papain-like, (ii) trypsin- like with preference for cleavage at the arginine residue, (iii) specific to glutamic acid, and (iv) others. Papain is the best-known cysteine protease.

9.3.2.3   Aspartic Endopeptidases

Aspartic acid proteases, commonly known as acidic proteases, are the endopeptidases that depend on aspartic acid residues for their catalytic activity. Acidic proteases have been grouped into three families, namely, pepsin (A1), retropepsin (A2) and enzymes from Pararetro viruses (A3) (Barett, 1995), and have been placed in clan AA.

9.3.2.4   Metallo Endopeptidases

Metallopeptidases are the most diverse of the catalytic types of proteases and are characterized by the requirement for a divalent metal ion for their activity (Barett, 1995). They include enzymes from a variety of origins such as collagenases from higher organisms, hemorrhagic toxins from snake venoms and thermolysins from bacteria (Weaver et al., 1977; Hibbs et al., 1985; Okada et al., 1986; Wilhelm et al., 1987). Based on the specificity of their action, metalloproteases can be divided into four groups, (i) neutral, (ii) alkaline, (iii) Myxobacter I, and (iv) Myxobacter II.

9.4   PROTEIN AND GENETIC ENGINEERING OF MICROBIAL PROTEASES

During the last decade, advances in molecular biology, genetic engineering and computer hardware and software have enhanced the scientist's ability to determine amino acid sequence, 3-dimensional structure of proteins and its encoding genes. Protein engineering improves the structure or function of a protein by altering the amino acid sequence (Outtrup and Boyce, 1990). The requirements for protein engineering of microbial proteases are: a cloned, sequenced and expressed gene coding for specific protein, a system for site-specific mutagenesis, a 3-D structure of the protein based on X-ray crystallography, a computer modeling program to predict 3-D structures from the amino acids sequences, to locate the active-site residues and/or to alter the enzyme properties to suit its commercial applications. Genetic engineering or gene cloning is a rapidly progressing technology for improving the structure-function relationship of genetic systems. It provides an excellent method for the manipulation and control of genes. More than 50% of the industrially important enzymes are now produced from genetically engineered microorganisms. To increase the enzyme production, over-expression of the gene is carried out. The overexpressed protein or enzyme is studied for immunogenic and pathogenic properties. Protease genes from bacteria, fungi, and viruses have been cloned and sequenced (Rao et al., 1998). Zhao et al. (2012), reported a draft genome sequence of *Bacillus pumilus* BA06, a producer of alkaline serine protease with leather-dehairing function.

9.4.1   BACTERIA

The objective of cloning bacterial protease genes has been mainly the over production of enzymes for various commercial applications in the food, detergent and pharmaceutical industries. The virulence of several bacteria is related to the secretion of several extracellular proteases. Gene cloning in these microbes was studied to understand the basis of their pathogenicity and to develop therapeutics against them. Proteases play an important role in cell physiology, and protease gene cloning, especially in *E. coli*, has been attempted to study the regulatory aspects of proteases. Many bacterial cultures like alkaliphilic *Bacillus* sp., *Lactococci* sp., *Streptomyces* sp., *Pseudomonas* sp., *Serratia* sp., *Aeromonas* sp., *Vibrio* sp., and *E. coli* were used for isolation of protease encoding gene and cloning of the same in a suitable host.

9.4.2   FUNGI

As in bacteria, cloning of the protease genes of fungi has been attempted from both the commercial and pathogenicity points of view. Fungi with protease encoding gene are *Mucor* sp., *Rhozopus* sp., *Aspergillus* sp., *Acremonium* sp., *Fusarium* sp., *Tritirachium* sp., *Saccharomycopsis* sp., *Candida* sp. and *Saccharomyces* sp. The protease-encoding genes from these fungi were cloned and sequenced.

9.4.3   VIRUSES

Gene cloning of viral proteases has been undertaken for the isolation and overexpression of the gene and for subsequent screening of inhibitory compounds that may be used in the development of chemotherapeutic agents. Viral protease is responsible for processing polyprotein precursors into the structural proteins of the mature virion. Animal viruses like Herpes viruses, Adenoviruses, Retroviruses, Picorna viruses, Human rhinovirus type 14, Foot-and-mouth disease virus, Encephalomyocarditis virus, Poliovirus and plant viruses like Bean yellow mosaic virus and Zucchini yellow mosaic virus (Singapore isolate) were used as sources for protease gene. To determine the functional domains of this key enzyme, protease genes from various types of viruses have been cloned and sequenced.

The potential contributions of protein and genetic engineering to mankind are enormous and will benefit agriculture, animal husbandry, environmental protection, food production and processing, human health care and manufacture of biochemicals. In general, the application of protein and genetic engineering to proteases will facilitate their use in industry and enable the development of therapeutic agents against the proteases that are important in the life cycle of organisms which cause serious diseases.

9.5   REVIEW ON LEATHER INDUSTRY

9.5.1   LEATHER SECTOR – WORLDWIDE AND INDIA

Leather sector has offered unique advantages in terms of raw material resources, manufacturing capacity, expertise and knowledge base, employment and export potentials (Luthra, 2006). The worldwide general situation of the leather industry sector is summarized in Table 9.3. Total animal population is of 3293 Million Heads (Bovine 1515 Million Heads, Sheep 1058 Million Heads, and Goat 720 Million Heads). Out of this 1192 Million Pieces (Bovine 322 Million Pieces, Sheep 530 Million Pieces, and Goat 340 Million Pieces) of hides and skins production was observed. India has a strong raw material base, ranking first in buffalo and cattle population. 21% of the world's cattle and buffalo and 11% of the world's goat and sheep population are located in India. The country has 189 million cattle (14%), 110 million buffalo (57%), 120 million goat (27%) and 55 million sheep population (6%). India has the manufacturing capacity to produce 10% of the global supply of finished leather (2 billion sq. ft.) and has potential for enhancing the capacity to 4 bn sq. ft. (National Manufacturing Competitiveness Council Report, Government of India, New Delhi, 2010).

The sub segments of leather industry are tanning, footwear and footwear components, leather garments, leather goods (bags, wallets, belts, gloves and accessories), saddlery and harness. India is the second largest producer of footwear. India's annual production is Rs. 25,000 crore, and annual export Rs. 10,700 crore. Annual export growth was 8.20% (2003) against 5.30% global trade growth (2003). The Indian share in global leather trade is of 5.21% of 88 billion US$. Major production centers for leather and leather products in India are Tamil Nadu (Chennai, Ambur, Ranipet, Vaniyambadi, Trichy, Dindigul, Erode), West Bengal (Kolkata), Uttar Pradesh (Kanpur, Agra, Noida), Punjab (Jallandhar), Karnataka (Bangalore),

TABLE 9.3   Worldwide Situation of the Leather Industry Sector

Sample	Bovine	Sheep	Goat	Total
Animal population (Million Heads)	1515	1058	720	3293
Hides and skins production (Million Pieces)	322	530	340	1192

Maharashtra (Mumbai), Andra Pradesh (Hyderabad), Haryana (Ambala, Gurgaon, Panchkula, Karnal), Delhi.

9.5.2   LEATHER INDUSTRY – A POLLUTING INDUSTRY

A raw-hide or skin undergoes a series of treatments prior to tanning and is finally converted into finished leather. The principle leather making protein collagen exists in association with various substances such as keratin, elastin, reticulum, albumin, globulin, mucoid, lipids, carbohydrates and mineral salts. During leather manufacture, the non-collagenous constituents are removed in the various beam house operation such as soaking, dehairing, liming, deliming, bating and pickling and the extent of removal of these constituents decides the characteristics of the final leather. The pollution causing chemicals are lime, sodium sulphide, salts, solvents which arise mainly from the pre-tanning phase of leather processing. Leather industry waste contributes to one of the major industrial pollution problems faced by the world. Blacksmith Technical Advisory Board has declared different worst polluted places at global level. Ranipet from Tamil Nadu state of India is one of them. All these places are shown in Table 9.4. It was suggested that to decrease the pollution load enzyme or enzyme assisted processes should be used in leather manufacture. Lots of activities were started in leather sector. But still Blacksmith advisory board suggests leather industry is one of the reasons of pollution (McCartor and Becker, 2010).

TABLE 9.4   World's Worst Polluted Places of Year 2006

Rank	Polluted places	Country
1	Chernobyl	Ukraine
2	Dzerzhinsk	Russia
3	Haina	Dominican Republic
4	Kabwe	Zambia
5	La Oroya	Peru
6	Linfen	China
7	MaiuuSuu	Kyrgyzstan
8	Norilsk	Russia
9	Ranipet	India
10	RudnayaPristan/Dalnegorask	Russia

9.6   ENZYMES FOR LEATHER PROCESSING

Proteases have a large variety of applications, mainly in the detergent and food industries. Marketed value of microbial protease was 89 % in detergent industry, 8% in rennet, about 1% each in baking, leather and in miscellaneous industries (Figure 9.4). Different leather processing steps are shown in Figure 9.5. Out of these, protease has an important role in beam house operations such as soaking, dehairing and bating processes. While in degreasing process lipase can be used.

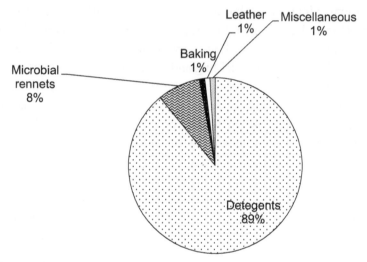

FIGURE 9.4   Pie chart for microbial protease market.

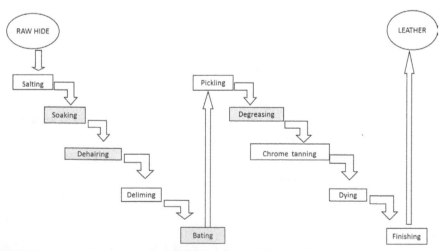

FIGURE 9.5   Steps in leather processing.

9.6.1 SOAKING

Soaking is the first important operation of leather processing. Hides and skins received into the tannery are in four conditions, as green or fresh, wet salted, dry salted or dried. Soaking is for removal of globular proteins, dirt, dung, blood and salts. Soaking agents fall into three categories: chemical agents, surface-active agents and enzymatic agents. Enzymatic agents are biocatalysts. Specific protease and lipase enzymes enhance water uptake by dissolving interfibrillary proteins that cement fibers together, and disperse fats and oils together with dirt and other contaminants present on skin. Proteases produced from *Aspergillus flavus*, *A. parasiticus*, *A. oryzae* and *Bacillus subtilis* were reported for soaking of animal skins. Some amylolytic enzymes from *A. awamori* have also shown soaking application (Kamini et al., 1999). A soaking method using proteolytic enzymes and carbohydrases in the pH range of 5.5 to 10 has been described by Kamini et al. (1999). Soaking enzymes are available in market, but in literature very limited reports are available. Most importantly, the soaking re-hydrates skins to bring it as far as possible back to state of green hides (Zambare et al., 2013). The enzyme treatment is responsible for improving softness, elasticity, area yield with reduction in time of process of leather production. Dry hide and skin require a longer period of soaking for complete rehydration. To minimize the soaking time, enzymes have an important role. The hide pieces were soaked in water bath with different enzyme concentrations and were observed periodically for its water holding capacity as well as the softening of the hide. This increased water holding capacity is due to the hydrolysis of the proteins especially globular proteins present in hide (Zambare et al., 2013). The soaking duration is very important and many researchers have reported it with different durations such as 36–48 h by Puvankrishnan and Dhar (1988), 20 h by Kamini et al. (1999) and 16–18 h by Zambare et al. (2013).

9.6.2 DEHAIRING

The conventional and most widespread way to remove hair from bovine hides is to use lime and sodium sulphide in a hair-burning process. They dissolve the hair and open up the fiber structure. Though the process is efficient for removal of hair, it leads to heavy pollution. Enzyme-assisted unhairing results in a cleaner grain surface and improved area yield and softness.

The use of a specific protease also offers tanneries a number of options. For instance, the sulphide and lime requirements can be reduced by as much as 40% while maintaining the same liming time. Alternatively tanners can shorten the liming time by at least half without any loss of the quality of leather. Another possibility is to avoid the use of amines, which can be converted into carcinogenic compounds. The hair-burning process is the most widespread but a better alternative to this is the hair-saving process, which is environment friendly, and where the hair is not dissolved but can be filtered out from the liming float. It can be used as agriculture fertilizer, soil conditioner, compost and poultry feed. It is possible to reduce the chemical oxygen demand (COD) by up to 50% and biological oxygen demand (BOD) by up to 30% in waste discharges (Zambare et al., 2011).

9.6.3   BATING

To make leather pliable, the hides and skins require an enzymatic treatment before tanning known as bating. During bating, scud is loosened and other unwanted proteins are removed. Bating de-swells swollen pelts and prepares leather for tanning. It makes the grain surface of the finished leather clean, smooth and fine (Nilegaonkar et al., 2006, 2007a, 2007b; Zambare et al., 2010). Traditional methods for bating employed manure of dog, pigeon or hen. These were very unpleasant, unreliable and slow methods (Puvankrishnan and Dhar, 1988). Bating with enzymes is an indispensable operation of leather processing to obtain best quality of leather and it cannot be substituted with a chemical process. Biotechnical developments in science have now completely replaced these methods with those using of industrial enzymes.

9.6.4   DEGREASING

Lipases are enzymes that specifically degrade fat. Lipases hydrolyze not just the fat on the outside of the hides and skins, but also the fat inside the skin structure. Once most of the natural fat has been removed, subsequent chemical treatments such as tanning, re-tanning and dyeing have a better effect. Lipases represent a more environmentally sound method of removing fat. For bovine hides, lipases allow tensides to be replaced completely. For sheepskins, which contain up to 40% fat, the use of solvents is very common and these can also be replaced with lipases and surfactants. Solvents tend to dry out the skin and give it a pale color. If surfactants are used for sheepskins, they are usually not as effective and may be harmful to the environment. Stronger surfactants such as nonylphenol ethoxylate have a better effect but they are more detrimental to the environment. When using lipases, the original surfactant dosage can be reduced by at least 50% in the case of both sheepskins and pigskins. In addition, nonylphenol ethoxylate can be substituted with more biodegradable surfactants.

9.6.5   GLOBAL IMPACT

The use of enzymes in the process of leather formation/production has reduced pollution. The use of sodium sulphide and lime is decreased and hair is salvaged. The use of enzymatic process in leather preparation has provided better quality leather and decreased toxins in the environment of workers. Leather industry waste leads to environmental pollution and health hazards (Malathi and Chakraborty, 1991). Major pollutants from the leather industry that may have significant environmental impact include lime, sulphide and chromium (Riffel et al., 2003). Huge amounts of lime sludge and total solid formation are the main drawbacks of lime (Thanikaivelan et al., 2003). Untreated sulphide can cause major problem in sewers. The severe alkaline condition is a health hazard for the workers (Purushotham et al., 1996). Sulphide also reduces the strength of hair, which directly hampers the recovery of this value added byproduct. The tannery effluent has high total dissolved solids (TDS), BOD and COD (Thanikaivelan et al., 2004; Zambare et al., 2011). Enzymatic dehairing is being increasingly looked upon as a reliable alternative to the conventional lime- sulphide process, avoiding the problems created by sulphide. The advantages of enzymatic dehairing are total elimination of lime and sulphide from the effluent, recovery of hair as a byproduct, elimination of bating process, during deliming. However, these benefits remain unfulfilled, since enzymes are more expensive than the conventional chemical process, and require careful control (Huang et al., 2003; Kanagaraj et al., 2006). The latter point has been identified as the main obstacle in the wider application of enzymes. Pal et al. (1996) reported an enzyme-assisted dehairing with lime and protease from *R. oryzae*, where lime acts as a swelling agent. The potential for commercial use of enzymes in leather production is considerable, because of their properties of being highly efficient and substrate specific. Thus, the substitution of chemical dehairing agents in the leather industry with proteolytic enzymes produced by microorganisms will have an important economic and environmental impact.

Earlier studies using commercial enzymes for sheepskins dehairing revealed a high correlation between dehairing activity and proteolytic activity (Huang et al., 2003). Some microorganisms producing extracellular enzyme with dehairing activity have been described, *Streptomyces* sp. isolated from soil, degraded human hair, chicken feather, silk, wool, and unhaired goatskin (Mukhopadhyay and Chandra, 1993). Among bacteria, strains of *B. subtilis* and *B. amyloliquefaciens* have been characterized for dehairing purposes (Varela et al., 1997; Nashy et al., 2005). However, many proteases are not suitable for dehairing, since they have collagen- degrading activity, which destroys the collagen structure of hide. Therefore, it is essential to explore proteases with dehairing activity but no collagenolytic activity (Huang et al., 2003). A novel protease from *P. aeruginisa* with non collagenase and non keratinase activity but with efficient dehairing activity was reported by Zambare et al. (2011).

9.7 DIFFERENT METHODS OF ENZYMATIC DEHAIRING

Three methods of application are commonly used in the dehairing process: paint, dip and spray method. In the paint method, the enzyme solution is mixed with an inert material like kaolin, made into a thin paste, adjusted to the required pH, applied on the flesh side of hides and skins, piled flesh to flesh, covered with polythene sheets and kept till dehairing takes place. The disadvantage of this method is that it involves a lot of labor. In the dip method of enzymatic dehairing, the hides or skins are kept immersed in the enzyme solution at the required pH in a pit or tub. The disadvantages encountered are the unavoidable dilution of the enzyme solution and un-uniform penetration of enzyme in backbone and neck region. Recently, a novel spraying technique has been adopted for the application of multienzyme concentrate in dehairing. The advantages of this method over the painting and dip methods are: even concentrated solutions can be sprayed, when the enzyme solution is sprayed on the flesh side with force, entry becomes easier, backbone and neck can be sprayed with more amount of enzyme thereby making the process quicker, there is less effluent arising out of this method and after dehairing, hair will be almost free from all the adhering skin tissues.

9.7.1 ENZYMATIC DEHAIRING

Proteolytic enzymes derived from various sources such as microbial, animal and plants have been applied individually or in combination to produce effective dehairing of hides/skins.

9.7.1.1 Microbial Proteases

Microbial proteases are produced from a wide variety of bacteria, fungi and actinomycetes (Table 9.5), (Kanagaraj, 2009; Jaouadi et al., 2011; Jatavathu et al., 2011; Saha et al., 2011; Khan, 2013). Most of the dehairing enzymes were from bacterial species such as *Bacillus* and fungal species such as *Aspergillus*. Extracellular proteases have attracted significant commercial interest than intracellular proteases because of easy recovery. The work on microbial dehairing enzymes was more focused after Blacksmith Advisory Board Report (Grant et al., 2006).

TABLE 9.5 Microbial Enzymes Reported for Dehairing Stage of Leather Processing

Microorganisms	Enzymes	Substrates	References
Alakaligen faecalis	Protease	Goat skin	Thangram and Rajkumar (2002)
Bacillus subtilis P6	Keratinase	Bovine skin	Giongo et al. (2007)
B. amyloliquefaciens	Keratinase	Bovine skin	Giongo et al. (2007)
B. vekesensis P11	Keratinase	Bovine skin	Giongo ct al. (2007)
B. cereus MCM-B326	Protease	Buffalo hide	Nilegaonkar et al. (2007); Zambare et al. (2007)
B. subtilis S14	Keratinase	Bovine hide	Alexander et al. (2005)
B. subtilis	Protease	Goat skin	Arunachalam and Saritha (2009)
B. subtilis	Protease	Cow hide	Prabhavathy et al. (2013)
B. licheniformis	Protease	Bovine hide	Nashy ct al. (2005)
B. licheniformis N-?	Protease	Goat skin	Nadeem et al. (2010)
Bacillus sp. BSA-26	Protease	Buffalo hide	Nilegaonkar et al. (2004)
B. pumilus	Protease	Goat skin	Huang et al. (2003); Wang et al. (2007); Bholay et al. (2012)
B. subilis	Protease	Sheep and goat skin	Varela et al. (1997); Mukhtarand Ul-Haq (2008)
B. subilis strain VV	Protease	Goat hide	Vijayaraghavan et al. (2012)
B. subilis	Protease	Cow skin	Mukesh Kumar et al. (2012)

TABLE 9.5   Continued

Microorganisms	Enzymes	Substrates	References
Euphorbia nivulia Buch.-Ham	Protease	Rat hide	Badgujar and Mahajan (2013)
B. amyloliquefaciens	Protease	Goat skin	George et al. (1995)
Bacillus sp. Kr10	Protease	Bovine hide	Riffel et al. (2003)
Bacillus spp.	Protease	Goat skin	Raju et al. (1996)
Bacillus spp.	Protease	Cow hide	Haile and Gessesse (2012)
Bacillus sp. JB-99	Protease	Goat skin	Shivasharana and Naik (2012)
Bacillus megaterium DSM 319	Protease	Cow hide	Wahyuntari and Hendrawati (2012)
Bacillus thuringiensis TS2	Keratinase	Goat skin	Sivakumar et al. (2013)
Nocardiopsis dassonvillei NRC2aza	Keratinase	Bovine hide	Abdel-Fattah (2013)
Bacillus subtilis	Protease	Buffalo hide	Sathiya (2013)
Euphorbia nivulia Buch.-Ham			
P. aeruginosa PD100	Protease	Cow hide	Najafi et al. (2005)
P. aeruginosa MCM B 327	Protease	Buffalo hide	Zambare et al. (2011a); Pandeeti et al. (2011)
P. aeruginosa MTCC 10501	Protease	Goat skin	Bhoopathy et al. (2013)
P. aeruginosa	Protease	Goat skin	Patre and Dawande (2010)
A. niger	Protease	Goat skin	Patre and Dawande (2010)
A. flavus	Protease	Goat skin	Purushotham et al. (1996)
A. flavus	Protease	Goat skin	Malathi and Chakraborty (1991)
A. fumigatus	Protease	Goat skin	Tharwat et al. (2014)
A. parasiticus	Protease	Sheep skin	Gillespie (1953)
Rhizopus oryzae	Protease	Goat and sheep skin	Pal et al. (1996)
Streptomyces griseus	Protease	Black Angus steer hide	Gehring et al. (2002)
Streptomyces	Proteinase	Goat skin	Mane and Mahadik (2013)
Streptomyces indicusvar	Protease	Goat skin	Guravaiah et al. (2012)
S. moderatus	Proteinase	Goat skin	Chandrashekaran and Dhar (1985)

9.7.1.2 Bacterial Proteases

Proteolytic enzymes derived from a large number of *Bacillus* species have been reported to be useful in dehairing of hides/skins in earlier times. The concrete mixture of dehairing enzymes from *Bacillus subtilis* and *B. cereus* with sodium carbonate, caustic soda and thioglycolic acid, is described in a patent (Monsheimer and Pfleiderer, 1976). The use of protease from *B. subtilis* and other *Bacillus* sp. protease have been reported for dehairing of hides and skins (Raju et al., 1996; Varela et al., 1997; Riffle et al., 2003; Nilegaonkar et al., 2004). Purification and properties of dehairing enzyme from *B. pumilus* have been described by Huang et al. (2003). Likewise, detailed study of dehairing enzyme (keratinase) derived from *B. subtilis* S14 has been extensively studied by Alexander et al. (2005). A lime and sulphide free dehairing of buffalo hide has been obtained with 1% *B. cereus* protease dispersions in water alone, at a room temperature of 28±2°C (Nilegaonkar et al., 2007). *P. aeruginosa* PD100 produced a strong collagenase type protease having application in dehairing of cow hide. In this case purification of the hide was observed at 3 hour of incubation. Because of high collagenase activity this enzyme may not have importance in leather industry (Najafi et al., 2005). A robust and cost efficient process for a paradigm shift from chemical to biological leather processing has been reported by *P. aeruginosa* protease (Nilegaonkar et al., 2006, 2007a, 2007b; Zambare et al., 2011a). This novel enzyme is non-collagenase, non-keratinase and leather specific, efficient at room temperature and has ability to do soaking, dehairing, bating in one step.

9.7.1.3 Fungal Proteases

The protease from *A. flavus* has been first observed to be suitable for dehairing (Malathi and Chakraborty, 1991; Purushotham et al., 1996) of animal hide. Gillespie (1953), has observed that the enzyme preparations from a culture of *A. parasiticus*, exhibit marked depilatory activity on sheep skins. Based on studies on the influence of several cultural conditions and important nutritional factors on the formation of proteases, processes have been patented. *A. parasiticus* on cheap indigenous material for the production of enzyme depilant and bates. Dayanandan et al. (2003, 2004) have reported the keeping quality of the enzyme depilant and other enzyme preparations from

A. tamurri. The qualities of the leathers manufactured by dehairing with the enzyme depilant developed from *A. tamurri* and by lime – sulphide method have been comparatively assessed with respect to their physical parameters. Pal et al. (1996) has recommended the use of enzyme preparations from *Rhizopus oryzae* for dehairing.

9.7.1.4   Actinomycetes Proteases

Use of protease in dehairing of Black Angus steer hide derived from a strain of *Streptomyces griseus* (Gehring et al., 2002) is reported. A multiple proteinase concentrate from the culture filtrate of *Streptomyces moderatus* has been developed and a novel spraying technique has been standardized in place of conventional dip method or painting method (Chandrasekaran and Dhar, 1985).

9.7.2   COMMERCIAL BIOPRODUCTS

Recently, a review on commercial proteases has been published stating the wide range of protease application; but leather industry is one of the main application areas in it (Li et al., 2013). Although studies on the use of enzymes for various stages of leather processing are numerous, the commercial production and application of enzymes in the leather industry is limited. In limited commercial enzyme products, the process, called enzyme-assisted process, needs either lime or sulphide in addition to the enzyme. All the commercial enzymatic products with their company name are summarized in Table 9.6.

TABLE 9.6   Commercial Bioproducts Available for Different Process of Leather Manufacturing

Enzyme	Company	Process	Reference
Nercozyme	—	Dehairing	Andrews and Dempsey (1967)
M-ZYME	—	Dehairing	Jones et al. (1968)
Clarizyme	CLRI	Dehairing	Purushotham et al. (1996)
Erhavit MC	TFL Italia S.p.A.	Dehairing	Cassano et al. (2000)
Pelvit SPH	Together for leather	Dehairing	Thanikaivelan et al. (2004)
Truponat HL	Trumpler	Dehairing	Thanikaivelan et al. (2004)
Mystozyme ECO-S	Catomance technologies	Dehairing	Thanikaivelan et al. (2004)
Basozym-L10	BASF Group	Dehairing	Thanikaivelan et al. (2004)

TABLE 9.6 Continued

Enzyme	Company	Process	Reference
Microdep C	Textan chemicals	Dehairing	Thanikaivelan et al. (2004)
Forezyme LM	La Forestal Tanica	Dehairing	Thanikaivelan et al. (2004)
Biodart	SPIC	Dehairing	Thanikaivelan et al. (2004)
Novolime	Novozyme	Dehairing	Thanikaivelan et al. (2004)
NOVOLime, NUE	Novozyme	Dehairing	www.novozymes.com
Greasex	Novozyme	Dehairing	www.novozymes.com
NOVOBate WB	Novozyme	Bating	www.novozymes.com
NOVOCor AB	Novozyme	Bating	www.novozymes.com
NOVOBate 100	Novozyme	Bating	www.novozymes.com
NOVOBate 115	Novozyme	Bating	www.novozymes.com
Pyrase	Novozyme	Bating	www.novozymes.com
Greasex	Novozyme	Soaking and degreasing	www.novozymes.com
Novocor S	Novozyme	Soaking	www.novozymes.com
NOVOCor ADL	Novozyme	Degreasing	www.novozymes.com
Palkobate	Maps (India)	Bating	www.mapsindia.com
Palkoacid	Maps (India)	Bating	www.mapsindia.com
Palcosoak	Maps (India)	Soaking	www.mapsindia.com
Palkosoak ACP	Maps (India)	Soaking	www.mapsindia.com
Palkodehair	Maps (India)	Dehairing	www.mapsindia.com
Palkodegrease	Maps (India)	Degreasing	www.mapsindia.com
Palkodegrease AL	Maps (India)	Degreasing	www.mapsindia.com
Neosoak	Biocon	Soaking	www.biocon.com
Neobate Alkali	Biocon	Bating	www.biocon.com
Neobate Acid	Biocon	Bating	www.biocon.com
Neodepilase	Biocon	Dehairing	www.biocon.com
Neodegresase	Biocon	Degreasing	www.biocon.com
Neomix WF	Biocon	Ant wrinkling	www.biocon.com
SEBSoak	Advance Enzyme	Soaking	www.enzymeindia.com
SEBate	Advance Enzyme	Bating	www.enzymeindia.com
SEBDgrease	Advance Enzyme	Degreasing	www.enzymeindia.com
SEBLime	Advance Enzyme	Dehairing	www.enzymeindia.com
Freesoak	Textan chemicals	Soaking	www.textanchem.com
Microbate R	Textan chemicals	Bating	www.textanchem.com
Microbate AB	Textan chemicals	Bating	www.textanchem.com

9.8 CONCLUSION

Leather industry waste leads to environmental pollution and health hazards. Major pollutants from the leather industry that may have significant environmental impact include lime, sulphide and chromium. Huge amounts of sludge and total solids formation are the main drawbacks of processing with lime. Untreated sulphide can cause major problem in sewers. The severe alkaline condition is a health hazard for the workers. Use of sulphide also reduces the strength of hair, directly hampering the recovery of this value added byproduct. The tannery effluent has high TDS, BOD and COD. To overcome the environmental pollution problem, use of enzymes as an alternative to the conventional method, would be a paradigm shift from chemical to biological process. Use of a single enzyme in soaking, dehairing and bating operations will be beneficial from industrial point of view. Developing, patenting and utilizing enzymes in the processing, is the need of the day for the leather industry.

KEYWORDS

- **Bating**
- **Dehairing**
- **Hide**
- **Leather**
- **Microbial enzyme**
- **Pollution**
- **Protease**
- **Skin**
- **Soaking**

REFERENCES

Abdel-Fattah, A. M. Novel keratinase from marine *Nocardiopsis dassonvillei* NRC2aza exhibiting remarkable hide dehairing. *Egyptian Pharm J* 2013, 12, 142–147.
Alexander, J. M., Beys de Silva, W. O., Gava, R., Driemeier, D., Henriques, J. A. P., Termignoni, C. Novel keratinase from *Bacillus subtilis* S14 exhibiting remarkable dehairing capabilities. *Appl Environ Microbiol* 2005, 71, 594–596.

Andrew, R. S., Dempsey, M. Some investigations into methods of unhairing II: experiments on enzyme unhairing. *J Am Leather Chem Assoc* 1967, 51, 247–258.

Anwar, A., Saleemuddin, M. Alkaline proteases. *Bioresour Technol* 1998, 64, 175–183.

Arunachalam, C., Saritha, K. Protease enzyme, an eco-friendly alternative for leather industry. *Indian J Sci Technol* 2009, 2, 29–32.

Atsumi, Y., Yamamoto, S., Morihara, K., Fukushima, J., Takeuchi, H., Mizuki, N., Kawamoto, S., Okuda, K. Cloning and expression of the alkaline proteinase gene from *Pseudomonas aeruginosa* IFO 3455. *J Bacteriol* 1989, 171, 5173–5175.

Badgujar, S. B., Mahajan, R. T. Characterization of thermo- and detergent stable antigenic glycosylated cysteine protease of *Euphorbia nivulia* Buch.-Ham. and evaluation of its ecofriendly applications. *The Sci World J* 2013, 1–12.

Barett, A. J. Proteolytic enzymes, aspartic and metallopeptidases. *Methods Enzym* 1995, 248, 183–200.

Barett, A. J. Proteolytic enzymes, serine and cysteine peptidases. *Methods Enzym* 1994, 244, 1–15.

Bholay, A. D., More, S. Y., Patil, V. B., Patil, N. Bacterial extracellular alkaline proteases and its industrial applications. *Int Res J Biol Sci* 2012, 1, 1–5.

Bhoopathy, N. R., Indhuja, D., Shrinivasan, K., Uthirappan, M., Gupta, R., Ramudu, K. N., Chellan, R. Statistical medium optimization of an alkaline protease from *Pseudomonas aeruginosa* MTCC 10501, its characterization and application in leather processing. *Indian J Expt Biol* 2013, 51, 336–342.

Binod, P., Palkhiwala, P., Gaikaiwari, R., Nampoothiri K. M., Duggal, A., Day, K., Pandey, A. Industrial enzymes- present status and future perspectives in India. *J Sci Ind Res* 2013, 72, 271–283.

Cassano, A., Drioli, E., Molinari, R., Grimaldi, D., Caha, F. L., Rossi, M. Enzymatic membrane reactor for eco-friendly goat skin unhairing. *J Soc Leather Technol Chem* 2000, 84, 205–211.

Cerny, G. Studies on the aminopeptidase test for the distinction of gram-negative from gram-positive bacteria. *Eur J App Microbiol Biotechnol* 1978, 5, 113–122.

Chandrasekaran, S., Dhar, S. C. Studies on the development of a multiple proteinases concentrate and its application in the depilation of skins. *Leather Sci* 1985, 32, 297–304.

Dayanandan, A., Kanagraj, J., Sounderraj, L., Govindaraju, R., Rajkumar, G. S. Application of an alkaline protease in leather processing, an ecofriendly approach. *J Cleaner Product* 2003, 11, 533–536.

Dayanandan, A., Marmer, W. N., Dubley, R. Enzymatic dehairing of cattlehide with an alkaline protease isolated from *Aspergillus temarii*. *J Am Leather Chem Asso* 2008, 103, 338–344.

Gehring, A., Dimaio, G. L., Marmer, W. N., Mazenko, C. E, Unhairing with proteolytic enzymes derived from *Streptomyces griseus*. *J Am Leather Chem Assoc* 2002, 97, 406–411.

George, S., Raju, V., Krishnan, M. R. V., Subramanian, T. V., Jayaraman, K. Production of protease by *Bacillus amyloliquefaciens* in solid-state fermentation and its application in the unhairing of hides and skins. *Process Biochem* 1995, 30, 457–462.

Gillespie, J. M. The depilation of sheepskins with enzymes. *J Soc Leather Technol Chem* 1953, 37, 344–353.

Giongo, J. L., Lucas, F. S., Casarin, F., Heeb, P., Brandelli, A. Keratinolytic protease of *Bacillus* species isolated from the Amazon basin showing remarkable dehairing activity. *World J Microbiol Biotechnol* 2007, 23, 375–382.

Grant, T., Bonomo, J., Block, M., Hanrahan, D., Spiegle, J. *The World's Worst Polluted Places: The Top Ten*. Blacksmith Institute: New York, USA, 2006.

Gupta, R., Beg, Q. K., Lorenz, P. Bacterial alkaline proteases, molecular approaches and industrial applications. *Appl Microbiol Biotechnol* 2002, 59, 15–32.

Guravaiah, M., Hatti, I., Prabhakar, T., Daniel, K., Sirisha, P., Nautha, M. T., Anusha, A., Jahnavi, N. Optimization, purification and characterization of alkaline protease enzyme from *Streptomyces indicus* var. GAS-4. *Indian J Eng* 2012, 1, 108–117.

Guzzo, J., Pages, J. M., Duong, F., Lazdunski, A., Murgier, M. *Pseudomonas aeruginosa* alkaline protease: evidence for secretion genes and study of secretion mechanism. *J Bacteriol* 1991, 173, 5290–5297.

Haile, G., Gessesse, A. Properties of alkaline protease c45 produced by alkaliphilic *Bacillus* sp. isolated from Chitu, Ethiopian Soda Lake. *J Biotechnol Biomater* 2012, 2, 1–4.

Hartley, B. S. Proteolytic enzymes. *Annual Review Biochem* 1996, 29, 45–72.

Hibbs, M. S., Hasty, K. A., Seyer, J. M., Kang, A. H., Mainardi, C. L. Biochemical and immunological characterization of the secreted forms of human neutrophil gelatinase. *J Biol Chem* 1985, 260, 2493–2500.

Huang, Q., Peng, Y., Li, X., Wang, H., Zhang, Y. Purification and characterization of an extracellular protease with dehairing function from *Bacilus pumilus*. *Current Microbiol* 2003, 46, 169–173.

International Union of Biochemistry. *Enzyme Nomenclature*. Academic Press, Inc.: Orlando, USA, 1992.

Jaouadi, B., Abdelmalek, B., Jaouadi, N. Z., Bejar, S. The bioengineering and industrial applications of bacterial alkaline proteases, the case of SAPB and KERAB. In *Progress in Molecular and Environmental Bioengineering – From Analysis and Modeling to Technology Applications*, Angelo, C., Ed., 2011, pp. 445–466.

Jatavathu, M., Jatavathu S., Raghavendra Rao, M. V., Sambasiva Rao, K. R. S. Efficient leather dehairing by bacterial thermostable protease. *Intern J Bio-Sci BioTechnol* 2011, 3, 11–26.

Jones, H. W., Cordon, T. C., Windus, W. Light leather from enzyme-unhaired hides. *J Am Leather Chem Assoc* 1968, 63, 480–485.

Kalaskar, V. V., Narayanan, K., Subrahmanyam, V. M., Rao, V. J. Partial characterization and therapeutic application of protease from a fungal species. *Indian Drugs* 2012, 49, 42–46.

Kanagaraj, J. Cleaner leather processing by using enzymes: a review. *Adv Biotech* 2009, 13–18.

Kanagraj, J., Velappan, K. C., Chandrababu, N. K., Sadulla, S. Solid waste generation in the leather industry and its utilization for cleaner environment - a review. *J Sci Ind Res* 2006, 65, 541–548.

Kamini, N. R., Hemachander, C., Geraldine, J., Mala, S., Puvanakrishnan, R. Microbial enzyme technology as an alternative to conventional chemical in leather industry. *Current Sci* 1999, 77, 80–86.

Khalil, A. A., Mohamed, S. S., Taha, F. S., Eva, N. K. Production of functional protein hydrolysates from Egyptian breeds of soybean and lupin seeds. *Afr J Biotechnol* 2006, 5, 907–916.

Khan, F. New microbial proteases in leather and detergent industries. *Innov Res Chem* 2013, 1, 1–6.

Kim, H. S., Yoon, H., Minn, I., Park, C. B., Lee, W. T., Zasloff, M., Kim, S. C. Pepsin-mediated processing of the cytoplasmic histone H2A to strong antimicrobial peptide buforin II. *J Immunol* 2000, 165, 3268–3274.

Kumar, C. G., Takagi, H. Microbial alkaline proteases, from a bioindustrial viewpoint. *Biotechnol Adv* 1999, 17, 561–594.

Li, Q., Yi, L., Marek, P., Iverson, B. L. Commercial protease: present and future. *FEBS Lett* 2013, 587, 1155–1163.

Luthra, Y. K. Indian leather industry. *Leather Age* 2006, 69–71.

Mc Cartor, A., Becker, D. *World's Worst Pollution Problems Report 2010*. Blacksmith Institute: New York, USA, 2010.

Malathi, S., Chakraborty, R. Production of alkaline protease by a new *Aspergillus flavus* isolate under solid substrate fermentation conditions for use as a depilant agent. *Appl Environ Microbiol* 1991, 57, 712–716.

Mane, M., Mahadik, K., Kokare, C. Purification, characterization and applications of thermostable alkaline protease from marine *Streptomyces* SP. D1. *Int J Pharm Bio Sci* 2013, 4, 572–582.

Menon, A. S., Goldberg, A. L. Protein substrates activate the ATP-dependent protease La by promoting nucleotide binding and release of bound ADP. *J Biol Chem* 1987, 262, 14929–14934.

Monsheimer, R., Pfleiderer, E. Method for preparing tenable pelts from animal skins and hides. US Patent 3966551, 1976.

Mukesh Kumar, D. J., Rajan, R., Lawrence, L., Priyadarshini, S., Sandhiya, C., Kalaichelvan, P. T. Destaining and dehairing capabilities of partially purified *Bacillus subtilis* protease from optimized fermentation medium. *Asian J Exp Biol Sci* 2012, 3, 613–620.

Mukhopadhyay, R. P., Chandra, A. Protease of a keratinolytic streptomycete to unhair goat skin. *Indian J Exp Biol* 1993, 31, 557–558.

Mukhtar, H., Ul-Haq, I. Production of alkaline protease by *Bacillus subtilis* and its application as a depilating agent in leather processing. *Pak J Bot* 2008, 40, 1673–1679.

Nadeem, M., Qazi, J. I., Baig, S. Enhanced production of alkaline protease by a mutant of *Bacillus licheniformis* N-2 for dehairing. *Braz Arch Biol Technol* 2010, 53, 1015–1025.

Najafi, M. F., Deobhagkar, D., Deobhagkar, D. Potential application of protease isolated from *Pseudomonas aeruginosa* PD 100. *Electronic J Biotechnol* 2005, 8, 197–203.

Nashy, E. H. A., Ismail, S. A., Ahmady, A. M., Fadaly, H. E., Sayed, N. H. Enzymatic bacterial dehairing of bovine hide by a locally isolated strain of *Bacillus licheniformis*. *J Soc Leather Technol Chem* 2005, 89, 242–249.

National Manufacturing Competitiveness Council Report, Government of India, New Delhi, 2010.

Nilegaonkar, S. S., Zambare, V. P., Kanekar, P. P. Extracellular protease from *Bacillus* sp. BSA-326, application in dehairing of buffalo hide. In *Biotechnological Approaches for Sustainable Development*, Reddy, M. S., Khanna, S. (Eds.), Allied Publishers Pvt. Ltd.: New Delhi, India, 2004, pp. 186–191.

Nilegaonkar, S. S., Zambare, V. P., Kanekar, P. P., Dhakephalkar, P. K., Sarnaik, S. S., Chandrababu, N. K., Rajaram, R., Ramanaiah, B., Ramasami, T., Saikumari Y. K., Balaram, P. A novel protease for industrial application. *Indian Patent* NO. 2471DEL2006, 2006.

Nilegaonkar, S. S., Zambare, V. P., Kanekar, P. P., Dhakephalkar, P. K., Sarnaik, S. S. Production and partial characterization of dehairing protease from *Bacillus cereus* MCM B-326. *Bioresource Technol* 2007, 98, 1238–1245.

Nilegaonkar, S. S., Zambare, V. P., Kanekar, P. P., Dhakephalkar, P. K., Sarnaik, S. S., Chandrababu, N. K., Rajaram, R., Ramanaiah, B., Ramasami, T., Saikumari Y. K., Balaram, P. A novel protease for industrial application. *German Patent* N102007013950.2, 2007a.

Nilegaonkar, S. S., Zambare, V. P., Kanekar, P. P., Dhakephalkar, P. K., Sarnaik, S. S., Chandrababu, N. K., Rajaram, R., Ramanaiah, B., Ramasami, T., Saikumari Y. K., Balaram, P. A novel protease for industrial application. *US Patent* 2008/0220499 A1, 2007b.

Okada, Y., Nagase, H., Harris, E. D. A metalloproteinase from human rheumatoid synovial fibroblasts that digests connective tissue matrix components. Purification and characterization. *J Biol Chem* 1986, 261, 14245–14255.

Outtrup, H., Boyce, C. O. L. Microbial proteinases and biotechnology. In *Microbial Enzymes and Biotechnology, 2nd edition*, Fogarty, W. M., Kelly, C. T. (Eds.), Elsevier Applied Science: London, England, 1990, pp. 227–253.

Pal, S., Banerjee, R., Bhattacharya, B. C., Chakraborty, R. Application of a proteolytic enzyme in tanneries as a depilant agent. *J Am Leather Chem Assoc* 1996, 91, 59–63.

Pandeeti, E. V. P., Pitchika, G. K., Jotshi, J., Nilegaonkar, S. S., Kanekar, P. P., Siddavattam, D. Enzymatic depilation of animal hide, identification of elastase (LasB) from *Pseudomonas aeruginosa* MCM B-327 as a depilating protease. *PLoS One* 2011, 6, 1–8.

Patick, A. K., Potts, K. E. Protease inhibitors as antiviral agents. *Clinical Microbiol Review* 1998, 11, 614–627.

Patre, P. K., Dawande, A. Y. Production and partial purification of protease from *Aspergillus niger* and *Pseudomonas aeruginosa*. *Asiatic J Biotechnol Resour* 2010, 3, 278–281.

Peng, Y., Yang, X. J., Xiao, L., Zhang, Y. Z. Cloning and expression of a fibrinolytic enzyme (subtilisin DFE) gene from *Bacillus amyloliquefaciens* DC-4 in *Bacillus subtilis*. *Res Microbiol* 2004, 155, 167–173.

Poza, M., Miguel, T., Sieirol, C., Villa, T. G. Characterization of a broad pH range protease of *Candida caseinolytica*. *J Appl Microbiol* 2001, 91, 916–921.

Prabhavathy, G., Pandian, M. R., Senthilkumar, B. Identification of industrially important alkaline protease producing *Bacillus subtilis* by 16s rRNA sequence analysis and its applications. *Int J Res Pharma Biomed Sci* 2013, 4, 332–338.

Priem, B., Gitti, R., Bush, C. A., Gross, K. C. Structure of ten free N-glycans in ripening tomato fruit arabinose is a constituent of a plant N-glycan. *Plant Physiol* 1993, 102, 445–458.

Purushotham, H., Malathi, S., Rao, P. V., Rai, C. L., Immanuel, M. M., Raghavan, K. V. Dehairing enzyme by solid-state fermentation. *J Soc Leather Technol Chem* 1996, 80, 52–56.

Puvankrishnan, R., Dhar, S. C. Enzymes in soaking, dehairing, bating and degreasing. In *Enzyme Technology in Beamhouse Practice*, Central Leather Research Institute: Madras, India, 1988.

Raju, A. A., Chandrababu, N. K., Samivelu, N., Rose, C., Rao, N. M. Eco-friendly enzymatic dehairing using extracellular proteases from a *Bacillus* species isolate. *J Am Leather Chem Assoc* 1996, 91, 115–119.

Rao, M. B., Tanksale, A. M., Ghatge, M. S., Deshpande, V. V. Molecular and biotechnological aspects of microbial proteases. *Microbiol Mol Biol Review* 1998, 62, 597–635

Riffel, A., Ortolan, S., Brandelli, A. Dehairing activity of extracellular proteases produced by keratinolytic bacteria. *J Chem Technol Biotechnol* 2003, 78, 855–859.

Saha, M. L., Hashina Begum, K. J. M., Khan, M. R., Gomes, D. J. Bacteria associated with the tannery effluent and their alkaline protease activities. *Plant Tissue Cult Biotech* 2011, 21, 53–61.

Sarah, S. M., Leary, H. L., Nichols, Jr. D. J. Milk protein partial hydrolysate and process for preparation there of. 1996, *US Patent No.* 6,351,867.

Sarrouh, B., Santos, T. M., Miyoshi, A., Dias, R., Azevedo, V Up to-date insight on industrial enzymes applications and global market. *J Bioprocess Biotech* 2012, S4, 002.

Sathiya, G. Production of protease from *Bacillus subtilis* and its application in leather making process. *Int J Res Biotechnol Biochem* 2013, 3, 7–10.

Shivasharana, C. T., Naik, G. R. Ecofriendly applications of thermostable alkaline protease produced from a *Bacillus* sp. JB-99 under solid state fermentation. *Int J Environ Sci* 2012, 3, 956–964.

Sivakumar, T., Balamurugan, P., Ramasubramanian, V. Characterization and applications of keratinase enzyme by *Bacillus thuringiensis* TS2. *Int J Future Biotechnol* 2013, 2, 1–8.

Thangam, E. B., Rajkumar, G. S. Purification and characterization of alkaline protease from *Alcaligenes faecali*. *Biotechnol Appl Biochem* 2002, 35, 149–154.

Thanikaivelan, P., Rao, J. R., Nair, B. U., Ramasami, T. Approach towards zero discharge tanning, role of concentration on the development of eco-friendly liming- reliming process. *J Cleaner Prod* 2003, 11, 79–90.

Thanikaivelan, P., Rao, J. R., Nair, B. U., Ramasami, T. Progress and recent treads in biotechnological methods for leather processing. *Trends Biotechnol* 2004, 22, 181–188.

Tharwat, N., Sayed, M. A., Fadel, H. M. Biochemical and molecular characterization of alkalo-thermophilic proteases purified from *Aspergillus fumigatus*. *J Biol Chem Res* 2014, 31, 236–252.

Traidej, M., Caballero, A. R., Marquart, M. E., Thibodeaux, B. A., O'Callaghan, R. J. Molecular analysis of *Pseudomonas aeruginosa* protease IV expressed in *Pseudomonas putida*. *Investigative Ophthalmol Visual Sci* 2003, 44, 190–196.

Varela, H., Ferrai, M. D., Belobrajdic, L., Vazquez, A., Loperena, M. L. Skin unhairing proteases of *Bacillus subtilis*, production and partial characterization. *Biotechnol Lett* 1997, 19, 755–758.

Vijayaraghavan, P., Vijayan, A., Arun, A., Jenisha, K. K., Vincent, S. G. P. Cow dung, a potential biomass substrate for the production of detergent-stable dehairing protease by alkaliphilic *Bacillus subtilis* strain VV. *Springer Plus* 2012, 1, 1–9.

Wahyuntari, B., Hendrawati. Properties of an extracellular protease of *Bacillus megaterium* DSM 319 as depilating aid of hides. *Microbiol* 2012, 6, 77–82.

Wang, H. Y., Liu, D. M., Liu, Y., Cheng, C. F., Ma, Q. Y., Huang, Q., Zhang, Y. Z. Screening and mutagenesis of a novel *Bacillus pumilus* strain producing alkaline protease for dehairing. *Lett Appl Microbiol* 2007, 44, 1–6.

Weaver, L. H., Kester, W. R., Matthews, B. W. A crystallographic study of the complex of phosphoramidon with thermolysin. A model for the presumed catalytic transition state and for the binding of structures. *J Mol Biol* 1977, 114, 119–132.

Wilhelm, S. M., Collier, I. E., Kronberger, A., Eisen, A. Z., Marmer, B. L., Grant, G. A., Bauer, E. A., Goldberg, G. I. Human skin fibroblast stromelysin, structure, glycosylation, substrate specificity and differential expression in normal and tumorigenic cells. *Proc Natl Acad Sci USA* 1987, 84, 6725–6729.

Zambare, V., Nilegaonkar, S., Kanekar, P. Production of an alkaline protease and its application in dehairing of buffalo hide. *World J Microbiol Biotechnol* 2007, 23, 1569–1574.

Zambare, V., Nilegaonkar, S., Kanekar, P. Application of protease from *Bacillus cereus* MCM B-326 as a bating agent in leather processing. *The IIOAB J* 2010, 1, 18–21.

Zambare, V., Nilegaonkar, S., Kanekar, P. A novel extracellular protease from *Pseudomonas aeruginosa* MCM B-327: Enzyme production and its partial characterization. *New Biotech* 2011a, 28, 173–181.

Zambare, V., Nilegaonkar, S., Kanekar, P. Use of agroresidues for protease production and application in degelatinazation. *Res J BioTechnol* 2011b, 6, 62–65.

Zambare, V., Nilegaonkar, S., Kanekar, P. Protease production and enzymatic soaking of salt-preserved buffalo hides for leather processing. *IIOAB Lett* 2013, 3, 1–7.

Zambare, V., Nilegaonkar, S., Kshirsagar, P., Kanekar, P. Scale up production of protease using *Pseudomonas aeruginosa* MCM B-327 and its detergent compatibility. *J Biochem Technol* 2014, 5, 698–707.

Zhao, C. W., Wang, H. Y., Zhang, Y. Z., Feng, H. Draft genome sequence of *Bacillus pumilus* BA06, a producer of alkaline on serine protease with leather-dehairing function. *J Bacteriol* 2012, 194, 66–68.

CHAPTER 10

GENETIC ENHANCEMENT OF *SACCHAROMYCES CEREVISIAE* FOR FIRST AND SECOND GENERATION ETHANOL PRODUCTION

FERNANDA BRAVIM, MELINA CAMPAGNARO FARIAS,
OEBER DE FREITAS QUADROS, and
PATRICIA MACHADO BUENO FERNANDES

Núcleo de Biotecnologia, Centro de Ciências da Saúde, Universidade Federal do Espírito Santo, Vitória, ES, Brazil

CONTENTS

10.1   INTRODUCTION

The use of ethanol as a fuel additive or directly as a fuel source has grown in popularity due to international governmental regulations and, in some cases, economic incentives based on environmental concerns, as well as a desire to reduce dependence on petroleum. As a consequence, several countries are interested in developing the domestic market for the use of this biofuel. Currently, almost all the produced ethanol is from sugar cane or grain and most often used as a motor fuel, mainly as an additive for gasoline (Mussatto et al., 2010). Bioethanol has aroused great interest due to the high prices and the environmental problems caused by fossil fuels. It is a renewable product, which contributes to the reduction of the greenhouse effect and substantially reduces air pollution, minimizing its impacts on public health. Ethanol is the main biofuel employed worldwide, corresponding to 10% of the world's energy. Moreover, it has been estimated that its use will increase to 27% by 2050 (Iea, 2013).

Consumption of biofuels is projected to rise from 1.3 million barrels of oil equivalent per day (mboe/d) in 2011 to 2.1 mboe/d in 2020, and 4.1 mboe/d in 2035 (Table 10.1). By 2035, biofuels will meet 8% of total road-transport fuel demand, up to 3% from today's use. Combined, the United States, Brazil, the European Union, China and India account for about 90% of world biofuels demand according to the *Outlook* journal, with government policies driving the expansion in these regions (Iea, 2013). The United States remains the largest biofuels market, spurred on by the Renewable Fuel Standard (RFS) through 2022 and assumed continuation of support thereafter, with

TABLE 10.1 Ethanol and Biodiesel Consumption in Road Transport by Region in the New Policies Scenario

Countries/ Regions	Ethanol (mboe/d)		Biodiesel (mboe/d)		Biofuels total (mboe/d)		Transport energy use (%)	
	2011	2035	2011	2035	2011	2035	2011	2035
OECD	0.7	1.5	0.2	0.8	0.9	2.3	4	12
Americas	0.6	1.3	0.1	0.3	0.7	1.6	4	13
United States	0.6	1.2	0.1	0.3	0.7	1.5	5	15
Europe	0	0.2	0.2	0.5	0.2	0.7	4	12
Non-OECD	0.3	1.4	0.1	0.4	0.4	1.8	2	5
E. Europe/Eurasia	0	0	0	0	0	0	0	4
Asia	0	0.7	0	0.1	0.1	0.8	1	4
China	0	0.4	0	0	0	0.4	1	4
India	0	0.2	0	0	0	0.2	0	4
Latin America	0.3	0.8	0.1	0.2	0.4	1	10	20
Brazil	0.2	0.6	0	0.2	0.3	0.8	19	30
World	1	2.9	0.4	1.1	1.3	4.1	3	8
Europe Union	0	0.2	0.2	0.5	0.2	0.7	5	15

consumption increasing from around 0.7 mboe/d to 1.5 mboe/d in 2035, by which time biofuels meet 15% of road-transport energy needs (Iea, 2013). Driven by blending mandates and strong competition between ethanol and gasoline, Brazil remains the second-largest market and continues to have a larger share of biofuels in its transport fuel consumption than any other country. In 2035, biofuels will meet 30% of Brazilian road-transport fuel demand; up to 19% from today's use. Supported by the Renewable Energy Directive and continued policy support, biofuels use in the European Union more than triples over the same period to 0.7 mboe/d in 2035, representing 15% of road-transport energy consumption (Iea, 2013).

The agreement implemented by PEA, Policy Energy Act, followed by EISA, Energy Independence and Security Act, crave that it in 2022 to obtain about 36 billion gallons of ethanol per year, as well as the European Union seeks the use of 10% second-generation biofuels in transport in 2020 (Porzio et al., 2012). In this context, modeling studies indicates that a pilot plant would produce 40,000 t ethanol/year from 240,000 t biomass/year (Porzio et al., 2012). When the subject is biofuels, Brazil stands out as the protagonist. There are more than 35 years of research and development of various technologies involved in the production and use of ethanol from sugar cane, here called first generation biofuel. Today, the country has the lowest production costs, is the largest exporter and the world's second largest producer of this biofuel (Bastos, 2012).

Until December 1st 2013, sugar cane production on Brazil southern region has confirmed the expectation of a record growth in the supply of ethanol for domestic market during the season 2013/2014. This extended offering was enough to cover the entire increase in Brazilian demand for fuels for light vehicles, but is unlikely to be repeated to fully meet the new increases in demand in the future (Brazilian Sugarcane Industry Association, 2014). The numbers confirm the projection released by UNICA (Brazilian Sugarcane Industry Association) at the beginning of October, which pointed to an increment of 3.68 billion liters in ethanol production in this vintage, with the total reaching 25.04 billion liters against 21.36 billion liters in season 2012/2013. Despite all these advantages, many criticisms are made not only on the production of Brazilian ethanol, but also to the other first-generation biofuels produced in the world. For some, the current production of biofuels has generated pressure on both the price of food/feed as on the native forest cover. Although such criticisms do not apply to the production of ethanol based on sugar cane, even so the sustainability of ethanol production in Brazil could be incremented (Bastos, 2012).

Fuel ethanol can be produced from direct fermentation of simple sugars, considered as first generation ethanol production, or polysaccharides, second generation production, like starch or cellulose that can be converted into sugars. Thus, carbohydrate sources can be classified into three main groups: (i) simple sugars: sugarcane (Macedo et al., 2008; Leite et al., 2009); sugar beet (Ogbonna et al., 2001; Içoz et al., 2009); sorghum (Mamma et al., 1995; Prasad et al., 2007; Cheng, 2008) whey (Domingues et al., 2001; Gnansounou et al., 2005; Silveira et al., 2005; Dragone et al., 2009) and molasses (Roukas, 1996); (ii) starches: grains such as maize (Gaspar et al., 2007; Persson et al., 2009); wheat (Nigam, 2001); root crops such as cassava (Amutha and Gunasekaran, 2001; Kosugi et al., 2009; Rattanachomsri et al., 2009); (iii) lignocellulosic biomass: woody material (Ballesteros et al., 2004), straws (Huang et al., 2009; Silva et al., 2010), agricultural waste (Lin and Tanaka, 2006), crop residues (Hahn-Hägerdal et al., 2006).

Fermentation of simple sugar, mainly glucose, to produce ethanol is largely conducted by *Saccharomyces cerevisiae* (Figure 10.1), one of the most studied microorganisms in biofuel research for its ability to quickly

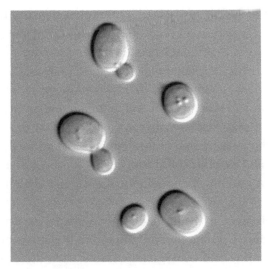

FIGURE 10.1 *Saccharomyces cerevisiae* cells; differential interference contrast (DIC) microscopy photograph of the yeast *S. cerevisiae* showing budded and unbudded cells (http://commons.wikimedia.org/wiki/File:S_cerevisiae_under_DIC_microscopy.jpg).

respond, adapt and develop tolerance to multiple simultaneous stress conditions by regulating the expression of genes involved in different cellular metabolic processes (Gasch et al., 2000; Fernandes, 2005; Bravim et al., 2010). This chapter examines some *S. cerevisiae* breeding technologies used nowadays for ethanol production, considering aspects related to the raw materials, processes and strain alterations.

10.2   FIRST GENERATION ETHANOL PRODUCTION

10.2.1   HISTORICAL BACKGROUND

Brazil is a pioneer in producing ethanol for cars, a program that started in 1927 when it was installed the first alcohol pump that continued to work until early 1930s (Balat and Balat, 2009; Mussatto et al., 2010). The market for fuel ethanol was revived in the 1970s when, for economic reasons, such as world oil crisis and problems with the international sugar market due to its overproduction, the National Alcohol Program (PROÁLCOOL) was created in Brazil. This program was based on the use of sugar cane as feedstock, and it was intended to focus more on the large-scale use of ethanol as a substitute for gasoline. With strong government intervention to increase supply and demand for ethanol, Brazil has developed institutional capacities and technologies for the use of renewable energy on a large scale. In 1984, most new cars sold in Brazil demanded hydrated bio-ethanol (96% bioethanol + 4% water) as fuel. As sugar/ethanol industry has matured and evolved Proálcool program was terminated in 1999, allowing more incentives for private investment and reducing government intervention in allocations and prices. Although Brazilians have driven some cars that run solely on ethanol from 1979, the introduction of new engines that allow drivers to switch between ethanol and gasoline transformed what was once an economic niche of the planet, to a leading example in the use of renewable fuels (Mussatto et al., 2010).

Brazil is a reference in the production of agroenergy. Programs such as ethanol and biodiesel production are attracting the attention of the world by offering economic and ecologically viable alternative in replacement for fossil fuels. Alcohol is the second main source of primary energy in Brazil, originating from agricultural products, which is less polluting and cheaper. Alcohol consumption exceeds the gasoline and biodiesel already has a significant share in the fuel matrix in the country in which the mixture of these fuels with gasoline is mandatory (until 25% ethanol) (Brazil, 2014). Alcohol is the second main source of primary energy in Brazil, originating from agricultural products, which is less polluting and cheaper. The energy content of sugarcane corresponds to 18% of the country's energy matrix and represents a higher share than hydroelectricity (Altieri, 2012). This value includes the solid part, mainly bagasse, which could be used as a substrate for second generation bioethanol production but is currently burned in boilers to generate electricity to the mills and, if there is a surplus, to the national grid. This sugar/energy market employs 4.3 million people (in both direct and indirect jobs) and has a turnover of more than 80 billion dollars annually.

Investment in research is the basis for the development of agricultural production technologies allowing the identification of the most suitable plants, more efficient production systems and regions with agricultural potential. Also, it is mandatory to seek for new yeast strains with better fermentation performance. The Agroenergy Brazilian National Plan explores the strategies and actions to organize and develop proposals for research, development, innovation and technology transfer. The goal is to ensure sustainability and competitiveness of bioenergy production chains (Brazil, 2014).

10.2.2 RAW MATERIAL AND PROCESSES

The biotechnological processes are responsible by the vast majority of ethanol currently produced. About 95% of the world ethanol is of agricultural products (Rossillo-Calle and Walter, 2006). The production of ethanol from crops, such as sugar cane and beet account for about 40% of the total produced ethanol, and approximately 60% corresponding to starch crops. Other more marginal feedstocks that are used or considered to produce first-generation bioethanol include but are not limited to whey, barley, potato wastes, and sugarbeets. Other raw materials more marginals that are used or considered to produce first generation bioethanol include but are not limited to whey, barley, the residues of potatoes, and beets (Lee and Lavoie, 2013). First-generation biofuels include ethanol and biodiesel and are directly related with a biomass that is many times more than edible. Ethanol is typically produced from the fermentation of sugars C6 (mainly glucose) using the classic or genetically modified strains of the yeast *Saccharomyces cerevisiae*. First generation ethanol production is usually performed in three steps: (i) obtainment of a solution of fermentable sugars, (ii) fermentation of sugars into ethanol and (iii) ethanol separation and purification, usually by distillation–rectification–dehydration (Demirbas, 2005).

During processing, the raw material is taken to a cleaning process, which can be by washing or dry cleaning. Then, it is directed to tillage equipment for broth extraction (Figure 10.2). Sugar crops need only a milling process for the extraction of sugars to fermentation (not requiring any hydrolysis step) as shown in Figure 10.2a, becoming a relatively simple process of sugar transformation into ethanol. In this process, ethanol can be fermented directly from cane juice or beet juice, or from molasses generally obtained as a byproduct after the extraction of sugar (Içoz et al., 2009).

Fermentation turns sugars into alcohol. The reactions occur in the fermentation vats, where it mixes the juice (must) and yeast. During the reaction, release of carbon dioxide and formation of side products such as higher alcohols, glycerol and aldehydes occurs. At the end of fermentation, 4–12 hours on average, the alcohol content on the vat is 7–10%, and the mixture is called fermented most. The most is then centrifuged to recover the yeast, which returns to the fermentation tank for a new cycle. And the centrifugation less dense phase is sent to distillation columns for ethanol recovery. Currently ethanol fermentation is carried out mainly by fed-batch processes

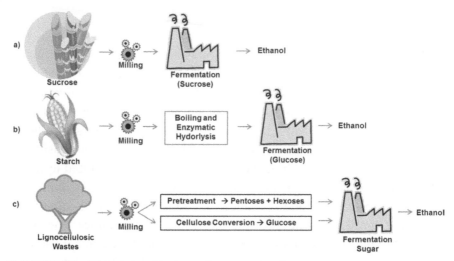

FIGURE 10.2 Flowchart with the main raw materials and processes used for ethanol production. Different types of feedstock, sucrose (A), starch (B) and agricultural waste (C) are processed to generate sugars. These sugars are subsequently converted into fuels and chemicals by a biocatalyst, for example, the yeast *S. cerevisiae*.

with cell recycle, and a small part is produced through multi-stage continuous fermentation with cell recycle (Bastos, 2007).

Although very advantageous for the producers, increases in the sugar price are a problem for the ethanol business. In August 2012, the price of raw sugar was close to US$0.20 per pound while the price for ethanol was US$2.59 per gallon (US$0.68/L). Production of 1 L of ethanol out of raw sugar should cost around US$0.30 to US$0.35, and therefore, the market favored production of raw sugar instead of ethanol (Lee and Lavoie, 2013).

Starch is the most utilized feedstock for ethanol production in North America and Europe. Yeast cannot directly use starch for ethanol production. Therefore, ethanol production from grains involves milling and hydrolysis of starch that has to be wholly broken down to glucose by combination of two enzymes, α-amylase and amyloglucosidase, before it is fermented by yeast to produce ethanol (Mussatto et al., 2010). In this process, starch feedstocks are grounded and mixed with water to produce a mash typically contained 15–20% starch. The mash is then cooked at or above its boiling point and treated subsequently with two enzyme preparation as shown in Figure 10.2b. The first enzyme, amylase, hydrolyzes starch molecules to short chains to glucose, the maltodextrin oligosaccharides. The dextrin and oligosaccharides are further hydrolyzed by other enzymes, pullulanase and glucoamylase, in a process known as saccharification. Saccharification converts all dextrans to glucose, maltose and isomaltose. The mash is then cooled to 28–30°C and yeast is added for fermentation (Lee et al., 2007).

The enzyme generally used for hydrolysis of starch, α-amylase, is rather inexpensive at US$0.04 per gallon of ethanol produced (McAloon et al., 2000). Corn market value in August 2012 was close to US$338/t, leading to the production of 400 to 450 L of ethanol, depending on the process efficiency. Moreover, the by-products value, like post-distillation spent grain used for livestock feed, is a net asset for the whole economical balance of the process (Lee and Lavoie, 2013).

Corn and wheat are mainly employed with these intentions. In tropical countries, other cultures of starch as the tubers (for example, cassava) can be used for commercial production of fuel ethanol (Cardona and Sanchez, 2007; Prasad et al., 2007). The processes of ethanol production using starchy crops are considered well established. Today, most fuel ethanol is produced from corn by either dry-grind (67%) or wet-mill (33%) process. Recent growth in the industry has been predominantly with dry-grind plants because of less capital costs per gallon and incentives for farmer-owned cooperatives (Bothast, 2005).

10.2.3 METABOLIC ENGINEERING OF YEAST FOR FIRST GENERATION ETHANOL PRODUCTION

Industrial application of microorganisms intended to biofuel production should take into account several considerations, including enhancement of product concentration, yield and productivity, simplification of downstream processes, and utilization of inexpensive substrates (Jang et al., 2012). Recently it has been reported the use of metabolically engineered microorganisms for the production of several different biofuels. General strategies for metabolic engineering of microorganisms and specific examples on the use of these strategies for the production of biofuels have been performed.

Owing to its role in bioethanol production, the yeast *S. cerevisiae* is already the most intensively applied microbial cell factory. An important reason for the applicability of *S. cerevisiae* within the field of biotechnology is its susceptibility to genetic modifications by recombinant DNA technology, which has been even further facilitated by the availability of the complete genome sequence of *S. cerevisiae*, published in 1996 (Goffeau et al., 1996). In addition, robustness under process conditions, genetic accessibility and a strong fundamental knowledge base in physiology and systems biology contribute to its current popularity as a 'general purpose' metabolic engineering platform (Krivoruchko et al., 2011; De Jong et al., 2012; Hong and Nielsen, 2012).

Strain improvement of baker's and brewer's yeasts has traditionally relied on random mutagenesis or classical breeding and genetic crossing of two strains followed by screening for mutants exhibiting enhanced properties of interest. However, the approach of random mutation and selection is difficult for further improvement of cellular performance due to the complexity associated with identifying modified gene as a consequence of random mutation (Jang et al., 2012). The developments of sophisticated methods in the field of recombinant DNA technology have enabled us to manipulate a given pathway of interest and hence to improve the cell by a more directed approach (Ostergaard et al., 2000). Thus, metabolic engineering aims at improving cellular performance by introducing of specific genetic perturbations, by performing deletions and/or overexpression of genes, or to introduce whole new genes or pathways into the cell based on engineering tools, which improved our knowledge of cellular physiology and our subsequent engineering results (Park and Lee, 2008; Lee et al., 2011).

Therefore, what distinguishes metabolic engineering from classical applied molecular biology is the use of the directed approach. This implies that it is necessary to have solid knowledge of the system being used considering careful analysis of the cellular system (the analysis part) and construction of the recombinant strain (the synthesis part) (Ostergaard et al., 2000). Usually metabolic engineer of *S. cerevisiae* for the production of first generation ethanol is used to increase their ethanol productivity and/or yield and to keep their highly desired tolerance traits, thus the overexpression and/ or deletion of genes in specific pathways are enough for this, while, the yeast metabolic engineer of yeast to the ethanol of the second generation of ethanol involves the introduction of whole new genes or pathways into the cells to change substrate utilization or the desired end product.

10.2.4 TOOLS AND APPROACHES EMPLOYED IN METABOLIC ENGINEERING FOR THE PRODUCTION OF FIRST GENERATION ETHANOL

In recent years a number of very powerful techniques have been developed that enable a far more in-depth analysis of the cellular physiology.

These include genome sequencing, genome-wide screening, DNA array technology for transcriptome analysis, two-dimensional gel electrophoresis for proteome analysis, gas chromatography-mass spectrometry (GC-MS), and liquid chromatography-mass spectrometry (LC-MS) methods for metabolome analysis, 13C-labeling experiments for metabolic network analysis (Gash et al., 2000; Causton et al., 2001; Förster et al., 2003; Kolkman et al., 2005; Bro et al., 2006; de Godoy et al., 2008; Argueso et al., 2009). Moreover, high-throughput technologies and parallel advances of computational and systems biology has enabled analyzing large amount of omics data for investigating cellular metabolism and physiology at systems-level.

Desirable traits to industrial *S. cerevisiae* strains to ethanol production comprise high ethanol titers (productivity and yield), reduced glycerol formation, as well as to high temperature, osmotic, oxidative stress and possible inhibitor tolerance. Having complete genome sequences and annotations for the organisms of interest, comparative genome analysis is possible to identify the genes or regulatory regions that need to be introduced, deleted, down- or up-regulated to attain a desired metabolic phenotype in yeasts. The comprehensive of all omics technologies coupled with suitable bioinformatics tools could propitiate the development of more robust yeast strains with improved stress tolerance and ethanol fermentation performance to be applied on the ethanol industry.

The detailed genomic structure of some of *S. cerevisiae* industrial strains has started to be determined (Argueso et al., 2009; Stambuk et al., 2009; Babrzadeh et al., 2012), revealing significant structural and sequence variations when compared to laboratory strains or others isolated from nature. These results have opened the door for applying advanced omics technologies and facilitating metabolic engineering strategies. These studies demonstrated for example that chromosomal rearrangements occur near chromosome ends (in regions that do not contain essential genes), they are unlikely to impair meiosis instead, they contain amplifications that improve fitness for industrial fermentations (Argueso et al., 2009; Stambuk et al., 2009). It was proposed that these polymorphisms could play a role in the higher adaptability of these variants throughout the fermentation process (Basso et al., 2008), contributing to their higher productivity.

Another important feature is the amplification of telomeric SNO and SNZ genes, shared by five industrial strains (PE-2, CAT–1, BG–1, SA–1, and VR–1) (Stambuk et al., 2009). These genes are involved in thiamine (vitamin B1) and pyridoxine (vitamin B6) biosyntheses, required for sugar catabolism by yeast. An increased copy number of these genes contributed for efficient growth under thiamine repression, when in medium lacking pyridoxine and with high sugar concentration (Della-Bianca et al., 2013). Transcriptome profiling allows examination of the genome-wide expression levels of mRNAs that can vary with genetic and environmental conditions using DNA microarrays. Reverse-transcribed cDNA is fluorescently labeled and then hybridized with the probes on the DNA microarray. Gene expression can be monitored by measuring the fluorescence intensities of the labeled cDNA that hybridized with the corresponding probes (DeRisi et al., 1997). Based on transcriptome profiling, potential target genes to be manipulated and strategies for strain improvement can be identified by comparing the expression levels of genes between the samples of same strain cultured under different environmental conditions or between strains of different genotypes under identical or different environmental conditions (Hibi et al., 2007; Hirasawa et al., 2010).

In the metabolic engineering field, systematic methodologies to identify the target of gene manipulation from DNA microarray data have been highly desired for conferring useful phenotypes including stress tolerance and substrate availability or improving target product productivity in industrial production (Hirasawa et al., 2010). Furthermore, it is often found that a single mutation (disruption or overexpression of a certain gene) results in a completely different expression pattern, and DNA array technology will therefore be a very powerful technique for analysis of the consequences of the individual genetic changes. However, transcriptional changes constitute only one domain of molecular organization and the complexity of dynamic cellular interactions between all levels including proteins, metabolites, and other molecular species has not yet been adequately addressed (Strassburg et al., 2010). Thus, another tool also used is the proteomic analysis, an effective method to analyze the cellular protein profile under different environments using two-dimensional gel electrophoresis (2DGE) or chromatography-coupled mass spectrometry (Kolkman et al., 2005; Pham et al., 2006; de Godoy et al., 2008). Comparative analysis of proteome profiles between different samples under genetically or environmentally different conditions can be used to identify those proteins showing altered expression levels and those proteins which are post-translationally modified (Lee and Lee, 2010), and has been employed for strain improvement (Han et al., 2003; Aldor et al., 2005; Han et al., 2005). Furthermore, often the pathway activity is directly correlated with the protein concentration, and when gene expression and/or protein-protein interactions are subjugated to metabolic regulation, it is important to quantify the protein levels in the different recombinant strains constructed. Clearly, a detailed proteome analysis may be valuable, but often it is sufficient to measure the levels of the proteins involved in the pathway studied and perhaps some of the regulatory proteins affecting the expression of the relevant genes (Ostergaard et al., 2000).

Metabolomics allows profiling metabolites, which are substrates, products, and intermediates of cellular metabolism, under desired culture conditions by using chromatography coupled with mass spectrometry and nuclear magnetic resonance (NMR). The profiles of metabolites represent the metabolic status of a cell and thus provide information on the physiological changes under genetic and environmental perturbations. Metabolic flux profiles, which can be considered as one of the ultimate phenotypes of a cell, are related closely with cellular metabolic performance (Wittmann and Heinzle, 2001; Liebeke et al., 2011). Fluxomics quantifies metabolic fluxes and collectively represent the metabolic characteristics of a cell under a given condition.

The study of flux control examines the effects of perturbations in the enzymatic activities on the systemic metabolic behavior in order to identify the best enzymatic target(s) for genetic manipulation. It is of interest to overexpress the specific gene(s) encoding the enzyme(s) that exerts the greatest control over flux through a pathway, since overexpression of a whole pathway may be very laborious and is often not possible because overexpression of several genes may have some negative metabolic consequences to the cell (Ostergaard et al., 2000).

10.2.5 IMPROVED SACCHAROMYCES CEREVISIAE TO THE PRODUCTION OF FIRST GENERATION ETHANOL

In the industrial processes using *S. cerevisiae* as host cells, it is known that cells encounter the environmental stresses, such as high osmotic pressure, high temperature, and high ethanol concentration. Since these stresses affect cell growth, substrate uptake, and yeast activity, tolerance to stresses is recognized as one of the useful phenotypes as host cells for industrial production. The genes and pathways involved in stress response and tolerance are good candidate targets for genetic engineering aiming at the development of improved strains.

Taking the above considerations together, many studies has showed successful performance by yeast metabolic engineering for ethanol production, both in laboratory and industrial strains. Many of these studies taken account the environmental stress response (ESR) in which yeast cells respond to a variety of stresses by inducing or repressing specific sets of genes which includes a stereotypical reaction common to all stresses (Gasch et al., 2000; Causton et al., 2001). The function of the expression of these ESR genes during early response to a single dose of stress was found not to be required for survival under that stimulus but it is instead required for the long-term cell protection against future stresses (Berry and Gasch, 2008). Some of examples describing genetically modified strains are listed below.

Our research group investigated recently whether comprehensive analysis of transcriptomic responses of wild yeast submitted to high hydrostatic pressure (HHP) stress might provide valuable information applicable to the broader biotechnology industries (Bravim et al., 2013). HHP is a stress that exerts broad effects on microorganisms with characteristics similar to those of other environmental stresses, such as high temperature, ethanol and oxidative stresses (Fernandes, 2008). The HHP response of wild *S. cerevisiae* shows high correlation with that resulting from increased ethanol concentration or high-temperature stresses, suggesting that HHP may be a useful stress model to employ for selecting suitable *S. cerevisiae* strains for industrial applications (Bravim et al., 2010b). Moreover, HHP treatment can elicit responses that provide cross-protection from multi-stresses. *S. cerevisiae* cells submitted to a mild sublethal pressure treatment (50 MPa for 30 min) followed by a short recovery at atmospheric pressure (0.1 MPa) acquire increased tolerance to heat, ultra-cold shock and high-pressure treatments (Palhano et al., 2004).

Thus, global transcriptional analysis identified genes induced by hydrostatic pressure and demonstrated that at least for one gene related to metabolism and stress response, its overexpression in the wild yeast strain enhanced ethanol production capacity, likely by increasing its tolerance to stress. Overexpression of *SYM1*, involved both in metabolism and tolerance to ethanol under high temperature, did in fact enhance the fermentative capacity of wild-type strain (Bravim et al., 2013). Trott and Morano (2004) previously reported that cells lacking *SYM1* displayed attenuated growth under the combined debilitating effects of both high temperature and high ethanol concentration. Thus, we speculate that overexpression of *SYM1* in the wild *S. cerevisiae* strain may help, to some degree, to boost tolerance to ethanol toxicity.

Corroborating with our results Cao et al. (2010) showed that cells deleted for *GPD2*, which encodes glycerol-3-phosphate dehydrogenase, but simultaneously overexpressing glutamate synthase (*GLT1*), glutamine synthetase (*GLN1*) and *SYM1*, produce approximately 14% more ethanol than the wild-type strain. The gpd2Δ strain is deficient in glycerol biosynthetic pathway and redirects carbon source flow from glycerol to ethanol synthesis; *GLT1* and *GLN1* overexpression reduces surplus nicotinamide adenine dinucleotide (NADH) and increases consumption of excess ATP in the ammonia assimilation pathway. It was speculated that *SYM1* overexpression might cause a general defect in NADH oxidation (NADH/NAD$^+$-ethanol/acetaldehyde shuttle), which occurs on the mitochondrial matrix (Bakker et al., 2001). To compensate for this impairment, NADH could be oxidized by formation of ethanol in the cytoplasm during fermentative metabolism to increase ethanol production.

Ethanol, which is the desired product of the fermentation, is also a particular stress for *S. cerevisiae* and its tolerance is intimately related to the ethanol productivity of this organism (Jones, 1989). It has been reported that yeast strain with the overexpressed *OLE1* gene, which encodes Δ-9 fatty acid desaturase, showed a higher growth yield and ethanol productivity than the control strains under aerobic conditions (Kajiwara et al., 2000). Moreover, another study using analysis of the expression dynamics allowed identification of candidate and key genes for the ethanol-tolerance and ethanol production under the stress, including heat shock proteins, trehalose-glycolysis-pentose phosphate pathways and pleiotropic drug resistance (PDR) gene family (Ma and Liu, 2010).

Another important trait for ethanol production is the thermotolerance. Growth improvement on high temperature is considered to be the crucial factor for the improvement of ethanol fermentation. Under ethanol fermentation conditions, the activity of trehalose-6-phosphate synthase and the accumulation of trehalose were significantly improved by overexpression of the trehalose-6-P synthase gene (*TPS1*). The ethanol fermentation performance of transformants with overexpression of the *TPS1* gene at 38°C was similar to that at 30°C, indicating that *TPS1* gene overexpression had remarkable effect in improving the fermentation capacity of the yeast strain at high temperatures (An et al., 2011). Moreover, it was reported for (Guo et al., 2011) that by overexpressing *TPS1* and *TPS2* (encoding trehalose-6-phosphate phosphatase), the osmotic stress tolerance, growth rate, and ethanol fermentation ability of the yeast strain were improved.

The genome-wide screening could be used together with DNA microarray data to determine the target genes for molecular breeding of yeast to have higher tolerance to sucrose stress. Ando et al. (2005) identified four genes by genome-wide screening related with the high sucrose sensitivity in laboratory yeast strain. The same study suggests that overexpression or up-regulation of these genes might increase the tolerance to high-sucrose stress in industrial process. In a microarray analysis of laboratory yeast strain exposed to 20% w/v glucose (Kaeberlein et al., 2002), an up-regulation of glycerol and trehalose biosynthetic genes was found. Moreover, DNA microarray analyzes of industrial yeast strains submitted to high sugar stress (40% (w/v) sugars) demonstrated an up-regulation of the glycolytic and pentose phosphate pathway and those involved in the formation of acetic acid from acetaldehyde (Erasmus et al., 2003). It is well know that under conditions of stress, acetate formation plays an important role in maintaining the redox balance in yeast cells since they require NAD^+ for this reaction to proceed (Nevoigt and Stahl, 1997).

Considerable efforts have been made to minimize or completely abolish formation of glycerol, the major by-product during current bioethanol production. During anaerobic growth of *S. cerevisiae*, glycerol serves as an essential electron sink for reoxidizing reduced redox cofactors (NADH) generated in biosynthesis. Glycerol formation can be prevented or reduced by deleting one or both genes encoding cytosolic NADH-dependent glycerol-3-phosphate dehydrogenases, *GPD1* and *GPD2* (Ansell et al., 1997). However, a double deletion renders cells unable to grow anaerobically. Deletion of, for example, *GPD2*, results in an increased ethanol yield and decreased glycerol formation, but severely hampers growth and ethanol productivity (Valadi et al., 1998). Alternative approaches aim at engineering cellular redox metabolism to reduce formation of cytosolic NADH. Nissen et al. (2000) deleted the NADPH-dependent glutamate dehydrogenase, *GDH1*, while overexpressing *GLN1* and *GLT1* (encoding glutamine synthetase and glutamate synthase), respectively. The resulting ammonium assimilation pathway consumed NADH as well as ATP and led to a reduction in glycerol yield by 38% while the yield of ethanol was increased by 10%.

A problem associated with reduced ability to produce glycerol in *S. cerevisiae* is that the osmosensitivity as well as the general robustness is reduced (Hohmann, 2002). Efforts have therefore been made to improve stress resistance even when glycerol formation is hampered. Guo et al. (2011) used a more targeted approach where they combined *GPD1* deletion with expression of NADP'-dependent counterpart (GAPN) and overexpression of the trehalose synthesis genes *TPS1* and *TPS2* and hereby obtained a robust, high-yielding strain. This could be of interest since trehalose is also a stress protectant and there is a correlation between accumulation of this compound and thermotolerance as well as ethanol tolerance of *S. cerevisiae* strains (Benjaphokee et al., 2012; Tao et al., 2012).

As previously mentioned the genome-wide expression profile show how the yeast cells respond to different environments stress at transcriptional level. However, increasing evidence shows that mRNA abundance is not always correlated with protein expression levels and enzyme activity as a result of posttranscriptional regulation and modifications (Gygi et al., 1999; Griffin et al., 2002). Therefore, it is essential to study yeast adaptation and other biological processes with proteome studies, which allow the analysis of all proteins present in a certain condition. In addition, all these studies validate the proteomic application for the identification of the molecular bases related with the environmental variations including the fermentative process stresses. A comparative proteomics study among different industrial strains reported a differential regulation of several proteins under various environmental challenges, including Arg1p, Sti1p and Pdc1p. It was speculated that these proteins might have important industrial implications for strain improvement and protection (Caesar et al., 2007). Another study, comparing the proteomic profiling of industrial yeast in realistic continuous and batch/fed-batch industrial fermentation, revealed dynamical changes of the expression levels of enzymes/proteins involved in various stress responses and metabolic pathways (pentose phosphate, glycolysis and gluconeogenesis, and glycerol biosynthetic process), providing a molecular understanding of physiological adaptation of industrial strain for optimizing the performance of industrial bioethanol fermentation (Cheng et al., 2008).

It has been identified features that could contribute to the adequate adaptation of yeast cells to fermentation conditions (Zuzuarregui et al., 2006). The proteomic analyzes of two commercial strains confirmed that the heat shock protein Hsp26p was differentially expressed in different strains and the major cytosolic aldehyde dehydrogenase (Ald6p) in the strain with better behavior during vinification. Moreover, higher levels of enzymes required for sulfur metabolism (Cys4p, Hom6p and Met22p) were observed, which could be related to the production of particular organoleptic compounds or to detoxification processes. Other studies validate the application of a proteomic approach for the identification of the molecular bases in high glucose stress (Pham and Wright, 2008) demonstrated an up-regulation of proteins involved in glycolysis and pentose phosphate pathways in yeast cells submitted to high glucose concentration. A similar approach has shown the overexpression of peroxiredoxin fewer than 20% w/v glucose, being this protein involved in protection against oxidative stress insult (Guidi et al., 2010).

The genome-scale metabolic network model for *in silico* aided metabolic engineering is a promising computational method for accelerating the design of cells with improved and desired properties (Förster et al., 2003). This approach was used to reconstruct metabolic network of *S. cerevisiae* to score a number of strategies for metabolic engineering of the redox metabolism to identify the most optimal strategy in terms of reducing glycerol formation and increasing the ethanol yield on glucose under anaerobic conditions. The best-scored strategy was the heterologous expression of $NADP^+$-dependent glyceraldehyde-3-phosphate dehydrogenase (GAPN) that successfully can be used to decrease the yield of glycerol in order to increase the yield of ethanol in 10% (Bro et al., 2006). Another strategy for ethanol production improvement was found in the study by Hjersted and Henson (2009) who used a different approach, also aided by genome-scale metabolic models, where ethanol production in batch culture is modeled both with steady-state Flux Balance Analysis (FBA) and dynamic FBA. In this study, a combined deletion/overexpression/insertion mutant with improved ethanol productivity capabilities was computationally identified by dynamically screening multiple combinations of the ten metabolic engineering strategies. In addition, computational modeling based on a cyclic approach based on genomics for strain improvement analyzed the whole gene expression profile and flux distribution in the pathway related to the central carbon metabolism under the high osmotic pressure conditions. The results concluded that the glycerol synthetic pathway is significantly responsible for high osmotic stress tolerance and genetically modified strain overexpressing *GPD1* gene showed to shorten the lag-time due to high osmotic stress (Shioya et al., 2007). More recently, a simplified metabolic network to estimate ethanol and yeast biomass in fed-batch production processes was proposed by Barrera-Martínez et al. (2011). The results of these study confirmed that increasing glucose uptake rates, controlled mainly by the glucose concentration in the input flow, produced an up-regulation in reductive catabolism, resulting in higher ethanol excretion. As previously described, starchy biomass is also a potential alternate energy source, however there are economic viability limitations of currents starch-to-ethanol process, including the energy-intensive liquefaction and substantial amounts of exogenous amylases (Viktor et al., 2013).

The yeast *S. cerevisiae* remains the preferred microorganism for ethanol production, but it lacks starch degrading enzymes required for the efficient utilization of starch, thus *S. cerevisiae* cannot utilize starchy materials without liquefaction and saccharification processes. To reduce the cost of ethanol production from starch, researchers have developed direct ethanol-fermenting recombinant *S. cerevisiae* capable of expressing amylolytic genes from yeast, bacteria and fungi (Nonato and Shishido, 1988; Steyn and Pretorius, 1991; Galdino et al., 2011; Yang et al., 2011; Favaro et al., 2012; Viktor et al., 2013). Nowadays, several research groups have made a concerted effort to reduced the cost-effective conversion of raw starch to biofuels through the expression of starch-hydrolyzing enzymes in a fermenting yeast strain to achieve liquefaction, hydrolysis and fermentation (Consolidated Bioprocessing, CBP) by a single organism (Van et al., 2012; Viktor et al., 2013). However, CBP of raw starch would require recombinant *S. cerevisiae* strains to produce sufficient quantities of raw starch degrading enzymes to ensure hydrolysis at a high substrate loading and at moderate temperatures (Robertson et al., 2006).

An interesting recent approach provided an effective strategy to improve protein secretion demonstrating an advance that can induce ER and cytosolic chaperones simultaneously. In this work, it was investigated the effect of heat shock response activation on recombinant protein secretion. *HSF1* gene, a heat shock transcription factor, which can constitutively activate HSR (Heat shock response), was overexpressed in a *S. cerevisiae* recombinant strain with potential production of α-amylase from *Aspergillus oryzae*. The results demonstrated that activation of HSR increased the yield of heterologous α-amylase (Hou et al., 2013).

10.3 PRODUCTION OF SECOND GENERATION ETHANOL

Second generation biofuels are produced from biomass in a more sustainable fashion, which is truly carbon neutral or even carbon negative in terms of its impact on CO_2 concentrations. In the context of biofuel production, the term 'plant biomass' refers largely to lignocellulosic material as this makes up the majority of the cheap and abundant nonfood materials available from plants (Gomez et al., 2008; Zabaniotou et al., 2008). Biomass used for the production of second generation biofuels is generally divided into three main categories: homogeneous, such as chips of white wood, almost homogeneous, such as agricultural and forest waste, and not homogeneous, including low-value feed stocks such as municipal solid waste. The price for this biomass is significantly less than the price for vegetable oil, corn, and sugarcane, which is an incentive. On the other hand, such biomass is generally more complex to convert to fuel and the production of ethanol from biomass is still dependent on new technologies to reduce costs (Lee and Lavoie, 2013).

Due to the wide availability of agricultural residues, it was estimated that 491 billion liters of biofuel can be generated from waste lignocellulosic biomass, expanding up to 16 times its annual production (Sarkar et al., 2012). In the United States, the residual biomass generated is estimated at around 1.4 billion tons of dry matter per year, 30% originated from forests. In this context, such biomasses can supply in large scale production of this fuel, using different agro-industrial residues (Cardona and Sanchez, 2007; Hu and Wen, 2008; Cardona et al., 2009). Currently, waste derived from lignocellulosic most promising materials to be used in bioprocesses are sugarcane bagasse, rice straw, corn and wheat, from South America, Asia, United States and Europe, respectively (Cheng et al., 2008). Asia generates about 667.6 million tons of rice straw and 145.2 million tons of wheat straw, while America produced 140.86 million tons of corn stover (Sarkar et al., 2012).

Open field burning is already banned in many countries in Western Europe and some other countries have considered it seriously. Less than 1% of corn straw is collected for industrial processing and about 5% is used as animal feed and bedding. More than 90% of corn straw in United States is left in the fields. Sugarcane bagasse has its prominent use as a fuel for boilers and for cogeneration of electricity (Banerjee et al., 2010). Additionally, the pulp industry also generates industrial waste containing high content of cellulose fibers (approximately 80 %) with sufficient potential to also become raw material. This is due, mainly, to the presence of these materials in abundance and the non-necessity of the steps of pretreatment for many of them, a time that the process is performed to the delignification of pulp, which allows the removal of a large part of the lignin to produce pulp (Silva et al., 2010).

As shown in Figure 10.2c, the basic process steps in producing ethanol from lignocellulosic biomass are: (i) pretreatment to render cellulose and hemicellulose more accessible to the subsequent steps; pretreatment generally involves a mechanical step to reduce the particle size and a chemical pretreatment (diluted acid, alkaline, solvent extraction, steam explosion among others) to make the biomass more digestible; (ii) acid or enzymatic hydrolysis to break down polysaccharides to simple sugars; (iii) fermentation of the sugars (hexoses and pentoses) to ethanol using microorganisms; and (iv) separating and concentrating the ethanol produced by distillation–rectification–dehydration (Sánchez and Cardona, 2008).

10.3.1 CHEMICAL CONSTITUTION

Cellulosic or lignocellulosic material is a generic term to describe the main constituents in most plants; in other words, cellulose, hemicellulose and lignin, which composition depends not only on the type of plant, but also growth conditions (Barl et al., 1991), the part of the selected plant (Brown, 1999), and the time of harvest. Cellulose is a linear homopolysaccharide composed of β-D-glucopyranose units, linked by β-$(1{\rightarrow}4)$-glycosidic bonds. Cellulose fibers provide wood's strength. Cellobiose is the smallest repetitive unit of cellulose and can be converted into glucose residues (Dogaris et al., 2013). There are different degrees of ordering in microfibrillar structure of cellulose with very ordered crystalline regions, and amorphous regions with the lowest order. The crystallinity is the linearity of cellulose molecules formed by the intermolecular hydrogen bonds (Delmer, 1999). The high crystallinity regions are difficult to reach by solvents and reagents. In contrast, amorphous regions are more accessible and more susceptible to chemical reagents.

Hemicellulose, a component that comes in a variety of forms, is a polymer of pentose and hexose sugars, as well as sugar acids (Saha, 2000; Saha, 2003). The most common hemicelluloses are xylan, mannan, and arabinofuranosyl, which differ in their composition and arrangement, but primarily consist of xylose, arabinose, and glucose (Nigam, 2001). Hemicellulose has glycosidic linkages of the type β-1, 3 and β-1, 4. The polymer branched and with a degree of polymerization less than 200, coats the cellulose to form a cellulose-hemicellulose cell wall domain. Hemicellulose prevent the parallel fibers of cellulose polymers to collapse, and also allows a weak interaction between one fiber and another, forming a network (Buckeridge et al., 2008). Arabinoxylans and xylans are the hemicellulose chains responsible for cellulose microfibrils and lignin connection.

The hemicellulose can suffer attacks at intermediate positions along the structure, releasing oligomers made of many sugar molecules, and these can be successively broken into smaller oligomers, until one molecule of a simple sugar is formed (Coughlan and Ljungdahl, 1988). Unlike cellulose, hemicelluloses are composed of various kinds of sugar units. These macromolecules are branched, have short chains and are soluble in strong alkali solutions. They are essentially amorphous, being more susceptible to chemical pretreatments. Lignin is a polyphenolic compound or phenylpropanoid, which is formed by polymerisation of phenyl propane (monolignols) units and its constituents are the three major compounds: p-coumaryl alcohol, coniferyl alcohol and sinapilic alcohol. The ratio of these three compounds results in different types of lignins, those formed by the combination of coniferyl alcohol and p-coumaryl alcohol have more complex structures than those formed by coniferyl alcohol and sinapilic alcohol. Cellulose microfibrils, hemicellulose and lignin form structures called macrofibrils, which are organized into fibrils forming fibers, which mediate the structural stability of plant cell wall (Rubin, 2008; Buranov and Mazza, 2009). It is known that polymers derived from sinapilic alcohol are primarily responsible for the linkage with the hemicellulose fraction contained in the biomass (Sun and Cheng, 2002).

10.3.2   PRETREATMENT

The most important processing challenge in the production of biofuel is the pretreatment of the biomass. Lignocellulosic biomass is composed of three main constituents namely hemicellulose, lignin and cellulose. Pretreatment methods refer to the solubilization and separation of one or more of these components of biomass. It makes the remaining solid biomass more accessible to further chemical or biological treatment (Demirbas, 2005). The lignocellulosic complex is comprised of a matrix of cellulose and hemicellulose chains linked to lignin. The pretreatment is done to break down the matrix in order to reduce the degree of crystallinity of the cellulose and to increase the fraction of amorphous cellulose, the most appropriate way to enzymatic attack (Sánchez and Cardona, 2008). The acid pretreatment promotes hemicellulose hydrolysis, while the alkali results in the removal of part of the lignin fraction, exposing the cellulose fibers and cellulose gets more accessible to enzymatic attack (Hahn-Hägerdal et al., 2006).

Pretreatment is undertaken to bring about a change in the macroscopic and microscopic size and structure of biomass as well as submicroscopic structure and chemical composition as shown in Figure 10.3. It makes the lignocellulosic biomass susceptible to quick hydrolysis with increased yields of monomeric sugars (Mosier et al., 2005). The steam explosion is a pretreatment technology successfully applied to various cellulosic biomass types (hard and soft woods, as well as agricultural residues) (Sun and Cheng, 2002). This process can occur with or without the presence of chemical catalyst (sulfuric acid, sulfur dioxide, sodium hydroxide

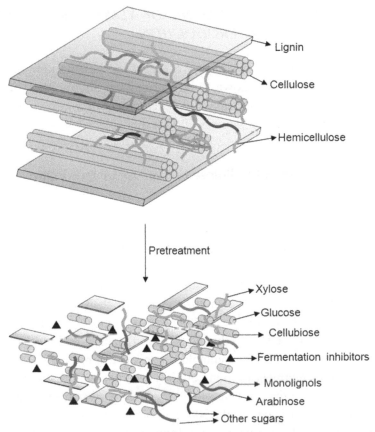

FIGURE 10.3 Lignocellulosic biomass is composed of three main constituents: lignin, cellulose and hemicellulose. Pretreatment methods lead to the solubilization and separation of one or more of these components of biomass.

and ammonia) (Öhgren et al., 2007) at high temperatures (160–290°C) for a certain period of time (a few seconds to several minutes) before pressure is released explosively. Goals of an effective pretreatment process are: (i) formation of sugars directly or subsequently by hydrolysis; (ii) to avoid loss and/or degradation of sugars formed; (iii) to limit formation of inhibitory products (iv); to reduce energy demands; and (v) to minimize costs. Physical, chemical, physicochemical and biological treatments are the four fundamental types of pretreatment techniques employed. In general, a combination of these processes is used in the pretreatment step (Sarkar et al., 2012). Figure 10.4 summarizes the advantages and disadvantages of each one.

10.3.3   METABOLIC ENGINEERING OF YEAST FOR SECOND-GENERATION ETHANOL PRODUCTION

The development of industrial strains of *S. cerevisiae* with the addition of new metabolic capabilities has shown to be a prominent alternative. Therefore, several groups have been pursuing the development of a new robust strain able of transform cellulosic biomass into ethanol.

10.3.3.1   Conversion of Cellulose to Glucose by Enzymatic Hydrolysis

Current technology for conversion of cellulose to ethanol requires chemical or enzymatic conversion of the substrate to fermentable sugars followed by fermentation by a microorganism such as *S. cerevisiae* (Lynd et al., 2005). The capacity to degrade cellulose implies the synthesis of the entire cellulolytic system. Cellulase is a multi-enzyme which is formed by several proteins. It catalyzes the conversion of cellulose to glucose in an enzymatic hydrolysis-based biomass to ethanol process (Maurya et al., 2012). One strategy for development of a suitable organism to produce ethanol from biomass involves engineering non-cellulolytic organisms that exhibit high product yields to produce a heterologous cellulase system enabling cellulose utilization (Den Haan et al., 2007). Various strains of bacteria (aerobic species such as *Pseudomonas* and *Actinomycetes*, facultative anaerobes such as *Bacillus* and *Cellulomonas* and strict anaerobes such as *Clostridium*) and fungi (*Trichoderma reesei, Aspergillus niger* and *Trichoderma viride*)

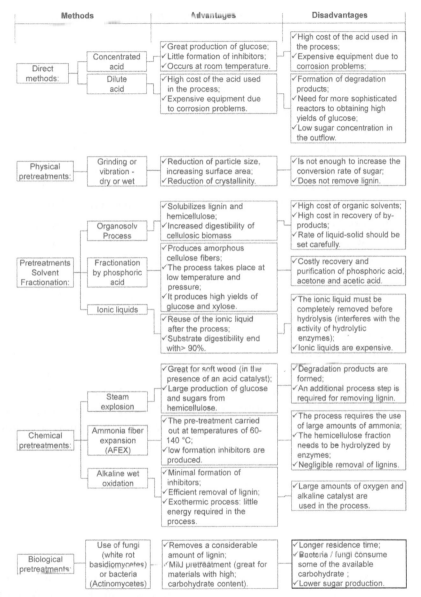

Methods		Advantages	Disadvantages
Direct methods:	Concentrated acid	✓ Great production of glucose; ✓ Little formation of inhibitors; ✓ Occurs at room temperature.	✓ High cost of the acid used in the process; ✓ Expensive equipment due to corrosion problems;
	Dilute acid	✓ High cost of the acid used in the process; ✓ Expensive equipment due to corrosion problems.	✓ Formation of degradation products; ✓ Need for more sophisticated reactors to obtaining high yields of glucose; ✓ Low sugar concentration in the outflow.
Physical pretreatments:	Grinding or vibration - dry or wet	✓ Reduction of particle size, increasing surface area; ✓ Reduction of crystallinity.	✓ Is not enough to increase the conversion rate of sugar; ✓ Does not remove lignin.
Pretreatments Solvent Fractionation:	Organosolv Process	✓ Solubilizes lignin and hemicellulose; ✓ Increased digestibility of cellulosic biomass	✓ High cost of organic solvents; ✓ High cost in recovery of by-products; ✓ Rate of liquid-solid should be set carefully.
	Fractionation by phosphoric acid	✓ Produces amorphous cellulose fibers; ✓ The process takes place at low temperature and pressure; ✓ It produces high yields of glucose and xylose.	✓ Costly recovery and purification of phosphoric acid, acetone and acetic acid.
	Ionic liquids	✓ Reuse of the ionic liquid after the process; ✓ Substrate digestibility end with > 90%.	✓ The ionic liquid must be completely removed before hydrolysis (interferes with the activity of hydrolytic enzymes); ✓ Ionic liquids are expensive.
Chemical pretreatments:	Steam explosion	✓ Great for soft wood (in the presence of an acid catalyst); ✓ Large production of glucose and sugars from hemicellulose.	✓ Degradation products are formed; ✓ An additional process step is required for removing lignin.
	Ammonia fiber expansion (AFEX)	✓ The pre-treatment carried out at temperatures of 60-140 °C; ✓ low formation inhibitors are produced.	✓ The process requires the use of large amounts of ammonia; ✓ The hemicellulose fraction needs to be hydrolyzed by enzymes; ✓ Negligible removal of lignins.
	Alkaline wet oxidation	✓ Minimal formation of inhibitors; ✓ Efficient removal of lignin; ✓ Exothermic process: little energy required in the process.	✓ Large amounts of oxygen and alkaline catalyst are used in the process.
Biological pretreatments	Use of fungi (white rot basidiomycetes) or bacteria (Actinomycetes)	✓ Removes a considerable amount of lignin; ✓ Mild pretreatment (great for materials with high; carbohydrate content).	✓ Longer residence time; ✓ Bacteria / fungi consume some of the available carbohydrate ; ✓ Lower sugar production.

FIGURE 10.4 Summary of advantages and disadvantages of the different methods of pretreatment and hydrolysis of cellulosic material.

produce cellulase. *T. reesei* has a long history of safe use in industrial scale enzyme production (Lejeune et al., 1995).

Efficient degradation of cellulosic biomass requires the synergistic action of the cellulolytic enzymes endoglucanase, cellobiohydrolase, and β-glucosidase. Endoglucanase first cuts the crystalline cellulose in an amorphous zone and cellobiohydrolase subsequently cleave these large insoluble chunks of cellulose into smaller, soluble cellodextrins which can be used by the cell. The most common cellodextrins are: cellobiose, cellotriose, cellotetraose, cellopentaose and cellohexaose. The β-glucosidase converts cellobiose to glucose residues or even waste cellooligosaccharide (Sandgren et al., 2013). Once *S. cerevisiae* cannot make use of cellulosic materials, these materials must first undergo saccharification to glucose, followed by ethanol production (Fujita et al., 2002). Generally, new pathways can be easily constructed in *S. cerevisiae* with well defined genetic engineering, to improve the efficiency and yield of direct ethanol production from cellulose. For example, heterologous expression of endoglucanase and β-glucosidase (Jeon et al., 2009); engineered minicellulosomes on the cell surface (Fan et al., 2012); co-expression of cellulases and cellodextrin transporter in the cell surface (Yamada et al., 2013).

Although there are many reports describing the use of recombinant yeast *S. cerevisiae* to express cellulolytic enzymes, the cellulose degradation efficiency has not been sufficiently improved for ethanol production when compared with low-cost first generation ethanol (Lynd et al., 2005; La Grange et al., 2010; Yamada et al., 2013). The cellulase used in simultaneous saccharification and fermentation is easily inhibited by fermentation products such as ethanol and fermentation conditions (pH and temperature), which would decrease the efficiency of ethanol production from cellulosic materials and increase the cost. Furthermore, some cellulosic materials such as rigid wood are very hard in hydrolyzing to fermentable sugar using only enzymes produced by microorganisms (Yu and Zhang, 2003).

10.3.3.2   Fermentation of Pentoses

The yeast *S. cerevisiae* is still the dominant organism in the production of ethanol by several factors, but unfortunately it is unable to efficiently metabolize pentoses (Demeke et al., 2013). In nature there are many species of bacteria, filamentous fungi and other species of yeasts which are naturally capable to metabolize pentoses. However none of these microorganisms has the robustness of *S. cerevisiae* (Weber et al., 2010). The development of strains of bacteria such as *Escherichia coli* and yeast as *Scheffersomyces stipitis* with ethanol rate production and tolerance still lag behind industrial strains of *S. cerevisiae* (Huang et al., 2009). Alternatively, breeding *S. cerevisiae* in order to metabolize pentoses has shown better results (Nevoigt, 2008).

Different pathways are available in nature for metabolism of arabinose and xylose; which are converted to xylulose 5-phosphate (intermediate compound) to enter the pentose phosphate pathway as shown in Figure 10.5. In yeasts, xylose is first reduced by xylose reductase to xylitol, which in turn is oxidized to xylulose by xylitol dehydrogenase. In bacteria and some anaerobic fungi, xylose isomerase is responsible for direct conversion of xylose to xylulose. Xylulose is finally phosphorylated to xylulose-5-phosphate by xylulokinase. In fungi, L-arabinose is reduced to L-arabitol (by arabinose reductase), L-xylulose (by arabitol dehydrogenase), xylitol (by L-xylulose reductase). Xylitol is finally converted to xylulose (by xylitol dehydrogenase), whose activity is also part of xylose utilization pathways. In bacteria, L-arabinose is converted to L-ribulose (by L-arabinose isomerase), L-ribulose-5-P (by L-ribulokinase) and finally D-xylulose-5-P (by L-ribulose-5-P 4-epimerase) (Bettiga et al., 2008).

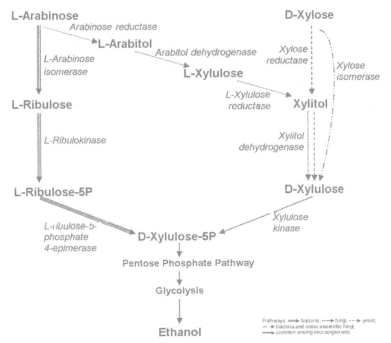

FIGURE 10.5 Different pathways available in nature for metabolism of arabinose and xylose.

The genes for the L-arabinose utilization pathway of *Lactobacillus plantarum*, were engineered overexpression of the *S. cerevisiae* genes, encoding the enzymes of the nonoxidative pentose phosphate pathway. The resulting *S. cerevisiae* strain showed high rates of arabinose consumption and a high ethanol yield, during anaerobic growth on L-arabinose as the sole carbon source. However, ethanol production from sugar mixtures containing glucose and arabinose are less effective; which is prejudicial for application in industrial ethanol production by lignocellulosic biomass (Wisselink et al., 2007).

Important progress has been made in engineering pentose fermentation capacity into the yeast *S. cerevisiae*, mainly on advances in research about heterologous xylose metabolism, reprogramming of the native pathways for xylose fermentation and simultaneous co-fermentation of sugars derived from lignocellulosic biomass (Ohgren et al., 2006). The strategies that are frequently used to metabolism engineering of xylose fermenting by *S. cerevisiae* have employed the pathways: xylose isomerase, xylose reductase and xylitol dehydrogenase. Both approaches require overexpression of xylulokinase, which connects xylulose to the endogenous pentose phosphate pathway of *S. cerevisiae* (Kim et al., 2013). However, this pathway represent a cofactor redox imbalance, because xylose reductase can use both NADPH and NADH, whereas xylitol dehydrogenase uses only NAD^+, elevating xylitol accumulation and, therefore, low ethanol productivity. Xylitol formation could be reduced by providing sufficient oxygen. However, aeration has high cost and lowers ethanol yield (Demeke et al., 2013). Despite decades of research, the use of xylose reductase and xylitol dehydrogenase for ethanol production is still not viable because this pathway cannot be operated optimally under anaerobic conditions. Studies for the optimization of the xylose metabolism are still ongoing; therefore, it is still difficult to determine which pathway is better for engineering *S. cerevisiae* (Wei et al., 2013).

10.3.4   FERMENTATION INHIBITORS

Lignocellulose waste hydrolysis to fermentable sugars still brings many challenges, because the degradation of cellulose and hemicellulose are hampered by the structure and composition of lignocellulosic biomass (Himmel et al., 2007). Due to its structural recalcitrance, a chemical and physicochemical pretreatment is a required stage to obtain potentially fermentable sugars (Agbor et al., 2011). However, the severe condition applied in the pretreatment liberates fermentation inhibitors, such as acetic acids, formic acids, furan derivatives, and phenolic compounds (Blaschek and Boateng, 2009). Acetic acid released during solubilization and hydrolysis of biomass is generally found at a high concentration. Formic acid is typically present at lower concentrations than acetic acid, but is more toxic to *S. cerevisiae* (Hasunuma et al., 2011a). Furan derivatives, such as 2-furaldehyde (furfural) and 5-hydroxymethyl-2-furaldehyde (5-HMF) are formed by dehydration of pentoses and hexoses, respectively. Due to lignin break, several phenolic compounds are generated, such as syringaldehyde, vanillin, 4-hydroxybenz-aldehyde and ferulic acid (Tomas-Pejo et al., 2010).

To enhance the production of ethanol in the presence of inhibitors of fermentation, several approaches to genetically engineering strains of *S. cerevisiae* were responsible for increased tolerance to inhibitors (Nevoigt, 2008). The overexpression of homologous or heterologous genes encoding these enzymes has been successfully applied for the detoxification of inhibi-tors, such as derivatives of furan and phenolic compounds (Hasunuma and Kondo, 2012). To identify genes involved in furfural tolerance, trials were made in a metagenomic library *S. cerevisiae* mutants which exhibit growth better in the presence of furfural. The overexpression of *ZWF1*, encoding glucose-6-phosphate dehydrogenase, in *S. cerevisiae* allowed growth at fur-fural concentrations that are normally toxic (Gorsich et al., 2006). Another strategy was to engineered overexpression of *ADH6*, encoding a strictly NADPH-dependent alcohol dehydrogenase. This genetic modification could increase the resistance against furfural, reducing the amount of 5-HMF in the crude extract of biomass (Petersson et al., 2006).

Treatment with the enzymes peroxidase and laccase, obtained from the ligninolytic fungus *Trametes versicolor*, has been shown to increase the maximum ethanol productivity in a hemicellulose hydrolysate (Palmqvist and Hahn-Hagerdal, 2000). To detoxify phenolic inhibitors, laccase from *T. versicolor* was expressed in *S. cerevisiae*. The laccase expressing strain had the ability to convert coniferyl aldehyde at a faster rate than a control strain not expressing laccase, which enabled faster growth and ethanol fer-mentation in the presence of coniferyl aldehyde (Larsson et al., 2001).

The overexpression of the *PAD1* (phenylacrylic acid decarboxylase gene) from *S. cerevisiae*, resulted in an ethanol productivity improvement in the presence of ferulic acid and cinnamic acid. The *PAD1* overexpressing strain also showed 22–25%, 24–29% and 40–45%, faster glucose consumption rate, ethanol production rate and mannose consumption rate, respectively, in a dilute acid hydrolysate of under aerobic and oxygen-limited conditions (Larsson et al., 2001). During the fermentation of xylose with different concentrations of formic acid, microarray analysis revealed the expression of seven genes from *S. cerevisiae* (*FDH1, FDH2, ALD5, HXK2, VCX1, GPD2,* and *HXT4*) is up regulated in response to increasing doses of formic acid. To reinforce the capability for formic acid breakdown, the *FDH1* gene was also overexpressed in the xylose-fermenting recombinant yeast strain. This modification allowed the yeast to rapidly decompose excess formic acid (Hasunuma et al., 2011b).

While there has been progress in increased resistance of *S. cerevisiae* of inhibiting substance isolated, is important to note that different inhibitory compounds are present in lignocellulosic hydrolysates, besides having additive or synergistic effects between them (Klinke et al., 2004). Furthermore, the relevant conditions in fermentations on an industrial scale can further increase the deleterious effects of the inhibitors (Graves et al., 2007). Another factor is the composition of hydrolyzate strongly depends on the source material and the pretreatment method. One strategy would the multigene cloning, but the peculiarity of each fermentation may fail in this context, since various metabolic pathways inserted within of *S. cerevisiae* may worsen the fermentation. Alternative approaches would be the isolation of new super-tolerant strains or alternative strategies to establish pretreatment production lower levels of toxic substances (Nevoigt, 2008).

10.4   CONCLUSION AND FUTURE PERSPECTIVES

Cost-effective production of ethanol from lignocellulosic materials remains a major challenge, primarily due to the high recalcitrance of these materials. Typically, high cost is associated with the large quantity of enzymes required for the complete hydrolysis of cellulose to ethanol. Thus, evolutionary engineering is a useful strategy to obtain a robust strain, while an important challenge for metabolic engineering is identification of gene targets that have importance for improvement of cellular robust. Tolerant yeast strains can be obtained by enhancing the genetic background (Hasunuma and Kondo, 2012). A systems biology approach using analysis such as transcriptomics, proteomics and metabolomics which present a dynamic view of gene expression, protein and metabolite biosynthesis in yeast stress responses, will facilitate metabolic engineering to optimize microorganisms for industrial ethanol production from lignocellulosic materials (Demeke et al., 2013).

KEYWORDS

- **Alcohol**
- **Biofuels**
- **Biomass**
- **Cellulose**
- **First generation ethanol**
- **Hemicelluloses**
- **High ethanol titers**
- **Laccase**
- **Lignocellulosic biomass**
- **Metabolic engineering of yeast**
- **Pentoses**
- **Pretreatment**
- ***Saccharomyces cerevisiae***
- **Second generation ethanol**
- **Stress tolerance**
- **Sugar**
- **Transcriptional analysis**
- **Xylose**

REFERENCES

Agbor, V. B., Cicek, N., Sparling, R., Berlin, A., Levin, D. B. Biomass pretreatment: Fundamentals toward application. *Biotechnol Adv.* 2011, 29, 675–685.

Aldor, I. S., Krawitz, D. C., Forrest, W., Chen, C., Nishihara, J. C., Joly, J. C., Champion, K. M. Proteomic profiling of recombinant *Escherichia coli* in high-cell-density fermentations for improved production of an antibody fragment biopharmaceutical. *Appl Environ Microbiol.* 2005, 71, 1717–1728.

Altieri, A. Overview and perspectives: Brazilian sugarcane industry. Brazilian Sugarcane Industry Association UNICA, 2012, 7, 10–11.

Amutha, R., Gunasekaran, P. Production of ethanol from liquefied cassava starch using co-immobilized cells of *Zymomonas mobilis* and *Saccharomyces diastaticus*. *J Biosci Bioeng*. 2001, 92, 560–564.

An, M. Z., Tang, Y. Q., Mitsumasu, K., Liu, Z. S., Shigeru, M., Kenji, K. Enhanced thermotolerance for ethanol fermentation of *Saccharomyces cerevisiae* strain by overexpression of the gene coding for trehalose-6-phosphate synthase. *Biotechnol Lett*. 2011, 33, 1367–1374.

Ando, A., Suzuki, C., Shima, J. Survival of genetically modified and self-cloned strains of commercial baker's yeast in simulated natural environments: Environmental risk assessment. *Appl Microbiol Biot*. 2005, 71, 7075–7082.

Ansell, R., Granath, K., Hohmann, S., Thevelein, J. M., Adler, L. The two isoenzymes for yeast NAD$^+$-dependent glycerol 3-phosphate dehydrogenase encoded by *GPD1* and *GPD2* have distinct roles in osmoadaptation and redox regulation. *EMBO J*. 1997, 16, 2179–2187.

Argueso, J. L., Carazzolle, M. F., Mieczkowski, P. A., Duarte, F. M., Netto, O. V., Missawa, S. K., Galzerani, F., Costa, G. G., Vidal, R. O., Noronha, M. F., Dominska, M., Andrietta, M. G., Andrietta, S. R., Cunha, A. F., Gomes, L. H., Tavares, F. C., Alcarde, A. R., Dietrich, F. S., McCusker, J. H., Petes, T. D., Pereira, G. A. Genome structure of a *Saccharomyces cerevisiae* strain widely used in bioethanol production. *Genome Res*. 2009, 19, 2258–2270.

Babrzadeh, F., Jalili, R., Wang, C., Shokralla, S., Pierce, S., Robinson-Mosher, A., Nyren, P., Shafer, R., Basso, L., de Amorim, H., de Oliveira, A., Davis, R., Ronaghi, M., Gharizadeh, B., Stambuk, B. Whole-genome sequencing of the efficient industrial fuel-ethanol fermentative *Saccharomyces cerevisiae* strain CAT–1. *Mol Genet Genomics*. 2012, 287, 485–494.

Balat, M., Balat, H. Recent trends in global production and utilization of bio-ethanol fuel. *Appl Energy* 2009, 86, 2273–2282.

Ballesteros, M., Oliva, J. M., Negro, M. J., Manzanares, P., Ballesteros, I. Ethanol from lignocellulosic materials by a simultaneous saccharification and fermentation process (SFS) with *Kluyveromyces marxianus* CECT 10875. *Process Biochem*. 2004, 39, 1843–1848.

Bakker, B. M., Overkamp, K. M., van Maris, A. J. A. Stoichiometry and compartmentation of NADH metabolism in *Saccharomyces cerevisiae*. *FEMS Microbiol Rev*. 2001, 25, 15–37.

Banerjee, S., Mudliar, S., Sen, R., Giri, B., Satpute, D., Chakrabarti, T. Commercializing lignocellulosic bioethanol: technology bottlenecks and possible remedies. *Biofuels, Bioprod. Biorefin*. 2010, 4, 77–93.

Barl, B., Biliaderis, C. G., Murray, D. M., MacGregor, A. W. Combined chemical and enzymic treatments of corn husk lignocellulosics. *J Sci Food Agric*. 1991, 56, 195–214.

Barrera-Martínez, I., González-García, R. A., Salgado-Manjarrez, E., Aranda-Barradas, J. S. A simple metabolic flux balance analysis of biomass and bioethanol production in *Saccharomyces cerevisiae* fed-batch cultures. *Biotech Bioprocess Eng*. 2011, 16, 13–22.

Basso, L. C., Amorim, H. V., Oliveira, A. J., Lopes, M. L. Yeast selection for fuel ethanol production in Brazil. *FEMS Yeast Res*. 2008, 8, 1155–1163.

Bastos, V. D. *Etanol, Alcoolquímica e Biorrefinarias*. BNDES Setorial: Rio de Janeiro. 2005, pp. 25–38.

Bastos, V. D. *Ethanol and biorefineries*. BNDES Setorial: Rio de Janeiro. 2012, pp. 5–48.

Benjaphokee, S., Koedrith, P., Auesukaree, C., Asvarak, T., Sugiyama, M., Kaneko, Y., Boonchird, C., Harashima, S. CDC19 encoding pyruvate kinase is important for high-temperature tolerance in *Saccharomyces cerevisiae*. *New Biotechnol.* 2012, 29, 166–176.

Berry, D. B., Gasch, A. P. Stress-activated genomic expression changes serve a preparative role for impending stress in yeast. *Mol Biol Cell*. 2008, 19, 4580–4587.

Bettiga, M., Hahn-Hagerdal, B., Gorwa-Grauslund, M. F. Comparing the xylose reductase/xylitol dehydrogenase and xylose isomerase pathways in arabinose and xylose fermenting *Saccharomyces cerevisiae* strains. *Biotechnol Biofuels*. 2008, 1, 16.

Blaschek, H., Boateng, A. New biofuels and biomass chemicals. *Appl Biochem Biotechnol*. 2009, 154, 180–181.

Bothast, R.J. New technologies in biofuel production. *Agricultural Outlook Forum* 2005, 10, 19–21.

Bravim, F., de Freitas, J. M., Fernandes, A. A., Fernandes, P. M. High hydrostatic pressure and the cell membrane: stress response of *Saccharomyces cerevisiae*. *Ann N Y Acad Sci*. 2010a, 1189, 127–132.

Bravim, F., Palhano, F. L., Fernandes, A. A., Fernandes, P. M. Biotechnological properties of distillery and laboratory yeasts in response to industrial stresses. *J Ind Microbiol Biotechnol*. 2010b, 37, 1071–1079.

Bravim, F., Lippman, S. I., da Silva, L. F., Souza, D. T., Fernandes, A. A., Masuda, C. A., Broach, J. R., Fernandes, P. M. High hydrostatic pressure activates gene expression that leads to ethanol production enhancement in a *Saccharomyces cerevisiae* distillery strain. *Appl Microbiol Biotechnol*. 2013, 97, 2093–2107.

Brazil, C. *Ministry of Agriculture, Livestock and Food Supply.* Mapa: Brasilia, 2014, pp. 47–49.

Bro, C., Regenberg, B., Förster, J., Nielsen, J. *In silico* aided metabolic engineering of *Saccharomyces cerevisiae* for improved bioethanol production. *Metab Eng*. 2006, 8, 102–111.

Brown, R. M. Cellulose structure and biosynthesis. *Pure Appl Chem*. 1999, 71, 767–775.

Buckeridge, M. S., Silva, G. B., Cavalari, A. A. Cell wall. Guanabara Koogan: Rio de Janeiro, 2008, pp. 165–181.

Buranov, A. U., Mazza, G. Extraction and purification of ferulic acid from flax shives, wheat and corn bran by alkaline hydrolysis and pressurized solvents. *Food Chem*. 2009, 115, 1542–1548.

Caesar, R., Palmfeldt, J., Gustafsson, J. S., Pettersson, E., Hashemi, S. H., Blomberg, A. Comparative proteomics of industrial lager yeast reveals differential expression of the cerevisiae and non-cerevisiae parts of their genomes. *Proteomics* 2007, 7, 4135–4147.

Cao, L., Kong, Q., Zhang, A., Chen, X. Overexpression of *SYM1* in a gpdDelta mutant of *Saccharomyces cerevisiae* with modified ammonium assimilation for optimization of ethanol production. *J Taiwan Inst Chem Eng*. 2010, 41, 2–7.

Cardona, C. A., Quintero, J. A., Paz, I. C. Production of bioethanol from sugarcane bagasse: status and perspectives. *Bioresour Technol*. 2009, 101, 4754–4766.

Cardona, C. A., Sánchez, J. Fuel ethanol production: process design trends and integration. *Bioresour Technol*. 2007, 98, 2415–2457.

Causton, H. C., Ren, B., Koh, S. S., Harbison, C. T., Kanin, E., Jennings, E. G., Lee, T. I., True, H. L., Lander, E. S., Young, R. A. Remodeling of yeast genome expression in response to environmental changes. *Mol Biol Cell*. 2001. 12, 323–337.

Cheng, J. S., Qiao, B., Yuan, Y. J. Comparative proteome analysis of robust *Saccharomyces cerevisiae* insights into industrial continuous and batch fermentation. *Appl Microbiol Biotechnol*. 2008, 81, 327–338.

Coughlan, M. P., Ljungdahl, L. G. *Biochemistry and genetics of cellulose degradation*. Academic Press: New York, 1988, pp. 11–15.

de Jong, B., Siewers, V., Nielsen, J. Systems biology of yeast: enabling technology for development of cell factories for production of advanced biofuels. *Curr Opin Biotechnol*. 2012, 23, 624–630.

de Godoy, L. M., Olsen, J. V., Cox, J., Nielsen, M. L., Hubner, N. C., Fröhlich, F., Walther, T. C., Mann, M. Comprehensive mass-spectrometry based proteome quantification of haploid versus diploid yeast. *Nature* 2008, 455, 1251–1254.

Della-Bianca, B. E., Basso, T. O., Stambuk, B. U., Basso, L. C., Gombert, A. K. What do we know about the yeast strains from the Brazilian fuel ethanol industry? *Appl Microbiol Biotechnol*. 2013, 97, 979–991.

Delmer, D. P. Cellulose biosynthesis: Exciting times for a difficult field of study. *Annu Rev Plant Physiol Plant Mol Biol*. 1999, 50, 245–276.

Demeke, M. M., Dietz, H., Li, Y., Foulquie-Moreno, M. R., Mutturi, S., Deprez, S., Den Abt, T., Bonini, B. M., Liden, G., Dumortier, F., Verplaetse, A., Boles, E., Thevelein, J. M. Development of a D-xylose fermenting and inhibitor tolerant industrial *Saccharomyces cerevisiae* strain with high performance in lignocellulose hydrolysates using metabolic and evolutionary engineering. *Biotechnol Biofuels*. 2013, 6, 89.

Demirbas, A. Bio-ethanol from cellulosic materials: a renewable motor fuel from biomass. *Energy Source* 2005, 27, 327–337.

Den Haan, R., McBride, J. E., La Grange, D. C., Lynd, L. R., Van Zyl, W. H. Functional expression of cellobiohydrolases in *Saccharomyces cerevisiae* towards one-step conversion of cellulose to ethanol. *Enzyme Microb Technol*. 2007, 40, 1291–1299.

DeRisi, J. L., Iyer, V. R., Brown, P. O. Exploring the metabolic and genetic control of gene expression on a genomic scale. *Science* 1997, 278, 680–686.

Dogaris, I., Diomi, M., Dimitris, K. Biotechnological production of ethanol from renewable resources by *Neurospora crassa*: an alternative to conventional yeast fermentations? *Appl Microbiol Biotechnol*. 2013, 97, 1457–1473.

Domingues, L., Lima, N., Teixeira, J. A. Alcohol production from cheese whey permeate using genetically modified flocculent yeast cells. *Biotechnol Bioeng*. 2001, 72, 507–514.

Dragone, G., Mussatto, S. I., Oliveira, J. M., Teixeira, J. A. Characterization of volatile compounds in an alcoholic beverage produced by whey fermentation. *Food Chem*. 2009, 112, 929–935.

Erasmus, D. J., van der Merwe, G. K., van Vuuren, H. J. Genome-wide expression analyzes: Metabolic adaptation of *Saccharomyces cerevisiae* to high sugar stress. *FEMS Yeast Res*. 2003, 3, 375–399.

Fan, L. H., Zhang, Z. J., Yu, X. Y., Xue, Y. X., Tan, T. W. Self-surface assembly of cellulosomes with two miniscaffoldins on *Saccharomyces cerevisiae* for cellulosic ethanol production. *Proc Natl Acad Sci USA*. 2012, 109, 13260–13265.

Favaro, L., Jooste, T., Basaglia, M., Rose, S. H., Saayman, M., Görgens, J. F., Casella, S., van Zyl, W. H. Codon-optimized glucoamylase sGAI of *Aspergillus awamori* improves starch utilization in an industrial yeast. *Appl Microbiol Biotechnol*. 2012, 95, 957–968.

Fernandes, P. M. How does yeast respond to pressure? *Braz J Med Biol Res*. 2005, 38, 1239–1245.

Fernandes, P. M. B. *Saccharomyces cerevisiae* response to high hydrostatic pressure. In *High-Pressure Microbiology*. Michiels, C., Bartlett, D. H., Aertsen, A. (Eds.), American Society of Microbiology: Washington, 2008, pp. 145–166.

Förster, J., Famili, I., Fu, P., Palsson, B. Ø., Nielsen, J. Genome-scale reconstruction of the *Saccharomyces cerevisiae* metabolic network. *Genome Res*. 2003, 13, 244–253.

Fujita, Y., Takahashi, S., Ueda, M., Tanaka, A., Okada, H., Morikawa, Y., Kawaguchi, T., Arai, M., Fukuda, H., Kondo, A. Direct and efficient production of ethanol from cellulosic material with a yeast strain displaying cellulolytic enzymes. *Appl Environ Microbiol*. 2002, 68, 5136–5141.

Galdino, A. S., Silva, R. N., Lottermann, M. T., Alvares, A. C., de Moraes, L. M., Torres, F. A., de Freitas, S. M., Ulhoa, C. J. Biochemical and structural characterization of amy1: an α-amylase from *Cryptococcus flavus* expressed in *Saccharomyces cerevisiae*. *Enzyme Res*. 2011, 157294.

Gasch, A. P., Spellman, P. T., Kao, C. M., Carmel-Harel, O., Eisen, M. B., Storz, G., Botstein, D., Brown, P. O. Genomic expression programs in the response of yeast cells to environmental changes. *Mol Biol Cell*. 2000, 11, 4241–4257.

Gaspar, M., Kálmán, G., Réczey, K. Cornfiber as a raw material for hemicellulose and ethanol production. *Process Biochem*. 2007, 42, 1135–1139.

Gnansounou, E., Dauriat, A., Wyman, C. E. Refining sweet sorghum to ethanol and sugar: economic trade-offs in the context of North China. *Bioresour Technol*. 2005, 96, 985–1002.

Goffeau, A., Barrell, B. G., Bussey, B., Davis, R. W., Dujon, B., Feldmann, H., Galibert, F., Hohcisel, J. D., Jacq, C., Johnston, M., Louis, E. J., Mewes, H. W., Murakami, Y., Philippsen, P., Tettelin, H., Oliver, S. G. Life with 6000 genes. *Science* 1996, 274, 546–567.

Gomez, L. D., Clare, G. S., McQueen-Mason, J. Sustainable liquid biofuels from biomass: the writing's on the walls. *New Phytol*. 2008, 178, 473–485.

Gorsich, S. W., Dien, B. S., Nichols, N. N., Slininger, P. J., Liu, Z. L., Skory, C. D. Tolerance to furfural-induced stress is associated with pentose phosphate pathway genes *ZWF1*, *GND1*, *RPE1*, and *TKL1* in *Saccharomyces cerevisiae*. *Appl Microbiol Biotechnol*. 2006, 71, 339–349.

Graves, T., Narendranath, N., Power, R. Development of a "Stress Model" fermentation system for fuel ethanol yeast strains. *J Inst Brew*. 2007, 113, 263–271.

Griffin, T. J., Gygi, S. P., Ideker, T., Rist, B., Eng, J., Hood, L., Aebersold, R. Complementary profiling of gene expression at the transcriptome and proteome levels in *Saccharomyces cerevisiae*. *Mol Cell Proteomics*. 2002, 1, 323–333.

Guidi, F., Magherini, F., Gamberi, T., Borro, M., Simmaco, M., Modesti, A. Effect of different glucose concentrations on proteome of *Saccharomyces cerevisiae*. *Biochim Biophys Acta*. 2010, 1804, 1516–1525.

Guo, Z., Zhang, L., Ding, Z., Shi, G. Minimization of glycerol synthesis in industrial ethanol yeast without influencing its fermentation performance. *Metab Eng*. 2011, 13, 49–59.

Gygi, S. P., Rochon, Y., Franza, B. R., Aebersold, R. Correlation between protein and mRNA abundance in yeast. *Mol Cell Biol*. 1999, 19, 1720–1730.

Hahn-Hägerdal, B., Galbe, M., Gorwa-Grauslund, M. F., Lidén, G., Zacchi, G. Bio-ethanol-the fuel of tomorrow from the residues of today. *Trends Biotechnol*. 2006, 24, 549–556.

Han, M. J., Jeong, K. J., Yoo, J. S., Lee, S. Y. Engineering *Escherichia coli* for increased productivity of serine-rich proteins based on proteome profiling. *Appl Environ Microbiol*. 2003, 69, 5772–5781.

Han, M. J., Yoon, S. S., Lee, S. Y. Proteome analysis of metabolically engineered *Escherichia coli* producing Poly(3-hydroxybutyrate). *J Bacteriol*. 2005, 183, 301–308.

Hasunuma, T., Kondo, A. Consolidated bioprocessing and simultaneous saccharification and fermentation of lignocellulose to ethanol with thermotolerant yeast strains. *Process Biochem*. 2012, 47, 1287–1294.

Hasunuma, T., Sanda, T., Yamada, R., Yoshimura, K., Ishii, J., Kondo, A. Metabolic pathway engineering based on metabolomics confers acetic and formic acid tolerance to a recombinant xylose-fermenting strain of *Saccharomyces cerevisiae*. *Microb Cell Fact*. 2011a, 10, 2.

Hasunuma, T., Sung K. M., Sanda, T., Yoshimura, K., Matsuda, F., Kondo, A. Efficient fermentation of xylose to ethanol at high formic acid concentrations by metabolically engineered *Saccharomyces cerevisiae*. *Appl Microbiol Biotechnol*. 2011b, 90, 997–1004.

Hibi, M., Yukitomo, H., Ito, M., Mori, H. Improvement of NADPH-dependent bioconversion by transcriptome-based molecular breeding. *Appl Environ Microbiol*. 2007, 73, 7657–7663.

Himmel, M. E., Ding, S. Y., Johnson, D. K., Adney, W. S., Nimlos, M. R., Brady, J. W., Foust, T. D. Biomass recalcitrance: Engineering plants and enzymes for biofuels production. *Science* 2007, 315, 804–807.

Hirasawa, T., Furusawa, C., Shimizu, H. *Saccharomyces cerevisiae* and DNA microarray analyzes: what did we learn from it for a better understanding and exploitation of yeast biotechnology? *Appl Microbiol Biotechnol*. 2010, 87, 391–400.

Hjersted, J. L., Henson, M. A. Steady-state and dynamic flux balance analysis of ethanol production by *Saccharomyces cerevisiae*. *IET Syst Biol*. 2009, 3, 167–179.

Hohmann, S. Osmotic stress signaling and osmoadaptation in yeasts. *Microbiol Mol Biol Rev*. 2002, 6, 300–372.

Hong, K. K., Nielsen, J. Metabolic engineering of *Saccharomyces cerevisiae*: a key cell factory platform for future biorefineries. *Cell Mol Life Sci*. 2012, 69, 2671–2690.

Hou, J., Osterlund, T., Liu, Z., Petranovic, D., Nielsen, J. Heat shock response improves heterologous protein secretion in *Saccharomyces cerevisiae*. *Appl Microbiol Biotechnol*. 2013, 97, 3559–3568.

Hu, Z., Wen, Z. Enhancing enzymatic digestibility of switch grass by microwave assisted alkali pretreatment. *Biochem Engin J*. 2008, 38, 369–378.

Huang, C. F., Lin, T. H., Guo, G. L., Hwang, W. S. Enhanced ethanol production by fermentation of rice straw hydrolysate without detoxification using a newly adapted strain of *Pichia stipitis*. *Bioresour Technol*. 2009, 100, 3914–3920.

Içoz, E., Tugrul, M. K., Saral, A., Içoz, E. Research on ethanol production and use from sugar beet in Turkey. *Biomass Bioenerg*. 2009, 33, 1–7.

International Energy Agency. International energy outlook. *OECD/IEA* 2013, 6, 204–210.

Jang, Y. S., Park, J. M., Choi, S., Choi, Y. J., Seungdo, Y., Cho, J. H., Lee, S. Y. Engineering of microorganisms for the production of biofuels and perspectives based on systems metabolic engineering approaches. *Biotechnol Adv*. 2012, 30, 989–1000.

Jeon, E., Hyeon, J. E., Eun, L. S., Park, B. S., Kim, S. W., Lee, J., Han, S. O. Cellulosic alcoholic fermentation using recombinant *Saccharomyces cerevisiae* engineered for the production of *Clostridium cellulovorans* endoglucanase and *Saccharomycopsis fibuligera* beta-glucosidase. *FEMS Microbiol Lett*. 2009, 301, 130–136.

Jones, R. P. Biological principles for the effects of ethanol. *Enzyme Microbiol Technol*. 1989, 11, 130–153.

Kaeberlein, M., Andalis, A. A., Fink, G. R., Guarente, L. High osmolarity extends life span in *Saccharomyces cerevisiae* by a mechanism related to calorie restriction. *Mol Cell Biol.* 2002, 22, 8056–8066.

Kajiwara, S., Aritomi, T., Suga, K., Ohtaguchi, K., Kobayashi, O. Overexpression of the *OLE1* gene enhances ethanol fermentation by *Saccharomyces cerevisiae. Appl Microbiol Biotechnol.* 2000, 53, 568–74.

Kim, S. R., Park, Y. C., Jin, Y. S., Seo, J. H. Strain engineering of *Saccharomyces cerevisiae* for enhanced xylose metabolism. *Biotechnol Adv.* 2013, 31, 851–861.

Klinke, H. B., Thomsen, A. B., Ahring, B. K. Inhibition of ethanol-producing yeast and bacteria by degradation products produced during pre-treatment of biomass. *Appl Microbiol Biotechnol.* 2004, 66, 10–26.

Kolkman, A., Slijper, M., Heck, A. J. Development and application of proteomics technologies in *Saccharomyces cerevisiae. Trends Biotechnol.* 2005, 23, 598–604.

Kosugi, A., Kondo, A., Ueda, M., Murata, Y., Vaithanomsat, P., Thanapase, W., Araia, T., Moria, Y. Production of ethanol from cassava pulp via fermentation with a surface-engineered yeast strain displaying glucoamylase. *Renew Energy* 2009, 34, 1354–1358.

Krivoruchko, A., Siewers, V., Nielsen, J. Opportunities for yeast metabolic engineering: lessons from synthetic biology. *Biotechnol J.* 2011, 6, 262–276.

La Grange, D. C., den Haan, R., van Zyl, W. H. Engineering cellulolytic ability into bioprocessing organisms. *Appl Microbiol Biotechnol.* 2010, 87, 1195–1208.

Larsson, S., Cassland, P., Jonsson, L. J. Development of a *Saccharomyces cerevisiae* strain with enhanced resistance to phenolic fermentation inhibitors in lignocellulose hydrolysates by heterologous expression of laccase. *Appl Environ Microbiol.* 2001, 67, 1163–1170.

Lee, J. W., Kim, T. Y., Jang, Y. S., Choi, S., Lee, S. Y. Systems metabolic engineering for chemicals and materials. *Trends Biotechnol.* 2011, 29, 370–378.

Lee, J. W., Lee, S. Y. Proteome-based physiological analysis of the metabolically engineered succinic acid producer *Mannheimia succiniciproducens* LPK7. *Bioprocess Biosyst Eng.* 2010, 33, 97–107.

Lee, R. A., Lavoie, J. M. From first- to third-generation biofuels: Challenges of producing a commodity from a biomass of increasing complexity. *Animal Frontiers* 2013, 3, 6–11.

Lee, S., Speight, J. G., Loyalka, S. K. In *Handbook of Alternative Fuel Technologies.* CRC Press: USA, 2007, pp. 26–28.

Leite, R. C. C., Leal, M. R. L. V., Cortez, L. A. B., Barbosa, L. A., Griffin, W. M., Scandiffio, M. I. G. Can Brazil replace 5% of the 2025 gasoline world demand with ethanol? *Energy* 2009, 34, 655–661.

Lejeune, R., Baron, G. V. Effect of agitation on growth and enzyme-production of *Trichoderma reesei* in batch fermentation. *Appl Microbiol Biotechnol.* 1995, 43, 249–258.

Liebeke, M., Dörries, K., Zühlke, D., Bernhardt, J., Fuchs, S., Pané-Farré, J., Engelmann, S., Völker, U., Bode, R., Dandekar, T., Lindequist, U., Hecker, M., Lalk, M. A metabolomics and proteomics study of the adaptation of *Staphylococcus aureus* to glucose starvation. *Mol Biosyst.* 2011, 7, 1241–1253.

Lin, Y., Tanaka, S. Ethanol fermentation from biomass resources: current state and prospects. *Appl Microbiol Biotechnol.* 2006, 69, 627–642.

Lynd, L. R., van Zyl, W. H., McBride, J. E., Laser, M. Consolidated bioprocessing of cellulosic biomass: an update. *Curr Opin Biotechnol.* 2005, 16, 577–583.

Ma, M., Liu, L. Z. Quantitative transcription dynamic analysis reveals candidate genes and key regulators for ethanol tolerance in *Saccharomyces cerevisiae*. *BMC Microbiol.* 2010, 10, 169.

Macedo, I. C., Seabra, J. E. A., Silva, J. E. A. R. Green house gases emissions in the production and use of ethanol from sugarcane in Brazil: the 2005/2006 averages and a prediction for 2020. *Biomass Bioenerg.* 2008, 32, 582–595.

Mamma, D., Christakopoulos, P., Koullas, D., Kekos, D., Macris, B. J., Kouki, E. An alternative approach to the bioconversion of sweet sorghum carbohydrates to ethanol. *Biomass Bioenerg.* 1995, 8, 99–103.

Maurya, D. P., Singh, D., Pratap, D., Maurya, J. P. Optimization of solid state fermentation conditions for the production of cellulase by *Trichoderma reesei*. *J Environ Biol.* 2012, 33, 5–8.

McAloon, A., Taylor, F., Yee, W., Ibsen, K., Wooley, R. Determining the cost of producing ethanol from corn starch and lignocellulosic feedstocks. *National Renewable Energy Laboratory* 2000, 17, 1–25.

Mosier, N., Wyman, C., Dale, B., Elander, R., Lee, Y. Y., Holtazapple, M. Features of promising technologies for pretreatment of lignocellulosic biomass. *Bioresource Technology* 2005, 96, 673–686.

Mussatto, S. I., Dragone, G., Guimarães, P. M., Silva, J. P., Carneiro, L. M., Roberto, I. C., Vicente, A., Domingues, L., Teixeira, J. A. Technological trends, global market, and challenges of bio-ethanol production. *Biotechnol Adv.* 2010, 28, 817–830.

Nevoigt, E. Progress in metabolic engineering of *Saccharomyces cerevisiae*. *Microbiol Mol Biol Rev.* 2008, 72, 379–412.

Nevoigt, E., Stahl, U. Osmoregulation and glycerol metabolism in the yeast *Saccharomyces cerevisiae*. *FEMS Microbiol Rev.* 1997, 21, 231–241.

Nigam, J. Ethanol production from wheat straw hemicellulose hydrolysate by *Pichia stipitis*. *J Biotechnol.* 2001, 87, 17–27.

Nissen, T. L., Kielland-Brandt, M. C., Nielsen, J., Villadsen, J. Optimization of ethanol production in *Saccharomyces cerevisiae* by metabolic engineering of the ammonium assimilation. *Metab Eng* 2, 69–77.

Nonato, R. V., Shishido, K. α-factor-directed synthesis of *Bacillus stearothermophilus* α-amylase in *Saccharomyces cerevisiae*. *Biochem Biophys Res Com.* 2000, 152, 76–82.

Ogbonna, J. C., Mashima, H., Tanaka, H. Scale up of fuel ethanol production from sugar beet juice using loofa sponge immobilized bioreactor. *Bioresour Technol.* 2001, 76, 1–8.

Öhgren, K., Bengtsson, O., Gorwa-Grauslund, M. F., Galbe, M., Hahn-Hagerdal, B., Zacchi, G. Simultaneous saccharification and co-fermentation of glucose and xylose in steam-pretreated corn stover at high fiber content with *Saccharomyces cerevisiae* TMB3400. *J Biotechnol* 2006, 126, 488–498.

Öhgren, K., Bura, R., Saddler, J., Zacchi, G. Effect of hemicellulose and lignin removal on enzymatic hydrolysis of steam pretreated corn stover. *Bioresour Technol.* 2007, 98, 2503–2510.

Ostergaard, S., Olsson, L., Nielsen, J. Metabolic engineering of *Saccharomyces cerevisiae*. *Microbiol Mol Biol Rev.* 2000, 64, 34–50.

Palhano, F. L., Gomes, H. L., Orlando, M. T., Kurtenbach, E., Fernandes, P. M. Pressure response in the yeast *Saccharomyces cerevisiae*: from cellular to molecular approaches. *Cell Mol Biol.* 2004, 50, 447–457.

Palmqvist, E., Hahn-Hagerdal, B. Fermentation of lignocellulosic hydrolysates. I: inhibition and detoxification. *Bioresour Technol.* 2000, 74, 17–24.

Park, J. H., Lee, S. Y. Towards systems metabolic engineering of microorganisms for amino acid production. *Curr Opin Biotechnol.* 2008, 19, 454–460.

Persson, T., Garcia, A., Paz, J., Jones, J., Hoogenbooma, G. Maize ethanol feedstock production and net energy value as affected by climate variability and crop management practices. *Agric Syst.* 2009,100, 11–21.

Petersson, A., Almeida, J. R. M., Modig, T., Karhumaa, K., Hahn-Hagerdal, B., Gorwa-Grauslund, M. F., Liden, G. A 5-hydroxymethyl furfural reducing enzyme encoded by the *Saccharomyces cerevisiae ADH6* gene conveys HMF tolerance. *Yeast* 2006, 23, 455–464.

Pham, T. K., Chong, P. K., Gan, C. S., Wright, P. C. Proteomic analysis of *Saccharomyces cerevisiae* under high gravity conditions. *J Proteome Res.* 2006, 5, 3411–3419.

Pham, T. K., Wright, P. C. Proteomic analysis of calcium alginate-immobilized *Saccharomyces cerevisiae* under high-gravity fermentation conditions. *J Proteome Res.* 2008, 7, 515–525.

Porzio, G. F., Prussi, M., Chiaramonti, D., Pari, L. Modeling lignocellulosic bioethanol from poplar: estimation of the level of process integration, yield and potential for co-products. *Journal of Cleaner Production* 2012, 34, 66–75.

Prasad, S., Singh, A., Jain, N., Joshi, H. C. Ethanol production from sweet sorghum syrup for utilization as automotive fuel in India. *Energ Fuel.* 2007, 21, 2415–2420.

Rattanachomsri, U., Tanapongpipat, S., Eurwilaichitr, L., Champreda, V. Simultaneous non-thermal saccharification of cassava pulp by multi-enzyme activity and ethanol fermentation by *Candida tropicalis. J Biosci Bioeng.* 2009, 107, 488–493.

Robertson, G. H., Wong, D. W., Lee, C. C., Wagschal, K., Smith, M. R., Orts, W. J. Native or raw starch digestion: a key step in energy efficient biorefining of grain. *J Agric Food Chem.* 2006, 54, 353–365.

Rossillo-Calle, F., Walter, A. Global market for bio-ethanol: historical trends and future prospects. *Energ Sustain Dev.* 2006, 10, 20–32.

Roukas, T. Ethanol production from non-sterilized beet molasses by free and immobilized *Saccharomyces cerevisiae* cells using fed-batch culture. *J Food Eng.* 1996, 27, 87–96.

Rubin, E. Genomics of cellulosic biofuels. *Nature* 2008, 454, 841–845.

Saha, B.C. Alpha-L-arabinofuranosidases: biochemistry, molecular biology and application in biotechnology. *Biotechnol Adv.* 2000, 18, 403–423.

Saha, B.C. Hemicellulose bioconversion. *J Ind Microbiol Biotechnol.* 2003, 30, 279–291.

Sánchez, O. J., Cardona, C.A. Trends in biotechnological production of fuel ethanol from different feedstocks. *Bioresour Technol.* 2008, 99, 5270–5295.

Sandgren, M., Wu, M., Karkehabadi, S., Mitchinson, C., Kelemen, B. R., Larenas, E. A., Stahlberg, J., Hansson, H. The structure of a bacterial cellobiohydrolase: the catalytic core of the *Thermobifida fusca* family GH6 cellobiohydrolase Cel6B. *J Mol Biol.* 2013, 425, 622–635.

Sarkar N., Ghosh, S. K., Bannerjee, S., Aikat, K. Bioethanol production from agricultural wastes: An overview. *Renewable Energy* 2010, 37, 19–27.

Shiyoa, S., Shimizu, H., Hirasawa, T., Nagahisa, K., Furusawa, C., Pandey, G., Katakura, Y. Metabolic pathway recruiting through genome data analysis for industrial application of *Saccharomyces cerevisiae. Biochem Eng J.* 2007, 36, 28–37.

Silva, J. P. A., Mussatto, S. I., Roberto, I. C. The influence of initial xylose concentration, agitation, and aeration on ethanol production by *Pichia stipitis* from rice straw hemicellulosic hydrolysate. *Appl Biochem Biotechnol*. 2010, 162, 1306–1315.

Silveira, W. B., Passos, F. J. V., Mantovani, H. C., Passos, F. M. L. Ethanol production from cheese whey permeate by *Kluyveromyces marxianus* UFV-3: a flux analysis of oxidoreductive metabolism as a function of lactose concentration and oxygen levels. *Enzyme Microb Technol*. 2005, 36, 930–936.

Stambuk, B. U., Dunn, B., Alves, S. L., Duval, E. H., Sherlock, G. Industrial fuel ethanol yeasts contain adaptive copy number changes in genes involved in vitamin B1 and B6 biosynthesis. *Genome Res*. 2009, 19, 2271–2278.

Steyn, A. J., Pretorius, I. S. Co-expression of a *Saccharomyces diastaticus* glucoamylase-encoding gene and a *Bacillus amyloliquefaciens* α-amylase-encoding gene in *Saccharomyces cerevisiae*. *Gene* 1991, 100, 85–93.

Strassburg, K., Walther, D., Takahashi, H., Kanaya, S., Kopka, J. Dynamic transcriptional and metabolic responses in yeast adapting to temperature stress. *OMICS* 2010, 14, 249–259.

Sun, Y., Cheng, J. Hydrolysis of lignocellulosic material for ethanol production: a review. *Bioresource Technology* 2002, 96, 673–686.

Tao, X., Zheng, D., Liu, T., Wang, P., Zhao, W., Zhu, M., Jiang, X., Zhao, Y., Wu, X. A novel strategy to construct yeast *Saccharomyces cerevisiae* for very high gravity fermentation. *PLoS One* 2012, 7, 312–335.

Tomas-Pejo, E., Ballesteros, M., Oliva, J. M., Olsson, L. Adaptation of the xylose fermenting yeast *Saccharomyces cerevisiae* F12 for improving ethanol production in different fed-batch SSF processes. *J Ind Microbiol Biotechnol*. 2010, 37, 1211–1220.

Trott, A., Morano, K. A. *SYM1* is the stress-induced *Saccharomyces cerevisiae* ortholog of the mammalian kidney disease gene Mpv17 and is required for ethanol metabolism and tolerance during heat shock. *Eukaryot Cell*. 2004, 3, 620–631.

Valadi, H., Larsson, C., Gustafsson, L. Improved ethanol production by glycerol-3-phosphate dehydrogenase mutants of *Saccharomyces cerevisiae*. *Appl Microbiol Biotechnol*. 1998, 50, 434–439.

Van Zyl, W. H., Bloom, M., Viktor, M. J. Engineering yeasts for raw starch conversion. *Appl Microbiol Biotechnol*. 2012, 95, 1377–1388.

Viktor, M. J., Rose, S. H., van Zyl, W. H., Viljoen-Bloom, M. Raw starch conversion by *Saccharomyces cerevisiae* expressing *Aspergillus tubingensis* amylases. *Biotechnol Biofuels* 2013, 6, 167.

Weber, C., Farwick, A., Benisch, F., Brat, D., Dietz, H., Subtil, T., Boles, E. Trends and challenges in the microbial production of lignocellulosic bioalcohol fuels. *Appl Microbiol Biotechnol*. 2010, 87, 1303–1315.

Wei, N., Quarterman, J., Kim, S. R., Cate, J. H. D., Jin, Y. S. Enhanced biofuel production through coupled acetic acid and xylose consumption by engineered yeast. *Nat Commun*. 2013, 4, 1.

Wisselink, H. W., Toirkens, M. J., Berriel, M. D. R. F., Winkler, A. A., van Dijken, J. P., Pronk, J. T., van Maris, A. J. A. Engineering of *Saccharomyces cerevisiae* for efficient anaerobic alcoholic fermentation of L-arabinose. *Appl Environ Microbiol*. 2007, 73, 4881–4891.

Wittmann, C., Heinzle, E. Modeling and experimental design for metabolic flux analysis of lysine-producing *Corynebacteria* by mass spectrometry. *Metab Eng*. 2001, 3, 173–191.

Yamada, R., Hasunuma, T., Kondo, A. Endowing non-cellulolytic microorganisms with cellulolytic activity aiming for consolidated bioprocessing. *Biotechnol Adv.* 2013, 31, 754–763

Yang, S., Jia, N., Li, M., Wang, J. Heterologous expression and efficient ethanol production of a Rhizopus glucoamylase gene in *Saccharomyces cerevisiae*. *Mol Biol Rep.* 2011, 38, 59–64.

Yu, Z. S., Zhang, H. X. Ethanol fermentation of acid-hydrolyzed cellulosic pyrolysate with *Saccharomyces cerevisiae*. *Bioresour Technol.* 2003, 90, 95–100.

Zabaniotou, A., Ioannidou, O., Skoulou, V. Rapeseed residues utilization for energy and 2nd generation biofuels. *Fuel* 2008, 87, 1492–502.

Zuzuarregui, A., Monteoliva, L., Gil, C., del Olmo, M. Transcriptomic and proteomic approach for understanding the molecular basis of adaptation of *Saccharomyces cerevisiae* to wine fermentation. *Appl Environ Microbiol.* 2006, 72, 836–847.

CHAPTER 11

BIOLOGICAL SYSTEM AS REACTOR FOR THE PRODUCTION OF BIODEGRADABLE THERMOPLASTICS, POLYHYDROXYALKANOATES

AKHILESH KUMAR SINGH[1] and NIRUPAMA MALLICK[2]

[1]Amity Institute of Biotechnology, Amity University Uttar Pradesh, Lucknow Campus, India

[2]Agricultural and Food Engineering Department, Indian Institute of Technology, Kharagpur, West Bengal, India

CONTENTS

11.1   INTRODUCTION

Plastics have unquestionably been the miracle materials and attained an inimitable place in contemporary material technology. These are omnipresent in the present day society, appearing in manners, which range from ordinary to high-tech, from essential to completely lavish. The way in which these materials are being utilized has molded our impression of plastics not only due to wonder materials but also as necessary evils.

Plastics have become attention-grabbing materials during 1940s, and since then these are substituting glass, wood and other construction materials, and even metals in many industrial, domestic and environmental applications (Lee et al., 1991; Cain, 1992; Poirier et al., 1995; Lee, 1996). These extensive applications of plastics are not only owing to their promising mechanical and thermal properties but also because of their stability, durability and low-cost.

The first thermoplastic resin, celluloid, was manufactured by the reaction of cellulose with nitric acid in the mid-19th century. Since then the number along with the kinds and qualities have significantly intensified, generating superior materials for example silicones, epoxies, teflon, polycarbonates, polysulfones, etc. Durability and resistance to degradation are desirable characteristics when plastics are in use. However, they cause problems for disposal when not in use. Globally, a wide variety of these polymers are manufactured to the tune of 140 million tons per year (approx.), but significant quantities of these materials are dumped into the ecosystem as industrial waste products (Shimao, 2001). Furthermore, plastics also display significant role in several 'short-live' applications for instance packaging and these denote the foremost part of plastic wastes (Rivard et al., 1995; Witt et al., 1997; Muller et al., 2001). The mammoth scale per capita utilization of these materials to the extent of 60 and 80 kg, respectively in Europe and USA has generated a prominent havoc to solid waste management programme (Kalia et al., 2000). Several goods composed of plastics are ended up after their useful life as discarded wastes, which accounting up to 20% by volume of landfills in developed countries (Stein, 1992). These conventional plastics can remain for several years because these materials are highly resistant to microbial attack owing to their high molecular mass, large number of aromatic rings, unusual bonds and/or halogen substitutions, once discarded in nature (Alexander, 1981). On account of these reasons, some communities are now susceptible to the effects of discarded conventional plastics on the environment, together with detrimental impacts on wildlife and on the aesthetic qualities of cities and forests.

11.2   BIODEGRADABLE POLYMERS

There have been increasing public and scientific interests over the past three decades in relation to the utilization and development of environmentally friendly polymers (biodegradable polymers) as an ecologically useful substitute to plastics, which must still preserve the preferred properties of synthetic conventional plastics; therefore presenting a solution for the persisting serious problem of plastic waste materials. Biodegradability, which is nothing more than catabolism measures the capacity of to be broken down, especially into nontoxic products, through the action of microbes. Amongst microorganisms, bacteria and fungi are the most important members in the biodegradation process. The disintegration of materials furnishes them with precursors for constituents of cell in addition to energy for energy-demanding processes. With the likely exclusion of coral reefs and other analogous structures, the biological world actively degrades what it synthesizes. Therefore, biodegradable materials are generally the product of life itself.

The exploration for such degradable polymers has focused for the introduction of three categories of degradable plastic materials: semi-biodegradable, photodegradable, and completely biodegradable. Semi-biodegradable plastics include the starch-incorporated plastics where starch is introduced to grasp together the small polyethylene fragments. The thought behind starch-incorporated plastics is that as soon as dumped into landfills, microorganisms live in the soil will attack the starch and liberate fragments of polymer, which can be degraded subsequently. Bacteria in fact attack the starch but are turned off through the polyethylene fragments, which as a consequence remain non-degradable (Johnstone, 1990). Photodegradable plastics possess light sensitive groups, which introduced as additives straightforwardly into the polymer backbone. Ultraviolet radiation is able to disrupt their polymeric structure rendering them open to further bacterial attack (Kalia et al., 2000). Nevertheless, they remain non-degraded as landfills deprive of sunlight. The third kinds of biodegradable plastics are quite new and promising owing to its real exploitation by bacteria. Included biodegradable plastics are poly-lactides (PLA), polyhydroxyalkanoates (PHAs), polysaccharides, aliphatic polyesters, co-polymers and/ or blends of the above. Of all these, the micro-bially-synthesized PHAs offer ample potential.

11.3   POLYHYDROXYALKANOATES

PHAs are a group macromolecules accumulated as water insoluble storage compounds in the cytoplasm of microbial cells. They have gained much con-cern over conventional plastics due to their biodegradability and geneses from renewable resources. Intracellular degradation of endogenous storage compounds, for example, PHAs involve their breakdown when supply of limiting nutrient is reestablished. In addition, a vast group of PHAs degrad-ing microorganisms occur in nature and is capable to degrade PHAs by uti-lizing secreted particular PHA hydrolases and PHA depolymerases, and the degradation products can be used as carbon and energy source (Madison and Huisman, 1999). The actions of these enzymes can differ and depend on the monomer units of the polymer, the dimensions of the sample, its physi-cal form, for example, crystalline or amorphous and prominently, the con-ditions of environment. The degradation rate with respect to a portion of poly-3-hydroxybutyrate (PHB) is normally on the order of a few months in case of anaerobic sewage to years in seawater (Madison and Huisman, 1999). Similarly, co-polymers incorporating 4-hydroxybutyrate (4HB) monomers reported to degrade more rapidly as compared to poly-3-hy-droxybutyrate (PHB) or co-polymer comprised of 3-hydroxybutyrate and 3-hydroxyvalerate [P(3HB-co-3HV)] (Williamson and Wilkinson, 1958). Apart from biodegradability, another important feature associated with PHAs is their synthesis, which is based on renewable resources, such as agri-cultural feedstocks resulting from CO_2 and water. After the transformation of agricultural feedstocks into biodegradable PHAs, the breakdown products are once again CO_2 and water. Therefore, while for certain applications the biodegradability is crucial, PHAs gain comprehensive consideration as they arc based on renewable resources instead of on non-renewable shrinking fossil fuel stocks (Williams and Peoples, 1996).

11.3.1 CHEMICAL STRUCTURE

The several diverse PHAs, which have been documented so far, are mainly linear, head-to-tail polymers made up of 3-hydroxy fatty acid monomers (Madison and Huisman, 1999). In these polyesters, the carboxyl group of one monomer establishes an ester bond with the hydroxyl group of the adjoining monomer being represented by the general chemical structure as exhibited in Figure 11.1, where 'X' can arrive up to 30,000 and R-pendant group comprises H-atom or a large varieties of C-chains.

For example, when

$n = 1$	R = hydrogen	Poly(3-hydroxypropionate)
	methyl	Poly(3-hydroxybutyrate)
	ethyl	Poly(3-hydroxyvalerate)
	propyl	Poly(3-hydroxyhexanoate)
	pentyl	Poly(3-hydroxyoctanoate)
	nonyl	Poly(3-hydroxydodecanoate)
$n = 2$	R = hydrogen	Poly(3-hydroxybutyrate)
$n = 3$	R = hydrogen	Poly(5-hydroxyvalerate)

So far, in all PHAs, which have been characterized the hydroxyl replaced carbon atom is of R configuration, except in certain special circumstances where there is no chirality. At the identical C-3 or β position an alkyl group that can differ from methyl to tridecyl, is placed. Nevertheless, this alkyl side chain is not essentially saturated; unsaturated, aromatic, epoxidized, halogenated and branched monomers have been reported too (Lageveen et al., 1988; Abe et al., 1990; Doi and Abe, 1990; Fritzsche et al., 1990a,b,c; Kim et al., 1991, 1992; Choi and Yoon, 1994; Hazer et al.,

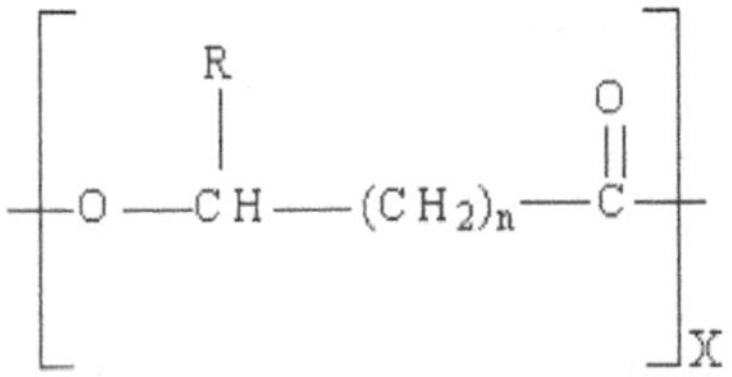

FIGURE 11.1 General structure of polyhydroxyalkanoates.

1994; Curley et al., 1996; Song and Yoon, 1996). In the side chains of PHAs, substituents can be altered chemically such as by cross-linking of unsaturated bonds (de Koning et al., 1994; Gagnon et al., 1994a,b). These differences in the length and composition of the side chains and the capacity to alter their reactive substituents are the foundation for the diversity of the PHA polymer family and their huge array of potential applications (Madison and Huisman, 1999).

11.3.2   TYPES OF PHAs

Polymers of 3-hydroxyalkanoates (3HAs) have been established in a wide range of prokaryotes and in several eukaryotic plants and animal cells. However, only prokaryotes are proficient of synthesizing high molecular weight PHAs in the form of amorphous granules with insignificant osmotic activity. To date, more than 300 species of microorganisms are identified to build and accumulate PHAs (Lee et al., 1999).

PHAs can be assigned into three key categories based on the number of carbons in the monomer units integrated into the polymer chain, for example, short-chain-length (SCL), medium-chain-length (MCL) and long-chain-length (LCL) PHAs composed of hydroxyacids with 3–5, 6–14 or more than 14 carbon atoms, respectively (Anderson and Dawes, 1990; Steinbuchel, 1992; Steinbuchel et al., 1992). Interestingly, it is not viable to produce PHAs chemically; therefore, microorganisms are the merely potential source for these polymers. PHAs are non-toxic, biocompatible, piezoelectric, thermoplastic and/or elastomeric, enantiomerically pure, insoluble in water and exhibit a high degree of polymerization with molecular weights of up to several million Da (Muller and Seebach, 1993; Hocking and Marchessault, 1994). Amongst the 150 diverse types of PHA monomer units known so far, the homopolymer of 3-hydroxybutyrate (PHB) is common in different taxonomic groups of prokaryotes (Vincenzini and De Philippis, 1999; Reddy et al., 2003). The monomer composition of the PHAs is regulated by two most important factors, for example, the type of microorganism used and the type of substrate exploit to grow the microorganism (Valappil et al., 2007). This generates an opportunity for synthesizing various types of PHAs with a widespread range of properties (Reddy et al., 2003). Because of the stereospecificity of biosynthetic enzymes, the PHAs monomer constituents are all in D(–) configuration (Reddy et al., 2003). PHAs can be advocated for various commercial applications, owing to resemblances of material properties with conventional plastics. The potential applications of the PHAs vary depending on material properties governed by monomer composition (Valappil et al., 2007). Poly-3-hydroxybutyrate (PHB), which is the common member of SCL-PHAs is quite rigid and crystalline, therefore making it problematical to process and be commercially exploited (Park and Lee, 2004). In contrast, MCL-/LCL-PHAs are elastomeric materials depicting poor tensile strength. Owing to these limitations, SCL-, MCL- and LCL-PHAs as such not feasible for various industrial applications. However, integration of MCL- or LCL-monomer units into the PHB backbone transforms the material properties of the polymer, making it appropriate for various commercial application (Matsusaki et al., 2000; Singh and Mallick, 2008, 2009a,b; Singh et al., 2013).

11.3.3 POTENTIAL APPLICATIONS FOR PHAs

The foremost attractive features of these polymers are their biodegradability together with hydrophobicity; even if economical biodegradable plastics are commercially available, merely PHAs have a good moisture resistance, so that their use in application for which these characteristics are required, for example, packaging of liquids and food materials, papers coating, etc. can be envisaged. The practicable applications of PHAs as plastics, related to their more important properties, and their real commercial applications (Vincenzini and Philippis, 1999) are in the areas of: (i) Agriculture: PHAs could be utilized for controlled release of pesticides, plant growth regulators, herbicides and fertilizers owing to their biodegradability coupled with retarding properties; (ii) Medical: PHAs biocompatibility and biodegradability together with piezoelectric properties lead to potential application as absorbable sutures, bone plates, film around bone fracture, surgical pins and staples; (iii) Pharmaceutical: Owing to their biodegradability and biocompatibility, PHAs could be employed as drug carrier and retarded drug release; (iv) Disposables: PHAs could be used as sole structural materials such as razors, tray for foods, utensils, etc. because of their biodegradability with good mechanical properties; (v) Packaging: PHAs could be exploited for manufacturing bottles and film for food packaging owing to their biodegradability, good mechanical properties, moisture resistance, hydrophobic and low oxygen permeability.

11.3.4 PHA BIOSYNTHETIC PATHWAYS

Bacterial species producing PHAs have been fundamentally subdivided into three groups (Nomura et al., 2005). One group together with the bacterium *Alcaligenes eutrophus*, now termed as *Cupriavidus necator*, yields SCL-PHA, whereas a second group in conjunction with a number of *Pseudomonads*, produces MCL-PHA (Steinbuchel, 1991). The third group, including the bacteria *Aeromonas hydrophila* and *Pseudomonas* sp. 61–3 accumulates SCL-MCL-PHA co-polymer composed of both SCL and MCL monomer units. In contrast, sludge-isolated *Pseudomonas aeruginosa* MTCC 7925 has been known so far to synthesize a novel SCL-LCL-PHA co-polymer composed of LCL 3-hydroxyhexadecanoate (3HHD) and 3-hydroxyoctadecanoate (3HOD) units with SCL 3-hydroxybutyrate (3HB) and 3-hydroxyvalerate (3HV), for example, P(3HB-*co*-3HV-*co*-3HHD-*co*-3HOD) as constituents of PHAs (Singh and Mallick, 2008, 2009a,b; Singh et al., 2013). This partition between SCL-PHA, MCL-PHA, SCL-MCL-PHA and LCL-MCL-PHA co-polymer is predominantly determined by the substrate specificity of the PHA synthase liable for the polymerization of the substrate *R*-3-hydroxyacyl-CoA to form PHAs.

PHB is the most common and comprehensively characterized PHA present in bacteria. It is synthesized as a carbon and/or energy storage material in a number of microorganisms generally under limiting nutritional conditions for instance N and P stresses and/or in presence of excess carbon sources (Steinbuchel, 1991; Byrom, 1994; Liebergesel et al., 1994; Yu, 2001). PHB biosynthesis and degradation are best studied in a few bacterial species that follow a cyclic process, first established in 1973 for *Azotobacter beijerinckii* and *Cupriavidus necator*. In these microorganisms along with many other bacteria when they are grown on carbohydrates or on acetyl-CoA producing carbon sources, the biosynthetic pathway for the production of PHB starts from acetyl-CoA and involves three consecutive, enzyme-mediated reactions, for example, the action of 3-ketothiolase, acetoacetyl-CoA reductase and PHA synthase (Oeding and Schlegel, 1973; Senior and Dawes, 1973). The first enzyme of the pathway, 3-ketothiolase, catalyzes the reversible condensation of two acetyl-CoA moieties to form acetoacetyl-CoA. Acetoacetyl-CoA reductase consequently reduces acetoacetyl-CoA to *R*-(–)-3-hydroxybutyryl-CoA, which is then polymerized by the action of a PHA synthase to form PHB (Figure 11.2). Nevertheless, slight modification of this pathway has been detected in *Rhodospirillum rubrum*, which needs

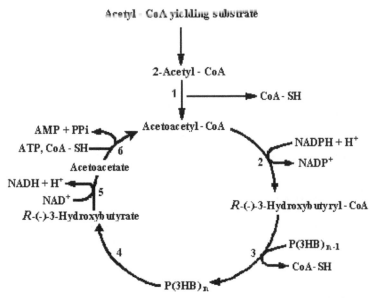

FIGURE 11.2 PHB biosynthesis in *Alcaligenes eutrophus* (*Cupriavidus necator*). Enzymes: (1) 3-Ketothiolase, (2) NADPH-linked acetoacetyl-CoA reductase, (3) PHA synthase, (4) PHA depolymerase, (5) 3-hydroxybutyrate dehydrogenase, and (6) Acetoacetyl-CoA synthase (Vincenzini and De Philippis, 1999).

the additional action of two enoyl-CoA hydratases to synthesize *R*-(–)-3-hydroxybutyryl-CoA (Moskowitz and Merrick, 1969), but without altering the basic rules.

On the basis of types of PHAs that are produced by *Cupriavidus necator* one general observation is that the integrated monomers always consist of only 3–5 carbons. The nature and ratio of these monomers are nevertheless, influenced by the type and the relative quantity of carbon sources supplemented to the growth media (Doi et al., 1988; Steinbuchel, 1991; Steinbuchel et al., 1993). It has been revealed that supplementation of propionic acid or valeric acid to the growth media with glucose results into production of P(3HB-*co*-3HV) co-polymer, the biosynthetic pathway of which is presented in Figure 11.3.

The PHA biosynthesis in *Pseudomonas oleovorans* and in many *Pseudomonads* including *Pseudomonas aeruginosa* belonging to the rRNA homology group I involves the biosynthetic route that includes the cyclic β-oxidation along with thiolytic cleavage of fatty acids, for example, 3-hydroxyacyl-CoA and intermediates of the β-oxidation pathways

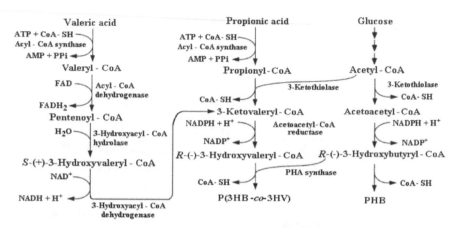

FIGURE 11.3   Biosynthetic pathway of PHB and P(3HB-*co*-3HV) in *Alcaligenes eutrophus* (*Cupriavidus necator*) (modified from Valentin and Steinbuchel, 1995; Suriyamongkol et al., 2007).

(Doi, 1990; Punrattanasin, 2001; Singh and Mallick, 2009a; Singh et al., 2013). These microorganisms synthesize MCL-PHAs and rarely LCL-PHAs from alkanes, alkanoates or alcohols. Contrary to *Cupriavidus necator*, these organisms normally do not synthesize PHAs comprising SCL monomers (SCL-PHAs). The β-oxidation of fatty acids could supply the foremost substrate flux for PHA polymerase (Lageveen et al., 1988). This cycle follows an enzymatic cascade of reactions where fatty acids are incorporated into the cell (Figure 11.4): they are activated to CoA thioesters, reduced to 2-*trans*-enoyl-CoA, and catalyzed by acyl-CoA dehydrogenase with FAD as a cofactor. These intermediates are transformed to *S*-(+)-3-hydroxyacyl-CoA and catalyzed by enoyl-CoA hydratase. The compound so generated is oxidized to 3-ketoacyl-CoA with the assistance of NAD dependent 3-ketoacyl-CoA dehydrogenase. Lastly, acetyl-CoA is cleaved from 3-ketoacyl-CoA and two carbons lesser fatty acid is synthesized. By reduction of 3-ketoacyl-CoA, a reaction catalyzed by a ketoacyl-CoA reductase, intermediate *R*-(−)-3-hydroxyacyl-CoA is produced. As the PHA synthase accepts merely the *R*-(−)-3-hydroxyacyl-CoA and the bacterial β-oxidation of fatty acids produces merely the *S*-(+)-3-hydroxyacyl-CoA, bacteria must have enzymes proficient of generating *R*-(−)-3-hydroxyacyl-CoA. One such enzyme is a 3-hydroxyacyl-CoA epimerase, mediating the reversible conversion of the *S* and *R* isomers of 3-hydroxyacyl-CoA. Thus, *S*-(+)-3-hydroxyacyl-CoA epimerizes to *R*-(−)-3-hydroxyacyl-CoA through 3-hydroxyacyl-CoA epimerase and the enoyl-CoA is converted to *R*-(−)-3-hydroxyacyl-CoA by

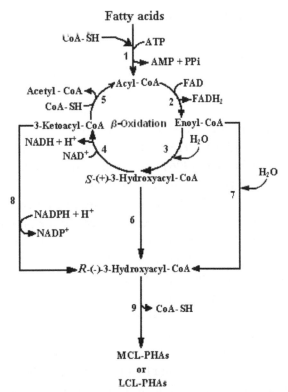

FIGURE 11.4 Biosynthetic pathway of MCL-PHAs and LCL-PHAs in *Pseudomonas oleovorans* and *Pseudomonas aeruginosa*, respectively using β-oxidation cycle. Enzymes: 1: Acyl-CoA synthase, 2: Acyl-CoA dehydrogenase, 3: Enoyl-CoA hydratase, 4: NAD dependent 3-hydroxyacyl-CoA dehydrogenase, 5: Ketoacyl-CoA thiolase (reductase), 6: 3-Hydroxyacyl-CoA epimerase, 7: Enoyl-CoA hydratase, 8: Ketoacyl-CoA reductase and 9: PHA polymerase (Lagaveen et al., 1988; Huisman et al., 1989; Singh and Mallick, 2009a; Singh et al., 2013).

enoyl-CoA hydratase activity. These are finally utilized by the PHA polymerase for the production of either MCL-PHAs or LCL-PHAs (Lageveen et al., 1988; Kraak et al., 1997, Singh and Mallick, 2009a).

In *P. aeruginosa* nevertheless, PHA is produced from acetyl-CoA through fatty acid biosynthetic pathway (Huijberts et al., 1995; Steinbuchel, 1996; Singh and Mallick, 2008, 2009b). Various *Pseudomonads* belonging to the rRNA homology group I too accumulate MCL-PHAs by this pathway. However, *Pseudomonas aeruginosa* MTCC 7925 produce a novel co-polymer composed of LCL 3HA units of C_{16} (3-hydroxyhexadecanoate, 3HHD) and C_{18} (3-hydroxyoctadecanoate, 3HOD) along with SCL 3HA units of 3-hydroxybutyrate (3HB) and 3-hydroxyvalerate (3HV). In these microorganisms the pathway operated is known as the *P. aeruginosa* PHA biosynthetic pathway. In most cases, the building blocks of PHAs are incorporated into the polymer only if the cells are cultivated on carbon source whose carbon skeleton is related to the structure of the constituent monomer unit (Huisman et al., 1989). In some cases, of course, the PHA monomers do not have similarity with the carbon sources; this was referred as the capability of the organism to synthesis PHA from unrelated substrates (Anderson and Dawes, 1990). Steinbuchel (1996) and Singh and Mallick (2008) reported that MCL- and LCL-PHAs synthesized by *P. aeruginosa* PHA biosynthetic pathway are from unrelated substrates, for example, glucose, ethanol, gluconate or acetate. These substrates are first used for the production of fatty acids, which then yield precursors for PHA polymerase to produce either MCL- or LCL-PHA polymer (Figure 11.5). The fatty acid synthesis pathway exhibits different biosynthetic pathways according to the growth substrates utilized by organisms. In this pathway, the carbon source is first converted to acetyl-CoA and then carboxylated to malonyl-CoA. A malonyl transacylase links the malonyl-CoA to an acyl carrier protein (ACP), liberating coenzyme A. Malonyl-ACP is further converted to 3-ketoacyl-ACP, *R*-(–)-3-hydroxyacyl-ACP, enoyl-ACP and finally acyl-ACP. The latter compound is then reacted to a malonyl-ACP to produce a new 3-ketoacyl-ACP molecule. In this pathway, *R*-(–)-3-hydroxyacyl-ACP is a presumed precursor for the production of PHA. As the fatty acid biosynthesis intermediate is in the form of *R*-(–)-3-hydroxyacyl-ACP, an additional biosynthetic step is required to transform it into the *R*-(–)-3-hydroxyacyl-CoA form. An enzyme identified as 3-hydroxyacyl-ACP-CoA transacylase, has been shown to be proficient of directing the intermediates of the fatty acid biosynthesis pathway to PHA biosynthesis. Nevertheless, *P. oleovorans* does not synthesize MCL-PHAs from these unrelated substrates. The difference between these two related organisms is that *P. oleovorans* degrades fatty acids to form acyl-CoA, the foremost substrate for the β-oxidation cycle, while *P. aeruginosa* produces acyl-ACP (acyl carrier protein), a precursor substrate for fatty acid production from unrelated substrates resulting in either MCL- or LCL-PHA (Figure 11.5).

As stated above, the majority of bacteria produce either SCL-, MCL- or rarely LCL-PHAs but not generally SCL-MCL or SCL-LCL co-polymers because of the mutual exclusivity of the SCL-, MCL- and LCL-PHA synthase enzymes (Kato et al., 1996; Ashby et al., 2002; Singh et al., 2013).

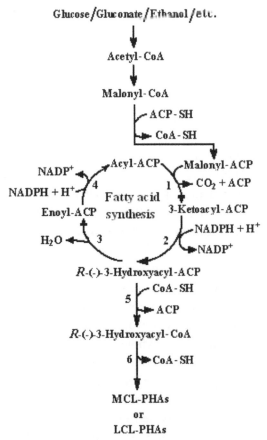

FIGURE 11.5 Biosynthetic pathway of MCL-PHAs and LCL-PHAs in *Pseudomonas aeruginosa* using fatty acid synthesis pathway. Enzymes: 1: Ketoacyl-ACP synthase, 2: Ketoacyl-ACP reductase, 3: 3-Hydroxyacyl-ACP dehydratase, 4: Enoyl-ACP reductase, 5: 3-Hydroxyacyl-ACP-CoA transacylase and 6: PHA polymerase (Timm and Steinbuchel, 1990; Steinbuchel, 1996; Singh and Mallick, 2008; Singh and Mallick, 2009b).

Biosynthesis of PHAs, in which SCL and MCL or SCL and LCL monomers are covalently linked in the same polyester molecules, for example, the real co-polymer of SCL-MCL or SCL-LCL needs the existence of a PHA synthase showing a substrate range combining those of SCL-PHA and MCL-PHA or SCL-PHA and LCL-PHA synthases. Interestingly, in some cases presence of SCL- and MCL-PHA, where the monomer units not connected in the same polyester molecule has also been detected (Steinbuchel and Wiese, 1992; Kato et al., 1996; Ashby et al., 2002). Simultaneous presence of a SCL-PHA synthase and a MCL-PHA synthase is the reason for formation of such mixture of SCL- and MCL-PHA rather than the real SCL-MCL co-polymer.

11.3.5 ACCUMULATION OF PHA CO-POLYMERS

As indicated in Table 11.1, some bacterial species such as *Aeromonas caviae, Rhodococcus* sp. NCIMB 40126, *Bacillus* sp. and some species of *Pseudomonas* (*P. fluorescens, Pseudomonas* sp. A33, *Pseudomonas* sp. 61–3, *P. marginalis* DSM50276 and *P. mendocina* DSM50017) are found to synthesize the real co-polymer of SCL-MCL-PHAs with a maximum carbon skeleton up to C_{14} (Haywood et al., 1991; Caballero et al., 1995; Doi et al., 1995; Lee et al., 1995a; Kato et al., 1996; Tajima et al., 2003; Nomura et al., 2005; Li et al., 2011; Gao et al., 2012; Phithakrotchanakoon et al., 2013). However, *Pseudomonas aeruginosa* MTCC 7925 produces a novel SCL-LCL-PHA co-polymer with incorporation of a maximum carbon skeleton up to C_{18}. Thus, for synthesis of the co-polymer of SCL-MCL-PHAs and SCL-LCL-PHAs, *Pseudomonads* are believed to be of the appropriate candidates, due to their inherent properties of synthesizing MCL-PHAs and LCL-PHAs (Huisman et al., 1989; Singh and Mallick, 2008, 2009a,b; Singh et al., 2013).

Bacterial system appropriate for biosynthesis of SCL-MCL-PHA and SCL-LCL-PHA co-polymer is of current research concern owing to their interesting material properties (Chen et al., 2001b; Lee and Park, 2002; Singh and Mallick, 2008, 2009a,b; Singh et al., 2013; Phithakrotchanakoon et al., 2013). The SCL-MCL-PHA and SCL-LCL-PHA co-polymers are found to have superior material properties as compared to those of PHAs comprising of SCL or MCL or LCL monomers only (Doi et al., 1995; Singh and Mallick, 2008, 2009a,b; Singh et al., 2013). It has been revealed that SCL-MCL-PHA co-polymers with a high mol percentage of SCL monomers along with a low mol fraction of MCL and SCL-LCL-PHA co-polymers with high mol percentage of SCL monomers in conjugation with low mol fraction of LCL have properties comparable to polypropylene (PP) and polyethylene (PE) (Matsusaki et al., 2000; Abe and Doi, 2002; Noda et al., 2004; Singh and Mallick, 2008, 2009a,b; Singh et al., 2013). Co-polymer of P(3HB-*co*-3HHx) containing 10–17 mol% of 3HHx retains higher elongation to break as compared to P(3HB-*co*-3HV) co-polymer containing 20 mol% of 3HV, commercialized under the trade name of BIOPOL® (Lee et al., 2000b). The co-polymer having 20 mol% of 3HHx exhibits properties akin to low-density

TABLE 11.1 Accumulation of Co-Polymer of PHA in Various Bacterial Species

Bacterium	Substrate	% PHA (dcw)	PHA composition	Reference
Alcaligenes eutrophus H16	Propionate	56	P(3HB-*co*-3HV)	Doi et al. (1987)
Alcaligenes feacalis NCIB 8156	Acetate + Propionate	14	P(3HB-*co*-3HV)	Haywood et al. (1989)
Pseudomonas extorquens MP4	Methanol + Propionate	26	P(3HB-*co*-3HV)	
*Rhodococcus sp. NCIMB 40126	Hexanoate	25	P(3HB-*co*-3HV-*co*-3HHx)	Haywood et al. (1991)
Methylobacterium extorquens	Methanol + Valerate	60–70	P(3HB-*co*-3HV)	Bourque et al. (1992)
Ralstonia eutropha	Glucose + Propionate	70–80	P(3HB-*co*-3HV)	Byrom (1992)
*Bacillus cereus ATCC14579	Hexanoate	2	P(3HB-*co*-3HHx)	Caballero et al. (1995)
*Pseudomonas sp. A33	1,3-butanediol	18	P(3HB-*co*-3HHx-*co*-3HO-*co*-3HD)	Lee et al. (1995a)
	3-hydroxybutyrate	18	P(3HB-*co*-3HHx-*co*-3HO-*co*-3HD)	
	Gluconate	38	P(3HB-*co*-3HHx-*co*-3HO-*co*-3HD-*co*-3HDD)	
*Pseudomonas fluorescens GK13	1,3-butanediol	15	P(3HB-*co*-3HHx-*co*-3HO-*co*-3HD)	
	3-hydroxybutyrate	18	P(3HB-*co*-3HHx-*co*-3HO-*co*-3HD)	
	Gluconate	38	P(3HB-*co*-3HHx-*co*-3HO-*co*-3HD-*co*-3HDD)	
*Pseudomonas marginalis DSM50276	3-hydroxybutyrate	12	P(3HB-*co*-3HHx-*co*-3HO-*co*-3HD)	
*Pseudomonas mendocina DSM50017	3-hydroxybutyrate	19	P(3HB-*co*-3HHx-*co*-3HO-*co*-3HD)	
*Aeromonas caviae	Dodecanoate	36	P(3HB-*co*-3HHx)	Doi et al. (1995)
*Pseudomonas sp. 61–3	Glucose	26	P(3HB-*co*-3HO-*co*-3HD) + PHB	Kato et al. (1996)

TABLE 11.1   Continued

Bacterium	Substrate	% PHA (dcw)	PHA composition	Reference
	Fructose	17	P(3HB-*co*-3HHx-*co*-3HO-*co*-3HD-*co*-3HDD)	
Aeromonas hydrophila	Lauric acid	50	P(3HB-*co*-3HHx)	Chen et al. (2001a)
Bacillus cereus UW85	ε-caprolactone + Glucose	9	P(3HB-*co*-3HV-*co*-6HHx)	Labuzek and Radecka (2001)
Bacillus sp. INT005	ε-caprolactone	23	P(3HB-*co*-3HHx-*co*-6HHx)	Tajima et al. (2003)
	4-hydroxybutyrate	55	P(3HB-*co*-4HB-*co*-3HHx)	
	Octanoate	65	P(3HB-*co*-3HHx)	
Recombinant Pseudomonas putida GPp104 PHA	Octanoate	40	P(3HB-*co*-3HHx-*co*-3HO-*co*-3HD)	Sheu and Lee (2004)
Recombinant Ralstonia eutropha PHB-4	Octanoate	45	P(3HB-*co*-3HHx-*co*-3HO)	
Recombinant Aeromonas hydrophila 4AK4	Lauric acid	53	P(3HB-*co*-3HHx)	Tian et al. (2005)
Recombinant Ralstonia eutropha PHB-4	Gluconate + Octanoate	34	P(3HB-*co*-3HHx-*co*-3HO-*co*-3HD)	
	Gluconate + Octanoate	36	P(3HB-*co*-3HHx-*co*-3HO-*co*-3HD-*co*-3HDD)	Luo et al. (2006)
Methylobacterium sp. GW2	Methanol	30	P(3HB-*co*-3HV)	Yezza et al. (2006)
Recombinant Aeromonas hydrophila CQ4	Dodecanoate + Gluconate + Yeast extract	21	P(3HB-*co*-3HHx-*co*-3HO-*co*-3HD-*co*-3HDD)	
Aeromonas hydrophila 4AK4	Dodecanoate + Yeast extract	39	P(3HB-*co*-3HHx)	Qin et al. (2007)
Recombinant Aeromonas hydrophila CQ4	Dodecanoate + Yeast extract	45	P(3HB-*co*-3HHx)	
Bacillus sp. 256	Mahua flowers	51	P(3HB-*co*-3HV)	Kumar et al. (2007)

TABLE 11.1 Continued

Bacterium	Substrate	% PHA (dcw)	PHA composition	Reference
**Pseudomonas aeruginosa* MTCC 7925	Ethanol	69	P(3HB-co-3HV-co-3HHD-co-3HOD)	Singh and Mallick (2008)
**Pseudomonas aeruginosa* MTCC 7925	Palm oil + Extract of palm oil cakes	75	P(3HB-co-3HV-co-3HHD-co-3HOD)	Singh and Mallick (2009a)
**Pseudomonas aeruginosa* MTCC 7925	Ethanol + Glucose	78	P(3HB-co-3HV-co-3HHD-co-3HOD)	Singh and Mallick (2009b)
*Recombinant *E. coli* DH5α	Decanoate + Glucose	7.3	P(3HB-co-3HHx-co-3HO -co-3HD)	Li et al. (2011)
*Recombinant *E. coli* LS521	Dodecanoate	4.0	P(3HB-co-3HHx-co-3HO-co-3HD-co-3HDD)	Gao et al. (2012)
	Gluconate + Dodecanoate	4.0	P(3HB-co-3HHx-co-3HO-co-3HD-co-3HDD)	
	Gluconate + Dodecanoate	2.1	P(3HB-co-3HO)	
**Pseudomonas aeruginosa* MTCC 7925	Palm oil + Extract of palm oil cakes	77	P(3HB-co-3HV-co-3HHD-co-3HOD)	Singh et al. (2013)
*Recombinant *E. coli*-ABC$_{Re}$J1$_{Pp}$	Glycerol + Bacto-yeast extract + Sodium dodecanoate	3.2	P(3HB-co-3HHx)	Phithakrotchana-koon et al. (2013)
*Recombinant *E. coli*-ABC2$_{Pp}$J4$_{Pp}$	Glycerol + Bacto-yeast extract + Sodium dodecanoate	0.1	P(3HB-co-3HHx-co-3HO -co-3HD)	
*Recombinant *E. coli*-ABC$_{Re}$J4$_{Pp}$	Glycerol + Bacto-yeast extract + Sodium dodecanoate	1.0	P(3HB-co-3HHx-co-3HO)	
Recombinant *Escherichia coli* XL1	Glucose	62	P(3HB-co-3HV)	Yang et al. (2014)

3HB: 3-hydroxybutyrate, 3HV: 3-hydroxyvalerate, 3HHx: 3-hydroxyhexanoate, 3HO: 3-hydroxyoc-tanoate, 3HD: 3-hydroxydecanoate, 3HDD: 3-hydroxydodecanoate, 6HHx: 6-hydroxyhexanoate, 3HHD: 3-hydroxyhexadecanoic acid, 3HOD: 3-hydroxyoctadecanoic acid.

*SCL-MCL co-polymer producers.

**SCL-LCL co-polymer producers.

polyethylene (LDPE). In contrast, P(3HB-*co*-3HV-*co*-3HHD-*co*-3HOD) co-polymer containing mol fraction of 1.8:1.5 of 3HHD:3HOD units, respectively exhibits significantly higher elongation to break with improve Young's modulus and tensile strength as compared to SCL-MCL-PHA co-polymer (Singh and Mallick, 2008, 2009a,b; Singh et al., 2013). Hence, incorporation of MCL or LCL monomer unit into the PHB polymer backbone not merely increases the ductility and processability but also the polymer become very much appropriate for commercial applications. The material properties of PHB, P(3HB-*co*-3HV), P(3HB-*co*-3HA), P(3HB-*co*-3HV-*co*-3HHD-*co*-3HOD), PP and LDPE are summarized in Table 11.2 for comparison.

11.3.6 COMMERCIAL DRAWBACK OF BACTERIAL PHAS

In spite of the 75 years, on and off, of investigation on PHAs and 20 years of profound industrial interest, PHAs still appear to be far reached from huge scale production. Currently, a range of P(3HB-*co*-3HV) co-polymer with 0–24 mol% 3HV produced from glucose and propionate by using the bacterium *C. necator* is marketed under the trade name BIOPOL® by Metabolix Inc. (Massachussets, USA). The foremost commercial disadvantage for the exploitation of bacterial PHAs as a source of biodegradable plastics for low-value commodity products, such as packaging materials and disposable items, is its high production cost, predominantly the cost of the substrates, causing them much more costly as compared to the conventional plastics. The price of PHA employing the natural accumulator *C. necator* is approximately US$16 per Kg, which is approximately 18 times more costly than polypropylene (Reddy et al., 2003). Consequently, to reduce this drawback, photoautotrophic hosts or alternate low-cost substrates are receiving the current attention.

11.3.6.1 Attempts for Low-Cost PHAs Production in Transgenic Plants

In photoautotrophic host system, synthesis of PHAs in agricultural crops has been regarded as a promising alternative (Poirier et al., 1995; Poirier, 1999, 2002). The first PHA synthesized in plants was a homopolymer of PHB. Expression of the last two enzymes, for example, an acetoacetyl-CoA reductase and a PHA synthase, from the bacterium *C. necator* into

TABLE 11.2 A Comparative Account on the Properties of PHB, P(3HB-*co*-3HV), P(3HB-*co*-3HA), P(3HB-*co*-3HV-*co*-3HHD-*co*-3HOD), PP and LDPE (Sudesh et al., 2000; Tsuge, 2002; Singh and Mallick, 2008, 2009a,b; Singh et al., 2013)

Property	PHB	P(3HB-*co*-3HV) (mol fraction 80:20)	P(3HB-*co*-3HA) (mol fraction 94:06)	P(3HB-*co*-3HV-*co*-3HHD-*co*-3HOD) (mol fraction 84.8:7.2:3.1:4.9–95.7:1.0:1.8:1.5)	PP	LDPE
T_m (°C)	180	145	133	115 to 131	176	130
T_g (°C)	4	–1	–8	–8 to –14	–10	–36
Crystallinity (%)	60	56	45	–	50 to 70	20 to 50
Young's modulus (Gpa)	3.5	0.8	0.2	0.2 to 0.3	1.7	0.2
Tensile strength (Mpa)	40	20	17	17 to 19	38	10
Elongation to break (%)	5	50	680	682 to 723	400	620

T_m: Melting temperature, T_g: Glass-transition temperature.

3HA: [3-hydroxydecanoate (3 mol%, 3-hydroxydodecanoate (3 mol%), 3-hydroxyoctanoate (<1 mol%) and 3-hydroxy-*cis*-dodecenoate (<1 mol%)].

the cytoplasm of *Arabidopsis thaliana* cells steered to production of PHB as intracellular inclusions up to 0.1% of shoot dry weight (Table 11.3). Growth of this transgenic plant was harshly reduced, probably triggered by the limitation of the acetyl-CoA pool available in the cytoplasm for isoprenoid and flavonoid biosynthesis (Poirier, 2002).

Higher production levels were achieved by directing the PHB pathway to the plastid. In a research where three *C. necator* genes altered for plastid-targeting were expressed in *A. thaliana* from separate T-DNA vectors, PHB production in shoots were stated to achieve up to 14% of shoot dry weight (Nawrath et al., 1994). Further, PHB synthesis up to 40% shoot dry weight was attained by expressing the similar genes on a single T-DNA vector (Bohmert et al., 2000). Nevertheless, plants producing PHB higher than approximately 5% of dry weight were chlorotic and exhibited a negative correlation between plant growth and PHB accumulation. These experiments established that although high level of PHB could be produced in the plastid, polymer accumulation away from some critical point had a negative impact on chloroplast function (Rezzonico et al., 2002).

In an attempt to move this technology to field, a group of scientist at Monsanto has discovered that transgenic rape expressing the PHB pathway in leucoplasts of developing embryos synthesized PHB up to 8% of dry weight in mature seeds of heterozygous plants (Houmiel et al., 1999). While the effects of PHB production on triacylglycerol synthesis has not been reported, no harmful impacts of PHB production on seed development or germination have been observed. The same group has also been capable to express the PHB biosynthetic pathway in the chloroplasts of stalks and leaves of corn, and reported a level of PHB only up to 6% (Mitsky et al., 2000). Nevertheless, as with *A. thaliana*, accumulation of PHB in corn leaves was associated with chlorosis.

As presented in Table 11.3, expression of PHB biosynthetic pathway either in tobacco, *Linum*, maize or sugarcane has so far produced only low amounts of polymer (Hahn et al., 1999; Wrobel et al., 2004; Lossl et al., 2003, 2005; Petrasovits et al., 2007; Matsumoto et al., 2009; Tilbrook et al., 2011; Schnell et al., 2012; Somleva et al., 2012; Petrasovits et al., 2013). It is estimated that these plants may either not express the genes from *C. necator* very well, or that the bacterial proteins are poorly active owing to instability or inactivation. PHB has been accumulated in cotton fiber cells in order to alter the physical properties of the fibers (John and Keller, 1996; John, 1997). Production of PHB upto 0.3% of shoot dry weight of the fiber was

TABLE 11.3 Summary of the Production of PHAs in Transgenic Plants

Sub-cellular compartment	Species	Tissue type	PHA composition	% PHA (dcw)	Reference
Cytoplasm	*Arabidopsis thaliana*	Shoot	PHB	0.1	Poirier et al. (1992)
Plastid	*Arabidopsis thaliana*	Shoot	PHB	14	Nawrath et al. (1994)
Cytoplasm	*Gossypium hirsutum*	Fiber	PHB	0.3	John and Keller (1996)
Plastid	*Gossypium hirsutum*	Fiber	PHB	0.05	John (1997)
Peroxisome	*Arabidopsis thaliana*	Seedling	MCL-PHA	0.6	Mittendorf et al. (1998, 1999)
Cytoplasm	*Nicotiana tabacum*	Shoot	PHB	0.01	Nakashita et al. (1999)
Plastid	*Brassica napus*	Seed	PHB	8	Houmiel et al. (1999)
Plastid	*Arabidopsis thaliana*	Shoot	P(3HB-*co*-3HV)	1.6	Slater et al. (1999)
Plastid	*Brassica napus*	Seed	P(3HB-*co*-3HV)	2.3	
Peroxisome	*Arabidopsis thaliana*	Seed	MCL-PHA	0.1	Poirier (1999)
Peroxisome	*Zea mays*	Cell suspension	PHB	2	Hahn et al. (1999)
Plastid	*Arabidopsis thaliana*	Shoot	PHB	40	Bohmert et al. (2000)
Plastid	*Zea mays*	Stalks and leaves	PHB	6	Mitsky et al. (2000)
Plastid	*Arabidopsis thaliana*	Shoot	PHB	8	Poirier and Gruys (2001)
Plastid	*Nicotiana tabacum*	Leaf	PHB	1.7	Lossl et al. (2003)
Plastid	*Beta vulgaris*	Hairy root	PHB	5.5	Menzel et al. (2003)
Plastid	*Linum usitatissimum*	Stem	PHB	0.0004*	Wrobel et al. (2004)
Plastid	*Nicotiana tabacum*	Shoot	PHB	0.14	Lossl et al. (2005)
Peroxisome	*Arabidopsis thaliana*	Leaves	P(3HB-*co*-3HA)	0.02	Matsumoto et al. (2006)
		Stems	P(3HB-*co*-3HA)	0.004	
Plastid	*Saccharum* spp.	Leaves	PHB	1.88	Petrasovits et al. (2007)

TABLE 11.3 Continued

Sub-cellular compartment	Species	Tissue type	PHA composition	% PHA (dcw)	Reference
Plastid	*Saccharum* spp.	Leaves	PHB	1.77	Purnell et al. (2007)
Plastid	*Elaeis guineensis*	Mesocarp	PHB	—	Omidvar et al. (2008)
Plastid	*Elaeis guineensis*	Leaves	PHB	—	Masani et al. (2009)
Plastid	*Arabidopsis thaliana*	Whole plants without roots	SCL-/ MCL- PHA	0.18**	Matsumoto et al. (2009)
Plastid	*Elaeis guineensis*	Mesocarp	PHB	—	Ismail et al. (2010)
Plastid	*Panicum virgatum*	Leaves	PHB	6.1**	Somleva and Ali (2010)
Plastid	*Camelina sativa*	Seeds	PHB	19.9**	Patterson et al. (2011a,b)
Peroxisome	*Arabidopsis thaliana*	Various plant developmental stages	SCL-PHA	1.8*	Tilbrook et al. (2011)
Vacuole	*Glycine max*	Seed coat	PHB	0.36**	Schnell et al. (2012)
Plastid	*Panicum virgatum*	Leaves	PHB	6.1**	Somleva et al. (2012)
Plastid	*Saccharum* spp.	Leaves	PHB	1.6**	Petrasovits et al. (2013)

*% Fresh weight; ** Dry weight.

stated. Thus, low expression level, lengthy growth period and problems in isolating PHB from other cellular components are the main hurdles in plant-based PHB production.

11.3.6.2 Attempts for Low-Cost PHAs Production in Cyanobacteria

Synthesis of PHAs in cyanobacteria has attracted the attention since Prof. N. G. Carr, University of Warwick, UK, reported the presence of PHB in a N_2-fixing cyanobacterium, *Chlorogloea fritschii* (Carr, 1966). Cyanobacteria are indigenously the sole prokaryotes that accumulate PHA by oxygenic photosynthesis. Several strains are found to accumulate PHA under photoautotrophic conditions and others only with the addition of acetate or other carbon sources. Nevertheless, the demand of exogenous carbon, is considerably lower (about 0.4%) when compared to heterotrophic bacteria, where supplementation of 4–5% carbon has been registered (Chen et al., 2001b; Shang et al., 2003). The maximum PHB accumulation reported for *Spirulina platensis* was 6% of dry cell weight (dcw) under photoautotrophic condition (Campbell et al., 1982). Nevertheless, the report of Nishioka et al. (2001) and the recent reports of Mallick and her group (Panda and Mallick, 2007; Bhati and Mallick, 2012; Samantaray and Mallick, 2012) demonstrated PHA accumulation of 85, 60, 55 and 38%, respectively in *Aulosira fertilissima*, *Nostoc muscorum* Agardh, *Synechococcus* sp. MA19 and *Synechocystis* sp. PCC 6803 under various specific conditions (Table 11.4). These reports are rather encouraging to viewed cyanobacteria as hosts for PHAs production. However, except *Aulosira fertilissima* (Table 11.4) the values seem quite low when compared to the PHB producing capability of bacteria, where a synthesis up to 80% (dcw) has been registered (Reddy et al., 2003). Interestingly, a comparative investigation conducted by Sudesh et al. (2001) with *Synechocystis* sp. PCC 6803 and *C. necator* having 11% and 50% PHB, respectively revealed comparable density while subjected to sucrose density gradient ultracentrifugation. Furthermore, in both the species the numbers and sizes of PHB inclusions too revealed close resemblance. These researchers inferred that the seemingly low PHB value as confirmed for the cyanobacterial cells might be owing to the larger size and mass of *Synechocystis* cells. In addition, the thicker cyanobacterial cell wall could hamper the elimination of water during drying, therefore could provide considerably

TABLE 11.4 Accumulation of PHAs in Cyanobacterial Species under Various Specific Conditions

Cyanobacterium	Condition	% PHA (dcw)	PHA composition	Reference
Synechococcus sp. MA19	Phosphate-deficiency	55	PHB	Nishioka et al. (2001)
Nostoc muscorum	Dark + Acetate	43	PHB	Sharma and Mallick (2005)
Synechocystis sp. PCC 6803	Phosphate-deficiency + gas-exchange limitation + Acetate	38	PHB	Panda and Mallick (2007)
Spirulina platensis UMACC 161	Acetate and CO_2	10	PHB	Toh et al. (2008)
Synechocystis sp. UNIWG	Acetate and CO_2	14	PHB	
Nostoc muscorum	Photoautotrophy	9	PHB	Bhati et al. (2010)
Nostoc muscorum Agardh	Phosphate-deficiency + Acetate + Valerate	58	P(3HB-*co*-3HV)	Bhati and Mallick (2012)
	Nitrogen-deficiency + Acetate + Valerate	60	P(3HB-*co*-3HV)	
Aulosira fertilissima	Phosphate-deficiency + Citrate + Dark	51	PHB	Samantaray and Mallick (2012)
	Acetate + Citrate	66	PHB	
	Phosphate-deficiency + Acetate	77	PHB	
	Citrate + Acetate and K_2HPO_4	85	PHB	
Recombinant *Synechococcus* sp. PCC 6803	Photoautotrophy + CO_2	1.4	PHB	Osanai et al. (2013)
Aulosira fertilissima CCC 444	Fructose + Valerate	77	P(3HB-*co*-3HV)	Samantaray and Mallick (2014)

larger mass for cyanobacteria that in turn give rise to low PHB values on dry weight basis.

Cyanobacteria, therefore, do have prospective for accumulation of PHB. Exploiting cyanobacteria is one of the most motivating eco-friendly approaches as the CO_2 is transformed into PHAs photosynthetically by using sunlight as the source of energy. Furthermore, maximum cyanobacteria are aquatic microorganisms; consequently, they do not need productive lands for their cultivation. Irrespective of these advantages, the foremost hurdle in photosynthetic production of PHB by cyanobacteria is the scarcity of an economically feasible mass cultivation system. Belay (2004) reported that ongoing mass cultures of cyanobacteria have productivity merely up to 10–15 g dry wt/m^2/day in sunlight receiving area. Moreover, the low-priced harvesting of these tiny cyanobacterial cells/filaments is another limitation linked to cyanobacterial accumulation of PHA.

11.3.6.3   Attempts for Low-Cost PHAs Production with Inexpensive Substrates

While considering the second option, for example, a search for low-cost alternative substrates for bacterial PHA production, the use of organic wastes such as swine waste liquor, palm oil mill effluents, and vegetable and fruit wastes is being considered as alternative substrates for reducing the production cost of PHA (Hassan et al., 1996; Hassan et al., 1997a,b; Meesters, 1998; Reis et al., 2003; Salehizadeh and van Loosdrecht, 2004). Wastewaters, spent washes, pressed-muds and sludges originating from food and brewery industries, municipal sewage, etc. contain considerably high amount of organic carbon compounds such as volatile fatty acids (VFAs). The VFAs such as acetic, propionic and butyric acids are precursors of PHAs (Ruan et al., 2003). If waste products can be used as substrates for production of PHAs, combined advantages of reducing disposal cost and production of value-added products can be envisaged. Consequently, different researchers have studied PHAs production from activated sludge supplemented with malt waste, acetate, volatile fatty acids, paper mill wastewater or rice grain-based distillery spent wash, municipal wastewaters alone as well as supplemented with carbon sources, fermented food waste, synthetic wastewater and also anaerobic wastewaters. Table 11.5 summarizes PHAs production from various wastes; a maximum of 71% (dcw) was recorded

TABLE 11.5 PHAs Production by Different Microorganisms from Wastes

Microorganism	Substrate	% PHA (dcw)	PHA composition	Reference
Activated sludge	Malt	43	PHB	Yu et al. (1999)
Municipal wastewater only	—	31	PHB	Chua et al. (2003)
Fermented food waste	—	51	PHB	Rhu et al. (2003)
Synthetic wastewater	—	66	PHB	Khardenavis et al. (2005)
Anaerobic wastewater	—	58	PHB	
Activated sludge	Rice grain-based distillery spent wash	67	PHB	Khardenavis et al. (2007)
Activated sludge	Paper mill wastewater	48	P(3HB-co-3HV)	Bengtsson et al. (2008)
Activated sludge	Volatile fatty acids	57	PHB	Mengmeng et al. (2009)
Activated sludge	—	50	PHB	Rodgers and Wu (2010)
Activated sludge	Sodium acetate	67	PHB	Liu et al. (2011)
Activated sludge	Volatile fatty acids	49	PHB	Mokhtarani et al. (2012)
Activated sludge	Municipal wastewater	34	PHB	Morgan-Sagastume et al. (2014)
Mixed culture	Synthetic wastewater	71	PHB	Vargas et al. (2014)

with mixed culture biomass grown in synthetic wastewater (Vargas et al., 2014). These studies demonstrated that PHA could be produced in mixed culture system. Apart from this, costs for process development using mixed cultures can be substantially reduced because cheap (or even free) substrates and simple, non-sterilizable reactors are used and less process control is needed when compared with a pure culture (Satoh et al., 1998; Reis et al., 2003; Salehizadeh and van Loosdrecht, 2004). Furthermore, utilization of open mixed microbial cultures facilitates the use of mixed substrates since the microbial population can adapt continuously to changes in substrate (Reis et al., 2003). Thus, using activated sludges and wastes enriched under adequate conditions may be a promising option for PHAs production, which needs further investigation as no studies on the optimization of nutrient removal and PHA production by mixed cultures are available.

Apart from this, several inexpensive carbon substrates such as molasses, whey, cellulose, plant oils, hydrolysates of starch (corn and tapioca), wheat bran hydrolysate and rice bran hydrolysate, can be excellent substrates for bacteria utilizing them to produce PHAs, which could lead to significant economical advantages. Several investigators have studied PHAs production by different bacterial strains such as *Azotobacter vinelandii, Ralstonia eutropha (Cupriavidus necator), Actinobacillus* sp., *Alcaligenes latus, Azotobacter chroococcum, Pseudomonas cepacia, Pseudomonas corrugate, Haloferax mediterranei, Methylobacterium* sp., recombinant *Escherichia coli, Bacillus* sp., *Rhizobium meliloti, Sphingomonas* sp., *Brevundimonas vesicularis, Sphingopyxis macrogoltabida* and *Herbaspirillum seropedicae, Bacillus tequilensis, Cupriavidus necator* IPT 026, *Pseudomonas putida* CA-3, *Hydrogenophaga pseudoflava, Aeromonas* sp., *Pseudomonas* sp., *Bacillus subtilis, Pseudomonas putrefaciens* and *Burkholderia cepaca* to convert inexpensive substrates into PHAs. Table 11.6 summarizes some of the important findings on the production of PHAs from inexpensive substrates by various microorganisms.

To date, whey has been the most extensively studied waste material for PHAs production by a variety of microorganisms and, so far, recombinant *Escherichia coli* has been reported to yield 80% PHB on dcw basis using whey as the substrate (Kim, 2000). *Azotobacter vinelandii* UWD was however, reported to produce P(3HB-*co*-3HV) co-polymer up to 34% of dcw, when grown in swine waste liquor, which was found to increase up to 58% (dcw) following addition of glucose (30 g L^{-1}, Cho et al., 1997). Alcoholic distillery wastewater rich in sugars and nitrogen compounds was

TABLE 11.6 PHAs Production from Inexpensive Substrates by Various Microorganisms

Microorganism	Substrate	% PHA (dcw)	PHA composition	Reference
Haloferax mediterranei	Starch	60	PHB	Lillo and Rodriguez-Valera (1990)
Azotobacter vinelandii UWD	Molasses	66	PHB	Page and Cornish (1993)
Pseudomonas cepacia	Lactose	56	PHB	Young et al. (1994)
Pseudomonas cepacia ATCC 17759	Xylose	60	PHB	Ramsay et al. (1995)
Ralstonia eutropha	Tapioca hydrolysate	58	PHB	Kim and Chang (1995)
Actinobacillus sp. EL-9	Alcoholic distillery wastewater	42	PHB	Son et al. (1996)
Azotobacter vinelandii UWD	Swine waste liquor	34	P(3HB-*co*-3HV)	Cho et al. (1997)
Methylobacterium sp. ZP24	Whey	60	PHB	Yellore and Desia (1998)
Alcaligenes latus DSM 1124	Malt waste	70	PHB	Yu et al. (1999)
Azotobacter chroococcum	Starch	74	PHB	Kim (2000)
Recombinant *E. coli*	Whey	80\	PHB	
Ralstonia eutropha	Food scraps	73	PHB	Du and Yu (2002)
Ralstonia eutropha	Soy cake + Sugar cane molasses	39	PHB	Oliveira et al. (2004)
Pseudomonas cepacia G13	Sugar beet molasses	70	PHB	Celik and Beyatli et al. (2005)

TABLE 11.6 Continued

Microorganism	Substrate	% PHA (dcw)	PHA composition	Reference
Pseudomonas corrugata	Soy molasses	17	MCL-PHA	Solaiman et al. (2006)
Recombinant *E. coli*	Whey + Corn steep liquor	73	PHB	Nikel et al. (2006)
Bacillus sp. Z56	Mahua flowers	51	P(3HB-*co*-3HV)	Kumar et al. (2007)
Rhizobium meliloti	Mahua flowers	31	P(3HB-*co*-3HV)	
Sphingomonas sp.	Mahua flowers	22	P(3HB-*co*-3HV)	
Brevundimonas vesicularis LMG P-23615	Acid-hydrolyzed sawdust	64	PHB	Silva et al. (2007)
Sphingopyxis macrogoltabida LMG 17324	Acid-hydrolyzed sawdust	72	PHB	
Recombinant *Herbaspirillum seropedicae* Z69	Lactose	36	PHB	Catalan et al. (2007)
Bacillus megaterium	Sugarcane molasses	43	PHB	Chaijamrus and Udpuay (2008)
Cupriavidus necator DSM 545	Waste glycerol	50	PHB	Cavalheiro et al. (2009)
Burkholderia cepaca	Waste glycerol	31	PHB	Zhu et al. (2010)
Pseudomonas putrefaciens	Corn cob	67	PHB	Ogunjobi et al. (2011)
Cupriavidus necator CCGUG 52238	Kitchen-waste derived organic acids	53	PHB	Omar et al. (2011)

TABLE 11.6 Continued

Microorganism	Substrate	% PHA (dcw)	PHA composition	Reference
Aeromonas sp.	Municipal waste water	11	PHB	Sangkharak and Prasertsan (2012)
	POME	23	PHB	
	Glycerol	20	PHB	
	Molasses	20	PHB	
Pseudomonas sp.	Municipal waste water	61	PHB	
	POME	60	P(3HB-*co*-3HV-*co*-HHx)	
	Glycerol	62	PHB	
	Molasses	60	PHB	
Bacillus subtilis	Municipal waste water	47	PHB	
	POME	50	P(3HB-*co*-3HV-*co*-3HHx)	
	Glycerol	19	PHB	
	Molasses	16	PHB	
Bacillus sp.	WBH + RBH*	59		Shamala et al. (2012)
Hydrogenophaga pseudoflava	Lactose	30	P(3HB-co-3HV-co-4HB)	Povolo et al. (2013)
	Whey	10	P(3HB-co-3HV-co-4HB)	
Pseudomonas putida CA-3	Volatile fatty acids	39	MCL-PHA	Cerrone et al. (2014)
Cupriavidus necator IPT 026	Crude glycerin	65	PHB	Campos et al. (2014)

*WBH: Wheat bran hydrolysate, RBH: Rice bran hydrolysate, POME: Palm oil mill effluent.

also found to be a potential feedstock for PHB production and *Actinobacillus* sp. EL-9 was found to accumulate up to 42% (dcw) PHB when grown in such wastewater (Son et al., 1996). Residual oil from biotechnological rhamnose production was used as a carbon source for growth as well as PHA production by *Ralstonia eutropha* (*Cupriavidus necator*) H16 and *Pseudomonas oleovorans*. Approximately 20–25% of the components of the residual oil were converted into PHAs (Fuechtenbusch et al., 2000). *Bacillus megaterium* was grown on various carbon sources such as date syrup, beet molasses, fructose, lactose, sucrose and glucose in mineral salts medium. Best results with regard to growth and PHAs production were obtained in the cheaper carbon sources such as date syrup and beet molasses (Omar et al., 2001). Kumar et al. (2007) reported that *Bacillus* sp. 256 accumulated 51% P(3HB-*co*-3HV) co-polymer in mahua flower feded cultures. Acid-hydrolyzed sawdust was also found to be a potential feedstock for PHAs production (Silva et al., 2007). Accumulation of PHB up to 50% was registered in *Cupriavidus necator* DSM 545 supplemented with waste glycerol (Cavalheiro et al., 2009). *Pseudomonas putrefaciens* under supplementation of corn cob accumulated 67% PHB (Ogunjobi et al., 2011). Shamala et al. (2012) reported that *Bacillus* sp. accumulated 59% P(3HB-*co*-3HV) co-polymer under wheat bran and rice bran hydrolysate-supplemented cultures. *Hydrogenophaga pseudoflava* depicted 30 and 10% P(3HB-*co*-3HV-*co*-4HB) co-polymer accumulation, respectively in lactose and whey-supplemented cultures (Povolo et al., 2013). *Cupriavidus necator* IPT 026 and *Bacillus tequilensis* registered PHAs up to 65 and 40%, respectively under supplementation with crude glycerin and acidogenic spent wash (Amulya et al., 2014; Campos et al., 2014). Thus, exploitation of bacterial strains for PHAs production using inexpensive substrates could be an ideal economic solution not only for low-cost polymer production but also for successful commercialization of PHAs.

11.4 CONCLUSION AND FUTURE PERSPECTIVES

PHA polymers have gained substantial interest in recent years as a result of rising environmental concerns over conventional plastics. Despite the basic attractiveness as an alternate for petroleum-derived polymers, the foremost obstacle facing triumphant commercialization of PHAs is the high price of bacterial fermentation, which brought photoautotrophic hosts and alternate inexpensive substrates into focus for low-cost PHAs production. Photoautotrophic hosts that do have the potential to produce PHAs also have some drawbacks, which makes the system less preferred for large-scale production. Presently, major attempt has been devoted to making PHAs production process cost-effectively more feasible by changing the substrate from expensive to cost-effective. The major challenge would be to improve the economical production together with expression level of PHAs either by exploiting an novel cost-effective carbon source, engineering efficient microorganisms, improving fermentation and separation processes or by applying mutational approaches/genetic engineering techniques, thus, paving the way for a greener, post-conventional plastic era.

KEYWORDS

- 3-Hydroxybutyrate
- 4-Hydroxybutyrate
- Biodegradability
- Biodegradable polymers
- Biosynthetic polymers
- Co-polymer
- Cyanobacteria
- Microbially-synthesized PHAs
- PHA depolymerases
- PHA hydrolases
- Polycarbonates
- Polyhydroxyalkanoates
- Polysulfones
- Thermoplastics
- Transgenic polymers

REFERENCES

Abe, C., Taima, Y., Nakamura, Y., Doi, Y. New bacterial copolyesters of 3-hydroxyalkanoates and 3-hydroxy-ω-fluoroalkanoates produced by *Pseudomonas oleovorans*. *Polym Commun* 1990, 31, 404–406.

Abe, H., Doi, Y. Side-chain effect of second monomer units on crystalline morphology, thermal properties, and enzymatic degradability for random copolyesters of (*R*)-3-hydroxybutyric acid with (*R*)-3-hydroxyalkanoic acids. *Biomacromolecules* 2002, 3, 133–138.

Alexander, M. Biodegradation of chemicals of environmental concern. *Science* 1981, 211, 132–138.

Amulya, K., Reddy, M. V., Mohan, S. V. Acidogenic spent wash valorization through polyhydroxyalkanoate (PHA) production using *Bacillus tequilensis*: Integration with fermentative biohydrogen production process. *Biores Technol* 2014, http://dx.doi.org/10.1016/j.biortech.2014.02.026.

Anderson, A. J., Dawes, E. A. Occurrence, metabolism, metabolic role, and industrial uses of bacterial polyhydroxyalkanoates. *Microbiol Rev* 1990, 54, 450–472.

Ashby, R. D., Solaiman, D. K. Y., Foglia, T. A. The synthesis of short and medium chain-length poly(hydroxyalkanoate) mixtures from glucose- or alkanoic acid-grown *pseudomonas oleovorans*. *J Ind Microbiol Biotechnol* 2002, 28, 147–153.

Belay, A. Mass culture of *Spirulina* outdoors - the Earthrise farms experience. In *Spirulina platensis (Arthrospira): Physiology, Cell-Biology and Biotechnology*, Vonshak, A. (Ed.), Taylor and Francis: London, 2004, pp. 131–158.

Bengtsson, S., Werker, A., Christensson, M., Welander, T. Production of polyhydroxyalkanoates by activated sludge treating a paper mill wastewater. *Biores Technol* 2008, 99, 509–516.

Bhati, R., Samantaray, S., Sharma, L., Mallick, N. Poly-β-hydroxybutyrate accumulation in cyanobacteria under photoautotrophy. *Biotechnol J* 2010, 5, 1181–1185.

Bhati, R., Mallick, N. Production and characterization of poly(3-hydroxybutyrate-*co*-3-hydroxyvalerate) co-polymer by a N_2-fixing cyanobacterium, *Nostoc muscorum* Agardh. *J Chem Technol Biotechnol* 2012, 87, 505–512.

Bohmert, K., Balbo, I., Kopka, J., Mittendorf, V., Nawrath, C., Poirier, Y., Tischendorf, G., Tretchewey, R. N., Willmitzer, L. Transgenic *Arabidopsis* plants can accumulate polyhydroxybutyrate up to 4% of their fresh weight. *Planta* 2000, 211, 841–845.

Bourque, D., Ouellette, B., Andre, G., Groleau, D. Production of poly-*β*-hydroxybutyrate from methanol: characterization of a new isolate of *Methylobacterium extorquens*. *Appl Microbiol Biotechnol* 1992, 37, 7–12.

Byrom, D. Production of poly-*β*-hydroxybutyrate and poly-*β*-hydroxyvalerate copolymers. *FEMS Microbiol Rev* 1992, 103, 247–250.

Byrom, D. Polyhydroxyalkanoates. In *Plastics from Microbes*: *Microbial Synthesis of Polymers and Polymer Precursors*, Mobley, D. P. (Ed.), Hanser: New York, 1994, pp. 5–33.

Caballero, K. P., Karel, S. F., Register, R. A. Biosynthesis and characterization of hydroxybutyrate-hydroxycaproate copolymers. *Int J Biol Macromol* 1995, 17, 86–92.

Cain, R. B. Microbial degradation of synthetic polymers. In *Microbial Control of Pollution*, *48th Symposium of the Society for General Microbiology*, University of Cardiff, 1992, pp. 293–338.

Campbell, J., Stevens, S. E. Jr.; Bankwill, D. L. Accumulation of poly-*β*-hydroxybutyrate in *Spirulina platensis*. *J Bacteriol* 1982, 149, 361–366.

Campos, M. I., Figueiredo, V. B. F., Sousa, L. S., Druzian, J. I. The influence of crude glycerin and nitrogen concentrations on the production of PHA by *Cupriavidus necator* using a response surface methodology and its characterizations. *Ind Crops Prod* 2014, 52, 338–346.

Carr, N. G. The occurrence of poly-*β*-hydroxybutyrate in the blue-green alga, *Chlorogloea fritschii*. *Biochem Biophys Acta* 1966, 120, 308–310.

Cavalheiro, J. M. B. T., de Almeida, M. C. M. D., Grandfils, C., da Fonseca, M. M. R. Poly(3-hydroxybutyrate) production by *Cupriavidus necator* using waste glycerol. *Process Biochem* 2009, 44, 509–515.

Cerrone, F., Choudhari, S. K., Davis, R., Cysneiros, D., O'Flaherty, V., Duane, G., Casey, E., Guzik, M. W., Kenny, S. T., Babu, R. P., O'Connor, K. Medium chain length polyhydroxyalkanoate (mcl-PHA) production from volatile fatty acids derived from the anaerobic digestion of grass. *Appl Microbiol Biotechnol* 2014, 98, 611–20.

Chaijamrus, S., Udpuay, N. Production and characterization of polyhydroxybutyrate from molasses and corn steep liquor produced by *Bacillus megaterium* ATCC 6748. *Agr Engg Int* 2008, 10(7), 30.

Chen, G. Q., Xu, J., Wu, Q., Zhang, Z., Ho, K. P. Synthesis of copolyesters consisting of medium-chain-length β-hydroxyalkanoates by *Pseudomonas stutzeri* 1317. *React Funct Polym* 2001a, 48, 107–112.

Chen, G. Q., Zhang, G., Park, S. J., Lee, S. Y. Industrial scale production of poly(3-hydroxy-butyrate-*co*-3-hydroxyhexanoate). *Appl Microbiol Biotechnol* 2001b, 57, 50–55.

Chiras, D. D. *Environmental Science*. The Benjamin/Cumming Publishing Company, Inc.: Redwood, California, 1994.

Cho, K. S., Ryu, H. W., Park, C. H., Goodrich, P. R. Poly(hydroxybutyrate-*co*-hydroxyvalerate) from swine waste liquor by *Azotobacter vinelandii* UWD. *Biotechnol Lett* 1997, 19, 7–10.

Choi, M. H., Yoon, S. C. Polyester biosynthesis characteristics of *Pseudomonas citronellolis* grown on various carbon sources, including 3-methyl-branched substrates. *Appl Environ Microbiol* 1994, 60, 3245–3254.

Chua, A. S. M., Takabatake, H., Satoh, H., Mino, T. Production of polyhydroxyalkanoates (PHA) by activated sludge treating municipal wastewater: effect of pH, sludge retention time (SRT) and acetate concentration in influent. *Water Res* 2003, 37, 3602–3611.

Curley, J. M., Hazer, B., Lenz, R. W., Fuller, R. C. Production of poly(3-hydroxyalkanoates) containing aromatic substituents by *Pseudomonas oleovorans*. *Macromolecules* 1996, 29, 1762–1766.

de Koning, G. J. M., van Bilsen, H. M. M., Lemstra, P. J., Hazenberg, W., Witholt, B., Preusting, H., van der Galien, J. G., Schirmer, A., Jendrossek, D. A biodegradable rubber by crosslinking poly(hydroxyalkanoate) from *Pseudomonas oleovorans*. *Polymer* 1994, 35, 2090–2097.

Doi, Y. Microbial Polyesters. VCH Publishers: New York, 1990.

Doi, Y., Abe, C. Biosynthesis and characterization of a new bacterial copolyester of 3-hydroxy-alkanoates and 3-hydroxy-ω-chloroalkanoates. *Macromolecules* 1990, 23, 3705–3707.

Doi, Y., Kitamura, S., Abe, H. Microbial synthesis and characterization of poly(3-hydroxybu-tyrate-*co*-3-hydroxyhexanoate). *Macromolecules* 1995, 28, 4822–4828.

Doi, Y., Kunioka, M., Nakamura, Y., Soga, K. Biosynthesis of copolyesters in *Alcaligenes eutrophus* H16 from ^{13}C-labeled acetate and propionate. *Macromolecules* 1987, 20, 2988–2991.

Fiechter, A. Plastics from bacteria and for bacteria: poly (β-hydroxyalkanoates) as natural, biocompatible, and biodegradable polyesters. Springer-Verlag: New York, 1990.

Fritzsche, K., Lenz, R. W., Fuller, R. C. Bacterial polyesters containing branched poly (β-hydroxyalkanoate) units. *Int J Biol Macromol* 1990a, 12, 92–101.

Fritzsche, K., Lenz, R. W., Fuller, R. C. Production of unsaturated polyesters by *Pseudomonas oleovorans*. *Int J Biol Macromol* 1990b, 12, 85–91.

Fritzsche, K., Lenz, R. W., Fuller, R. C. An unusual bacterial polyester with a phenyl pendant group. *Macromol Chem* 1990c, 191, 1957–1965.

Fuchtenbusch, B., Wullbrandt, D., Steinbuchel, A. Production of polyhydroxyalkanoic acids by *Ralstonia eutropha* and *Pseudomonas oleovorans* from oil remaining from biotechnological rhamnose production. *Appl Microbiol Biotechnol* 2000, 53, 167–172.

Gagnon, K. D., Lenz, R. W., Farris, R. J., Fuller, R. C. Chemical modification of bacterial elastomers: 1. Peroxide crosslinking. *Polymer* 1994a, 35, 4358–4367.

Gagnon, K. D., Lenz, R. W., Farris, R. J., Fuller, R. C. Chemical modification of bacterial elastomers: 2. Sulfur vulcanization. *Polymer* 1994b, 35, 4368–4375.

Gao, X., Yuan, X. X., Shi, Z. Y., Shen, X. W., Chen, J. C., Wu, Q., Chen, G. Q. Production of copolyesters of 3-hydroxybutyrate and medium-chain-length 3-hydroxyalkanoates by *E. coli* containing an optimized PHA synthase gene. *Microb Cell Fact* 2012, doi: 10.1186/1475–2859-11-130.

Hahn, J. J., Eschenlauer, A. C., Sletyr, U. B., Somer, D. A., Srienc, F. Peroxisomes as site for synthesis of polyhydroxyalkanoates in transgenic plants. *Biotechnol Prog* 1999, 15, 1053–1057.

Hassan, M. A., Shirai, N., Kusubayashi, N., Abdul Karim, M. I., Nakanishi, K., Hashimoto, K. Effect of organic acid profiles during anaerobic treatment of palm oil mill effluent on the production of polyhydroxyalkanoates by *Rhodobacter sphaeroides*. *J Ferment Bioengg* 1996, 82, 151–156.

Hassan, M. A., Shirai, N., Kusubayashi, N., Abdul Karim, M. I., Nakanishi, K., Hashimoto, K. The production of polyhydroxyalkanoates from anaerobically treated palm oil mill effluent by *Rhodobacter spheroides*. *J Ferment Bioengg* 1997a, 83, 485–488.

Hassan, M. A., Shirai, N., Kusubayashi, N., Abdul Karim, M. I., Nakanishi, K., Hashimoto, K. Acetic acid separation from anaerobically treated palm oil mill effluent for the production of polyhydroxyalkanoate by *Alcaligenes eutrophus*. *Biosci Biotechnol Biochem* 1997b, 61, 1465–1468.

Haywood, G. W., Anderson, A. J., Dawes, E. A. A survey of the accumulation of novel polyhydroxyalkanoates by bacteria. *Biotechnol Lett* 1989, 11, 471–476.

Haywood, G. W., Anderson, A. J., Williams, D. R., Dawes, E. A. Accumulation of a poly(hydroxyalkanoate) copolymer containing primarily 3-hydroxyvalerate from simple carbohydrate substrates by *Rhodococcus* sp. NCIMB 40126. *Int J Biol Macromol* 1991, 13, 83–88.

Hazer, B., Lenz, R. W., Fuller, R. C. Biosynthesis of methyl-branched poly (β-hydroxyalkanoate) s by *Pseudomonas oleovorans*. *Macromolecules* 1994, 27, 45–49.

Hocking, P. J., Marchessault, R. H. Biopolyesters. In *Chemistry and Technology of Biodegradable Polymers*, Griffin, G. J. L. (Ed.), Chapman and Hall: London, 1994, pp. 48–96.

Houmiel, K. L., Slater, S., Broyles, D., Casagrande, L., Colburn, S., Gonzalez, K. Poly(β-hydroxybutyrate) production in oilseed leucoplasts of *Brassica napus*. *Planta* 1999, 209, 547–550.

Huijberts, G. N. M., de Rijk, T. C., de Ward, P., Eggink, G. [13]C nuclear magnetic resonance study of *Pseudomonas putida* fatty acid metabolic routes involved in poly(3-hydroxyalkanoate) synthesis. *J Bacteriol* 1995, 176, 1661–1666.

Huisman, G. W., de Leeuw, O., Eggink, G., Witholt, B. Synthesis of poly-3-hydroxyalkanoates is a common feature of fluorescent *Pseudomonads*. *Appl Environ Microbiol* 1989, 55, 1949–1954.

Ismail, I., Iskandar, N. F., Chee, G. M., Abdullah, R. Genetic transformation and molecular analysis of polyhydroxybutyrate biosynthetic contents of polyhydroxybutyrate are associated with growth reduction. *Plant Cell Rep* 2010, 21, 891–899.

John, M. E. Cotton crop improvement through genetic engineering. *Crit Rev Biotechnol* 1997, 17, 185–208.

John, M. E., Keller, G. Metabolic pathway engineering in cotton: biosynthesis of polyhydroxybutyrate in fiber cells. *Proc Natl Acad Sci* 1996, 93, 12768–12773.

Johnstone, B. A throw away answer. *Far East Econ Rev* 1990, 147, 62–63.

Kalia, V. C., Raizada, N., Sonakya, V. Bioplastics. *J Sci Ind Res* 2000, 59, 433–445.

Kato, M., Bao, H. J., Kang, C. K., Fukui, T., Doi, Y. Production of novel copolyester of 3-hydroxybutyric acid and medium-chain-length 3-hydroxyalkanoic acids by *Pseudomonas* sp. 61–3 from sugars. *Appl Microbiol Biotechnol* 1996, 45, 363–370.

Khardenavis, A., Kumar, M. S., Mudliar, S. N., Chakrabarti, T. Biotechnological conversion of agro-industrial wastewaters into biodegradable plastic, poly β-hydroxybutyrate. *Biores Technol* 2007, 98, 3579–3584.

Kim, D. Y., Kim, Y. B., Rhee, Y. H. Evaluation of various carbon substrates for the biosynthesis of polyhydroxyalkanoates bearing functional group by *Pseudomonas putida*. *Int J Biol Macromol* 2000, 28, 23–29.

Kim, Y. B., Lenz, R. W., Fuller, R. C. Preparation and characterization of poly (β-hydroxyalkanoates) obtained from *Pseudomonas oleovorans* grown with mixtures of 5-phenylvaleric acid and n-alkanoic acids. *Macromolecules* 1991, 24, 5256–5360.

Kim, Y. B., Lenz, R. W., Fuller, R. C. Poly(β-hydroxyalkanoate) copolymers containing brominated repeating units produced by *Pseudomonas oleovorans*. *Macromolecules* 1992, 25, 1852–1857.

Kraak, M. N., Kessler, B., Witholt, B. *In vitro* activities of granule-bound poly[(R)-3-hydroxyalkanoate] polymerase Cl of *Pseudomonas oleovorans*: development of an activity test for medium-chain-length-poly(3-hydroxyalkanoate) polymerases. *Eur J Biochem* 1997, 250, 432–439.

Kumar, P. K. A., Shamala, T. R., Kshama, L., Prakash, M. H., Joshi, G. J., Chandrashekar, A., Kumari, K. S. L., Divyashree, M. S. Bacterial synthesis of poly(hydroxybutyrate-co-hydroxyvalerate) using carbohydrate-rich mahua (*Madhuca* sp.) flowers. *J Appl Microbiol* 2007, 103, 204–209.

Labuzek, S., Radecka, I. Biosynthesis of PHB tercopolymer by *Bacillus cereus* UW85. *J Appl Microbiol* 2001, 90, 353–357.

Lageveen, R. G., Huisman, G. W., Preusting, H., Ketelaar, P., Eggink, G., Witholt, B. Formation of polyesters by *Pseudomonas oleovorans*: effect of substrates on formation and composition of poly(*R*)-3-hydroxyalkanoates and poly(*R*)-3-hydroxyalkenoates. *Appl Environ Microbiol* 1988, 54, 2924–2932.

Lee, B., Pometto, A. L. III., Fratzke, A., Bailey, T. B. Biodegradation of degradable plastic polyethylene by *Phanerochaete* and *Streptomyces* species. *Appl Environ Microbiol* 1991, 57, 678–685.

Lee, E. Y., Jendrossek, D., Schirmer, A., Choi, C. Y., Steinbuchel, A. Biosynthesis of copolyesters consisting of 3-hydroxybutyric acid and medium-chain-length 3-hydroxyalkanoic acids from 1,3-butanediol or from 3-hydroxybutyrate by *Pseudomonas* sp. A33. *Appl Microbiol Biotechnol* 1995a, 42, 901–909.

Lee, S. H., Oh, D. H., Ahn, W. S., Lee, Y., Choi, J., Lee, S. Y. Production of poly(3-hydroxybutyrate-*co*-3-hydroxyhexanoate) by high-cell-density cultivation of *Aeromonas hydrophila*. *Biotechnol Bioengg* 2000b, 67, 240–244.

Lee, S. Y. Bacterial polyhydroxyalkanoates. *Biotechnol Bioeng* 1996, 49, 1–4.

Lee, S. Y., Choi, J., Wong H. W. Recent advances in polyhydroxyalkanoate production by bacterial fermentation: mini-review. *Int J Biol Macromol* 1999, 25, 31–36.

Lee, S. Y., Park, S. J. Biosynthesis and fermentative production of SCL-MCL-PHAs. In *Biopolymers, Vol. 3a,* Doi, Y., Steinbuchel, A. (Eds.), Wiley/VCH: Weinheim, 2002, pp. 317–336.

Li, Q. A., Chen, Q. A., Li, M. J., Wang, F. S., Qi, Q. S. Pathway engineering results the altered polyhydroxyalkanoates composition in recombinant *Escherichia coli*. *New Biotech* 2011, 28, 92–95.

Liebergesell, M., Sonomoto, K., Madkour, M., Mayer, F., Steinbuchel, A. Purification and characterization of the poly(hydroxyalkanoic acid) synthase from *Chromatium vinosum* and localization of the enzyme at the surface of poly(hydroxyalkanoic acid) granules. *Eur J Biochem* 1994, 226, 71–80.

Liu, Z., Wang, Y., He, N., Huang, J., Zhu, K., Shao, W., Wang, H., Yuan, W., Li, Q. Optimization of polyhydroxybutyrate (PHB) production by excess activated sludge and microbial community analysis. *J Hazard Mater* 2011, 185, 8–16.

Lossl, A., Bohmert, K., Harloff, H. J., Eibl, C., Muhlbauer, S., Koop, H. U. Inducible transactivation of plastid transgenes: expression of the *R. eutropha* phb operon in transplastomic tobacco. *Plant Cell Physiol* 2005, 46, 1462–1471.

Lossl, A., Eibl, C., Harloff, H. J., Jung, C., Koop, H. U. Polyester synthesis in transplastomic tobacco (*Nicotiana tabacum* L.): significant contents of polyhydroxybutyrate are associated with growth reduction. *Plant Cell Rep* 2003, 21, 891–899.

Luo, R., Chen, J., Zhang, L., Chen, G. Polyhydroxyalkanoate copolyesters produced by *Ralstonia eutropha* PHB-4 harboring a low-substrate-specificity PHA synthase PhaC2Ps from *Pseudomonas stutzeri* 1317. *Biochem Engg J* 2006, 32, 218–225.

Madison, L. L., Huisiman, G. W. Metabolic engineering of poly(3-hydroxyalkanoates): from DNA to plastic. *Microbiol Mol Biol Rev* 1999, 63, 21–53.

Masani, M. Y., Parveez, G. K., Izawati, A. M., Lan, C. P., Siti Nor, A. Construction of PHB and PHBV multiple-gene vectors driven by an oil palm leaf-specific promoter. *Plasmid* 2009, 62, 191–200.

Matsumoto, K., Arai, Y., Nagao, R., Murata, T., Takase, K., Nakashita, H., Taguchi, S., Shimada, H., Doi, Y. Synthesis of short-chain-length/medium-chain-length polyhydroxyalkanoate (PHA) copolymers in peroxisome of the transgenic *Arabidopsis thaliana* harboring the PHA synthase gene from *Pseudomonas* sp. 61–3. *J. Polym Environ* 2006, 14, 369–374.

Matsumoto, K., Murata, T., Nagao, R., Nomura, C. T., Arai, S., Arai, Y., Takase, K., Nakashita, H., Taguchi, S., Shimada, H. Production of short-chain-length/medium-chain-length polyhydroxyalkanoate (PHA) copolymer in the plastid of *Arabidopsis thaliana* using an engineered 3-ketoacyl-acyl carrier protein synthase III. *Biomacromolecules* 2009, 10, 686–690.

Matsusaki, H., Abe, H., Doi, Y. Biosynthesis and properties of poly (3-hydroxybutyrate-*co*-3-hydroxyalkanoates) by recombinant strains of *Pseudomonas* sp. 61–3. *Biomacromolecules* 2000, 1, 17–22.

Meesters, K. H. P. *Production of poly(3-hydroxyalkanoates) from waste streams*. Technical University of Delft: Netherlands, 1998, p 247.

Mengmeng, C., Hong, C., Qingliang, Z., Shirley, S. N., Jie, R. Optimal production of polyhydroxyalkanoates (PHA) in activated sludge fed by volatile fatty acids (VFAs) generated from alkaline excess sludge fermentation. *Bioresour Technol* 2009, 100, 1399–1405.

Menzel, G., Harloff, H. J., Jung, C. Expression of bacterial poly(3-hydroxybutyrate) synthesis genes in hairy roots of sugar beet (*Beta vulgaris* L.). *Appl Microbiol Biotechnol* 2003, 60, 571–576.

Mitsky, T. A., Slater, S. C., Reiser, S. E., Hao, M., Houmiel, K. L. Multigene expression vectors for the biosynthesis of products via multienzyme biological pathways. PCT application WO00/52183, 2000.

Mittendorf, V., Bongcam, V., Allenbach, L., Coullerez, G., Martini, N., Poirier, Y. Polyhydroxyalkanoate synthesis in transgenic plants as a new tool to study carbon flow through β-oxidation. *Plant J* 1999, 20, 45–55.

Mittendorf, V., Robertson, E. J., Leech, R. M., Krüger, N., Steinbuchel, A., Poirier, Y. Synthesis of medium-chain-length polyhydroxyalkanoates in *Arabidopsis thaliana* using intermediates of peroxisomal fatty acid β-oxidation. *Proc Natl Acad Sci* 1998, 95, 13397–13402.

Mokhtarani, N., Ganjidoust, H., Farahani, E. V. Effect of process variables on the production of polyhydroxyalkanoates by activated sludge. *Iran J Environ Health Sci Eng* 2012, 9, 1–6.

Morgan-Sagastume, F., Valentino, F., Hjort, M., Cirne, D., Karabegovic, L., Gerardin, F., Johansson, P., Karlsson, A., Magnusson, P., Alexandersson, T., Bengtsson, S., Majone, M., Werker, A. Polyhydroxyalkanoate (PHA) production from sludge and municipal wastewater treatment. *Water Sci Technol* 2014, 69, 177–184.

Moskowitz, G. J., Merrick, J. M. Metabolism of poly-β-hydroxybutyrate II. Enzymatic synthesis of D(–)-β-hydroxybutyryl coenzyme-A by an enoylhydrases from *Rhodospirillum rubrum*. *Biochemistry* 1969, 8, 2748–2755.

Muller, H. M., Seebach, D. Poly(hydroxyfettsaureester), eine funfte klasse von physiologisch bedeutsamen organischen biopolymeren? *Angew. Chem.* 1993, 105, 483–509.

Muller, R. J., Kleeberg, I., Deckwer, W. D. Biodegradation of polyesters containing aromatic constituents. *J Biotechnol* 2001, 86, 87–95.

Nakashita, H., Arai, Y., Yoshioka, K., Fukui, T., Doi, Y., Usami, R. Production of biodegradable polyester by a transgenic tobacco. *Biosci Biotechnol Biochem* 1999, 63, 870–874.

Nawrath, C., Poirier, Y., Somerville, C. Targeting of the polyhydroxybutyrate biosynthetic pathway to the plastids of *Arabidopsis thaliana* results in high levels of polymer accumulation. *Proc Natl Acad Sci* 1994, 91, 12760–12764.

Nishioka, M., Nakai, K., Miyake, M., Asada, Y., Taya, M. Production of poly-β-hydroyxybutyrate by thermophilic cyanobacterium, *Synechococcus* sp. MA19, under phosphate limitation. *Biotechnol Lett* 2001, 23, 1095–1099.

Noda, I., Satkowski, M. M., Dowrey, A. E., Marcott, C. Polymer alloy of Nodax co-polymer and poly (lactic acid). *Macromol Biosci* 2004, 4, 269–275.

Nomura, C. T., Taguchi, K., Gan, Z., Kuwabara, K., Tanaka, T., Takase, K., Doi, Y. Expression of 3-ketoayl-acyl carrier protein reductase (*fabG*) genes enhanced production of polyhydroxyalkanoates copolymer from glucose in recombinant *Escherichia coli* JM109. *Appl Environ Microbiol* 2005, 71, 4297–4306.

Oeding, V., Schlegel, H. G. Beta-ketothiolase from *Hydrogenomonas eutropha* HI6 and its significance in the regulation of poly-beta-hydroxybutyrate metabolism. *Biochem J* 1973, 134, 239–248.

Ogunjobi, A. A., Ogundele, A. O., Fagade, O. E. Production of polyhydroxyalkanoates by *Pseudomonas putrefaciens* from cheap and renewable carbon substrates. *Electron J Environ Agr Food Chem* 2011, 10, 2806–2815.

Omar, F. N., Aini, N., Rahman, A., Hafid, H. S., Mumtaz, T., Yee, P. L., Hassan, M. A. Utilization of kitchen waste for the production of green thermoplastic polyhydroxybutyrate (PHB) by *Cupriavidus necator* CCGUG 52238. *Afr J Microbiol Res* 2011, 5, 2873–2879.

Omar, S., Rayes, A., Eqaab, A., Voss, I., Steinbuchel, A. Optimization of cell growth and poly(3-hydroxybutyrate) accumulation on date syrup by a *Bacillus megaterium* strain. *Biotechnol Lett* 2001, 23, 1119–1123.

Omidvar, V., Siti Nor, A., Marziah, M., Maheran, A. A transient assay to evaluate the expression of polyhydroxybutyrate genes regulated by oil palm mesocarp-specific promoter. *Plant Cell Rep* 2008, 27, 1451–1459.

Osanai, T., Numata, K., Oikawa, A., Kuwahara, A., Iijima, H., Doi, Y., Tanaka, K., Saito, K., Hirai, M. Y. Increased bioplastic production with an RNA polymerase sigma factor SigE during nitrogen starvation in *Synechocystis* sp. PCC 6803. *DNA Res* 2013, 20, 525–535.

Panda, B., Mallick, N. Enhanced poly-β-hydroxybutyrate accumulation in a unicellular cyanobactrium, *Synechocystis* sp. PCC 6803. *Lett Appl Microbiol* 2007, 44, 194–198.

Park, S. J., Lee, S. Y. Biosynthesis of poly (3-hydroxybutyrate-co-3-hydroxyalkanoates) by metabolically engineered *Escherichia coli* strains. *Appl Biochem Biotechnol* 2004, 113⁄116, 335–346.

Patterson, N., Tang, J., Cahoon, E. B., Jaworski, J. G., Wang, W., Peoples, O. P., Snell, K. D. Generation of high polyhydroxybutryate producing oilseeds. International Patent Application WO/2011/034945, 2011a.

Patterson, N., Tang, J., Han, J., Tavva, V., Hertig, A., Zhang, Z., Ramseier, T. M., Bohmert-Tatarev, K., Peoples, O. P., Snell, K. D. Generation of high polyhydroxybutryate producing oilseeds. International Patent Application WO/2011/034946, 2011b.

Petrasovits, L. A., McQualter, R. B., Gebbie, L. K., Blackman, D. M., Nielsen, L. K., Brumbley, S. M. Chemical inhibition of acetyl coenzyme A carboxylase as a strategy to increase polyhydroxybutyrate yields in transgenic sugarcane. *Plant Biotechnol J* 2013, 11, 1146–1151.

Petrasovits, L. A., Purnell, M. P., Nielsen, L. K., Brumbley, S. M. Production of polyhydroxybutyrate in sugarcane. *Plant Biotechnol J* 2007, 5, 162–172.

Phithakrotchanakoon, C., Champreda, V., Aiba, S., Pootanakit, K., Tanapongpipat, S. Engineered *Escherichia coli* for short-chain-length medium-chain-length polyhydroxyalkanoate copolymer biosynthesis from glycerol and dodecanoate. *Biosci Biotechnol Biochem* 2013, 77, 1262–1268.

Poirier, Y. Production of new polymeric compounds in plants. *Curr Opin Biotechnol* 1999, 10, 181–185.

Poirier, Y. Polyhydroxyalkanoate synthesis in plants as a tool for biotechnology and basic studies of lipid metabolism. *Progr Lip Res* 2002, 41, 131–135.

Poirier, Y., Dennis, D. E., Klomparens, K., Somerville, C. Polyhydroxybutyrate, a biodegradable thermoplastic, produce in transgenic plants. *Science* 1992, 256, 520–523.

Poirier, Y., Gruys, K. J. Production of polyhydroxyalkanoates in transgenic plants. In *Biopolyester*, Doi, Y., Steinbuchel, A. (Eds.), Wiley VCH: Weinheim, 2001, pp. 401–435.

Poirier, Y., Nawrath, C., Somerville, C. Production of polyhydroxyalkanoates, a family of biodegradable plastics and elastomers, in bacteria and plants. *Biotechnology* 1995, 13, 142–150.

Povolo, S., Romanelli, M. G., Basaglia, M., Ilieva, VI., Corti, A., Morelli, A., Chiellini, E., Casella, S. Polyhydroxyalkanoate biosynthesis by *Hydrogenophaga pseudoflava*

DSM1034 from structurally unrelated carbon sources. *New Biotechnol* 2013, 30, 629–634.

Punrattanasin, W. *The Utilization of Activated Sludge Polyhydroxyalkanoates for the Production of Biodegradable Plastics*. Virginia Polytechnic Institute and State University: Blacksburg, Virginia, USA, PhD Dissertation, 2001.

Purnell, M. P., Petrasovits, L. A., Nielsen, L. K., Brumbley, S. M. Spatio-temporal characterization of polyhydroxybutyrate accumulation in sugarcane. *Plant Biotechnol J* 2007, 5, 173–184.

Qin, L. F., Gao, X., Liu, Q., Wu, Q., Chen, G. Q. Biosynthesis of polyhydroxyalkanoate copolyesters by *Aeromonas hydrophila* mutant expressing a low-substrate-specificity PHA synthase PhaC2Ps. *Biochem Eng J* 2007, 137, 144–150.

Reddy, C. S. K., Ghai, R., Rashmi, Kalia, V. C. Polyhydroxyalkanoates: an overview. *Biores Technol* 2003, 87, 137–146.

Reis, M. A. M., Serafim, L. S., Lemos, P. C., Ramos, A. M., Aguiar, F. R., Van Loosdrecht, M. C. M. Production of polyhydroxyalkanoates by mixed microbial cultures. *Biopro Biosyst Engg* 2003, 25, 377–385.

Rezzonico, E., Moire, L., Poirier, Y. Polymers of 3-hydroxyacids in plants. *Phytochem Rev* 2002, 1, 87–92.

Rhu, D. H., Lee, W. H., Kim, J. Y., Choi, E. Polyhydroxybutyrate (PHA) production from waste. *Water Sci Technol* 2003, 48, 221–228.

Rivard, C., Moens, L., Roberts, K., Brigham, J., Kelley, S. Starch esters as biodegradable plastics: effects of ester group chain length and degree of substitution on anaerobic biodegradation. *Enzyme Microb Technol* 1995, 17, 848–852.

Rodgers, M., Wu, G. Production of polyhydroxybutyrate by activated sludge performing enhanced biological phosphorus removal. *Biores Technol* 2010, 101, 1049–1053.

Ruan, W., Chen, J., Lun, S. Production of biodegradable polymer by *A. eutrophus* using volatile fatty acids from acidified wastewater. *Process Biochem.* 2003, 39, 295–299.

Salehizadeh, H., Van Loosdrecht, M. C. M. Production of polyhydroxyalkanoates by mixed culture: recent trends and biotechnological importance. *Biotechnol Adv* 2004, 22, 261–279.

Samantaray, S., Mallick, N. Production and characterization of poly-β-hydroxybutyrate (PHB) polymer from *Aulosira fertilissima*. *J Appl Phycol* 2012, 24, 803–814.

Samantaray, S., Mallick, N. Production of poly(3-hydroxybutyrate-*co*-3-hydroxyvalerate) co-polymer by the diazotrophic cyanobacterium *Aulosira fertilissima* CCC 444. *J Appl Phycol* 2014, 26, 237–245.

Sangkharak, K., Prasertsan, P. Screening and identification of polyhydroxyalkanoates producing bacteria and biochemical characterization of their possible application. *J Gen Appl Microbiol* 2012, 58, 173–182.

Satoh, H., Iwamoto, Y., Mino, T., Matsuo, T. Activated sludge as a possible source of biodegradable plastic. *Water Sci Technol* 1998, 38, 103–109.

Schnell, J., Treyvaud-Amiguet, V., Arnason, J., Johnson, D. Expression of polyhydroxybutyric acid as a model for metabolic engineering of soybean seed coats. *Transgenic Res* 2012, 21, 895–899.

Senior, P. J., Dawes, E. A. The regulation of poly-β-hydroxybutyrate metabolism in *Azotobacter beijerinckii*. *Biochem J* 1973, 134, 225–238.

Shamala, T. R., Vijayendra, S. V., Joshi, G. J. Agro-industrial residues and starch for growth and co-production of polyhydroxyalkanoate copolymer and α-amylase by *Bacillus* sp. CFR-67. *Braz J Microbiol* 2012, 43, 1094–1102.

Shang, L., Jiang, M., Chang, H. N. Poly(3-hydroxybutyrate) synthesis in fed-batch culture of *Ralstonia eutropha* with phosphate limitation under different glucose concentrations. *Biotechnol Lett* 2003, 25:1415–1419.

Sharma, L., Mallick, N. Accumulation of poly-β-hydroxybutyrate in *Nostoc muscorum*: regulation by pH, light-dark cycles, N and P status and carbon sources. *Biores Technol* 2005, 96, 1304–1310.

Sheu, D. S., Lee, C. Y. Altering the substrate specificity of polyhydroxyalkanoate synthase 1 derived from *Pseudomonas putida* GPo1 by localized semirandom mutagenesis. *J Bacteriol* 2004, 186, 4177–4184.

Shimao, M. Biodegradation of plastics. *Curr Opin Biotechnol* 2001, 12, 242–247.

Silva, J. A., Tobella, L. M., Becerra, J., Godoy, F., Martínez, M. A. Biosynthesis of poly-β-hydroxyalkanoate by *Brevundimonas vesicularis* LMGP-23615 and *Sphingopyxis macrogoltabida* LMG 17324 using acid-hydrolyzed sawdust as carbon source. *J Biosci Bioengg* 2007, 103, 542–546.

Singh, A. K., Bhati, R., Samantaray, S., Mallick, N. *Pseudomonas aeruginosa* MTCC 7925: Producer of a novel SCL-LCL-PHA co-polymer. *Curr Biotechnol* 2013, 2, 81–88.

Singh, A. K., Mallick, N. Enhanced production of SCL-LCL-PHA co-polymer by sludge-isolated *Pseudomonas aeruginosa* MTCC 7925. *Lett Appl Microbiol* 2008, 46, 350–357.

Singh, A. K., Mallick, N. Exploitation of inexpensive substrates for production of a novel SCL-LCL-PHA co-polymer by *Pseudomonas aeruginosa* MTCC 7925. *J Ind Microbiol Biotechnol* 2009a, 36, 347–354.

Singh, A. K., Mallick, N. SCL-LCL-PHA co-polymer production by a local isolate, *Pseudomonas aeruginosa* MTCC 7925. *Biotechnol J* 2009b, 4, 703–711.

Slater, S., Mitsky, T. A., Houmiel, K. L., Hao, M., Rieser, S. E., Taylor, N. B., Tran, M., Valentin, H. E., Rodriguez, D. J., Stone, D. A., Padgette, S. R., Kishore, G., Gruys, K. J. Metabolic engineering of *Arabidopsis* and *Brassica* for poly (3-hydroxybutyrate-co-3-hydroxyvalerate) copolymer production. *Nature Biotechnol* 1999, 17, 1011–1016.

Somleva, M., Ali, A. Propagation of transgenic plants. International Patent Application WO/2010/102220, 2010.

Somleva, M., Chinnapen, H., Ali, A., Snell, K. D., Peoples, O. P., Patterson, N., Tang, J., Bohmert-Tatarev, K. Increasing carbon flow for polyhydroxybutyrate production in biomass crops. US Patent Application 2012/0060413, 2012.

Son, H., Park, G., Lee, S. Growth associated production of poly-β-hydroxybutyrate from glucose or alcoholic distillery wastewater by *Actinobacillus* sp. EL-9. *Biotechnol Lett* 1996, 18, 1229–1234.

Song, J. J., Yoon, S. C. Biosynthesis of novel aromatic copolyesters from insoluble 11-phenoxyundecanoic acid by *Pseudomonas putida* BMO1. *Appl Environ Microbiol* 1996, 62, 536–544.

Stein, R. S. Polymer recycling: opportunities and limitations. *Proc Natl Acad Sci* 1992, 89, 835–838.

Steinbuchel, A. Polyhydroxyalkanoic acids. In *Biomaterials: Novel Materials from Biological Sources*, Byrom, D. (Ed.), Stockton: New York, 1991, pp. 124–213.

Steinbuchel, A. Biodegradable plastics. *Curr Opin Biotechnol* 1992, 3, 291–297.

Steinbuchel, A. PHB and other polyhydroxyalkanoic acids. In *Biodegradable Plastics and Polymers*, Doi, Y., Fukuda, K. (Eds.), Elsevier Science: New York, 1996, pp. 362–364.

Steinbuchel, A., Debzi, E. M., Marchessault, R. H., Timm, A. Synthesis and production of poly(3-hydroxyvaleric acid) homopolymer by *Chromobacterium violaceum*. *Appl Microbiol Biotechnol* 1993, 39, 443–449.

Steinbuchel, A., Hustede, E., Liebergesell, M., Pieper, U., Timm, A., Valentin, H. Molecular basis for biosynthesis and accumulation of polyhydroxyalkanoic acids in bacteria. *FEMS Microbiol Rev* 1992, 103, 217–230.

Steinbuchel, A., Wiese, S. A. *Pseudomonas* strain accumulating polyesters of 3-hydroxybutyric acid and medium-chain-length 3-hydroxyalkanoic acids. *Appl Microbiol Biotechnol* 1992, 37, 691–697.

Sudesh, K., Abe, H., Doi, Y. Synthesis, structure and properties of polyhydroxyalkanoates: biological polyesters. *Prog Polym Sci* 2000, 25, 1503–1555.

Sudesh, K., Taguchi, K., Doi. Y. Can cyanobacteria be a potential PHA producer? *RIKEN Rev* 2001, 42, 75–76.

Suriyamongkol, P., Weselake, R., Narine, S., Moloney, M., Shah, S. Biotechnological approaches for the production of polyhydroxyalkanoates in microorganisms and plants - a review. *Biotechnol Adv* 2007, 25, 148–175.

Tajima, K., Igari, T., Nishimura, D., Nakamura, M., Satoh, Y., Munekata, M. Isolation and characterization of *Bacillus* sp. INT005 accumulating polyhydroxyalkanoate (PHA) from gas field soil. *J Biosci Bioengg* 2003, 95, 77–81.

Tian, S. J., Lai, W. J., Zheng, Z., Wang, H. X., Chen, G. Q. Effect of over-expression of phasin gene from *Aeromonas hydrophila* on biosynthesis of copolyesters of 3-hydroxybutyrate and 3-hydroxyhexanoate. *FEMS Microbiol Lett* 2005, 244, 19–25.

Tilbrook, K., Gebbie, L., Schenk, P. M., Poirier, Y., Brumbley, S. M. Peroxisomal polyhydroxyalkanoate biosynthesis is a promising strategy for bioplastic production in high biomass crops. *Plant Biotechnol J* 2011, 9, 958–969.

Timm, A., Steinbuchel, A. Formation of polyesters consisting of medium-chain-length 3-hydroxyalkanoic acids from gluconate by *Pseudomonas aeruginosa* and other fluorescent *Pseudomonads*. *Appl Environ Microbiol* 1990, 56, 3360–3367.

Toh, P. S. Y., Jau, M. H., Yew, S. P., Abed, R. M. M., Sudesh, K. Comparison of polyhydroxyalkanoates biosynthesis, mobilization and the effects on cellular morphology in *Spirulina platensis* and *Synechocystis* sp. UNIWG. *J Biosci* 2008, 19:21–38.

Tsuge, T. Metabolic improvements and use of inexpensive carbon sources in microbial production of polyhydroxyalkanoates. *J Biosci Bioengg* 2002, 94, 579–584.

Valappil, S. P., Boccaccini, A. R., Bucke, C., Roy, I. Polyhydroxyalkanoates in Gram-positive bacteria: insights from the genera *Bacillus* and *Streptomyces*. *Antonie van Leeuwenhoek* 2007, 91, 1–17.

Valentin, H. E., Steinbuchel, A. Accumulation of poly(3-hydroxybutyric acid-co-3-hydroxyvaleric acid-co-4-hydroxyvaleric acid) by mutants and recombinant strains of *Alcaligenes eutrophus*. *J Environ Polym Degr* 1995, 3, 169–175.

Vargas, A., Montano, L., Amaya, R. Enhanced polyhydroxyalkanoate production from organic wastes via process control. *Biores Technol* 2014, 156, 248–255.

Vincenzini, M., De Philippis, R. Polyhydroxyalkanoates. In *Chemicals from Microalgae*, Cohen, Z. (Ed.), Taylor and Francis Inc.: USA, 1999, pp. 292–312.

Williams, S. F., Peoples, O. P. Biodegradable plastics from plants. *Chemtech* 1996, 26, 38–44.

Williamson, D. H., Wilkinson, J. F. The isolation and estimation of the poly-β-hydroxybutyrate inclusions of *Bacillus* species. *J Gen Microbiol* 1958, 19, 198–209.

Witt, U., Muller, R. J., Deckwer, W. D. Biodegradation behavior and material properties of aliphatic/aromatic polyesters of commercial importance. *J Environ Polym Degrad* 1997, 15, 81–89.

Wrobel, W., Zebrowski, J., Szopa, J. Polyhydroxybutyrate synthesis in transgenic flax. *J Biotechnol* 2004, 107:41–54.

Yang, J. E., Choi, Y. J., Lee, S. J., Kang, K. H., Lee, H., Oh, Y. H., Lee, S. H., Park, S. J., Lee, S. Y. Metabolic engineering of *Escherichia coli* for biosynthesis of poly(3-hydroxy-butyrate-*co*-3-hydroxyvalerate) from glucose. *Appl Microbiol Biotechnol* 2014, 98, 95–104.

Yezza, A., Fournier, D., Halasz, A., Hawari, J. Production of polyhydroxyalkanoates from methanol by a new methylotrophic bacterium *Methylobacterium* sp. GW2. *Appl Microbiol Biotechnol* 2006, 73, 211–218.

Yu, J. Production of PHA from starch wastewater via organic acids. *J Biotechnol* 2001, 86, 105–112.

Yu, P. H. F., Chua, H., Huang, A. L., Lo, W. H., Ho, K. P. Transformation of industrial food wastes into PHA. *Water Sci Technol* 1999, 40, 367–370.

Zhu, C., Nomura, C. T., Perrotta, J. A., Stipanovic, A. J., Nakas, J. P. Production and characterization of poly-3-hydroxybutyrate from biodiesel-glycerol by *Burkholderia cepacia* ATCC 17759. *Biotechnol Prog* 2010, 26, 424–430.

CHAPTER 12

HYDROTHERMAL AND THERMOCHEMICAL SYNTHESIS OF BIO-OIL FROM LIGNOCELLULOSIC BIOMASS: COMPOSITION, ENGINEERING AND CATALYTIC UPGRADING

SONIL NANDA,[1] PRAVAKAR MOHANTY,[2] JANUSZ A. KOZINSKI,[3] and AJAY K. DALAI[1]

[1]*Department of Chemical and Biological Engineering, University of Saskatchewan, Saskatchewan, Canada*

[2]*Science and Engineering Research Board, Department of Science and Technology, Government of India, New Delhi, India*

[3]*Lassonde School of Engineering, York University, Ontario, Canada*

CONTENTS

12.1   INTRODUCTION

There is a wide scale increase in the demand for energy and its resources today due to rapid increase in industrialization, urbanization and population outgrowth. The discovery of crude oil in the 19th century supplied an inexpensive liquid fossil fuel source that helped industrialize the world and improved standards of living. Since then, various fossil fuels and their derivatives have remained the most favorite and widely exploited resource of energy for the world. The global use of petroleum and other liquid fossil fuels was 85.7 million barrels per day in 2008 with a projection for increase to 97.6 million barrels per day in 2020 and 112.2 million barrels per day in 2035 (USEIA, 2011). With these concerns of amplifying demand for petroleum resources by emerging economies, their declining reserves, rising fuel prices combined with environmental apprehensions of global warming, it has become imperative to develop ecofriendly and energy-efficient processes for the sustainable production of fuels and chemicals. The total world energy consumption in 2008 was 505 quadrillion British thermal units (BTU); however the consumption is expected to rise to 619 quadrillion BTU in 2020 and 770 quadrillion BTU in 2035 (USEIA, 2011).

In this scenario, waste plant biomass seems to be the only current renewable source of organic carbon, and fuels derived from their processing could be the only alternative liquid fuels to replace the conventional fossil fuels.

Moreover, biofuels are considered to be carbon neutral as the net amount of CO_2 released from their combustion is consumed by the plants during photosynthesis to produce new biomass. These waste biomass are lignocellulosic in composition, for example, they contain two major structural carbohydrates, namely cellulose and hemicellulose bound together by the polymeric lignin. These three biopolymers are highly energy rich components as they store the energy from the sun in their molecules (Huber et al., 2006). Lignocellulosic biomasses are potential feedstocks for the production of biofuels as they are inexpensive, available in abundant supplies worldwide and retain no competence towards the food supply. For instance, the annual lignocellulosic biomass availability in USA and Canada ranges up to 256 and 1000 million dry tons, respectively (Gronowska et al., 2009). The worldwide biomass energy potential in 2050 has been estimated to be in the range of 150–450 exajoule/year with an energy equivalence of 25×10^9 to 76×10^9 barrels of oil energy equivalent (Huber et al., 2006).

Lignocellulosic biomasses are obtained as non-edible residues from agriculture, forestry, municipal, industrial and urban refuse. These organic residues are suitable raw material for biofuel production through efficient conversion technologies employing physico-chemical, thermochemical or biochemical routes. The worldwide production of biomass from terrestrial plants is $170–200 \times 10^9$ tons, with an estimated 70% made of plant cell walls (Pauly and Keegstra, 2008). From an average estimate of 146 billion metric tons of world biomass production per annum, about 20 metric tons per acre of biomass are produced from farm crops and trees (Demirbas, 2007). Exploring the potential of lignocellulosic biomass for biofuel production is essential to minimize the fossil fuel consumption and supplement the increasing energy demand.

The energy contained in the biomass can be recovered through a variety of physical, thermochemical, hydrothermal and biochemical routes (Figures 12.1 and 12.2). Thermochemical conversion is the most widely used biomass conversion pathway which involves pyrolysis, gasification and liquefaction (Demirbas, 2009). Thermochemical decomposition can be utilized for most biomass, however biomass with low moisture content are typically preferred as moisture may affect the overall energy balance and compromise with the product quality (Nanda et al., 2013). Through gasification, the biomass is converted to hydrocarbons and synthesis gas (or syngas), while pyrolysis and liquefaction directly convert biomass at high temperatures to bio-oils, gases and char. Furthermore, there is a flexibility

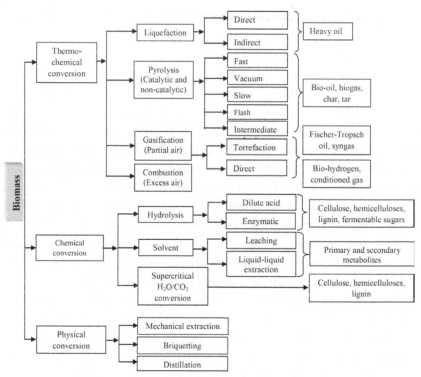

FIGURE 12.1 Different physical, chemical and thermochemical conversion routes of biomass to biofuels (adapted from Mohanty et al., 2014).

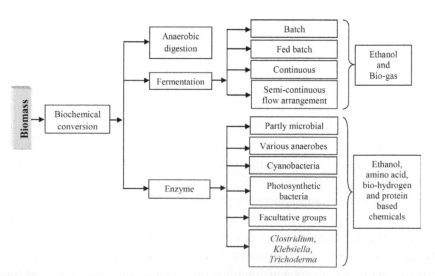

FIGURE 12.2 Different biochemical conversion routes of biomass to biofuels (adapted from Mohanty et al., 2014).

to choose between different process parameters to engineer the end product quality and properties. The biochemical processes mainly involve microbial or enzymatic digestion and fermentation (Figure 12.1). In contrast to thermochemical processes, biochemical processes mostly prefer high moisture-containing biomasses where water acts as a nutrient for microorganisms for their metabolic reactions.

Pyrolysis is a thermochemical process that converts lignocellulosic materials at high temperatures in absence of air to produce fuels, chemicals and biomaterials. The process results in energy-rich bio-oils, carbon-rich biochar, condensable organic liquids, non-condensable gases, acids (e.g., acetic acid, carboxylic acid), solvents (e.g., acetone), alcohols (e.g., phenol, methanol) (Joshi and Lawal, 2012). Thermochemical conversion is attractive because it converts solid biomass into energy-densified liquid products with an ease of transport, storage, combustion and marketing (Demirbas, 2007). Bio-oil is a synthetic fuel obtained from the pyrolysis of biomass that contains 70% of the energy of the biomass feed (French and Czernik, 2010). They are complex mixtures of alcohols, acids, aldehydes, esters, ketones and many other aromatic compounds. Bio-oils have found wide array of applications as a sustainable fuel in industries' boilers for power generation and in synthesis of chemicals. Upon upgrading, these bio-oils can be used as transportation fuels.

In this chapter, the main technical aspects of different thermochemical processes for biomass conversion are discussed. In addition, various hydrothermal treatments applied to lignocellulosic biomass for the production of fermentable sugars, hydrogen or other value-added products are presented. This chapter also describes about pyrolysis along with its various parameters that influence conversion product yields. Bio-oil, which is a complex mixture or both aqueous and organic biomass components, is discussed in details along with its catalytic upgrading for use as transportation fuel.

12.2 LIGNOCELLULOSIC BIOMASS

Lignocellulosic biomasses are inexpensive resources that are abundantly available throughout the world and have the tendency to continually supply biofuels through efficient conversion processes. Lignocellulosic materials are usually categorized into agricultural and forage residues (e.g., sugarcane bagasse, corn stover, wheat straw, rice straw, rice husk), dedicated energy crops (e.g., switch grass, timothy grass, hybrid poplar), forest wood residues (e.g., saw mill residues, paper mill discards) and municipal paper waste. Typically, lignocellulosic biomass contains 35–55% cellulose, 20–40% hemicellulose and 10–25% lignin (Sukumaran et al., 2010). About 90% of dry matter in lignocellulosic biomass comprises of cellulose, hemicelluloses and lignin, whereas the rest consists of extractives and ash (Balat, 2011). With ash representing the mineral components in biomass, extractives include structural biopolymers such as terpenoids, steroids, resin acids, fats, lipids, waxes and phenolic components.

Cellulose is a glucose polymer consisting of β (1, 4) linked D-glucose subunits, whereas hemicellulose is a mixture of polysaccharides composed of C_5 (e.g., xylose and arabinose) and C_6 (e.g., galactose, glucose and mannose) sugars as well as sugar acids such as methylglucuronic and galaturonic acids. In the polymeric form, cellulose consists of linear chains of 1, 4-D-glucopyranose units with an average molecular weight of around 100,000 (Goyal et al., 2008). In the monomeric form, hemicellulose consists of D-glucose, D-mannose, D-galactose, D-xylose, L-arabinose, L-rhamnose, D-glucuronic acid and D-galacturonic acid, with an average molecular weight less than 30,000. On the other hand, lignin is a phenyl propane polymer linked with ester bonds that holds cellulose and hemicellulose together as a matrix. The lignin in biomass is highly branched, substituted, mononuclear aromatic polymer and consists of macromolecules containing phenolic groups. The average molecular weight of lignin is around 20,000.

The lignin in biomass makes it recalcitrant to degradation by enzymes, acids, insects and microorganisms. Moreover, the presence of lignin makes it difficult to obtain cellulose and hemicellulose to produce fermentable sugars during bioconversion. In addition, during the biomass pretreatment, lignin can form furan compounds such as furfural and hydroxymethyl furfural that could inhibit fermentation (Zaldivar et al., 1999). In contrast, thermochemical conversion on biomass yields various degradation products from lignin that have high industrial values in the production of aromatic compounds, flavoring agents, food preserving agents and other chemicals (Ayyachamy et al., 2013). The major structural components of lignin are the monolignols (e.g., hydroxyl cinnamyl alcohols), such as coniferyl alcohols (e.g., guaiacyl propanol), coumaryl alcohols (e.g., p-hydroxyphenyl propanol) and sinapyl alcohols (e.g., syringyl propanol).

During thermochemical conversions, the polymeric components of biomass such as cellulose, hemicellulose and lignin undergo a series of complex thermal degradation reactions. In most cases, cellulose and hemicellulose undergo cyclore version and dehydration followed by transglycosylation reaction. This series of reactions results in the yield of both low and high molecular weight compounds in the bio-oil fractions. This makes bio-oil a heterogeneous mixture of thermochemical derivatives arising from cellulose, hemicellulose and lignin. The typical degradation products of cellulose and hemicellulose in bio-oils include acids, esters, alcohols, ketones, aldehydes, sugars, furans and oxygenates, while lignin derivatives include phenols, guaiacols and syringols (Joshi and Lawal, 2012).

The primary products of cellulose and hemicellulose decomposition are condensable vapors (yields to liquid products) and gases. Typically, lignin decomposes to liquid, gas and solid char products. Extractives contribute to liquid and gas products either through simple volatilization or decomposition. Minerals in general remain in the char comprising their ash content. This distribution of components into products is shown schematically in Figure 12.3. Vapors formed by primary decomposition of biomass components can be involved in secondary reactions in the gas phase forming soot, and/or at hot char surfaces where a secondary char is formed (Antal and Gronli, 2003). This is particularly important in understanding the differences between slow, intermediate and fast pyrolysis and the factors affecting bio-oil, gas and biochar yields. In most cases, the inorganic metals in biomass, particularly the alkali metals, can have a catalytic effect on pyrolysis reactions leading to increased biochar yields in some circumstances in addition to the effect of ash contributing directly to char yield. Minerals also affect the reactivity and ignition properties of biochars (Antal and Gronli, 2003).

12.3 HYDROLYSIS OF LIGNOCELLULOSIC BIOMASS

Biomass is resistant to chemicals at ambient temperature. Hydrolysis reactions using acid or base are very complex mainly because the substrate is in solid phase and the catalyst in liquid phase. The reaction rate of hydrolysis depends on a number of variables such as temperature, acid concentration, residence time, substrate concentration and substrate composition.

In acidic reaction media, the monosaccharides are rapidly formed from the depolymerization of hemicelluloses at lower temperature (less than 120°C). These monosaccharides are subsequently converted to furfural, hydroxymethyl furfural and acid derivatives (Chheda et al., 2007b). The mild hydrolysis conditions do not significantly affect cellulose. The acid hydrolysis of the glycosidic bonds only takes place when the temperature is increased beyond 200–220°C and depolymerization is achieved rapidly at low acid concentrations. At 160–180°C, in weak acidic media such as sulphite salts, ether groups of lignin are cleaved forming benzylium ions. Sulfuric and hydrochloric acids are the most commonly used catalysts for hydrolysis of lignocellulosic residues (Kumar et al., 2009). In contrast to these acids, phosphoric acid can be more advantageous for hydrolysis.

In basic media, hydrolysis reactions of carbohydrates are slow, but with increase in temperature beyond 140°C the cleavage of glycosidic bonds becomes significant. Hydroxyl groups promote delignification by cleavage of ether linkages. This contributes to the dissolution of lignin and lignin-carbohydrate complexes. Thermal decomposition of lignocellulosic biomass begins at 200°C and is particular to carbohydrate degradation at 250°C and lignin degradation at 280°C. Homolytic cleavage of C–O and C–C bonds gives low molecular weight products and gases by further decomposition. However, depolymerization and char formation compete with each other through dehydration and decarboxylation. These secondary reactions tend to diminish the oxygen content of the bio-oil produced, tending to create unsaturated and polycyclic products (Chheda and Dumesic, 2007). In contrast to dehydration, decarboxylation reactions maintain the hydrogen/carbon ratio of the produced bio-oil. The overall mechanism is unselective and results in a range of complex oils.

Hydrolysis of α-glucose with H_2SO_4 was studied and a three step mechanism was proposed by Xiang et al. (2003). The reaction starts with a proton from acid interacting rapidly with the glycosidic oxygen, linking two sugar units and forming a conjugate acid. Further, the cleavage of the C–O bond and breakdown of the conjugate acid to the cyclic carbonium ion takes place, which adopts a half-chair conformation. After a rapid addition of water, free sugars and a proton are liberated. The effects of temperature and acid concentration have also been studied by few authors. The X-ray diffraction studies for the fibers indicate the highly crystalline structure of untreated cellulose is totally disrupted and a completely different diffraction pattern with near zero crystallinity appeared after dissolution with sulfuric acid. The scanning electron microscopy (SEM) studies also indicate the breakdown of fibers into smaller fragments with acid treatment. The original fibrous form of cellulose disappeared and changed into a gel-like substance.

In a study by Rahman et al. (2007), the selectivity and xylose yield were developed to study individual and combined effect of parameters such as temperature, retention time and acid concentration at five different levels. The feedstock under study was oil palm empty fruit bunch fiber which is a lignocellulosic waste from palm oil mills and a potential source of xylose that can be used as a raw material for production of xylitol. It was found that under optimum conditions (i.e., temperature: 119°C, reaction time: 60 min and H_2SO_4 concentration: 2%), the xylose yield and selectivity were 91.3% and 18 g/g, respectively. The combined effect of temperature with reaction time and temperature with acid concentration was found to be more prominent than combined effect of reaction time and acid concentration.

Table 12.1 summarizes the acid hydrolysis studies of various lignocellulosic biomasses such as corn cob, olive wood, potato peel, rice husk, sunflower seed hull and tobacco stalk. In most cases, H_2SO_4 is used in hydrolysis, although HCl, CH_3COOH and H_3PO_4 have also been investigated in biomass digestion. The hydrolysis temperature ranges from 50 to 220°C, depending on the residence time and acid concentration. Usually, higher acid concentrations require lower temperatures and short residence time, and *vice versa*.

TABLE 12.1   Acid Hydrolysis of Lignocellulosic Biomass

Biomass	Hydrolysis conditions	Remarks	Reference
Corn cob and sunflower seed hull	H_2SO_4: 0.2–1 N Temperature: 98–130°C Reaction time: 0–3.3 h	A first order hemicelluloses degradation mechanism was proposed including process parameters, for example, temperature and acid concentration in Arrehenius-type equation. The slow hydrolysis step was more acid concentration-dependent than the fast hydrolysis step.	Eken-Saracoglu et al. (1998)
Olive wood	H_2SO_4: 0–32% w/w Temperature: 60–90°C Reaction time: 0–4 h	The temperature and the acid concentration influenced hydrolysis. Arrhenius-type equation enabled the evaluation of activation energy values of 26.4 and 25.9 kJ/mol for fiber hydrolysis and sugar generation, respectively.	Romero et al. (2010)

TABLE 12.1   Continued

Biomass	Hydrolysis conditions	Remarks	Reference
Potato peel	H_3PO_4: 2.5–10% w/w Temperature: 135–200°C	An overall sugar yield of 82.5% was achieved under optimum conditions. This optimum yield was obtained at 135°C and 10% w/w acid concentration.	Lenihan et al. (2010)
Rice husk	H_2SO_4: 0.18 N Temperature: 140–220°C	With increase in temperature from 160–200°C, the total sugar concentration increased from 0.16–0.22 mol/L.	Megawati et al. (2010)
Sunflower stalk	H_2SO_4, HCl, CH_3COOH and maleic acid: 0.5–20% w/w Temperature: 50–150°C	Maximum (19 wt. %) sugar yield at 120°C in 60 min with solid to acid ratio of 1:30.	Du et al. (2012)
Sunflower stalk	H_2SO_4: 0.7–7.3% Temperature: 87–153°C	Optimum reaction conditions for maximum sugar recovery were 120°C, 30 min and 4% H_2SO_4 concentration.	Akpinar et al. (2011)
Tobacco stalk	Residence time: 5–55 min	Optimum reaction conditions for maximum sugar recovery were 133°C, 27 min and 4.9% H_2SO_4 concentration.	

12.4   HYDROTHERMAL CONVERSION OF BIOMASS

When biomass hydrolysis is performed in presence of hot compressed water, the complex molecules of cellulose, hemicelluloses and lignin depolymerize to produce liquid fuel in the range of gasoline hydrocarbons. The process proceeds through a series of structural and chemical transformations that involve: (i) solvolysis of biomass resulting in micellar-like structure, (ii) depolymerization of cellulose, hemicelluloses and lignin, and (iii) chemical and thermal decomposition of sugar monomers to smaller molecules.

When alkalis are added to the process, a better liquid yield is achieved. The reaction proceeds by the formate ions derived from the alkali carbonates which react with the hydroxyl group of biomass to cause decarboxylation. The alkali carbonates act as catalyst and convert biomass macro-molecules into smaller fragments. The micellar-like fragments thus produced are degraded to smaller compounds by dehydration, dehydrogenation, deoxygenation and decarboxylation. These compounds further rearrange through condensation, cyclization and polymerization, leading to newer compounds (Hong-Yu et al., 2008).

Hot compressed water can be generally described as water at temperature above 150°C and various pressures. Depending upon the temperature and pressure, hot compressed water can exhibit stimulating physical and chemical properties. Water plays a very active role as its dielectric constant sharply decreases and helps the reaction to proceed. Low relative dielectric constant in this state enhances the ionic reaction suitable for a variety of syntheses or some degradation reactions. Depending on the temperature and pressure, hot compressed water supports either free radical or ionic reactions. At high pressures and below the critical temperature, ionic reactions dominate and thus ionic and polar species of biomass are extracted. At high temperatures and low pressures, free-radical reactions are superior and non-polar substances are readily dissolved and extracted (Mohanty et al., 2014).

Table 12.2 gives the properties of sub- and supercritical water. Subcritical water is the water that is in a state under a pressurized condition at temperatures above its boiling point under ambient pressure and below the critical point ($Tc = 374°C$; $Pc = 22.1$ MPa, $\rho c = 320$ kg/cm^3). The dielectric constant of liquid water decreases with increasing temperature (Nanda et al., 2014b). At temperatures from 277 to 377°C, the dielectric constant becomes as low as those of polar organic solvents. The ionic product of water is maximized at temperatures between 227 and 372°C depending upon the pressure (Kruse and Dinjus, 2007). Thus, subcritical water acts as acid and/or base catalysts for reactions, such as hydrolysis of ether/ester bonds, and also as a solvent for the extraction of low molecular mass products (Brunner, 2009).

TABLE 12.2 Properties of Water at Normal, Subcritical and Supercritical Conditions

Property	Normal water	Subcritical water	Supercritical water
Temperature (°C)	25	250	400
Pressure (MPa)	0.1	5	25
Density (g/cm^3)	1.0	0.8	0.2
Dielectric constant, ε	78.5	27.1	5.9
pK$_w$	14	11.2	19.4
Heat capacity, Cp (kJ/Kg/K)	4.2	4.9	13
Heat conductivity, λ (mW/m/K)	608	620	160

Hot compressed water hydrolysis with $ZnCl_2$, NaH_2PO_4 and H_3BO_3 as catalysts for glucose, cellulose, methyl and phenyl glucosides, glucitol, levoglucosenone and hydroxymethyl furfural degradation have been studied (Miller and Saunders, 1987). The reactions with $ZnCl_2$ was carried out with hydroxymethyl furfural, which is an intermediate product formed from hydrolysis of glucose. The results showed that hydroxymethyl furfural does not form appreciable amounts of light hydrocarbons, indans or tetralins, whereas reactions with NaH_2PO_4 produced three main products including 5-methyl-2-furaldehyde. The unknowns of molecular weight 210 and 258 were formed from hydroxymethyl furfural and saccharides in $ZnCl_2$ catalyzed reactions. The results with H_3BO_3 were similar to those of $ZnCl_2$ except for lesser amount of heavy hydrocarbons production.

Calvo and Vallejo (2002) proposed the conversion mechanism of various organic acids such as formic, glycolic and lactic acids for hydrolysis in sub- and supercritical water. The decomposition of lactic acid in high temperature water followed three major pathways such as hydrolysis/thermal degradation, dehydration and oxidation. In the presence of an oxidant, the route leading to formation of acetic acid predominates. This route begins with the hydrogen abstraction step followed by hydroxylation to form acetic acid and formic acid, which requires relatively high activation energy. Formic acid decomposes readily under subcritical conditions. The decomposition of formic acid follows both dehydration route (producing CO and H_2O) and decarboxylation route (producing CO_2 and H_2). The thermal degradation, hydrolysis and/or dehydration of lactic acid in high temperature water may occur simultaneously with or without oxygen. In absence of oxygen, acetaldehyde was found to be a major intermediate. This suggested that hydrolysis and thermal decomposition of lactic acid may involve an acetaldehyde precursor, such as pyruvic acid. However, acetaldehyde was relatively stable in high-temperature water without the presence of an oxidant.

Switchgrass conversion in high compressed water was studied by Cheng et al. (2009). The GC/MS analyzes mainly showed methane-soluble fractions in the following categories: sugar compounds such as furfural and furans, lignin compounds such as phenol compounds, and some fatty acid components. The water-soluble fractions, analyzed by HPLC, mostly contained oligosaccharides, cellobiose, glucose, xylose, furfural and hydroxymethyl furfural. The SEM of char showed swollen structures and small-pored surfaces. This was due to the high decomposition rate of lignocellulosic structure and high diffusivity of water molecules at higher temperatures that led to a more significant release of decomposition products from the internal zones.

Variation in liquid yields from poplar leaves, bark and wood in an airproof stainless steel reactor at different temperatures was studied by Wu et al. (2009). The bio-oil yields from leaves, bark and wood were high at 350, 400 and 450°C, respectively. Various properties such as carbon, oxygen, hydrogen/carbon ratio and high heating value of poplar leaves bio-oil was found to be comparable with diesel. The bio-oils from leaves and barks contained light and long-chain alkanes. Recent studies by Yang et al. (2013) demonstrate the characterization of pyrolysis oils derived from sewage sludge and de-inking sludge for use in diesel engines. Compared with de-inking sludge pyro-oil, sewage sludge pyro-oil showed higher heating value but suffered from higher corrosiveness and viscosity. The results showed that although both pyro-oils could provide sufficient heat when used in diesel engines, yet the issues of poor combustion and carbon deposition may be encountered. However, blending of these pyro-oils with diesel could overcome such issues.

A hydrothermal hydrolysis study on rice bran in the temperature range of 100–360°C for 5 min gave a high content of water soluble fraction that contained glucose, sucrose, fructose and glyceraldehydes (Pourali et al., 2009). The liquid fraction obtained was separated into hexane, acetone and water soluble components. As the temperature increased, hexane and acetone soluble content also increased, whereas the solid content sharply decreased. Highest total organic carbon value was achieved at 250°C and total sugar content was maximum at 150°C. The presence of acetic, formic, glycolic and levulinic acids at temperatures above 190°C was observed.

Hydrolysis of palm fruit bunch in different base medium, for example, NaOH, KOH and K_2CO_3 was studied (Akhtar et al., 2010). The conversion of palm fruit bunch into polysaccharides, oligosaccharides and monosaccharides was among the initial hydrolysis reactions. Subsequent decarboxylation, dehydration and fragmentation of saccharides produced volatiles and char. The gases were produced mainly due to decarboxylation of hemicellulose and cellulose. High stability of OH-functional groups under operating conditions suggested the presence of high amount of phenols and alcohols in liquid fraction and solid residue. A similar hydrolysis and decomposition mechanism was proposed for lignin degradation. The high re-polymerization tendency in lignin produced more char than liquids. The secondary decomposition was low as negligible amounts of gases were obtained. Moreover, major compounds produced were phenols, esters and carboxylic acid.

Glucose conversion as a model product was studied in presence of $ZnSO_4$ and hot compressed water. A scheme for the decomposition was proposed by Kruse and Vogel (2010). At low temperature of 250°C glucose isomerized to fructose and mannose and dehydrated to 1,6-anhydroglucose. In addition, glucose, fructose and mannose dehydrated to hydroxyl methyl furfural. Erythrose, glycolaldehyde and glyceraldehydes were also formed via retro-aldol reactions. Erythrose undergoes further fragmentation to glycolaldehyde via a retro-aldol reaction; further glyceraldehyde reacts via pyruvaldehyde to lactic acid and hydroxyacetone.

Liquefaction of microalgae has also been attempted for the production of bio-oils. Hydrothermal liquefaction has also shown to convert triglycerides from microalgae to fatty acids and alkanes with the aid of certain heterogeneous catalysts. Investigation by Biller et al. (2011) has shown the comparison between the composition of lipids and free fatty acids from solvent extraction to those from hydrothermal processing. The initial decomposition products included free fatty acids and glycerol. The results indicated that the bio-crude yields from the liquefaction of *Chlorella vulgaris* and *Nannochloropsis occulta* increased slightly with the use of heterogeneous catalysts but the higher heating value and de-oxygenation level increased by up to 10%.

The conversion of microalgae *Nannochloropsis* in hot water at 350°C in presence of six different heterogeneous catalysts such as Pd/C, Pt/C, Ru/C, Ni/SiO$_2$-Al$_2$O$_3$, CoMo/γ-Al$_2$O$_3$ and zeolite was studied by Brown et al. (2010). In the presence of high pressure H$_2$, the bio-crude yields were largely insensitive to the presence of catalyst. The bio-crude yields ranged from a low of 35% from uncatalyzed liquefaction without H$_2$ to a high of 57% from Pd/C catalyzed liquefaction without H$_2$. The bio-oil from Pd/C contained large quantities of fatty acids, for example, palmitic acid, palmitoleic acid, some phenolic compounds and a number of different long-chain alkanes. It also contained aliphatic methyl and methylene carbon atoms.

High pressure hydrothermal conversion of *Spirulina* was studied with iron as catalyst (Matsui et al., 1997). It showed that the bio-oil yield increased linearly from 54.4 to 63.7 wt. % with increasing amount of Fe(CO)$_5$-S from 0 to 1 mmol. The conversion and gas yield were nearly constant. In a similar study, brown macroalga *Laminaria saccharina* was hydrothermally liquefied to bio-crude in a batch reactor (Anastasakis and Ross, 2011). A maximum bio-crude yield of 19.3 wt. % was obtained with a biomass to water ratio of 1:10 at 350°C and 15 min of residence time. The solid residue contained large proportion of calcium and magnesium, whereas the liquid phase was rich in sugars, ammonium, potassium and sodium.

The hydrolysis of *Paulownia* in hot compressed water was studied with Fe and Na_2CO_3 (Sun et al., 2010). Both the catalysts were found effective for enhancing the formation of heavy oil products, although significantly promoting the formation of gases within the range of 280–360°C. The yield of water soluble fraction decreased with increase in the reaction temperature irrespective of the presence of catalyst. The yield of heavy oil increased from 21.9 to 36.3 wt. % with the increment of temperature from 280 to 340°C in presence of iron. As the temperature increased, the isomerization and rearrangement of carbohydrates, alcohols, aldehydes produced by biomass hydrolysis occurred due to the endothermic nature of the thermal reaction. With the Fe catalyst, the aldehydes were converted to acids followed by transformation into ketones. The highest yield of total oil (53 wt. %) was produced at the lowest temperature, while the maximum yield of heavy oil (36.3 wt. %) was obtained at 340°C.

Baby food decomposition (as model biomass for protein and carbohydrate containing biomass) in near critical water (i.e., 375°C and 24 MPa) was studied with K_2CO_3, ZSM-5 and Ni by Sinag et al. (2010). The SEM images of char produced without Ni showed macropores, whereas biochars produced with Ni on silica and ZSM-5 gave both macro and micro-sized pores. However, K_2CO_3 resulted in only microporous chars. In presence of Ni on SiO_2 catalyst, the amount of hydroxylmethyl furfural increased. The amounts of acetic acid and acetaldehyde were considerably higher under acidic conditions, whereas under basic condition (in the presence of K_2CO_3), formic acid and glycolic acid concentrations increased in the aqueous phase.

Hydrothermal conversion of several model components, microalgae and cyanobacteria were studied at various concentrations of HCOOH and Na_2CO_3 (Biller and Ross, 2011). The microalgae used in the study were *Chlorella vulgaris, Nannochloropsis occulata* and *Porphyridium cruentum*, and the cyanobacteria used was *Spirulina*. The study on model compound was used to predict the liquid yield from microalgae. The bio-crude from *Chlorella* with Na_2CO_3 contained phenols and piperdine-derived compounds and alkanes. Water and HCOOH processing resulted in aliphatic amides such as dimethyldecanamide, dodecamine and fatty acid octanoic acid. *Nannochloropsis*, which had high lipid content, resulted in the bio-crude containing large amounts of fatty acids and heterocycles such as indole. The HCOOH processing resulted in the formation of aliphatic amide hexadacamide and fatty acid tetradecanoic acid. The higher fraction of nitrogen heterocycles, pyrroles and indoles in the bio-crude was observed with hydrolysis of algal feedstock and Na_2CO_3. The use of Na_2CO_3 increased the formation of phenolic compounds and stimulated the breakdown of lipids to alkanes, while water and HCOOH resulted in the lipids breakdown to fatty acids. Table 12.3 lists a few waste biomasses and their hot compressed water conversion.

TABLE 12.3   Hydrothermal Conversion of Biomass in Hot Compressed Water

Biomass	Operating conditions	Remarks	Reference
Catechol (lignin model compound)	Temperature: 370–420°C Water density: 0.2–0.5 g/cm³ Pressure: 25–40 MPa Residence time: 50–250 min	With increase in the reaction time, amount of higher molecular weight compounds increased. The conversion of catechol and the formation of phenol increased with an increase in water density.	Wahyudiono et al. (2008)
Cattle manure	Temperature: 260–355°C Pressure: 0–0.7 MPa Residence time: 0–40 min Biomass/water: 0.5–2 Base: NaOH	The maximum bio-oil yields were 48.8, 44.7, 38.5 and 28 wt.% with CO, H_2, N_2 and air, respectively at 310°C.	Yin et al. (2010)
Cellulose (with phenol)	Temperature: 130°C Catalyst: 0.4 g H_2SO_4 (1.0 wt. % based on phenol) Cellulose/phenol ratio: 1/3	An acid-catalyzed hydrolysis pathway of cellulose was proposed. The product composition was dependent on phenol/cellobiose ratio, catalyst concentration, temperature and reaction time.	Lin et al. (2004)
Glucose, wood and pyro-oil	Temperature: 300–350°C Residence time: 0–60 min Catalyst: $CuCl_2$, $CuSO_4$, Na_2SO_4, NaCl, $NiSO_4$, MoO_3, TiO_2, ZrO_2, CuO, Ru/TiO_2, Ru/Al_2O_3 and Ru/C	Compared to non-catalytic reactions, the tested catalyst increased the gas yield, decreased the char yields and had insignificant effects on the bio-oil yields.	Knezevic et al. (2010)
Japanese beech wood	Temperature: 170–290°C Residence time: 30 min Pressure: 20 MPa	In both batch and semi-batch conversions, saccharides yield obtained from cellulose and hemicelluloses was maximum.	Lu and Saka, 2010

TABLE 12.3   Continued

Biomass	Operating conditions	Remarks	Reference
Kenaf stem	Temperature: 100–250°C Pressure: 8–28 MPa	The formation of carbonic acid in reaction media played a catalytic role and led to the effective solubilization of the lignocellulosic material.	Ozturk et al. (2010)
Maize straw, sawdust, rice husk and wheat bran	Metal ions: Zn(II), Ni(II), Co(II) and Cr(III) Concentration: 0–800 ppm Temperature: 300°C Residence time: 120 s	The yield of lactic acid decreased as the concentration of Cr(III) and Ni(II) increased for maize straw, sawdust and rice husk. The increasing Cr(III), Zn(II) and Ni(II) concentrations led to a decrease in lactic acid yield from wheat bran.	Kong et al. (2008)
Palm fruit press fiber	Temperature: 207–350°C Residence time: 2–30 min Catalyst: ZnCl$_2$ Base: NaOH, Na$_2$CO$_3$	As the temperature increased, the gas yield decreased from 40–25% and liquid yield increased from 20–45%. With ZnCl$_2$, the gas yield increased and liquid yield decreased.	Mazaheri et al. (2010)
	Temperature: 220–320°C Residence time: 7.5–85 min Particle size: 250–710 μm Base: NaOH	As temperature increased from 276 to 307°C, the liquid yields decreases. An increase in residence time (up to 45 min) favored the liquid product yield.	
Wood	Temperature: 280°C Residence time: 15 min Base: NaOH, Na$_2$CO$_3$, KOH and K$_2$CO$_3$	The catalytic hydrothermal treatment of biomass produced mainly phenolic compounds. The catalytic activity can be ranked as follows: water (58%) < NaOH (86%) < Na$_2$CO$_3$ (88%) < KOH (91%) < K$_2$CO$_3$ (96%).	Karagoz et al. (2005)

12.5 THERMOCHEMICAL CONVERSION OF BIOMASS

Through a variety of processes, lignocellulosic wastes can be converted into solid, liquid and gaseous fuels. Some of the widely used conversion processes are combustion, pyrolysis, gasification, alcoholic fermentation and liquefaction as shown in Figure 12.3. The main products of biomass conversion are energy (e.g., thermal, steam and electricity), solid fuels (e.g., char and combustibles) and synthetic fuels (e.g., methanol, methane and hydrogen gas) (Nanda et al., 2014b). As a thermochemical conversion, combustion is widely practiced in rural areas and industries. In addition, gasification attracts a great deal of interest as it offers higher efficiencies compared to combustion. Nonetheless, pyrolysis is interesting as a high energy-rich liquid product is obtained that offers various advantages in storage and transport. Although it is still at an early stage of development, yet it has received special attention for its direct biomass conversion to solid, liquid and gaseous products by thermal decomposition in the absence of oxygen.

Combustion is based on a rapid oxidation of biomass to obtain energy mostly in the form of heat. During combustion, the biomass is burnt in presence of air for the conversion of chemical energy stored in biomass into

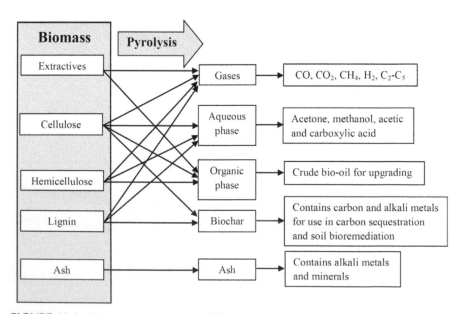

FIGURE 12.3   Biomass components and their pyrolytic conversions.

heat, electricity or mechanical power. However, the biomass requires to be dried (moisture < 50%), chopped and ground which might increase financial costs and energy expenditure in its large scale application. Since combustion takes place in the presence of excess air, CO_2 and water are the pivotal components. The flame temperature during combustion can go beyond 2000°C, depending on the heating value and the moisture content of the fuel, amount of air used to burn the fuel, and construction of the furnace. During combustion, a variety of combustors are used as the devices to convert the chemical energy of fuels into high temperature exhaust gases. The equations 1–3 are of relevant interest during combustion process.

In gasification, biomass is converted into a combustible synthesis gas mixture (mostly CO and H_2) by its partial oxidation at temperature in the range of 500–800°C. The thermal decomposition of the organic material results in major gases such as CO, H_2 and CH_4 as shown in the equations 1–4 (Goyal et al., 2008). In addition to biomass conversion, gasification is suitable for treating other organic matter including municipal solid wastes or hydrocarbons such as coal. It involves partial combustion of biomass in a gas flow containing a controlled level of oxygen at relatively high temperatures yielding a main product of combustible syngas with some biochar (Brewer et al., 2009).

$$C + O_2 \rightarrow CO_2 \tag{1}$$

$$C + 0.5\ O_2 \rightarrow CO \tag{2}$$

$$CO + 0.5\ O_2 \rightarrow CO_2 \tag{3}$$

$$CO_2 + C \longleftrightarrow 2CO \tag{4}$$

$$CO_2 + 4H_2 \longleftrightarrow CH_4 + 2H_2O \tag{5}$$

Hydrothermal carbonization is a thermochemical process involving the conversion of carbohydrate components (i.e., cellulose and hemicellulose) of biomass into carbon-rich solids in water at elevated temperature and pressure (Titirici et al., 2007b). Under acidic conditions with catalysis by iron salts, the reaction temperature during carbonation may be as low as 200°C (Titirici et al., 2007a). Iron oxide nanoparticles and iron ions were found to be effective in catalyzing hydrothermal carbonization of starch and rice grains under mild temperatures of $\leq 200°C$ and gave attractive nanostructures (Cui et al., 2006).

The process is suitable for biomass with high moisture content that would otherwise require drying before pyrolysis, making it complementary to pyrolysis and a potential alternative to anaerobic digestion. In contrast, flash carbonization involves partial combustion of biomass in a packed bed pressurized reactor with a controlled air supply. A high yield of char and gas are formed with no liquid product under these reaction conditions. Antal et al. (2003) describe the flash carbonation process as ignition and control of a flash fire at elevated pressure (~1 MPa) within a packed bed of biomass. As a result, the fire moves upward through the packed bed against the downward flow of air, thus initiating the conversion of biomass into gases at elevated pressure. The process also generates char, characterized by high fixed-carbon content, in less than 30 min of reaction time. The technology is currently being commercialized by Carbon Diversion Inc., USA.

Torrefaction is a thermochemical processing of biomass in the temperature range of 200–300°C (Chen et al., 2011b). It is a kind of mild pyrolysis process that improves the fuel properties of the biomass product. Torrefaction is carried out under atmospheric conditions and but in absence of oxygen. During the process, the moisture and superfluous volatiles contained in the biomass are removed, which partly decomposes the biopolymers (i.e., cellulose, hemicellulose and lignin) and releases various types of volatile matter (Chen et al., 2011a). The final product is a solid, dry, blackened material which is referred to as torrefied biomass or bio-coal (van der Stelt et al., 2011).

Torrefaction effectively lowers the O/C ratio of biomass enhancing its heating value. Torrefied products and volatiles are formed resulting in hardened, dried and more volatile free solid product. Torrefied product has much higher energy density than the raw biomass. This enhances the energy value of the product considering the distance of biomass transportation to biorefineries for use or processing because of its relative lower weight and volume. Torrefied biomass is also hydrophobic which makes it advantageous to store in the open space for long periods, similar to the infrastructures used for coal (Chen et al., 2011b). Moreover, torrefied biomass requires less energy to crushing and pulverizing which adds positively to its energy balance.

Liquefaction of biomass is mostly performed at low temperature and high pressure in presence of hydrogen and a catalyst. The process results in a liquid product along with considerable amount of tars which makes the process expensive for their removal. Autothermal reforming is an attractive alternative process to steam reforming of bio-oils. It is a combination of steam reforming and partial oxidation of the hydrocarbons to produce CO, CO_2 and H_2.

On the other hand, pyrolysis is the thermal decomposition of biomass occurring in the absence of oxygen at temperatures starting from 350–500°C upto 700°C or higher. This results in a formation of a liquid product, solid char and a mixture of gases. There are different modes of pyrolysis with variable operating conditions known today, a few of which are tabulated in Table 12.4. Among all, fast, intermediate, flash, slow and catalytic pyrolysis processes have substantial importance in the conversion of biomass to different liquid and gaseous products. The detailed summary of pyrolysis operating parameters and typical product yield ranges are given in Table 12.5 for slow, fast, intermediate and flash pyrolysis processes.

The gas product from pyrolysis is typically composed of a mixture of CO_2 (9–55 vol. %), CO (16–51 vol. %), H_2 (2–43 vol. %), CH_4 (4–11 vol. %) and small amounts of higher hydrocarbons. The gases are usually present with N_2 introduced to the reactor for creating an inert atmosphere during the pyrolysis equipment. The CO_2 and N_2 provide no energy value to the gas product in combustion, although the other gases are flammable and provide

TABLE 12.4 Summary of Different Pyrolysis Methods and Their Operating Conditions (Demirbas, 2009; Amonette and Joseph, 2009; Mohan et al., 2006; Maschio et al., 1992)

Pyrolysis method	Residence time	Heating rate (°C/s)	Typical temperature (°C)	Major products
Slow	2–30 min	0.1–1	300–700	Biochar
Fast	0.5–5 s	10–200	500–700	Bio-oil
Flash	< 0.5	> 1000	800–1000	Bio-oil, gases
Flash-liquid	< 1 s	> 1000	< 650	Bio-oil
Flash-gas	< 1 s	High	< 650	Liquid phase*, gases
Ultra	< 0.5 s	Very high	1000	Liquid phase*, gases
Intermediate	5–15 min	10–100	320–500	Bio-oil, biochar, gases
Vacuum	2–30 s	Medium	400	Bio-oil
Hydro-pyrolysis	< 10 s	High	< 500	Bio-oil
Methano-pyrolysis	< 10 s	High	> 700	Liquid phase*
Torrefaction	5–30 min	< 0.8	200–300	Biochar, gases
Carbonization	Days	Low	400	Biochar

* Liquid phase mostly contains organic chemicals.

TABLE 12.5 Typical Operating Conditions and Product Yields from Various Pyrolysis Methods (Maschio et al., 1992; Mohan et al., 2006; Demirbas, 2009)

Pyrolysis process		Slow	Intermediate	Fast	Flash
Temperature (°C)		300–700	320–500	500–700	800–1000
Residence time		2–30 min	5–15 min	0.5–5 s	< 0.5 s
Heating rate (°C/s)		0.1–1	10–100	10–200	> 1000
Biomass particle size (mm)		5–50	5–50	< 1	< 0.2
Yield (wt. %)	Char	29–38	30–40	22–34	20–35
	Liquid	28–35	35–45	46–64	46–50
	Gas	25–36	19–32	11–16	11–28

energy value in proportion to their individual properties. No special consideration of the CO_2 in the pyrolysis gas is required as it is not additional to what would result from biomass decomposition. On the other hand, biomass is a carbon-neutral energy resource.

The following equations represent the characteristic reaction scheme during pyrolysis, where $C_nH_mO_k$ is the biomass feedstock. The enthalpy (kJ/gmol) in each step is given at the reference temperature of 27°C and $n = 6$ (Tanksale et al., 2010). Equations 8–10 represent the scheme for partial oxidations in pyrolysis.

$$C_nH_mO_k \rightarrow (1 - n)\ CO + (m/2)\ H_2 + C \qquad\qquad 180\ (KJ/gmol) \quad (6)$$

$$C_nH_mO_k \rightarrow (1 - n)\ CO + \{(m - 4)/2\}\ H_2 + CH_4 \qquad 300\ (KJ/gmol) \quad (7)$$

$$C_nH_mO_k + 0.5\ O_2 \rightarrow nCO = (m/2)\ H_2 \qquad\qquad 71\ (KJ/gmol) \quad (8)$$

$$C_nH_mO_k + O_2 \rightarrow (1 - n)\ CO + CO_2 + (m/2)\ H_2 \qquad -213\ (KJ/gmol) \quad (9)$$

$$C_nH_mO_k + 2O_2 \rightarrow (n/2)\ CO + (n/2)\ CO_2 + (m/2)\ H_2 \quad -778\ (KJ/gmol) \quad (10)$$

The liquid fraction of the pyrolysis products consists of two phases: an aqueous phase and an organic phase. The aqueous phase contains a wide range of organo-oxygenates compounds of low molecular weight, whereas the non-aqueous phase contains insoluble organics (mainly aromatics) of higher molecular weight. The phase is called bio-oil which is the product of great interest in commercial thermochemical conversions. The ratios of acetic acid, methanol and acetone in aqueous phase are mostly higher than in the organic phase. The purpose of pyrolysis process is to maximize the yield of liquid products resulting from biomass pyrolysis with a low temperature, high heating rate and short gas residence time. However, for char production, a low temperature and low heating rate pyrolysis is preferred. In addition, for maximum flue gas production, a high temperature, low heating rate and long residence time process is ideal. The distinct component involvement during pyrolysis process is summarized in Figure 12.2.

As a result of thermochemical decomposition of biomass materials, biochar is produced in variable amounts depending on the process. Biochar is an interesting byproduct as it makes the process carbon negative by sequestering the CO_2 emitted through biomass combustion and other thermochemical processes. Biochar is capable in capturing this CO_2 in the soil leading to the reduction in net greenhouse gas emission and enriching the soil carbon pool. In addition to carbon sequestration, biochar is advantageous in increase soil fertility, water retention and nutrient availability to plants and decreasing soil acidity (Nanda et al., 2014a). In addition, biochar is mostly briquetted in single composition or in combination with biomass to be used as a high efficiency solid fuel in boilers of biorefineries. Recently, biochar have found application in value added products such as activated carbon (Azargohar and Dalai, 2006), carbon nanotubes and carbon fibers (Ozcimen and Ersoy-Mericboyu, 2010). However, it can also be used further for the gasification process to obtain H_2 rich gas by thermal cracking (Li et al., 2006).

12.5.1 FAST PYROLYSIS

Fast pyrolysis is a process in which an organic material is rapidly heated to high temperatures in the absence of oxygen. Main features of fast pyrolysis are high heating rates (about 300°C/min) and short residence time (0.5–5 s) (Nanda et al., 2014b). It generally requires a feedstock prepared with small particle sizes (< 1.0 mm) and a design that removes the vapors quickly from the presence of the hot solids (i.e., char) (Maschio et al., 1992). There are a number of different reactor configurations that can be used for fast pyrolysis operations this including fluidized beds, ablative systems, stirred or moving beds and vacuum pyrolysis systems (Bridgwater and Peacocke, 2000). In all fast pyrolysis reactor systems, a typical temperature of around 500±10°C is mostly preferred.

Currently fast pyrolysis for liquids production is attracting worldwide interest. Fast pyrolysis occurs in a time of few seconds or less. Hence, the overall chemical reaction kinetics, heat and mass transfer processes and phase transition phenomena play vital roles in determining the yield and property of the products. The critical issue is to bring the reacting biomass particle to an optimum process temperature and minimize its exposure to the intermediate and lower temperature regions that favor char formation. However, this is achieved by using small particle size biomass which at higher temperatures and high heating rates is decomposed to generate vapors, aerosols and some char particles. As a result of cooling and condensation of the vapors, a dark brown liquid is formed with a heating value about half that of conventional fuel oil.

Although evolved from the traditional (slow) pyrolysis processes for making biochar, fast pyrolysis is an advanced process with carefully controlled parameters to give higher yields of liquids. The essential features of a fast pyrolysis process that favor production of bio-oils are: (i) very high heating and heat transfer rates at the reaction interface, (ii) requirement of small particle size biomass feed, (iii) controlled reaction temperature of around 450–600°C, (iv) vapor phase temperature of 400–450°C, (v) short vapor residence times of typically < 2 s, and (vi) rapid cooling of the vapors to produce the bio-oil. The main product (bio-oil) is obtained in yields of up to 75 wt. % on dry feed basis together with biochar and gases which are used within the process so there are no waste stream other than flue gas and ash.

12.5.2   SLOW PYROLYSIS

Slow pyrolysis of biomass operates at relatively low heating rates (0.1–2°C/s) and longer solid and vapor residence time (2–30 min) to favor biochar yield (Nanda et al., 2014b). Slow pyrolysis operates at temperature lower than that of fast pyrolysis, typically 400±10°C and has a gas residence time usually > 5 s. Slow pyrolysis is similar to carbonization (for low temperatures and long residence times). During conventional pyrolysis, biomass is slowly devolatilized facilitating the formation of chars and some tars as the main products. This process yields different range of products with their properties dependent on temperature, inert gas flow rate and residence time.

12.5.3 INTERMEDIATE PYROLYSIS

Intermediate pyrolysis is used to describe biomass pyrolysis in a certain type of commercial screw-type pyrolyser or the Haloclean reactor (Balabanovich et al., 2005). This pyrolysis has operating conditions for preventing formation of high molecular weight tar and enhancing the quality, for example, dryness and brittleness of biochar which can be further utilized for the purpose of soil fertilization and carbon sequestration. In this case, mechanical briquetting of biomass is not necessary as larger sized biomass can be directly feed into the reactor. This helps in separating char from vapors easily. The intermediate pyrolysis reactors are mostly designed for disposal and useful processing of biomass and other organic wastes as well as electronic waste. Sewage sludge and municipal solid wastes can be pyrolysed under these conditions with provisions of gasification where the producer gas could be recycled for heating purpose. Although it is similar to slow pyrolysis, yet the process is relatively faster.

Some recent developments have led to slow or intermediate pyrolysis technologies of most interest for wide pyrolysis product distribution in wide ranges. These are generally based on a horizontal tubular kiln where the biomass is moved at a controlled rate through the kiln. These include agitated drum kilns, rotary kilns and screw pyrolysers (Brown, 2009). In several cases these have been adapted for biomass pyrolysis from original uses such as the coking of coal with production of town gas or the extraction of hydrocarbons from oil shale, for example, Lurgi twin-screw pyrolyser (Henrich et al., 2007). Town gas, a flammable gas usually produced by the destructive distillation of coal, contains calorific gases (e.g., H_2, CO, CH_4 and other volatile hydrocarbons) and small quantities of non-calorific gases (e.g., CO_2 and N_2), and has similar uses as natural gas. Although some of the pyrolysis technologies have well-established commercial applications, yet there is little commercial use with biomass in biochar production (Dahman et al., 2007).

12.5.4 CATALYTIC PYROLYSIS

Pyrolysis oil consists of a complex mixture of aliphatic and aromatic oxygenates and particulates. Also known as crude bio-oil, it is very viscous, acidic and unstable liquid with relatively low-energy density compared to conventional fossil oil. This nature of the bio-oil requires costly post-treatment processes that make the commercial pyrolysis economically less attractive. However, the presence of suitable catalysts during the pyrolysis process can affect the network of reactions (e.g., deoxygenation) and allow in situ upgrading of the bio-oil. A tremendous advantage of catalytic pyrolysis is that the expensive additional deoxygenation step for bio-oil processing can be avoided. In addition, it also improves the efficiency of pyrolysis process by providing better contact between the solid catalyst and solid biomass or waste. Lower pyrolysis temperature is crucial in this step to maximize the bio-oil yield and quality (Babich et al., 2011).

12.6 FACTORS AFFECTING BIOMASS PYROLYSIS

12.6.1 MOISTURE

Moisture content can have different effects on pyrolysis product yields depending on the operating conditions (Antal and Gronli, 2003). In general, fast pyrolysis requires a fairly dry biomass feed with around 10% moisture (Bridgwater and Peacocke, 2000). This prerequisite does not restrict the increase in rate of temperature with the evaporation of water. Slow pyrolysis processes are comparatively tolerant of moisture, although the main issue being the effect on process energy requirement. For char production, wood moisture contents of 15–20% are typical (Antal and Gronli, 2003). Moisture in the reaction affects char properties and this has been used to produce activated carbons through pyrolysis of biomass (Schroder et al., 2007). In all pyrolysis processes, water is a significant byproduct and is usually collected together with other condensable vapors in the liquid phase.

12.6.2 BIOMASS PARTICLE SIZE

Feed particle size can significantly affect the balance between char and liquid yields. Larger particle sizes tend to produce more char by restricting the rate of release of primary vapor products from the hot char particles, hence increasing the scope for secondary char forming reactions (Antal and Gronli, 2003). Hence, larger particles are beneficial in thermochemical processes targeting char production and smaller particles are preferred to maximizing bio-oil yields. A range of biomass particle size requirements for slow, fast, intermediate and flash pyrolysis processes are given in Table 12.5.

12.6.3 PYROLYSIS TEMPERATURE AND RESIDENCE TIME

The temperature profile is the most important aspect of operational control for pyrolysis processes. Material flow rates, both solid and gas phase, together with the reactor temperature control the key parameters of heating rate, highest process temperatures, residence time of solids and contact time between solid and gas phases. These factors affect the product distribution and the product properties. Solid residence time is another important factor in the bio-oil yields. A short residence time enhances bio-oil yields, while a longer residence time increases char production (Antal and Gronli, 2003).

For fast pyrolysis, a rapid heating rate and a rapid rate for cooling primary vapors are required to minimize the extent of secondary reactions. These reactions not only reduce the liquid yield but also tend to reduce its quality, giving a more complex mixture with an increased degree of polymerization and higher viscosity (Bridgwater and Peacocke, 2000). In contrast, slow pyrolysis employs slow heating rates that lead to higher char yields (Mohanty et al., 2013). Peak temperature also affects the yield and properties of bio-oil and biochar. Higher temperatures lead to lower char yield with high bio-oil yields in all pyrolysis reactions. This is because temperature is a main controlling variable of pyrolysis reaction kinetics. The high yields of bio-oil at higher temperatures are because of reduced char yield. The char yield is reduced because of the forced removal of volatile materials from the biomass at elevated temperatures that condense as bio-oils; although this increases the proportion of carbon in the char (Antal and Gronli, 2003).

Liquid yields are higher with increased pyrolysis temperatures, usually at 400–550°C but dependent on other operating conditions. Temperatures higher than 550°C result in secondary reactions causing vapor decomposition to be more dominant, reducing the condensed liquid yields. For fast pyrolysis the highest bio-oil yields are generally obtained at a temperature near to 500°C (Czernik and Bridgwater, 2004). However, highest bio-oil yields for slow pyrolysis are more variable. Demirbas (2001) reported highest pyrolysis liquid yields of 28–41 wt. % at temperatures between 377 and 577°C, depending on feedstock when using a laboratory scale slow pyrolysis reactor. The Haloclean process has bio-oil yields of 42–45 wt. % at temperatures of 385–400°C with different straw feeds (Balabanovich et al., 2005).

12.6.4   GAS ENVIRONMENT

The inert gas and the product gas phase during pyrolysis have profound influence on product distributions and on the thermodynamics of the reaction. Most of the effects can be understood by considering the secondary char-forming reactions between primary vapor products and the hot char (Antal and Gronli, 2003). Gas flow rate through the reactor affects the contact time between primary vapors and hot char influencing the degree of secondary char formation. Low flow rate of the inert gas favors char yield which is typically preferable for slow pyrolysis. On the other hand, high gas flow rates are used in fast pyrolysis for enhancing bio-oil yields through effectively stripping off the vapors as soon as they are formed.

12.6.5   PRESSURE

High pressure increases the activity of vapors within and at the surfaces of char particles which increases secondary char formation. The effect is most marked at pressures up to 0.5 MPa. Conversely, pyrolysis under vacuum yields less char, thus favoring liquid products. For pyrolysis under pressure, moisture in the vapor phase can steadily increase the yield of char. This is due to an autocatalytic effect of water that reduces the activation energy for pyrolysis reactions. Thermodynamics of pyrolysis are also influenced by the gas environment. The reaction is more exothermic at higher pressures and low gas flow rates. This is rationalized as being due to the greater degree of secondary char forming reactions. Hence, higher char yields are associated with conditions where pyrolysis is exothermic and such conditions tend to favor the overall energy balance of processes targeting char as product.

In summary, any factor of pyrolysis conditions that increases the contact between primary vapors and hot char, including high pressure, low gas flow, large particles or slow heating is likely to favor char formation at the expense of bio-oil yield. The total mass and carbon content of the pyrolysis products obtained could be in equivalence to the mass and carbon content of the feed material, if all the above factors are properly accounted. However, some energy is inevitably lost as heat from the process indicating the total energy value in the products being less than the feedstock.

12.7 BIO-OIL CHARACTERISTICS

Bio-oil is not a product of thermodynamic equilibrium during pyrolysis or liquefaction (Zhang et al., 2007). It is a result of rapid cooling or quenching of organic vapors obtained from the cracking of biomass components at high temperatures and short reaction times. The condensate that is produced following the quenching process is also not at thermodynamic equilibrium at storage temperatures. Hence, the chemical composition of bio-oil always tends to change toward thermodynamic equilibrium during storage, which makes it unstable unless upgraded.

The characteristics of bio-oil largely depend on the production technology, type of feedstock used, production parameters and the reactor type. The energy content of bio-oils is between 72,000 and 80,000 BTU/gallon. Towards the higher ranges of energy content of bio-oils, there are typically certain amounts of suspended char in the liquid phase. The conventional heating oil has an energy content of about 138,500 BTU/gallon; hence bio-oil has about 52–58% as much energy as heating oil per gallon. However, it is interesting to note that bio-oil weighs about 40% more per gallon than the conventional heating oil. Bio-oils are heterogeneous mixtures of bio-polymeric components derived from depolymerization and fragmentation of cellulose, hemicellulose and lignin (Demirbas, 2004). Therefore, the chemical composition and properties of bio-oil often tend to differ based on the feedstock type and production process (Nanda et al., 2014b). The typical physicochemical properties of bio-oil are tabulated in Table 12.6. The table also makes a comparison between bio-oils and the fuel properties of No. 2 diesel and heavy fuel oil.

Bio-oil is typically a dark brown liquid with a smoky acrid smell. Upon its generation through pyrolysis, it tends to have relatively high water content, typically in the range of 15–30 wt. %. This water content in bio-oil is derived from the original moisture in the feedstock and the product of dehydration during the pyrolysis reaction and storage. The water also originates from the condensation of some volatiles during the pyrolytic conversion process. The water content in the bio-oil tends to be entirely miscible in its organic phase, although at higher moisture levels, the water is found to be separated from the organic phase. It is desirable to have the biomass feedstock dried to 10% of its moisture level or less before it is fed into the pyrolysis reactor.

As the water content in the bio-oil increases, the energy content (or higher heating value) and flame temperature of the bio-oil decreases. In addition,

TABLE 12.6   Typical Fuel Properties of Bio-Oil with Reference to No. 2 Diesel and Heavy Fuel Oil (Mohan et al., 2006; Zhang et al., 2007; Demirbas, 2009)

Property	Bio-oil	No. 2 Diesel	Heavy fuel oil
Appearance	Reddish-brown to dark green	Creamish-white	Brownish-black
Density	~ 1.2 kg/L	6.4–7.4 lb/gal	—
Flash point (°C/min)	48	52	—
Pour point (°C)	–33	–30 to –40	–18
Specific gravity	1.2	0.8–0.9	0.9
Moisture content (wt.%)	15–30	—	0.1
Solid sediment (wt.%)	0.2–1	0.05	1
Distillation residue (wt.%)	< 50	—	1
Kinematic viscosity at 40°C (mm²/s)	25–1000	1.3–4.1	—
Viscosity at 50°C (cP)	40–100	—	180
Carbon (wt.%)	54–58	84–87	~85
Hydrogen (wt.%)	5.5–7	33–16	~11
Oxygen (wt.%)	35–40	0	~1
Nitrogen (wt.%)	< 0.2	—	~0.3
Sulfur (wt.%)	0.001–0.02	0.05	—
Ash (wt.%)	< 0.2	0.01	0.1
Carbon residue (wt.%)	0.001–0.02	0.3	—
Higher heating value (MJ/kg)	16–19 MJ/kg	43	40
Heat of combustion (BTU/lb)	7100	19,2000–20,000	17,600
Heating of combustion (MJ/L)	19.5	—	39.4
Miscibility	Miscible with polar solvents such as methanol and acetone	Miscible in ethanol	Miscible in ethanol
Cetane number	48–65	40–55	—

water reduces the viscosity but enhances the fluidity, which is good for the atomization and combustion of bio-oil in the engine (Zhang et al., 2007). The viscosity of bio-oil tends to be slightly higher than the conventional no. 2 fuel oil. The viscosities of bio-oils greatly vary depending on the biomass feedstocks and conversion process. Boucher et al. (2000) tested the fuel performance of bio-oil with methanol addition for gas turbine applications.

It was found that methanol reduced the density and viscosity of the bio-oil and increased its stability. However, the methanol blending caused a lowering in the bio-oil's flash point.

Bio-oils show a wide range of boiling point temperatures due to their complex composition. During distillation, bio-oils start boiling below 100°C due to the slow heating rate that induces polymerization of some reactive components (Zhang et al., 2007). However, polymerization reactions cease and bio-oil stops boiling at 250–280°C, leaving behind 35–50 wt. % solid residues. Hence, bio-oils cannot be subjected to evaporation before combustion.

The oxygen content of bio-oils is usually 35–40% (Table 12.7) due to the occurrence and distribution of a variety of oxygenated components derived from biomass. These oxygenates also occur depending on the biomass source and conversion process parameters such as temperature, heating rate and residence time. The presence of oxygen is one of the major factors of distinction between bio-oils and hydrocarbon fuels such as gasoline, diesel, etc. The high moisture and oxygen levels lower the bio-oil's energy density by 50% compared to that of conventional fuels. The low nitrogen content of bio-oil should help reduce NO_x emissions. For example, biofuel

TABLE 12.7 Density and Volumetric Energy Content of Various Solid and Liquid Fuels

Fuel	Density (kg/m³)	Volumetric energy content (GJ/m³)
Ethanol	790	23
Methanol	790	18
Diesel	850	39
Biodiesel	900	36
Gasoline	740	36
Bio-oil	1280	11
Agricultural residue	50–200	0.8–3.6
Bagasse	160	2.8
Baled straw	160–300	2.6–4.9
Nut shells	64	1.3
Rice hulls	130	2.1
Hardwood	280–480	5.3–9.1
Softwood	200–340	4–6.8
Coal	600–900	11–33

tests in combustion engine showed half the NO_x emissions than from diesel fuel (Boehman et al., 2004). The low sulfur content of bio-oil also result in reduced SO_x emissions compared to those from petroleum-based fuel oil or diesel fuel (Palash et al., 2013).

Bio-oil is moderately acidic, having a pH between 2.5 and 3.0 (similar to the acidity of acetic acid). Bio-oils comprise substantial amounts of carboxylic acids, such as acetic and formic acids. The pH of softwood bio-oils is lower compared to that of hardwood bio-oils. For instance, bio-oil of pinewood has a pH of 2.6, while that of hardwood has 2.8 (Sipila et al., 1998). The strong acidity makes bio-oil very corrosive and extremely unstable, which prompts special requirements on the construction materials of storage or transportation vessels. This raises a significant issue regarding the use of bio-oil in existing commercial installations since most of the existing fuel storage tanks used for heating oil are likely to be made of plain mild steel or stainless steel that is vulnerable to corrosion from bio-oil. This makes it necessary for the storage tanks for bio-oils made of materials resistant to acidic corrosion.

The presence of ash (> 0.1 wt. %) in bio-oil also contributes to corrosion and fuel ignition problems when used in the engines. These issues are attributed due to the occurrence of alkali metals in the bio-oils. While, sodium, potassium and vanadium are responsible for corrosion, calcium is responsible for hard deposit formation on engine surface (Zhang et al., 2007). Producing bio-oil with lower ash content and/or lower water content helps prolong the stability of bio-oil with lower corrosion problems.

The higher heating value (HHV) properties of bio-oils depend on factors such as biomass type, production temperature and heating rate. As discussed earlier, bio-oils with a high moisture and oxygen content has a lower HHV. On the other hand, Nanda et al. (2014a) suggest that bio-oils with high carbon and hydrogen content, and low oxygen usually tend to have a higher HHV due to more amount of energy contained in C–C and C–H bonds than in C–O and O–H bonds. Bio-oil produced from wood has a high HHV than that of bio-oils from straws and grasses. For example, the HHV of pinewood bio-oil was 27.1 MJ/kg, while that of wheat straw bio-oil and timothy grass bio-oil were 20.8 and 20.5 MJ/kg, respectively (Mohanty et al., 2013). In the same study, Mohanty et al. (2013) also indicated that bio-oils obtained from fast pyrolysis had a relatively higher HHV than that of slow pyrolysis bio-oils. Moreover, bio-oils from oil-based plants have a higher HHV compared to that from straws and wood. For instance, the HHV of safflower seed bio-oil, for example, 11 MJ/kg (Beis et al., 2002) is relatively higher than bio-oils from forest and agricultural biomass, for example, 21–27 MJ/kg (Mohanty et al., 2013).

In addition, bio-oil is a complex mixture containing carbon, hydrogen and oxygen in the form of acids, alcohols, aldehydes, esters, ethers, ketones, sugars, phenols, guaiacols, syringols, furans, phenols, terpenes, terpenoids, etc. (Nanda et al., 2014b). Bio-oil does not naturally blend with conventional petroleum fuel. It may be possible to add a solvent or to emulsify mixtures of bio-oil and fuel oil in order to get homogeneous blends. The density and volumetric energy content of various solid and liquid fuels are tabulated in Table 12.7. The biomass feed, char and liquid products have energy values roughly related to their carbon contents. Release of this energy by combustion can again be considered as renewable and is largely carbon neutral. The carbon returned to the atmosphere as CO_2 is the same as would otherwise have resulted from biomass decomposition. If the char product is not burnt, but retained in a way that the carbon in it is stable, then that carbon can be equated to CO_2 removed from the atmosphere and sequestered.

12.8 UPGRADING OF BIO-OIL

Regardless of the production process (i.e., thermochemical or hydrothermal conversion), bio-oils obtained from biomass could not be directly used as transportation fuel. As this bio-crude oil contains high oxygen and water content, it needs to be upgraded to enhance its fuel properties. The bio-oils in their crude form are also found to be less stable and less miscible than conventional fuels (Goyal et al., 2008). Various catalysts have found profound applications in improving the bio-oil's fuel properties. This is achieved mostly in two ways: (i) addition of catalysts during biomass pyrolysis or liquefaction to improve the quality of the oil, and (ii) addition of catalysts to the crude bio-oil post-pyrolysis or liquefaction for upgrading.

The oil obtained by catalytic biomass conversion usually does not require expensive pre-upgrading techniques involving condensation and re-evaporation. The limiting properties of bio-oil such as high viscosity, instability and corrosiveness present many difficulties for their direct use as a substitute to fossil fuels (Chheda et al., 2007a). Hence, an upgrading process is required to reduce their oxygen and water content, and enhance their fuel properties for a wide scale of applications as shown in Figure 12.4.

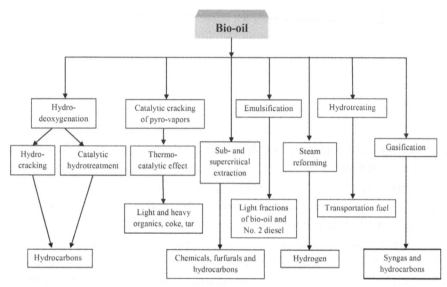

FIGURE 12.4   Integrated routes for upgrading of bio-oil.

The most commonly used bio-oil upgrading techniques can be categorized as: (i) hydrodeoxygenation, (ii) emulsification, and (iii) steam reforming.

Mostly related to each other, hydrodeoxygenation and hydrodesulphurization are used for the removal of oxygen and sulfur from the bio-oil, respectively (Mortensen et al., 2011). Both hydrodeoxygenation and hydrodesulphurization use H_2 for elimination of heteroatoms from bio-oils in the form of H_2O and H_2S, respectively. The hydrodeoxygenation is performed in presence of H_2 providing solvents activated by catalysts such as Co–Mo, Ni–Mo and their oxides (Zhang et al., 2007). These catalysts are also used as being loaded on Al_2O_3 under pressurized H_2 and/or CO atmosphere. Oxygen is removed as H_2O and CO_2 from the bio-oil, thus elevating its energy density.

Bio-oils could be blended with the conventional diesel fuels to be used as a transportation fuel. Since bio-oils are immiscible in hydrocarbons, they can be emulsified in presence of a surfactant. The viscosity and corrosiveness of emulsified bio-oils has been found to be much lower than that of crude bio-oil (Ikura et al., 2003). The high cost and energy input during emulsification are major factors for consideration in bio-oil upgrading.

Steam reforming is an upgrading process where hydrocarbons react with high temperature steam and are converted to CO and H_2. In general, the steam reforming of bio-oil or its model compounds removes the oxygenated organic compound ($C_nH_mO_k$) by the following reactions, where the enthalpy (kJ/gmol) of each step is given at reference temperature at 27°C and n = 6. The steam reforming is normally followed by water-gas shift (equation 13) and the methanation reaction (equation 15). The equations 11 and 12 are accompanied by water-gas shift reaction. However, equation 14 represents the overall reforming process. The steam reforming reaction is endothermic, while water-gas shift and methanation reactions are exothermic in behavior (Trane et al., 2012). While water-gas shift reaction alone is favored at high temperatures, the overall reforming equilibrium is favored at high temperatures and low pressures (Rostrup-Nielsen et al., 2002).

$$C_nH_mO_k + H_2O \rightarrow nCO + mH_2 \qquad\qquad 310 \text{ KJ/gmol} \quad (11)$$

$$C_nH_mO_k + nH_2O \rightarrow xCO + (n-x)\,CO + mH_2 \qquad 230 \text{ KJ/gmol} \quad (12)$$

$$CO + H_2O \longleftrightarrow CO_2 + H_2 \qquad\qquad (13)$$

$$C_nH_mO_k + (2n-k)\,H_2O \rightarrow nCO_2 + \{(2n+m)/(2-k)\}\ H_2\ 64 \text{ KJ/gmol} \quad (14)$$

$$CO + .3H_2O \longleftrightarrow CH_4 + H_2O \qquad\qquad (15)$$

The liquid bio-oil product from pyrolysis has many considerable advantages not only as a fuel but also as a potential source for numerous value added chemicals. Upgrading bio-oil to the quality of transportation liquid fuel possesses several technical challenges which adds to the overall process economics. Some chemicals, especially those produced from the bio-oils or its major fractions such as liquid smoke or from wood resins offer more interesting commercial opportunities. There are many challenges to overcome before bio-oil finds large-scale acceptance as fuel: (i) cost of bio-oil, which is 10 to 100% more than fossil fuel in energy terms, (ii) availability of bio-oil for application development remains a problem and there are limited supplies for testing, (iii) lack of established standards for use and distribution of bio-oil and inconsistent quality inhibits wider usage, (iv) considerable investigation is required to characterize and standardize these pyrolysis liquids and develop a wider range of energy applications, (v) incompatibility of bio-oil with conventional fuels and therefore need for dedicated fuel handling systems, (vi) environmental health and safety issues need to be completely resolved during bio-oil production, storage and handling, and (vii) more research and development is needed in the field of fast pyrolysis and bio-oil testing to develop large-scale commercial applications.

There are various upgrading processes such as hydrotreating and reforming but *in situ* conversion of biocrude vapors in presence of HZSM-5 is a promising bio-oil upgrading process. Zeolite and other mesoporous catalyst are shape selective and allow specific components to reach to catalyst surface. They provide acidic sites for the reaction such as deoxygenation, decarboxylation and decarbonylation of the bio-oil components. Reactions such as cracking, alkylation, isomerization, cyclization, oligomerization and aromatization are catalyzed by acidic sites of the zeolite by a carbonium ion mechanism. Some undesirable products such as tars and coke are also obtained during the process. Table 12.8 summarizes a few studies on the catalytic upgrading of bio-oils using alumina and zeolites.

The process of upgrading by Me-Al-MCM-41 has been studied with Cu, Zn and Fe by Nilsen et al. (2007). Their findings showed that Fe and Zn led to higher phenol productions. The presence of Al-MCM-41 showed a decrease in the fraction of undesirable oxygenated components. The Al-MCM-41 catalysts with SAR (silica-alumina ratio) of 20 produced fractions rich in hydrocarbons

TABLE 12.8   Upgrading of Bio-Oils Obtained from Biomass Pyrolysis with Alumina and Zeolites

Raw material	Catalyst	Remarks	Reference
Aspen wood	ZSM-5, Y and SAPO M ZSM-5 (Co, Fe, Ni, Ce, Ga, Cu and Na)	The best-performing catalysts belonged to the ZSM-5 group. The highest yields of hydrocarbons from wood, 16 wt. % including 3.5 wt. % of toluene was obtained from Ni-substituted ZSM-5 zeolite.	French and Czernik (2010)
Cassava rhizome	Al-MCM-41 (SAR 40), ZSM-5 (SAR 50)	ZSM-5 resulted in higher aromatic hydrocarbons compared to MCM-41. MCM gave a higher yield of syringyl lignin derivatives, whereas ZSM-5 produced higher amount of guaicyl-lignin derivatives. ZSM-5 gave more aldehydes and ketones products but lesser carbonyls containing hydroxyl groups.	Pattiya et al. (2008)
Cellulose	MCM-41 with Al, Mg, Ti, Sn, Zr and Si	All the examined meso-structured solids decreased the yields of levoglucosan with respect to uncatalysed cellulose. They increased the production of levoglucosenone and hydroxylactone.	Torri et al. (2009)

TABLE 12.8 Continued

Raw material	Catalyst	Remarks	Reference
Corn cob	Al$_2$O$_3$	Aliphatics were obtained in larger quantities compared to the non-catalytic reactions. The total amount of phenolic compounds decreased and the amount of polycyclic aromatic hydrocarbons increased at high pyrolysis temperatures without catalyst or at moderate pyrolysis temperatures with a catalyst.	Ates and Isikdag (2009)
Corncob pyro-vapors	MCM-41, CaO	The molality of carbonyl compounds decreased by 10.2% with MCM-41, while that of phenols, hydrocarbons and CH$_4$ increased by 15.3%, 4.3% and 10.2%, respectively compared with non-catalytic reactions.	Wang et al. (2010)
Lignin	Sand, HZSM-5, Al-MCM-41	With sand, the char, liquid and gas yields were 41, 38 and 21 wt. %. The sand, liquid and gas yields were 30 and 36, 48 and 40 as well as 22 and 24 wt. % with HZSM-5 and MCM-41, respectively.	Jackson et al. (2009)
Maple wood	HZSM-5 (SAR 56), H-Y (SAR 6), H-mordenite (SAR 14)	The yields of hydrocarbons were 27.9, 14.1 and 4.4 wt. % with HZSM-5, H-Y and H-mordenite, respectively.	Adjaye et al. (1995)
Miscanthus giganteus	Activated Al$_2$O$_3$	The H-NMR spectra of bio-oils indicated higher aromaticity. The yield of gases and solids decreased as the catalyst loading increased from 10 to 100 wt. %. The liquid product yields were not affected significantly with catalyst loading.	Yorgun and Simsek (2008)
Pine oil	HZSM-5 with SAR of 50 and 80	Highly deoxygenated bio-oils were obtained with aromatics. Better quality bio-oil was obtained at 450°C with HZSM-5/50 catalyst.	Vitolo et al. (2001)
Pinewood	β, Y, ZSM-5, mordenite, quartz	Maximum yield was obtained in presence of quartz, whereas zeolite Y yielded minimum bio-oil.	Aho et al. (2008)

TABLE 12.8   Continued

Raw material	Catalyst	Remarks	Reference
	Various proton and Fe-modified zeolites, for example, ferrierite, Y and β	The yield of bio-oil ranged from 43.5 to 52.7 wt. %. The water yield and formation of CO increased over all zeolites, while CO_2 production increased only to certain extent. β-zeolite was most active in de-oxygenation reactions followed by Y and ferrierite.	Aho et al. (2010)
Soybean oil	HZSM-5 (with SAR 28, 40 and 180), Me-MCM-41 (Ga, Al and Cu)	Ga-MCM-41 and HZSM-5 (SAR 28) gave highest liquid yield of about 77 and 71%, respectively. The main liquid products with H-ZSM-5 catalysts were aromatics such as benzene, toluene and xylene. Ga-MCM-41 catalysts gave a mixture of alkanes, alkenes, alkadienes, aromatics and carboxylic acids.	Ngo et al. (2010)
Spruce wood pyro-vapors	Al and Cu (SAR 20)	An increase of acetic acid and furans was observed with decrease in higher molecular mass phenols.	Adam et al. (2005)
Tong oil, palm oil, curcas oil and *Schisandra wilsoniana* Sojak oil	MCM-41, Al_2O_3	Cracking with MCM and Al_2O_3 yielded 71 and 75% of oil and 9% and 8.5% coke, respectively.	Junming et al. (2010)

as compared to SAR of 40. Investigation on the catalytic conversion of ligno-cellulosic biomass with two mesoporous aluminosilicate materials from the MSU family, for example, MSU-S/H with hexagonal mesopores and MSU-S/W with a wormhole-like structure and high textural porosity has been done (Triantafyllidis et al., 2007). The MSU-S catalysts led to a significantly lower quantity of organic phase in the obtained bio-oil, but had higher coke and char yields compared to Al-MCM-41. The MSU-S catalysts were selective towards polycyclic aromatic hydrocarbons and heavy fractions, whereas they produced only small amounts of acids, alcohols, carbonyls, and phenols. The MSU-S type materials appeared to possess stronger acidic sites than Al-MCM-41, resulting in enhanced yields of aromatics, polycyclic aromatic hydrocarbons, and coke along with propene in the pyrolysis gases.

Recent studies on catalytic conversion of Japanese larch by MCM-41 showed a higher amount of phenolics (Park et al., 2012). The bio-oil with enhanced stability was produced applying these mesoporous materials for transfer of oxygen, which is known to produce instability in bio-oils. The catalytic activity of Al-MCM-41 for bio-oil upgrading was higher than that of siliceous MCM-41 because of the larger number of acidic sites.

12.9 CONCLUSION AND FUTURE PERSPECTIVES

There is not only a need but also an urge to use waste biomass resources in the production of biofuels, due to the many environmental and economic impacts from the conventional fossil-based transportation fuels. The conversion routes applied to biomass for fuel production widely include thermochemical, hydrothermal and biochemical. All the three conversion methods are well-suited to achieve the energy requirements for being ecofriendly processes. However, in the present context both thermochemical and hydrothermal conversion are found effective to produce an energy dense liquid "bio-oil" that could not only be used as a transportation fuel but also for heat and power generation.

Pyrolysis and liquefaction have received significant amount of interest as they generate bio-oils of better quality and yield compared to other thermochemical and hydrothermal processes. It should be noted that the operating conditions during these processes along with the indigenous nature of the biomass feedstock affect the bio-oil yield and quality to a large extent. The high heating rates and short residence times of fast and flash pyrolysis result in high yields of bio-oils, whereas the slow heating rates and longer residence times of slow pyrolysis lead to a high amount of char products. The hydrothermal liquefaction, on the other hand, results in a very high yield of bio-oils although the process is performed with hot compressed water which adds significant energy and cost.

Compared to this, fast and flash pyrolysis processes are found to be good substitutes with high bio-oil yields. The bio-oils produced from pyrolysis cannot be used directly as transportation fuel due to their heterogeneous chemical and elemental composition. The bio-oil in crude form is highly oxygenated with high moisture content, corrosiveness, thermally unstable and acidic. In order to improve its fuel qualities, these bio-oils necessitate upgrading through hydrodeoxygenation, emulsification or steam reforming, with the aid of suitable catalysts. Catalytic biomass pyrolysis is another way to upgrade the quality of bio-oils, where the catalysts are added with the feedstock prior to pyrolysis reactions.

With the growing industrial interest in the production of biofuels, the most important areas that need attention and research seem to be: (i) scale-up, (ii) cost efficiency, (iii) better fuel properties, (iv) norms and standards for producers and end-users, (v) environment health and safety issues in biomass/biofuel handling, transportation and usage, (vi) encouragement to implement thermochemical processes and applications, (vii) efficient utilization of byproducts for value-added chemical or material production, and (viii) information dissemination. The biomass conversion process can be economically viable if used in an integrated manner for generation of other marketable co-products in addition to the primary biofuel product, thus contributing to sustainable development.

KEYWORDS

- Bio-oil
- Biochar
- Bioconversion
- Biomaterials
- Catalytic upgrading
- Cellulose
- Depolymerization
- Energy-efficient processes
- Enzymatic digestion
- Fatty acids
- Gasification
- Hydrolysis
- Hydrothermal synthesis
- Lignocellulosic biomass
- Liquefaction
- Microalgae
- Polycyclic products
- Pyrolysis
- Residence time
- Solvolysis
- Switchgrass
- Thermochemical synthesis

REFERENCES

Adam, J., Blazso, M., Meszaros, E., Stocker, M., Nilsen, M. H., Bouzga, A., Hustad, J. E., Gronli. M., Oye, G. Pyrolysis of biomass in the presence of Al-MCM-41 type catalysts. *Fuel* 2005, 84, 1494–1502.

Adjaye, J. D., Katikaneni, S. P. R., Bakhshi, N. N. Catalytic conversion of a biofuel to hydrocarbons: effect of mixtures of HZSM-5 and silica-alumina catalysts on product distribution. *Fuel Process Technol* 1996, 48, 115–143.

Aho, A., Kumar, N., Eranen, K., Salmi, T., Hupa, M., Murzin, D. Y. Catalytic pyrolysis of woody biomass in a fluidized bed reactor: influence of the zeolite structure. *Fuel* 2008, 87, 2493–2501.

Aho, A., Kumar, N., Lashkul, A. V., Eranen, K., Ziolek, M., Decyk, P., Salmi, T., Holmbom, B., Hupa, M., Murzin, D. Y. Catalytic upgrading of woody biomass derived pyrolysis vapors over iron modified zeolites in a dual-fluidized bed reactor. *Fuel* 2010, 89, 1992–2000.

Akhtar, J., Kuang, S. K., Amin, N. S. Liquefaction of empty palm fruit bunch (EPFB) in alkaline hot compressed water. *Renew Energ* 2010, 35, 1220–1227.

Akpinar, O., Levent, O., Sabanci, S., Uysal, R. S., Sapci, B. Optimization and comparison of dilute acid pretreatment of selected agricultural residues for recovery of xylose. *Bioresources* 2011, 6, 4103–4116.

Amonette, J. E., Joseph, S. Characteristics of biochars: microchemical properties. In *Biochar for Environmental Management: Science and Technology*, Lehmann, J., Joseph, S. (Eds.), Earthscan: Virginia, USA, 2009, pp. 33–52.

Anastasakis, K., Ross, A. B. Hydrothermal liquefaction of the brown macro-alga *Laminaria saccharina*: effect of reaction conditions on product distribution and composition. *Bioresource Technol* 2011, 102, 4876–4883.

Antal, M. J. Jr., Gronli, M. The art, science, and technology of charcoal production. *Ind Eng Chem Res* 2003, 42, 1619–1640.

Antal, M. J. Jr., Mochidzuki, K., Paredes, L. S. Flash carbonization of biomass. *Ind Eng Chem Res* 2003, 42, 3690–3699.

Ates, F., Isikdag, M. A. Influence of temperature and alumina catalyst on pyrolysis of corncob. *Fuel* 2009, 88, 1991–1997.

Ayyachamy, M., Cliffe, F. E., Coyne, J. M., Collier, J., Tuohy, M. G. Lignin: untapped biopolymers in biomass conversion technologies. *Biomass Conv Bioref* 2013, 3, 255–269.

Azargohar, R., Dalai, A. K. Biochar as a precursor of activated carbon. *Appl Biochem Biotechnol* 2006, 129–132, 762–773.

Babich, I. V., van der Hulst, M., Lefferts, L., Moulijn, J. A., O'Connor, P., Seshana, K. Catalytic pyrolysis of microalgae to high-quality liquid bio-fuels. *Biomass Bioenerg* 2011, 35, 3199–3207.

Balabanovich, A. I., Hornung, A., Luda, M. P., Koch, W., Tumiatti, V. Pyrolysis study of halogen-containing aromatics reflecting reactions with polypropylene in a post treatment decontamination process. *Environ Sci Technol* 2005, 39, 5469–5474.

Balat, M. Production of bioethanol from lignocellulosic materials via the biochemical pathway: a review. *Energ Convers Manage* 2011, 52, 858–875.

Beis, S. H., Onay, O., Kockar, O. M. Fixed-bed pyrolysis of safflower seed: influence of pyrolysis parameters on product yields and compositions. *Renew Energ* 2002, 26, 21–32.

Biller, P., Riley, R., Ross, A. B. Catalytic hydrothermal processing of microalgae: decomposition and upgrading of lipids. *Bioresource Technol* 2011, 102, 4841–4848.

Biller, P., Ross, A. B. Potential yields and properties of oil from the hydrothermal liquefaction of microalgae with different biochemical content. *Bioresource Technol* 2011, 102, 215–225.

Boehman, A. L., Morris, D., Szybist, J., Esen, E. The impact of the bulk modulus of diesel fuels on fuel injection timing. *Energ Fuel* 2004, 18, 1877–1882.

Boucher, M. E., Chaala, A., Roy, C. Bio-oils obtained by vacuum pyrolysis of softwood bark as a liquid fuel for gas turbines. Part I: properties of bio-oil and its blends with methanol and a pyrolytic aqueous phase. *Biomass Bioenerg* 2000, 19, 337–350.

Brewer, C. E., Schmidt-Rohr, K., Satrio, J. A., Brown, R. C. Characterization of biochar from fast pyrolysis and gasification systems. *Environ Prog Sustain Energy* 2009, 28, 386–396.

Bridgwater, A. V., Peacocke, G. V. C. Fast pyrolysis processes for biomass. *Renew Sust Energ Rev* 2000, 4, 1–73.

Brown, D., Gassner, M., Fuchino, T., Marechal, F. Thermo-economic analysis for the optimal conceptual design of biomass gasification energy conversion systems. *Appl Ther Eng* 2009, 29, 2137–2152.

Brown, T. M., Duan, P., Savage, P. E. Hydrothermal liquefaction and gasification of *Nannochloropsis* sp. *Energ Fuel* 2010, 24, 3639–3646.

Brunner, G. Near critical and supercritical water. Part I. Hydrolytic and hydrothermal processes. *J Supercrit Fluids* 2009, 47, 373–381.

Calvo, L., Vallejo, D. Formation of organic acids during the hydrolysis and oxidation of several wastes in sub- and supercritical water. *Ind Eng Chem Res* 2002, 41, 6503–6509.

Chen, W. H., Kuo, P. C. Isothermal torrefaction kinetics of hemicellulose, cellulose, lignin and xylan using thermogravimetric analysis. *Energy* 2011a, 36, 6451–6460.

Chen, W. H., Kuo, P. C. Torrefaction and co-torrefaction characterization of hemicellulose, cellulose and lignin as well as torrefaction of some basic constituents in biomass. *Energy* 2011b, 36, 803–811.

Cheng, L., Ye, X. P., He, R., Liu, S. Investigation of rapid conversion of switch grass in subcritical water. *Fuel Process Technol* 2009, 90, 301–311.

Chheda, J. N., Dumesic, J. A. An overview of dehydration, aldol-condensation and hydrogenation processes for production of liquid alkanes from biomass-derived carbohydrates. *Catal Today* 2007, 123, 59–70.

Chheda, J. N., Huber, G. W., Dumesic, J. A. Liquid-phase catalytic processing of biomass-derived oxygenated hydrocarbons to fuels and chemicals. *Angew Chem Int Ed* 2007a, 38, 7164–7183.

Chheda, J. N., Roman-Leshkova, Y., Dumesic, J. A. Production of 5-hydroxymethylfurfural and furfural by dehydration of biomass-derived mono- and poly-saccharides. *Green Chem* 2007b, 9, 342–350.

Cui, X. J., Antonietti, M., Yu, S. H. Structural effects of iron oxide nanoparticles and iron ions on the hydrothermal carbonization of starch and rice carbohydrates. *Small* 2006, 2, 756–759.

Czernik, S., Bridgwater, A. V. Overview of applications of biomass fast pyrolysis oil. *Energ Fuel* 2004, 18, 590–598.

Dahman, N., Dinjus, E., Henrich, E. Synthesis gas from biomass – problems and solutions *en route* to technical realization. *Oil Gas European Magazine* 2007, 1, 31–34.

Demirbas, A. Yields of hydrogen-rich gaseous producto via pyrolysis from selected biomass samples. *Fuel* 2001, 80, 1885–1891.

Demirbas, A. Recent advances in waste processing technologies for upgrading of synthetic fuels. *Energy Edu Sci Technol* 2004, 13, 1–12.

Demirbas, A. Biorefineries: current activities and future developments. *Energ Convers Manage* 2009, 50, 2782–2801.

Demirbas, A. The influence of temperature on the yields of compounds existing in bio-oils obtained from biomass samples via pyrolysis. *Fuel Process Technol*. 2007, 88, 591–597.

Du, W., Ren, X., Xu, M., Zhou, A. Influencing factors in hydrolysis of sunflower stalks by using dilute acid. *Energy Procedia* 2012, 17, 1468–1475.

Eken-Saracoglu, N., Mutlu, S. F., Dilmac, G., Cavusoglu, H. A comparative kinetic study of acidic hemicellulose hydrolysis in corn cob and sunflower seed hull. *Bioresource Technol* 1998, 65, 29–33.

French, R., Czernik, S. Catalytic pyrolysis of biomass for biofuels production. *Fuel Process Technol* 2010, 91, 25–32.

Goyal, H. B., Seal, D., Saxena, R. C. Bio-fuels from thermochemical conversion of renewable resources: a review. *Renew Sust Energ Rev* 2008, 12, 504–517.

Gronowska, M., Joshi, S., MacLean, H. L. A review of U.S. and Canadian biomass supply studies. *BioResources* 2009, 4, 341–369.

Henrich, E., Dahmen, N., Raffelt, K., Stahl, R., Weirich, F. The Karlsruhe "Bioliq" process for biomass gasification. *2nd European Summer School on Renewable Motor Fuels*, Warsaw: Poland, 2007, 1–23.

Hong-Yu, L., Yong-Jie, Y., Zheng-Wei, R. Online upgrading of organic vapors from the fast pyrolysis of biomass. *J Fuel Chem Technol* 2008, 36, 666–671.

Huber, G. W., Iborra, S., Corma, A. Synthesis of transportation fuels from biomass: chemistry, catalysts, and engineering. *Chem Rev* 2006, 106, 4044–4098.

Ikura, M., Stanciulescu, M., Hogan, E. Emulsification of pyrolysis derived bio-oil in diesel fuel. *Biomass Bioenerg* 2003, 24, 221–232.

Jackson, M. A., Compton, D. L., Boateng, A. A. Screening heterogeneous catalysts for the pyrolysis of lignin. *J Anal Appl Pyrolysis* 2009, 85, 226–230.

Joshi, J., Lawal, A. Hydrodeoxygenation of pyrolysis oil in a microreactor. *Chem Eng Sci* 2012, 74, 1–8.

Junming, X., Jianchun, J., Jie, C., Yunjuan, S. Biofuel production from catalytic cracking of woody oils. *Bioresource Technol* 2010, 101, 5586–5591.

Karagoz, S., Bhaskar, T., Muto, A., Sakata, Y., Oshiki, T., Kishimoto, T. Low-temperature catalytic hydrothermal treatment of wood biomass: analysis of liquid products. *Chem Eng J* 2005, 108, 127–137.

Knezevic, D., van Swaaij, W., Kersten, S. Hydrothermal conversion of biomass. II. Conversion of wood, pyrolysis oil, and glucose in hot compressed water. *Ind Eng Chem Res* 2010, 49, 104–112.

Kong, L., Li, G., Wang, H., He, W., Ling, F. Hydrothermal catalytic conversion of biomass for lactic acid production. *J Chem Technol Biotechnol* 2008, 83, 383–388.

Kruse, A., Dinjus, E. Hot compressed water as reaction medium and reactant: properties and synthesis reactions. *J Supercrit Fluids* 2007, 39, 362–380.

Kruse, A., Vogel, G. H. Chemistry in near- and supercritical water. *Handbook of Green Chemistry* 2010, 4, 457–475.

Kumar, P., Barrett, D. M., Delwiche, M. J., Stroeve, P. Methods for pretreatment of ligno-cellulosic biomass for efficient hydrolysis and biofuel production. *Ind Eng Chem Res* 2009, 48, 3713–3729.

Lenihan, P., Orozco, A., O'Neill, E., Ahmad, M. N. M., Rooney, D. W., Walker, G. M. Dilute acid hydrolysis of lignocellulosic biomass. *Chem Eng J* 2010, 156, 395–403.

Li, X., Hayashi, J., Li, C. Z. Volatilization and catalytic effects of alkali and alkaline earth metallic species during the pyrolysis and gasification of Victorian brown coal. Part VII. Raman spectroscopic study on the changes in char structure during the catalytic gasifi-cation in air. *Fuel* 2006, 85, 1509–1517.

Lin, L., Yao, Y., Yoshioka, M., Shiraishi, N. Liquefaction mechanism of cellulose in the pres-ence of phenol under acid catalysis. *Carbohydr Polym* 2004, 57, 123–129.

Lu, X., Saka, S. Hydrolysis of Japanese beech by batch and semi-flow water under subcritical temperatures and pressures. *Biomass Bioenerg* 2010, 34, 1089–1097.

Maschio, G., Koufopanos, C., Lucchesi, A. Pyrolysis, a promising route for biomass utiliza-tion. *Bioresource Technol* 1992, 42, 219–231.

Matsui, T. O., Nishihara, A., Ueda, C., Ohtsuki, M., Ikenaga, N. O., Suzuki, T. Liquefaction of micro-algae with iron catalyst. *Fuel* 1997, 76, 1043–1048.

Mazaheri, H., Lee, K. T., Bhatia, S., Mohamed, A. R. Sub/supercritical liquefaction of oil palm fruit press fiber for the production of bio-oil: effect of solvents. *Bioresource Tech-nol* 2010, 101, 7641–7647.

Miller, I. J., Saunders, E. R. Reactions of acetaldehyde, acrolein, acetol, and related con-densed compounds under cellulose liquefaction conditions. *Fuel* 1987, 66, 130–135.

Megawati, A., Sediawan, W. B., Sulistyo, H., Hidayat, M. Kinetics of sequential reaction of hydrolysis and sugar degradation of rice husk in ethanol production: effect of catalyst concentration. *Bioresource Technol* 2011, 102, 2062–2067.

Mohan, D., Pittman, C. U. Jr., Steele, P. H. Pyrolysis of wood/biomass for bio-oil: a critical review. *Energ Fuel* 2006, 20, 848–889.

Mohanty, P., Nanda, S., Pant, K. K., Naik, S., Kozinski, J. A., Dalai, A. K. Evaluation of the physiochemical development of biochars obtained from pyrolysis of wheat straw, timothy grass and pinewood: effects of heating rate. *J Anal Appl Pyrolysis* 2013, 104, 485–493.

Mohanty, P., Pant, K. K., Mittal, R. Hydrogen generation from biomass materials: challenges and opportunities. *Energ Environ* 2015, 4, 139–155.

Mortensen, P. M., Grunwaldt, J.-D., Jensen, P. A., Knudsen, K. G., Jensen, A. D. A review of catalytic upgrading of bio-oil to engine fuels. *Appl Catal A: Gen* 2011, 407, 1–19.

Nanda, S., Azargohar, R., Kozinski, J. A., Dalai, A. K. Characteristic studies on the pyrolysis products from hydrolyzed Canadian lignocellulosic feedstocks. *Bioenerg Res* 2014a, 7, 174–191.

Nanda, S., Mohammad, J., Reddy, S. N., Kozinski, J. A., Dalai, A. K. Pathways of lignocellu-losic biomass conversion to renewable fuels. *Biomass Conv Bioref* 2014b, 4, 157–191.

Nanda, S., Mohanty, P., Pant, K. K., Naik, S., Kozinski, J. A., Dalai, A. K. Characterization of North American lignocellulosic biomass and biochars in terms of their candidacy for alternate renewable fuels. *Bioenerg Res* 2013, 6, 663–677.

Ngo, T. -A., Kim, J., Kim, S. K., Kim, S. -S. Pyrolysis of soybean oil with H-ZSM5 (Proton-exchange of Zeolite Socony Mobil #5) and MCM41 (Mobil Composition of Matter No. 41) catalysts in a fixed-bed reactor. *Energy* 2010, 35, 2723–2728.

Nilsen, M. H., Antonakou, E., Bouzga, A., Lappas, A., Mathisen, K., Stocker, M. Investiga-tion of the effect of metal sites in Me–Al-MCM-41 (Me = Fe, Cu or Zn) on the catalytic behavior during the pyrolysis of wooden based biomass. *Micropor Mesopor Mat* 2007, 105, 189–203.

Ozcimen, D., Ersoy-Mericboyu, A. Characterization of biochar and bio-oil samples obtained from carbonization of various biomass materials. *Renew Energ* 2010, 35, 1319–1324.

Ozturk, I., Irmak, S., Hesenov, A., Erbatur, O. Hydrolysis of kenaf (*Hibiscus cannabinus* L.) stems by catalytical thermal treatment in subcritical water. *Biomass Bioenerg* 2010, 34, 1578–1585.

Palash, S. M., Kalam, M. A., Masjuki, H. H., Masum, B. M., Fattah, I. M. R., Mofijur, M. Impacts of biodiesel combustion on NO_x emissions and their reduction approaches. *Renew Sust Energ Rev* 2013, 23, 473–490.

Park, H. J., Park, K. -H., Jeon, J. -K., Kim, J., Ryoo, R., Jeong, K. -E., Park, S. H., Park, Y. -K. Production of phenolics and aromatics by pyrolysis of miscanthus. *Fuel* 2012, 97, 379–384.

Pauly, M., Keegstra, K. Cell-wall carbohydrates and their modification as a resource for bio-fuels. *Plant J* 2008, 54, 559–568.

Pattiya, A., Titiloye, J. O., Bridgwater, A. V. Fast pyrolysis of cassava rhizome in the presence of catalysts. *J Anal Appl Pyrolysis* 2008, 81, 72–79.

Pourali, O., Asghari, F. S., Yoshida, H. Sub-critical water treatment of rice bran to produce valuable materials. *Food Chem* 2009, 115, 1–7.

Rahman, S. H. A., Choudhury, J. P., Ahmad, A. L., Kamaruddin, A. H. Optimization studies on acid hydrolysis of oil palm empty fruit bunch fiber for production of xylose. *Bioresource Technol* 2007, 98, 554–559.

Romero, I., Ruiz, E., Castro, E., Moya, M. Acid hydrolysis of olive tree biomass. *Chem Eng Res Des* 2010, 88, 633–640.

Rostrup-Nielsen, J. R., Sehested, J., Norskov, J. K. Hydrogen and synthesis gas by steam- and CO_2 reforming. *Adv Catal* 2002, 47, 65–139.

Schroder, E., Thomauske, K., Weber, C., Hornung, A., Tumiatti, V. Experiments on the generation of activated carbon from biomass. *J Anal Appl Pyrolysis* 2007, 79, 106–111.

Sinag, A., Gulbay, S., Uskan, B., Canel, M. Biomass decomposition in near critical water. *Energ Convers Manage* 2010, 51, 612–620.

Sipila, K., Kuoppala, K., Fagernas, L., Oasmaa, A. Characterization of biomass–based flash pyrolysis oils. *Biomass Bioenerg* 1998, 14, 103–113.

Sukumaran, R. K., Surender, V. J., Sindhu, R., Binod, P., Janu, K. U., Sajna, J. V., Rajasree, K. P., Pandey, A. Lignocellulosic ethanol in India: prospects, challenges and feedstock availability. *Bioresource Technol* 2010, 101, 4826–4833.

Sun, P., Heng, M., Sun, S. Direct liquefaction of paulownia in hot compressed water: influence of catalysts. *Energy* 2010, 35, 5421–5429.

Tanksale, A., Beltramini, J. N., Lu, G. M. A review of catalytic hydrogen production processes from biomass. *Renew Sust Energ Rev* 2010, 14, 166–182.

Titirici, M. M., Thomas, A., Antonietti, M. Back in the black: hydrothermal carbonization of plant material as an efficient chemical process to treat the CO_2 problem? *New J Chem* 2007a, 31, 787–789.

Titirici, M. M., Thomas, A., Antonietti, M. Replication and coating of silica templates by hydrothermal carbonization. *Adv Funct Mater* 2007b, 17, 1010–1018.

Torri, C., Lesci, I. G., Fabbri, D. Analytical study on the pyrolytic behavior of cellulose in the presence of MCM-41 mesoporous materials. *J Anal Appl Pyrolysis* 2009, 85, 192–196.

Trane, R., Dahl, S., Skjoth-Rasmussen, M. S., Jensen, A. D. Catalytic steam reforming of bio-oil. *Int J Hydrogen Energ* 2012, 37, 6447–6472.

Triantafyllidis, K. S., Iliopoulou, E. F., Antonakou, E. V., Lappas, A. A., Wang, H., Pinnavaia, T. J. Hydrothermally stable mesoporous aluminosilicates (MSU-S) assembled from zeolite seeds as catalysts for biomass pyrolysis. *Micropor Mesopor Mat* 2007, 99, 132–139.

USEIA, U.S. Energy Information Administration. *Independent Statistics and Analysis.* International Energy Outlook 2011, www.eia.gov/ieo, Accessed on 30th June, 2013.

van der Stelt, M. J. C., Gerhauser, H., Kiel, J. H. A., Ptasinski, K. J. Biomass upgrading by torrefaction for the production of biofuels: a review. *Biomass Bioenerg* 2011, 35, 3748–3762.

Vitolo, S., Bresci, B., Seggiani, M., Gallo, M. G. Catalytic upgrading of pyrolytic oils over HZSM-5 zeolite: behavior of the catalyst when used in repeated upgrading–regenerating cycles. *Fuel* 2001, 80, 17–26.

Wahyudiono, T., Sasaki, M., Goto, M. Recovery of phenolic compounds through the decomposition of lignin in near and supercritical water. *Chem Eng Process* 2008, 47, 1609–1619.

Wang, D., Xiao, R., Zhang, H., He, G. Comparison of catalytic pyrolysis of biomass with MCM-41 and CaO catalysts by using TGA–FTIR analysis. *J Anal Appl Pyrolysis* 2010, 89, 171–177.

Wu, L., Guo, S., Wang, C., Yang, Z. Production of alkanes (C_7–C_{29}) from different part of poplar tree via direct deoxy-liquefaction. *Bioresource Technol* 2009, 100, 2069–2076.

Xiang, Q., Kin, J. S., Lee, Y. Y. A comprehensive kinetic model of dilute-acid hydrolysis of cellulose. *Appl Biochem Biotechnol* 2003, 105–108, 337–352.

Yang, Y., Brammer, J. G., Ouadi, M., Samanya, J., Hornung, A., Xu, H. M., Li, Y. Characterization of waste derived intermediate pyrolysis oils for use as diesel engine fuels. *Fuel* 2013, 103, 247–257.

Yin, S., Dloan, R., Harrison, M., Tan, Z. Subcritical hydrothermal liquefaction of cattle manure to bio-oil: effects of conversion parameters on bio-oil yield and characterization of bio-oil. *Bioresource Technol* 2010, 101, 3657–3664.

Yorgun, S., Simsek, Y. M. Catalytic pyrolysis of *Miscanthus × giganteus* over activated alumina. *Bioresource Technol* 2008, 99, 8095–8100.

Zaldivar, J., Martinez, A., Ingram, L. O. Effect of selected aldehydes on the growth and fermentation of ethanologenic *Escherichia coli*. *Biotechnol Bioeng* 1999, 65, 24–33.

Zhang, Q., Chang, J., Wang, T., Xu, Y. Review of biomass pyrolysis oil properties and upgrading research. *Energ Convers Manage* 2007, 48, 67–92.

CHAPTER 13

ADVANCES IN ECO-FRIENDLY PRE-TREATMENT METHODS AND UTILIZATION OF AGRO-BASED LIGNOCELLULOSES

MOHD AZMUDDIN ABDULLAH,[1] MUHAMMAD SHAHID NAZIR,[2] HUMA AJAB,[3] SAFOURA DANESHFOZOUN,[4] and SAKINATU ALMUSTAPHA[4]

[1]*Institute of Marine Biotechnology, Universiti Malaysia Terengganu, 21030 Kuala Terengganu, Terengganu, Malaysia*

[2]*Department of Chemical Engineering, COMSATS, Institute of Information Technology, Lahore, 54000 Punjab, Pakistan*

[3]*Department of Chemistry, COMSATS Institute of Information Technology, 22060 Abbottabad, Pakistan*

[4]*Department of Chemical Engineering, Universiti Teknologi Petronas, 32610 Seri Iskandar, Perak, Malaysia*

CONTENTS

13.1 INTRODUCTION

Biomass contributes 12% of total world energy demand, which may go as high as 40–50% in developing countries (Thiam and Bhatia, 2008).

This covers both the animals and plants origin and their remains. Agro-biomass or lignocelluloses constitute virgin wood, agricultural residues, food wastes and agro-industrial wastes. The total world biomass production of lignocellulosic material including herbaceous and woody crops is approximately 10 Mg ha^{-1} per year, with the temperate and sub-tropical rain forests contributing approximately 8 to 10–22 Mg ha^{-1} per year (Perlack et al., 2005). In the US, the annual biomass generated in millions tonnes on dry weight basis are: agricultural wastes (428), forest residues (370), power crops (377), cereals (87), oil seeds (48), municipal and industrial wastes (58). Wood or forest residues account for 30% of this total biomass production (Nieuwolt et al., 1982). The soft woody evergreen plants, classified based on gymnosperm, possess light fibers with low densities and fast growth. The hard woody plants, classified based on angiosperm, have comparatively higher-density fibers, and found commonly in deciduous habitat.

Lignocellulose is compact where cellulose is the backbone structure with parallel-rod like reinforcing material along with the deposition of hemicellulose and lignin. Total cellulose production makes up 10^{11}–10^{12} tonnes per year (Yuan et al., 2010), and is annually consumed at 7.5×10^{10} tonnes (Jacqueline and Kroschwitz, 2001). The special properties of cellulose include hydrophilicity, chirality, biodegradability, capacity for broad chemical modification, and the formation of versatile semicrystalline fiber morphologies (Klemm et al., 2005; Zhu et al., 2006). Celluloses and hemicelluloses are the more high valued lignocellulosic fractions as cheaper sources for biorefineries. Being one of the oldest polymers, cellulose has been involved in industrial production of textiles, plastics and food additives. It's most important application is in the holding of water with increasing importance for bioethanol, pulp and paper and biobleaching (Nazir, 2013). Commercialized cellulose and cellulose products from microcrystalline or nanofibrils include cellulose acetate, butyrate or propionate, methyl-, ethyl, hydroxyethyl-, ethyl hydroxyethyl-, cetyl hydroxyethyl-, hydroxyethyl methyl-, hydroxypropyl-, hydroxypropyl methyl-, methyl hydroxyethyl-, and nitrocelluloses. These have applications in many fields such as biomedical applications for cardiovascular surgery and bandages; for cancer therapies and hyperlipidemia; dermatological drugs, skin care cosmetics, and surgical implants; manufacturing of polymers such as acetate yarn, fibers, cellulosics and non-wovens; paint and coating such as methacrylates for high performance emulsifier, anticake agent, stabilizer, dispersing agent, gelling agent, fire retardant and thickener; civil and structural materials and composites such as shingle roofing, metal flashing, and automotive flooring (Nazir, 2013). Cellulose-modified carbon electrode enhances the detection of heavy metal ions (Almustapha et al., 2014).

Hemicellulose is considerably cross linked and is a combination of several polysaccharides with the extent of polymerization and orientation less than cellulose (Beg et al., 2001). The applications include in the production of resins and plastics, pharmaceuticals, furfurylalcohol, monochloroacetic acid, herbicides, tetrahydrofufurylic alcohol and maleic anhydride. Some co-products like xylans, mannans and galactans extracted from hemicelluloses are used as thickeners, stabilizers and emulsifiers, and as gels in the food, pharmaceutical and paper industries (Reddy and Yang, 2005). Different types of lignocellulosic fibers such as long, short and particulate fibers have been blended with starch, polylactic acid (PLA), polyhydroxyl alkanoates (PHA) and polyhydroxyl butyrates (PHB), polyolefin and polypropylene (Yu, 2009). Composites fabrication by compression and injection molding have been reported for wood fibers and cellulose in polypropylene especially fibers of jute, kenaf, wheat straw, bamboo, sugar cane bagasse, coir, rice husks, pineapple leaf and oil palm (Malkapuram et al., 2009).

Agro-based industries also produce 5×10^6 metric tonnes of lignin per year, making it the second main structural unit after cellulose. The major fraction of isolated lignin is burnt as fuel for energy (Fitz Patrick et al., 2010). Lignin is considered a low valued fraction due to low reactivity of lignin, a result of complex structure and variety of monomer units (Fitz Patrick et al., 2010). Lignin is normally discarded from wood during pulp and papermaking operation although it is a co-product. Depending on the plant resources and also the isolation protocol, different types of lignin exist with many industrial applications (Stewart, 2008). Lignin can be applied as binder, dispersant agent for pesticides and herbicides, emulsifier, as heavy metal sequestrate or curative substance of concrete, antioxidant, asphalt, carbon fiber, board binder, foams – plastics/polymers, battery, fuel and grease. For chemical applications, phenols, acetic acid, charcoal and ethylene can be produced from lignin. Lignin depolymerization provides routes to cresols, catechols, resorcinols, quinones, vanillin and guaiacols (Lora and Glasser, 2002).

13.2 THE CHEMISTRY OF LIGNOCELLULOSES

Complex lignocellulosic composite structure of cellulose, hemicellulose and lignin are held together through carbon-carbon (C-C), ether (C-O-C), ester and hydrogen (H)-bonding. Intra-lignocellulose bonding plays an important role in interacting and holding cellulose, hemicellulose and lignin as shown in Figure 13.1 (Douglas et al., 2012). Each type of interacting force is a characteristic of polymer molecule such as hemicellulose possessing C-O-C bond, lignin showing C-C and C-O-C bond, and cellulose exhibiting H-bond. There are three types of polymeric chain having different types of bond connectivity: (i) the interaction between hemicelluloses-lignin polymeric chain with C-O-C and ester linkage; (ii) cellulose-lignin polymeric chain with C-O-C linkage; and (iii) cellulose-hemicellulose, lignin-hemicellulose and cellulose-lignin with H-bonding (Faulon et al., 1994).

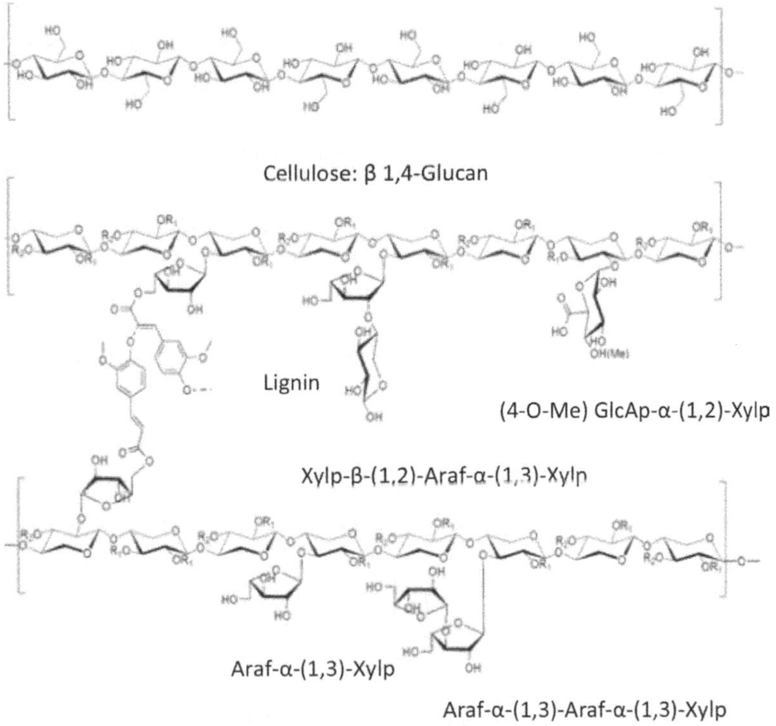

Cellulose: β 1,4-Glucan

Lignin

(4-O-Me) GlcAp-α-(1,2)-Xylp

Xylp-β-(1,2)-Araf-α-(1,3)-Xylp

Araf-α-(1,3)-Xylp

Araf-α-(1,3)-Araf-α-(1,3)-Xylp

FIGURE 13.1   Lignocellulose composite structure (Faulon et al., 1994).

13.2.1   HEMICELLULOSE

Hemicelluloses are non-cellulosic parts at a much lower molecular weight than cellulose. The branched chain, proposed by Schulze in 1891, is synthesized by the biosynthetic pathway different from that of cellulose (Sjöström, 1981). These are short-chained polysaccharides with 80–200 monomer units (Figure 13.2) (Balat et al., 2009), serving as cementing materials to bind cellulose, lignin and protein in the cell wall through ionic, covalent and H-bonding interactions (Sjöström, 1981). Hemicelluloses are the second most abundant organic chemicals after cellulose, composed of pentoses (xylose, arabinose), hexoses (mannose, glucose, galactose) and 4-O-ethylglucronic acid, and galacturonic acid residues (Balat et al., 2009). The quantitative and qualitative distribution of hemicellulose vary in different species and even in different parts of the same species. The typical formula of hardwood hemicellulose is described by gluconoxylene (O-acetyl-4-O-methylglucurono-β-D-xylan) and its content may vary from 15–30% of the material dry weight (Sjöström, 1981). Glucomannan (GM) makes up 2–5% of hemicelluloses, linked by β-glucopyranose through (1→4)-bonds.

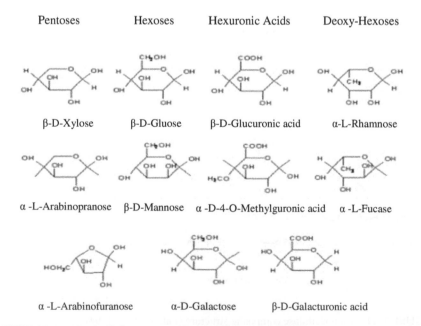

FIGURE 13.2   Hemicellulose monomer units (Balat et al., 2009).

The ratio of glucose and mannose in mannan varies at 1:1 or 1:2, depending on the source of the extraction, but galactose is totally absent (Sjöström, 1981). Xyloglucans are structural parts of the primary cell wall of hemicelluloses in some higher plants and constitute 2–5% in lower plants such as grasses (Alen, 2000). The function of xyloglucans is to bind and generate cross links between cellulose microfibrils, through H-bonding between resident hydroxyl groups (Carpita and Gibeaut, 1993).

The soft-wood hemicelluloses are composed of acetylated galactoglucomannans as major component of about 10% on a dry basis. In skeletal hemicelluloses, the polymeric chain carbon hydroxyl groups at C-2 and C-3 are partially substituted by acetyl functionality. The acetylgalactoglucomannans are classified based on the galactose constituents – at 5–8% of wood dry mass considered as low galactose; and at 10–15% of wood dry mass as high galactose. The degree of polymerization varies from 100–150, corresponding to 16000–24000 molecular weight (Scheller and Ulvskov, 2010). Arabinoglucuronoxylan constitutes 5–10% of non-woody materials. The main polymeric skeleton is composed of 4-O-methyl-D-glucuronic acid and an α-L-arabinofuranosyl unit at carbons C2, C3 through (1→4)-D-xylopranose. The furanosidic side chain is instantly hydrolyzed by acidic reagents, but arabinose and uronic acid protect xylan from alkali attack (Sjöström, 1981). Arabinogalactan is highly branched and the low viscosity makes the larches polysaccharide water soluble with the molecular weight ranges from 10000 to 120000 Da (Sjöström, 1981).

13.2.2 LIGNIN

The word 'lignin' comes from the latin word *lignum* which means wood. The English name is invented by Imison in 1822 for non-divisible wood samples after consecutive boiling in water and alcohol. In 1832, de Candolle uses the french term 'ligneuse' as a composite for cellulose matrix in lignin. In 1838, Payen reports that the lignocellulosic biomass can be segregated into isomeric starch (cellulose) and a carbon-enriched (lignin) fraction by several combinations of ammonia with sulphuric acid. The isolated carbon-enriched fraction is first described as lignin by Schulze in 1856 (Lundqvist et al., 2002). Lignin is defined as a condensed polymeric cementing material made up of C9-phenylpropane (*a*) monomeric unit derivatives such as syringyl alcohol (*s*), guaiacyl alcohol (*g*) and *p*-coumaryl alcohol (*p*) (Figure 13.3).

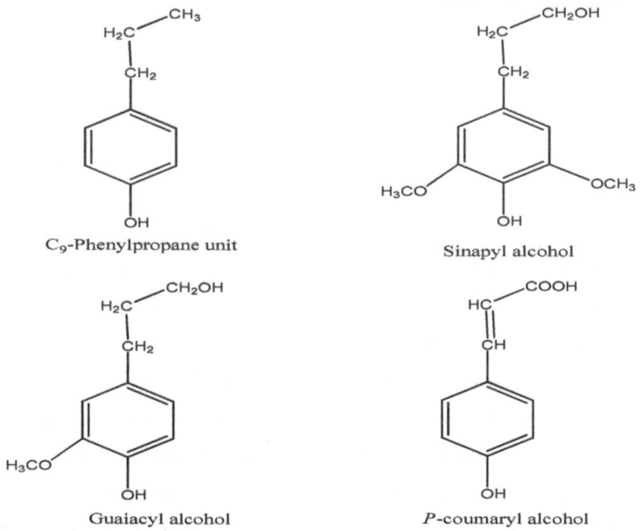

FIGURE 13.3   The C$_9$-phenylpropane (*a*), syringyl alcohol (*s*), guaiacyl alcohol (*g*) and *p*-coumaryl alcohol (*p*).

Other monomer units include sinapyl *p*-hydroxybenzoate (*SP*), β-aryl ether (*β-O-4*), phenyl coumaran (*β-5*), resinol (*β-β*), biphenyl ether (*4-O-5*), a phenolic group and a cinnamyl alcohol end group (Figure 13.4) (Stewart et al., 2009).

The three dimensional (3D) network of a lignin molecule is linked through C-C and C-O-C bond associated with hemicelluloses and celluloses in the polysaccharides. Lignin linkage starts from a methoxy phenolic monomer chain catenation with other monomeric units by condensation polymerization after the removal of water molecule and ends up with cinnamyl alcohol unit (Figure 13.4). The bonding lines in the model, shown as black bold line, depict radical joining during lignification in the cell wall, whilst the light grey lines suggest the rearrangement and rearomatization reaction. These polymeric chains result in the presence of many isomers (Ralph et al., 2001), although the sequence of the monomer units may not be strictly obeyed (Stewart et al., 2009).

13.2.3   CELLULOSE

Cellulose is the predominant constituent of lignocellulosic material and forms structural part of organic polymeric macromolecules present in prokaryotic

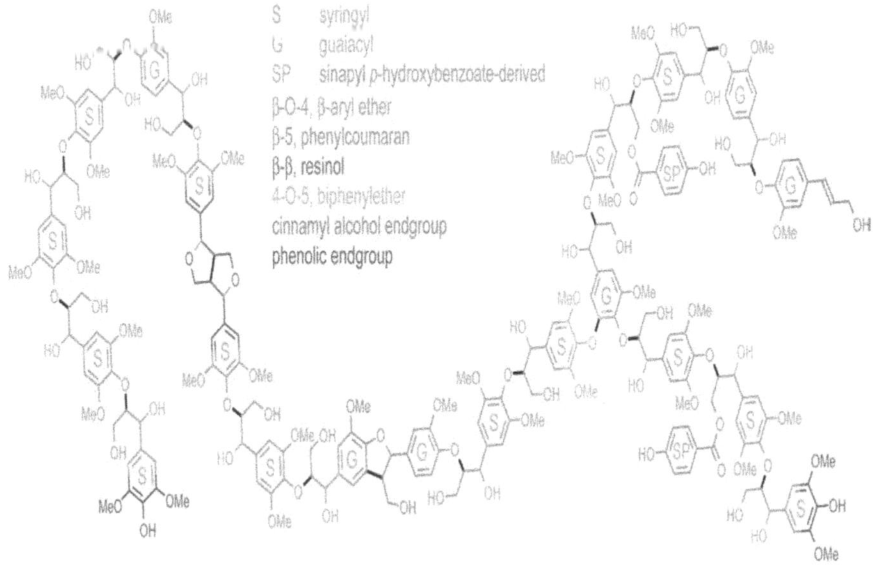

FIGURE 13.4 Lignin polymeric chain (Stewart et al., 2009).

and eukaryotic organisms. The molecular formula is $(C_6H_{10}O_5)_n$, having a linear, long chain of a hundred to a thousand 1–4-β-D-anhydroglucose units or glucosidic linkage (Abdullah et al., 2011). Cellulose from the lignocellulosic biomass has been first extracted by French Chemist, Anselme Payen (1795–1871) in 1838, where he reports its monomer composition in 1842. Hexose, a monomer unit of cellulose, is made up of six carbon atom rings, with atom numbering from carbon 1→ 6 (Figure 13.5), bearing three hydroxyl groups at C_2, C_3 and C_6 (Jean-Luc et al., 2010). The ring forming glucose, glucopyranose, has both α and β anomeric forms, coexisting at equilibrium. The α-glucose isomer carbon 1–OH stays above the ring and the β-glucose isomer –OH resides below the ring. The three –OH of the anhydroglucose molecules are responsible for the hydrophilic nature and receive more strength by intermolecular and intramolecular H-bonding. The cellulose chains are twisted and directed to certain planes and held tightly through the H-bond which is responsible for its crystallinity (Jean-Luc et al., 2010). Cellulose distribution is not only limited to photosynthetic or non-photosynthetic (protistan), but also found in animal-like Ascidians (Jean-Luc et al., 2010). Cellulose synthesis increases in human's skin during scleroderma, in diseased condition (Hall et al., 1960). Plant cellulosic fibers can be divided into lower plant and higher plant fibers (Figure 13.6).

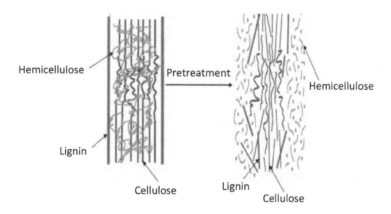

FIGURE 13.5   (a) Cellulose polymer chain, (b) Monomer unit with carbon numbering.

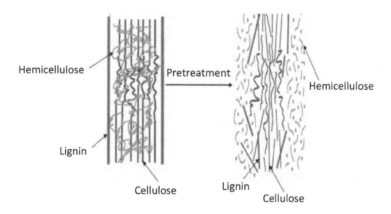

FIGURE 13.6   Pre-treatment of lignocellulosic biomass (Lejeune and Deprez, 2010).

13.2.3.1   Lower Plant Celluloses

The lower plant cellulosic fibers are obtained from bacteria, fungi and algae. Bacterial Celluloses (BC) are synthesized by strains such as *Sarcina*, *Pseudomonas*, *Aerobacter*, *Alcaligenes*, *Acetobacter*, *Rhizobium*, *Achromobacter* and *Zoogloea*. Prokaryotic, primitive *Eubacteria*, the gram positive *Sarcina*, synthesizes cellulose II as reported in 1961. All purple bacteria such as *Acetobacter*, *Rhizobium* and *Agrobacterium*, considered as the most advanced group and the genetic ancestor of eukaryotes (cells with defined mitochondria), synthesize only cellulose I, with the exception of mutant strain of *Acetobacter* which synthesizes cellulose II

(Lejeune and Deprez, 2010; Jean-Luc et al., 2010). *Acetobacter xylinum* is a strictly aerobe, non-photosynthetic bacteria where cellulose synthesis is in aerobic environment. Synthesis in algal species are not much known but group such as *Pyrrophyta, Chrysophyceae, Xanthophyceae, Phaeophyta,* and the *Chlorophyta* have been reported to produce cellulose (Perasso et al., 1989). Eukaryotic algal group including the *Oomycete, Achlya* and bisexual, show great diversity in producing cellulose synthase and cellulose. Their 18 S rRNA genetic sequences are proven by comparing with chitin-producing *Ochromonas danica* species (Gunderson et al., 1987).

Cellulose can be synthesized *in vitro* by using specific cellulose activator and those synthesized in cell membrane are secreted extracellularly through linear terminal complexes (TCs) from the bacterium outer-covering. synthesized microfibril particles are linearly arranged in parallel fashion. Minimum three particles are necessary to unite into subfibril which originate tangentially from the microfibril surface and a number of such fibrils are assembled, twisted and liaised in specific patterns to form ribbon-like structure. hydrated BC ribbons may have length up to 500Å due to several aggregations of para-crystalline fibrils at ~10 × 160 Å cross section (Jean-Luc et al., 2010). Cellulose synthesized by *Rhizobium* (the nitrifying bacterium) adhere at the tip of root meristem and causes infection. BCs may also have potential medical applications due to high porosity, with inter-pore distance ranging from 50–150 μm. Skin cells can be fabricated within the cellulose scaffold, to replace damaged blood vessels or used as template for cartilage regeneration. BC scaffold provides attachment to adhere chondrocytes for its multiplication as collagen type II substrate, assisting in the differentiation of chondrocytes and giving mechanical support for the normal cartilage growth (Lejeune and Deprez, 2010). Temporary BC mesoporous dressing keeps wound moist for quick healing.

13.2.3.2 Higher Plant Celluloses

The higher plant cellulosic fiber is sub-divided into non-woody fibers (straw, grass, bast, trunk, stem, frond, leaf, seed and fruit) and woody fibers (pine, rubber, *Acacia*, teak) (Nazir, 2013). Different plant biomass possesses different composition of cellulose (g/g) such as jute (61–71.5%), flax (71%), hemp (60.2–74.4%), ramie (68.6–76.2%), kenaf (31–39%), sisal (67–78%), pineapple leaf fiber (70–82%), henequen (77.6%), cotton seed (82.7%), rice

(28–48%), wheat (29–51%), bagasse (32–48%), bamboo (26–43%), esparto (33–38%), *Communis* (44–46%), abaca (56–63%), coniferous (40–45%), and deciduous (38–49%) (Long, 2009).

13.3 DELIGNIFICATION AND CELLULOSE EXTRACTION

The functional groups in the lignocelluloses are (a) lignins: composed of aromatic, –OH, C–C, R–O–R; (b) hemicelluloses: H-bond, RCOOR, R–O–R; and (c) cellulose: R–O–R, H-bond. The –OH groups are responsible for the substitution and enhance solubility. The H-bonding exists in cellulose polymer chain to give cellulose stability in aqueous medium, but very limited in hemicellulose due to unavailability of primary –OH group. Though considered as a non-functional group in organic compound and may not affect the molecular structure, the H-bonding has influence on solubility. Cellulose separation can be achieved by dislocating composite packed original structure, by breaking the H-bonding using physical treatment or addition of group that scavenges the H-atom. The presence of lignin covalent bond and hemicelluloses–cellulose H-bonding in lignocellulose has been reported as a weak H-bonding (Faulon et al., 1994), that hydrolysis and depolymerization of lignin could release the celluloses. Energy required to cleave the H-bond in cellulose at 8–15 kJmol^{-1} is less than to break H-bonding in water molecule at 18–21 kJmol^{-1} (Bochek, 2003).

Although the complex structure of lignin provides extra mechanical and chemical protection to the polysaccharides of the plant cell wall, it creates difficulty in hydrolyzing lignocellulosic biomass to pulp or biofuel (Vanholme et al., 2010). Within polymeric chain, the presence of the ortho and para reactive sites make it difficult to attack and open the structure. Lignin can be hydrolyzed by depolymerization into monomer and dimer units and also by reacting active sites with a phenolation mechanism (Vanholme et al., 2010). Lignin aromatic skeletal ring can be oxidized to dicarboxylic acids with oxidizing agents such as oxygen, chlorine oxides (ClOx) where side chains are also fragmented into smaller alkane with low carbon number (Jacqueline and Kroschwitz, 2001). The C-O-C bond holds glucose monomers in cellulose through glucosidic linkage and also present in lignin polymer chain. The breaking of C-O-C bond results in delignification and later the hydrolysis of cellulose to glucose. Ether bond cleavage can be initiated in acidic or basic medium. In acidic conditions, the protonation of ethereal oxygen is converted through sequential changes from hydroxyl, carbonyl or carboxylic compounds (Krassig and Schurz, 2002). basic condition provides the stability to aromatic ring and side chain of lignin which can be recovered after complete hydrolysis (Lin and Lin, 2002). The basic medium follows S_N2 mechanism for hydrogen substitution through the formation of epoxide intermediate (Solomon, 1988). The ester bond between hemicellulose–lignin, cellulose–lignin or hemicellulose–cellulose is not fully understood (Faulon et al., 1994). The cleavage products of ester bond are –COOH and –OH and the reaction is energetically favorable by alkaline hydrolysis called saponification (Solomon, 1988).

Physically and chemically compact structure needs stringent approaches to break the bonds from biomass and set free the cellulose, hemicellulose and lignin for further process. Pretreatment is the most important step in breaking down lignocellulose into fermentable constituents (Carvalheiro et al., 2008; Taherzadeh and Karimi, 2008a; Hendriks and Zeeman, 2009b; Alvira et al., 2010). Pretreatment of lignocellulosic composite breaks up cellulose, resulting in dissolution of amorphous cellulose, hemicellulose and lignin and affecting the surface area, delignification and depolymerization of hemicellulose, and the crystallinity of cellulose (Figure 13.6) (Lejeune and Deprez, 2010). It opens up and exposes lignocellulose for easy enzymatic (Mosier et al., 2005) and chemical attack (Mason et al., 1996). The pretreatment methods can be broadly divided into physical, chemical, biological and combination of methods.

13.3.1 PHYSICAL PRE-TREATMENT

Physical pre-treatment processes are employed without chemical agent and microorganisms (Zheng et al., 2014). The common methods are mechanical machines, irradiation and extrusion (Karimi et al., 2013) with the objectives to open up lignocelluloses, reducing crystallinity, and increasing surface area and porosity, for more attacking availability of post-treatment (Behera et al., 2014).

13.3.1.1 Milling

Milling cuts lignocellulosic biomass into smaller and smaller fragments using hammer mill or ball mill to get access to underneath biomass layers

(Behera et al., 2014) to ease bond breaking. Although eco-friendly in nature as no chemicals are used, milling is costly as it consumes high amount of energy to sustain the operation of machines during working hours (Chandel et al., 2013).

13.3.1.2   Irradiation

Irradiation brings the biomass into the path of ultrasound, microwave and gamma rays. Radiation reduces the bigger chain molecule to smaller molecule by breaking the inter and intra-molecular bonding and reducing crystallinity (Chandel et al., 2013) with increased biomass digestibility (Karimi et al., 2013). Microwaves are radio waves that have wavelengths of 1 m to 1 mm, with frequencies between 300 MHz to 300 GHz. It saves the reaction time (Zheng et al., 2014), but needs moisture content to be present in the biomass for the propagation of electromagnetic waves. Thermal and non-thermal effects generated by microwaves in aqueous environments are used for microwave pretreatment. The heating effect occur at the point when the waves interact with organic matters, where they get assimilated with water, fats and sugars and the energy gets transferred to organic molecules producing vast amount of heat (Banik et al., 2003). Radiations of the polar bonds take place in the surrounding aqueous medium and the biomass and within inhomogeneous material, a hot spot is created. This unique heating feature results in an explosion effect among the particles and improves the disruption of recalcitrant structures of lignocellulose (Hu and Wen, 2008). Microwave oven pre-treatment of lignocellulosic biomass is a suitable method and simple utilizing microwave oven high heating efficiency (Sarkar et al., 2012). Thermal pretreatment also provides an acidic environment for autohydrolysis by releasing acetic acid from the lignocellulosic materials.

In the non-thermal, electron beam irradiation (EBI) method, the vibration of polar bonds occur, because they are aligned to a continuously changing magnetic field. Disruption and shock to the polar bonds accelerate chemical, biological and physical processes (Sarkar et al., 2012). Cellulosic biomass undergoes changes such as an increase of specific surface area, a decrease of degree of polymerization and crystallinity of cellulose, hydrolysis of hemicellulose and partial depolymerization of lignin, due to high energy radiation (Intanakul et al., 2003). The pretreatment of rice straw by EBI at 80 kGy, 0.12 mA and 1 MeV and later hydrolyzed with 60 FPU of cellulase and

30 CBU of β-glucosidase achieve the glucose yield of 52.1% of theoretical maximum after 132 h of hydrolysis as compared to 22.6% with untreated rice straw (Bak et al., 2009). The combined method of gamma irradiation with acid hydrolysis can also significantly improve the saccharification process for bioethanol production from *Undaria* marine algae where gamma irradiation causes structure breakage of the cell wall (Yoon et al., 2012).

Ultrasound is the sound frequency higher than 18 kHz which can penetrate and be transmitted through solid, liquid and gases. The source produces vibration that travels through the medium and hits the sample which in case of gas, starts oscillating in the direction of wave, while solid and liquid oscillate normal to the wave propagation resulting in enhanced penetration (Luque de and Priego, 2007). Ultrasonic wave causes disturbance in the medium which travels as compression and expansion. Compression causes the medium molecules pushing closer to each other, while expansion pulls them away. The expansion cycles in liquid provide outward pressure with high energy that produces bubbles or cavities. When outward pressure at certain point on liquid surface increases, it breaks the surface inwardly to create cavitation. These cycles are executed continuously to generate cavitations. Adiabatic compression of gases and vapors generate high temperature and pressure at the hot spot of cavitation or bubble, and this may be at 5000°C which is equivalent to temperature on the Sun surface, and 2000 atm, the pressure equivalent to Marianas Trench (the deepest oceanic point). However, the heat is generated and dissipates instantaneously with no significant effect on the overall system that cavitation sometimes is referred to as cold boiling. The initially smaller bubble is generated in the medium and propagated in cyclic fashion to create maximum size before collapsing. When bubble collapses abruptly, it releases energy instantaneously and becomes cool at the rate of 10 billion °C per second, which is one million times faster than the red hot iron plunged into water for cooling (Luque de and Priego, 2007).

Cavitations play significant roles in the pretreatment of lignocelluloses. The cell disruption by cavitation (Figure 13.7a-d) involves steps such as (a) cavitation coming closer to the cell wall, the solid-liquid interaction takes place, (b) energetically unstable cavitation abruptly ruptures into protruded jet shape, with liquid thrust speed onto the wall surface estimated approximately at 400 kmh⁻¹, (c) and the strong effect of jet speed breaking the lignocellulose wall (Luque de and Priego, 2007). These lead to depolymerization in the form of successive cavitation formation and collapsing, generating negative high pressure onto the polymeric chain and opposite pulling effect

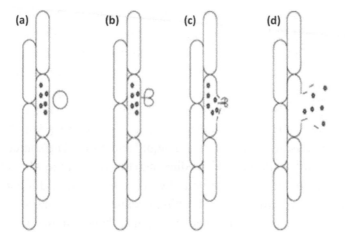

FIGURE 13.7   Cavitation (a) coming close to a plant cell wall, (b) jet release, (c) penetration into the cell, (d) rupturing the cell wall (Chemat et al., 2011).

to break the chain. The chemical pathway follows the formation of free radicals (OH•) which are energetically favorable at high temperature of cavitation. Ultrasonic treatment for depolymerization of carbohydrates such as hyaluronan and xylo-glucane is employed more frequently than the microwave technology and γ-radiation (Luque de and Priego, 2007).

13.3.2   CHEMICAL PRE-TREATMENT

Chemical pre-treatment loosens up the biomass structure under the effect of chemicals (Karimi et al., 2013), dissolving the crystalline region of cellulose and converting into amorphous cellulose (Tadesse and Luque, 2011); or removing the amorphous regions by strong acids which cleave the glycosidic bonds, to release the individual crystallites (Li et al., 2009; Spagnola et al., 2012). The method is adopted by many chemical industries like paper and pulp industries using inorganic acids and bases, catalyst, ammonia, ionic liquid, and buffer water treatment (Narayanaswamy et al., 2013).

13.3.2.1   Acid Pre-Treatment

In acid treatment, polysaccharides are hydrolyzed to monosaccharides which opens up the compact chain and allows enzyme for further fermentation

(Mood et al., 2013). The use of inorganic (mineral) acids include dilute sulphuric acid, nitric acid, hydrochloric acid, phosphoric acid (Brink, 1996; Herrera et al., 2003; Vázquez et al., 2007; Ballesteros et al., 2008) and organic acids such as peracetic acid (Zhao et al., 2007; Zhao et al., 2008), formic acid (Yang et al., 2012; Zhao and Liu, 2012), acetic (CH_3COOH) and maleic ($C_4H_4O_4$) acids. Sulphuric acid is mostly used (Zheng et al., 2014) which produce theoretically 80–90% of xyloses (Schell et al., 1992; Torget and Teh-An, 1994). The hydrolysis of biomass could be carried out at low and high temperature, using low and high concentration of acid. While effective for maximum yield, the process can be very expensive and produces hazardous by-products (Chiaramonti et al., 2012). Sulphonic acid hydrolysis for example may decrease the solubility of lignin with the formation of lignosulfonates (Jacqueline and Kroschwitz, 2001). Dilute acid is more favorable for cost effectiveness, but optimization is required for the right concentration at the right temperature for appropriate biomass.

13.3.2.2 Alkaline Pretreatment

Alkaline pre-treatment uses sodium hydroxide (NaOH), calcium hydroxide ($Ca(OH)_2$), potassium hydroxide (KOH), ammonia ($NH_{3(aq)}$), ammonium hydroxide (NH_4OH), or NaOH in combination with hydrogen peroxide (H_2O_2) (Ragauskas and Huang, 2013). These are effective for lignocellulosic material having low lignin content (Mahdy et al., 2014). Alkaline reagents could alter or open the lignocellulosic structure and produce pores which increase swelling and dissolution of lignin (Nazir et al., 2013) and freeing the cellulose and hemicelluloses for enzymatic attack (Lin et al., 2014).

13.3.2.3 Catalyzed Steam Explosion

Steam explosion increases the accessibility of cellulose but also rendering the production of compounds that may inhibit the activity of catalysts such as H_2SO_4, SO_2, and NaOH (Zheng et al., 2014). Hemicelluloses will be hydrolyzed into acetic acid due to the presence of acetyl group which could be further processed for the production of ethanol and syngas (Hendriks and Zeeman, 2009a).

13.3.2.4   Wet Oxidation

Wet oxidation uses oxidative compounds such as atomic or molecular oxygen, H_2O_2 (Zheng et al., 2014), chlorates and peroxychlorates, in aqueous medium at temperature ranging from 125 to 300°C and pressure from 0.5 to 20 MPa (Behera et al., 2014). Hydrolytic cleavage takes place at low temperature, and oxidative cleavage at high temperature (Karp et al., 2013). Pretreatment of softwood lignocellulose has been reported where 6 g of biomass milled up to 2 mm size is added into 1 L of water and heated to 170–200°C at 10–12 bar pressure for 20 min (Ragauskas and Huang, 2013).

13.3.2.5   Ozonolysis

Ozone treatment is effective for lignin dissolution from biomass and could be performed at room temperature and pressure, but requires optimum conditions of ozone concentration, feedstock particle size and flow rate, (Behera et al., 2014). Ozonolysis changes the morphology and chemical structure of biomass including peanut, cotton, sawdust, wheat straw and bagasse (Harmsen et al., 2010).

13.3.2.6   Organosolv Process

Organosolv method uses volatile organic solvents namely acetone, methyl alcohol, ethyl alcohol or non-volatile solvent such as ethylene glycol. Biomass is soaked in solvent or combination of solvents at different ratios to produce enzymatic or microbial accessible lignin-free feedstock for fermentation (Harmsen et al., 2010). Solvents mostly are recovered and reused to reduce the cost (Behera et al., 2014).

13.3.3   PHYSICO-CHEMICAL PRE-TREATMENT

Physico-chemical pretreatment is a synergistic effect of chemical reagent and physical approach. Lignin hydrolysis using NaOH is made easier with the assistance of physical treatment. Physico-chemical treatment of lignocelluloses depolymerizes and hydrolyzes hemicelluloses to monosaccharide. On continuous degradation with uncontrolled or prolonged treatment, xyloses will be generated (Yi et al., 2009). After depolymerization, mono-meric and oligomeric units show equal accessibility of attacks onto the ortho and meta carbon positions (Yuan et al., 2010). The chemical reagents used include hot water or ammonia along with heat treatment or freeze explosion (Galbe and Zacchi, 2012).

13.3.3.1 Steam Explosion

Water treatment at high temperature of 160–260°C and pressure of 0.69–4.83 MPa for a few minutes, when drops abruptly to a low temperature, causes the biomass to undergo explosion with the removal of lignin whilst the cellulose is made accessible for microbial fermentation (Kumar et al., 2009). Effectiveness depends on factors such as reaction time, process temperature, particle size of feedstock and moisture content of biomass. It removes hemicelluloses which otherwise would have become an enzyme inhibitor during fermentation (Hendriks and Zeeman, 2009a). steam explosion can be improved with the addition of acid or alkali reagents (Harmsen et al., 2010).

13.3.3.2 Freeze Explosion

In freeze explosion, NH_3 is used as a selective, non-corrosive and volatile reagent. Ammonia could be easily recycled after the completion of the treatment time due to its high volatility (Hu and Ragauskas, 2012). Biomass is dipped in liquefied anhydrous ammonia at 70–200°C and 100–400 psi. The pressure is made to undergo sudden elevation after the required time to create pores and break up the bonding in the lignocellulosic chains. This causes depolymerization, reduces the degree of crystallization and releases the hemicelluloses and lignin from the compact structure of biomass (Karp et al., 2013).

13.3.3.3 Hydrothermal Pre-Treatment

In hydrothermal process, hot water under high temperature and pressure penetrates into the compact structure of biomass and hydrolyzes the lignin, hemicelluloses and celluloses by breaking the H-bond and set the constituents free (Taherzadeh and Karimi, 2008). Hydrothermal pre-treatment is considered effective to achieve the required cellulose microbial/enzymatic digestibility due to enhanced surface area (Zheng et al., 2014).

13.3.3.4 Autoclave Treatment

The autoclave pre-treatment at 121°C and 15 psi with different chemicals and for different durations, disassemble lignocelluloses (Foston and Ragauskas, 2010; Gabriele et al., 2010; Hsu et al., 2011; Kim and Kim, 2013). Effective autoclaving treatment not only disinfects, but also opens up lignocelluloses for penetration of reagents and for hydrolysis to take place. Steam produced provides surface preparation of substrate for reagent attack, and is considered effective for further synthesis. Autoclaving with reactive chemicals may produce significant effect for sterilization and for partial and complete depolymerization of lignocelluloses to monomer constituents. Treatment duration and reagents concentration could determine the end products.

13.3.4 BIOLOGICAL PRE-TREATMENT

Biomass dissolution can be achieved in a cost-effective and less energy intensive manner by biological treatment utilizing bacteria, algae or fungi and/or enzymes which are secreted out. (Narayanaswamy et al., 2013). Fungal pre-treatment utilizes *Aspergilli* or *Neurospora, Crysosporium, Cinnabarinus, Podospora anserine*, and enzymes from fungi such as endoglucanase, cellobiohydrolase, polysaccharide monooxygenase, β-glucosidase, xylanase, β-xylosidase, acetyl esterase, feruloyl esterase, α-galactosidase (Geoffrey, 2014) to attack the biomass and selectively degrade lignin, hemicellulose, and cellulose with increased digestibility of cellulose (Sánchez and Montoya, 2013; Zheng et al., 2014). Once microorganisms such as brown-, white-, and soft-rot fungi start growing on the surface of lignocellulosic wood, degradation of celluloses, hemicelluloses and partial lignin ensues ((Potumarthi et al., 2013; Geoffrey, 2014). The rate limiting step in lignocellulosic materials fermentation is to break the lignin barrier and to disrupt the crystalline cellulose structure (Watanabe, 2013). Hence, the major drawback is the long duration to complete the solid-state fermentation process. Attention is now given to explore ways to improve the growth time, secondary metabolite production and digestibility of hemicelluloses and cellulose, and also through improved enzyme activity (Galbe and Zacchi, 2012). Lignocelluloses could also be treated with microbial consortium and microbes screened from natural environments which selectively attack and dissolve the celluloses and hemicelluloses in biomass (Zheng et al., 2014).

13.4 NON-GREEN AND GREEN EXTRACTION TECHNIQUES

13.4.1 CHLORITE METHOD

Acidified sodium chlorite ($NaClO_2^+$ H_3O^+) is used as a standard reagent for delignification and extraction of cellulose from wood materials (Wise, 1946). Chlorite (ClO_2) may produce chlorine radical, $Cl^\bullet$, which reacts and fragments the lignocellulosic material into highly toxic organochlorine compounds. Several studies have reported cellulose extraction by chlorites, in combination with alkali and heating at different duration (Table 13.1).

TABLE 13.1   Non-Green Cellulose Extraction from Various Feed Stocks

Materials	Methods	References
Soy hull	Treated with 2 % (w/v) sodium hydroxide at 100°C for 4 h with constant stirring, fiber bleaching with 1.7% (w/v) sodium chlorite in acetate buffer at 80°C for 4 h	Flauzino Neto et al. (2013)
Mengkuang leaves	Soaked in water for three days, boiling for 15 min and sun-dried. The dried fibers are ground and suspended in 4% (w/v) NaOH at 125°C for 2 h, the leave fibers further bleached with 1.7% (w/v) sodium chlorite at 125°C and pH 4.5 for 4 h	Sheltami et al. (2012)
Cocos nucifera L.	Dewaxed with toluene/ethanol (2:1 v/v) for 6 h in soxhlet. Fibers are bleached with $NaClO_2$ at 100°C for 2 h and pH adjusted to 4.0–4.2. Delignified fibers are washed with 2% $NaHSO_4$, water and ethanol several times. Hemicellulose is removed by treating with NaOH at 20°C for 45 min.	Uma Maheswari et al. (2012)
Kenaf fiber	Delignified at 80°C with 4% (w/w) NaOH for 3 h and further bleached with 1.7% acidified chlorite	Kargarzadeh et al. (2012)

TABLE 13.1 Continued

Materials	Methods	References
Rice husk	Depolymerization of hemicellulose, 4% NaOH for 2 h and delignified with 1.7% (w/w) chlorite under acetate buffer at 100–130°C for 4h	Johar et al. (2012)
Rubber saw dust	Treated with NaOH and sodium chlorite at 70°C for 2 h at pH 4.9 adjusted by 0.1M sodium acetate buffer	Kamphunthong et al. (2012)
Hardwood	NaBr, 4-acetamido-TEMPO and NaClO$_2$, with ultrasound 68, 170 kHz, 1000 W, TEMPO-chlorite at 70°C for 2 h	Shree et al. (2012)
Sugarcane	Delignification with acidified sodium chlorite at 70°C for 2 h and crude cellulose is extracted with 10% KOH for 10 h at 20–50°C KOH	Bian et al. (2012)
	Delignification with 0.7% (w/v) acidified sodium chlorite at 5 h and depolymerization of hemicellulose obtained with 17.5% (w/v) NaOH for 5 h, the white cellulose sonicated for 5 min to separate out nano-fibers	Mandal and Chakrabarty (2011)
Wood, flax, wheat straw and bamboo	Benzene/ethanol (2:1 v/v) in soxhlet for 6h. Delignification with chlorite at 75°C for 1 h and hemicellulose removed with 2% (w/v) KOH at 90°C for 2 h with ultrasonic treatment at 20–25 kHz and 1000 W	Chen et al. (2011)
Cotton	Chlorite/acetic acid, fibers are suspended in 6.5 M sulphuric acid at 45°C and 60°C for 14–20 min and further dialysis upto pH 6–7	Martins et al. (2011)
Coconut husks	Benzene/ethanol (1:2 v/v) and soxhlet-refluxed for 48 h, lignin removed with NaClO$_2$ at pH 4–5, 70°C for 1 h, hemicellulose depolymerized with 6% (w/v) KOH for 24 h, extracted cellulose dispersed in deionized water for 20 min with 200 W ultrasonication	Fahma et al. (2010)
Luffa cylindrica	Pretreatment with 4 % (w/v) NaOH at 80°C for 2 hours, bleaching with 1.7% (w/v) acidified NaClO$_2$ at 80° for 2 h	Gilberto et al. (2010)

13.4.2 GREEN METHOD

Green solvents, catalysts and media for chemical reactions and processes have become major research themes especially with regards to the reaction efficiency minus the use of toxic reagents, wastes reduction and effective utilization of resources (Disale et al., 2012). Green cellulose extraction involves the use of eco-friendly chemicals, at low temperature and pressure to make the process more economical and environmentally-friendly. Table 13.2 shows different techniques used for green and semi-green cellulose extraction. Among advanced green approaches explored are the use of supercritical fluid, ionic liquid and deep eutectic solvent.

13.4.2.1 Supercritical Fluid

Supercritical fluid is a substance at a temperature and pressure above its critical point where distinct liquid and gas phases do not exist. Supercritical fluid technology offers an interesting option for sustainable utilization of

TABLE 13.2 Green or Semi-Green Extraction of Cellulose From Various Feed Stocks

Materials	Methods	References
Oil palm empty fruit bunch	Use of 2.5 N HCl at 105°C for 30 min, neutralized with 5% NH_4OH and dried at 105°C	Haafiz et al. (2013)
	Microcrystalline cellulose extraction involving pre-treatment with soda anthraquinone method, and bleached with ozone and peroxide, later hydrolysis with 2.5 M HCl at 105°C for 15 min.	Wanrosli et al. (2011)
	Acid hydrolysis (2% H_2SO_4) with ultrasonication at 20 kHz and 2 kW with different amplitudes at 15 %, 60 % and 90 % for 15, 45 and 60 min, later acid hydrolyzed at 140°C and 2 bar.	Yunus et al. (2010)
	The 2% alkali treatment at 95°C for 60 min to yield 98% cellulose and 97.4% α-cellulose	Leh et al. (2008)
	Soda pulping of oil palm and evaluating pulping variables achieving 60% cellulose yield	Wanrosli et al. (2004)

TABLE 13.2   Continued

Materials	Methods	References
Juncus acutus	Extraction with NaOH at 6–7 M for 3 h and bleaching with H_2O_2 at 95°C for 45 min at pH 11	Ghali et al. (2012)
Rice husk	Pre-treatment with hexane/ethanol/water in soxhlet, delignified with 5% (w/v) in autoclave at 121°C. The husk is oxidized with 2% (v/v) H_2O_2 and 0.2% (v/v) (teraacetylethylenediamine) (TAED) for 12 h at 48°C. Cellulose further treated with 80% (v/v) acetic acid and 70% (v/v) HNO_3 at 120°C for 30 min. and cellulose dispersed in deionized water by ultrasonic treatment.	Rosa et al. (2012)
Bamboo	Delignification by formic acid (88% v/v), combination of formic acid (88%)-HCl (1%) v/v and formic acid (88%)-H_2O_2 (3%) v/v. Delignification achieved highest at 87.9% with formic acid-H_2O_2.	Li et al. (2012)
Switchgrass	Depolymerization, hydrodeoxygenation with 20 % (w/w) formic acid over Pt supported on carbon and H_2SO_4 hydrolysis	Xu et al. (2012); Yat et al. (2008)
Miscanthus x giganteus	Chopped feedstock treated with $Fe_2(SO_4)_3$ solution in NaOH plus H_2O_2 in formic acid at 19 bar and 35 bar pressure	Haverty et al. (2012)
Cotton fibers	Microcrystalline cellulose prepared with 2.5 to 15% HCl concentrations and refluxed for 1 h.	Nada et al. (2009)
Sisal fibers	Fibers suspended in 0.1M NaOH and ethanol at 45°C for 3 h, delignified with H_2O_2 (1–3 %) at pH 11.5 and 45°C. Hemicellulose is removed with 10% NaOH and 1% $Na_2B_4O7.10H_2O$ at room temperature for 15 h. Treated with 70% HNO3 and 80% CH_3COOH at 120°C for 15 min, filtered, washed and dried at 60°C.	Morán et al. (2008)
Barley straw	Dewaxing and alkaline extraction with 1.5% (v/v) H_2O_2 at 40°C for 14 h, further treated with 70% (v/v) HNO_3/80% CH_3COOH (1:10) at 120°C for 15 min.	Sun et al. (2005)

TABLE 13.2 Continued

Materials	Methods	References
	Pre-treatment with 1.5% (v/v) H_2O_2 at 45°C for 14 h and enzyme xylanase inoculated at 35°C for 24 h. Extracted fibers treated further with 10% NaOH at 25°C for 6 h, and crude cellulose collected.	
Wheat straw	Pre-treatment with toluene/ethanol (2:1 v/v) for 6 h and sonication treatment with 0.5 M KOH and 20 kHz for 0–35 min at 35°C and 100 W.	Run Cang and Jeremy (2002)
	Cellulose residue further bleached with 2 % (v/v) H_2O_2/0.2 % TAED at 48°C and pH 11.8 for 12 h. Cellulose purified with 70% (v/v) HNO_3/80% CH_3COOH (1:10) at 120°C for 15 min, filtered and washed at 60°C for 16 h.	Sun et al. (2004)

biomass as an alternative to conventional solvents. It presents unique physicochemical properties because of the duality between liquid and pure gas. Supercritical fluid behaves like a liquid with the viscosity of a gas, where it can diffuse through solids like a gas and dissolves material like a liquid. The thermo-physical properties such as dielectric constant, diffusivity, density or viscosity can be adjusted by altering the operating temperature and/or pressure (Knez et al., 2014) for high diffusion coefficient, high solvation power, high degree of selectivity and easy solvent separation. Carbon dioxide and water are the most commonly used supercritical fluids for various food and non-food applications. CO_2 becomes supercritical above 31°C and 73 bars, water above 374°C and 218 bar (Harmsen et al., 2013). Figure 13.8 shows the pressure-temperature diagram in the critical region of pure carbon dioxide. Supercritical carbon dioxide (SC-CO_2) is commonly used because of its non-toxicity, easy recovery after extraction, low cost, non-flammability, and environmentally-friendly (Rostagno et al., 2014). SC-CO_2 extraction displays gas-like mass transfer properties and considered as a possible pre-treatment route for lignocellulosic material. Carbon dioxide molecules are comparable in size to those of water and ammonia, and should be able to penetrate small pores accessible to water and ammonia molecules. Co-solvents such as ethanol–water or acetic acid–water, with carbon dioxide at high pressures, can be used to increase lignin removal. The co-solvent however should be used at low concentrations to a maximum of 10% (v/v) (Pasquini et al., 2005).

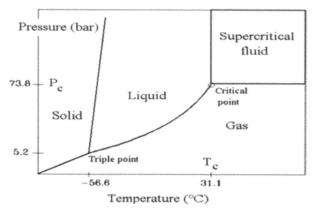

FIGURE 13.8   The phase diagram of carbon dioxide (Laitinen, 1999).

Water as a supercritical fluid is "green" with liquid-like density and gas-like transport properties, and behaves very differently than it does at room temperature. At pressure higher than its vapor saturation pressure and temperature around 423 K to 647 K (critical temperature), it becomes subcritical. The structure of water changes significantly near the critical point because of the breakage of infinite network of H-bonds where water exists as separate clusters with a chain structure. At supercritical conditions, the hydrogen bonds of water are weakened, and its dielectric constant decreases from about 78 at 25°C to the range of 2 to 20 near the critical point, which is similar to that of polar organic solvents at room temperature. It becomes highly non-polar, permitting complete solubilization of most organic compounds and oxygen (Cardenas-Toro et al., 2014). As compared to ambient water, supercritical water has 100-fold higher diffusivity but 100-fold lower viscosity, although the density may range between 0.2–0.9 g/cm^3 (Hendry, 2012).

Integrated supercritical fluid extraction and subcritical water hydrolysis for the recovery of bioactive compounds from pressed palm fiber (Cardenas-Toro et al., 2014), and the recovery of fermentable sugars with subcritical water and carbon dioxide from palm fiber and grape seed (Rostagno et al., 2014) have been reported. In supercritical water, celluloses and hemicelluloses are converted into monosaccharides, oligosaccharides and degradation products. As there is a competition between hydrolysis and product degradation rates, the composition of the raw material is of fundamental importance as it will determine the operational conditions to allow rapid conversion of polysaccharides into high saccharide production with low degradation products In lignocellulosic materials such as corn stover and sugar cane bagasse, lignin is present in a complex structure with cellulose and hemicellulose, resulting in increased resistance to the hydrolysis processes, and therefore requiring more aggressive conditions (Hendry, 2012; Farias-Campomanes et al., 2013).

13.4.2.2   Ionic Liquid

Ionic liquids (ILs) have been touted as effective and green solvents, primarily due to their high thermal and chemical stability, and miscibility with many other solvent systems. ILs are defined as organic salts that have melting points below 100°C, and are also liquids at low temperatures (Socha et al., 2014). Although often described as having non-flammable nature, ILs are known to thermally decompose at varying temperatures (Diallo et al., 2012; Hayyan et al., 2013). ILs are usually composed of large organic cations and small inorganic anions, with immeasurable combinations of anions and cations (Kubisa, 2009; da Costa et al., 2013). With uniquely beneficial and tunable properties such as negligible vapor pressure, high thermal stability, hydrophobicity, polarity and solvent power, ILs are recognized as enablers for more green applications in reactions and separations (Curnow, 2012). The ionic and non-coordinating nature of ILs allow them to dissolve combinations of organic and inorganic compounds, which aid diverse types of chemical transformation. The disintegration of lignocellulose in ILs distorts the primary bonds among cellulose, hemicellulose and lignin, providing more substrate accessibility to hydrolytic enzymes. The pre-treatment is dependent on the IL and the biomass (type, moisture, size and load), temperature, time of pre-treatment and precipitating solvent used (Gunny et al., 2013). The main procedure is by heating the lignocelluloses in IL at 100°C for a certain duration. Dissolution time depends on the IL used and the structure of the cellulose (Holm and Lassi, 2011). Anti-solvents such as water, methanol, ethanol or acetone can be used to precipitate out the dissolved cellulose in the IL and later separated by filtration. The filtered IL can be recovered and reused through the distillation of the anti-solvent (Ninomiya et al., 2015).

The mechanism for the dissolution of cellulose in ILs (Figure 13.9) suggests the interactions between the cation and anion of IL with H- and O-atoms of cellulose-OH leading to the dissolution of cellulose (Feng and

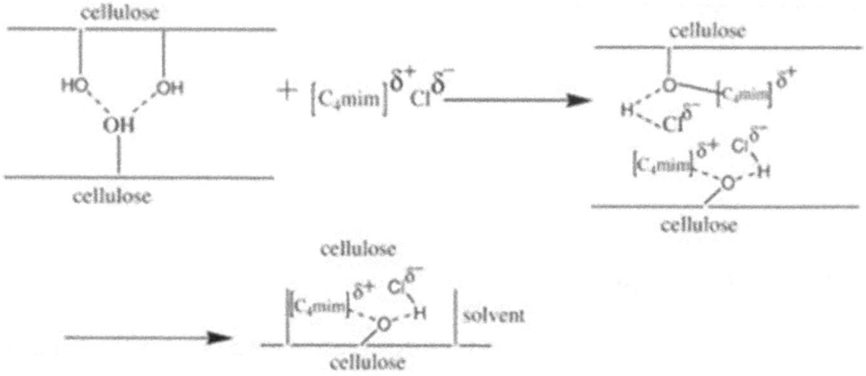

FIGURE 13.9   Dissolution mechanism of cellulose by ionic liquids (Feng and Chen, 2008).

Chen, 2008). The H-atoms act as electron acceptors and cellulose atoms serve as electron pair donors. The H- and O-atoms from hydroxyl groups are separated, upon interaction, which open up the H-bonds between the molecular chains of cellulose and as a result, the cellulose dissolves. The anion and cation side could be changed and tuned to enhance the pretreatment. Reducing the alkyl chain length in the cation side and introducing higher H-bond basicity and dipolarity on the anion side could increase the dissolution of cellulose in ILs (Gunny et al., 2013). The earlier report on cellulose dissolution in ILs combines 1-butyl-3-methyl imidazolium cation with different anions (Swatloski et al., 2002). Common examples for cellulose dissolution and biomass pretreatment include salts of organic cations such as 1-alkyl-3-methylimidazolium $[C_n\text{mim}]^+$, 1-alkyl-2,3-dimethylimidazolium $[C_n\text{mmim}]^+$, 1-allyl-3-methylimidazolium $[\text{Amim}]^+$, 1-allyl-2,3-dimethylimidazolium $[\text{Ammim}]^+$, 1-butyl-3-methylpyridinium $[C_4\text{mP}_y]^+$, and tetrabutylphosphonium $[\text{Bu4P}]^+$ (n = number of carbons in the alkyl chain) (Lee et al., 2014). The main focus for rational design of ILs with effective cellulose dissolution should be on anions having strong hydrogen bond suitability and the cations including strong acidic protons without having high electronegativity atoms such as oxygen and the bulky groups that can generate stearic hindrance. Acidic protons on the heterocyclic rings increase markedly the solubility by forming hydrogen bonds with hydroxyl and ether oxygen of cellulose (Lu et al., 2014). The existence of electronegative atoms on the cation decreases the acidity of the protons triggering a decrease in the solvation competency.

Efficient dissolution has been reported for the ones with halide counter ions like 1-butyl-3-ethyl-imidazolium chloride (BMIMCl) and other counter anions such as phosphate, formate and acetate (Swatloski et al., 2002; Gericke et al., 2012). ILs having imidazolium or pyridinium cations paired with chloride, formate, acetate or alkylphosphonate anions dissolve cellulose fibers through strong hydrogen bond basicity (Sathitsuksanoh et al., 2013). For example, 1-butyl-3-methylimidazolium acetate exhibits comparatively high solubility of 23 g/mol solubility at 40°C, but altering the cationic structure to 1-methoxyethyl-3-methylimidazolium reduces solubility dramatically to 8 g/mol solubility at the same temperature. Maximum solubility of cellulose has been established at 14.5 wt % for 1-allyl-3-methylimidazolium chloride at 80°C (Zhang et al., 2005) and 16 wt % for 1-ethyl-3-methylimidazolium acetate at 90°C (Zavrel et al., 2009), which can be improved up to 25 wt % with microwave heating. Aqueous solutions of 1-butyl-3-methylimidazolium methanesulfonate ([BMIM][MeSO$_3$]) are effective in the pre-treatment of switch grass (*Panicum virgatum*), resulting in fast saccharification of both cellulose and hemicellulose. Mixed solvents or ILs being considered for optimized biomass dissolution and conversion efficiency include cholineacetate ([Ch]Ac) and tributylmethylammonium chloride ([TBMA]Cl) (Zhang et al., 2012). The addition of an organic co-solvent such as DMSO can be utilized to boost the solvent power of the IL by reducing the time needed for dissolution, even at low temperatures (Andanson et al., 2014; Ries et al., 2014). The capability of ILs to dissolve cellulose (Swatloski et al., 2002) enables alteration of the physicochemical properties and the extraction of specific macromolecular component with different fractionation approaches after biomass dissolution. Figure 13.10 shows the possible chemical modifications that can be explored with cellulose dissolved in ILs (Isik et al., 2014). The chemical flexibility of both cellulose and ILs could pave the way for new materials and new processing technologies related to electrospinning, functionalization methods and new cellulose-based materials such as fibers, composites, blends, and ion gels (Isik et al., 2014). The major challenges that need to be overcome include fractionation and recoverability and the fact that ILs can be toxic to eco-system, microbes and marine organisms (Wood et al., 2011; Ventura et al., 2012). ILs are also expensive, difficult to synthesize, generally not biodegradable or renewable, and made from petrochemical resources (Kroon et al., 2013).

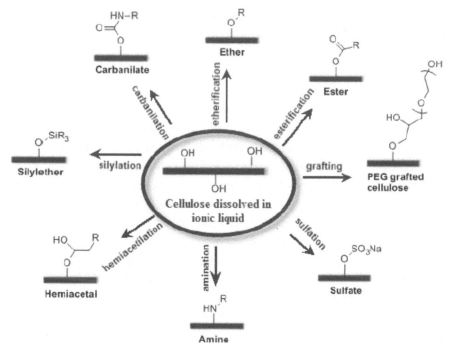

FIGURE 13.10   Possible chemical modifications to cellulose dissolved in ionic liquids (Isik et al., 2014).

13.4.2.3   Deep Eutectic Solvent

Conventional methods for biomass conversion into cellulose, hemicelluloses and lignin may require extreme and expensive techniques (steam explosion, high temperatures, strong acids/bases) but with potential product degradation and undesired side reactions such as hydroxymethylfurfural synthesis (Kroon et al., 2013). Low Transition Temperature Mixtures (LTTMs) or Deep Eutectic Solvents (DESs) are seen as alternatives to conventional ILs for lignin-containing biomass processing. The main constituents of the eutectic mixtures are solids with high melting points that show strong hydrogen bonding inter-actions (Kroon et al., 2013). The mixture can be two or more compounds with melting point lower than that of either of its components (Abbott et al., 2004; Hayyan et al., 2010). DESs can be considered as environmentally benign sol-vents and show low volatility, wide liquid range, water-compatibility, unusual low transition temperatures, non-flammability, non-toxicity, biocompatibility and biodegradability. The physical properties can be customized by choosing the right DES constituents in terms of chemical nature, relative compositions or water content. They can be prepared from readily available materials at high purities and low cost. The salt components and hydrogen bond donor (HBD)/complexing agent can be easily mixed and converted to DES without the need for further purification. The first reported DES was a mixture of urea and choline chloride. DESs can also be formed by mixing hydrogen bond acceptors such as amino-acids, salts, organic salts or natural salts (quaternary ammonium or phosphonium salt) acting with hydrogen bond donors such as acids, alcohols, amines, carbohydrates, urea, organic acids, alcohols, polyols, aldehydes, carbohydrates or saccharides. The potential applications are in food processing, in the removal of residual palm oil-based biodiesel catalyst and glycerine from palm-oil based biodiesel, and as solvents or catalysts for chemical reactions or biotransformations, metal electro-deposition, synthesis of nanoparticles, liquid and gas separations and heat transfer fluids (Abbott et al., 2004; Hayyan et al., 2010; Shahbaz et al., 2011; Hayyan et al., 2012; Singh et al., 2012; Kroon et al., 2013; Hayyan et al., 2014).

Low transition temperature mixtures (LTTMs) or solvents can be used in methods and systems to dissolve and hydrolyze lignin at mild conditions so that further degradation is prevented. LTTMs are cheap, renewable with non-toxic food ingredients and can dissolve lignin selectively and efficiently up to 90% lignin recovery. The remaining cellulose is also of higher quality and the much less water needed means less energy to evaporate large quantities of water. Cellulose shows much lower solubility in the LTTM type of solvents, and therefore can be separated out from the soluble lignin (Kroon et al., 2013). Among the possible mixtures are combinations of salts with organic acids or amino acids (e.g., choline chloride + malic acid), organic acids with amino acids (e.g., proline + malic acid), salts with alcohols or aldehydes (e.g., choline chloride + glycerol) and organic acids or amino acids with alcohols, carbohydrates or aldehydes (e.g., fructose + glucose + malic acid). The solvents dissolve lignin selectively at about 60°C, or 60–100°C for optimal dissolution kinetics without the risk of LTTM decomposition (Kroon et al., 2013).

13.5 OIL PALM LIGNOCELLULOSIC BIOMASS

13.5.1 RESOURCES AND UTILIZATION

The stable agricultural practices and well-established agro-based industries in Malaysia generate biomass totaling 77 million tonnes annually

(Foo et al., 2011). The relative percentage are: municipal solid waste (MSW) (9.5), wood powder (3.7), rice husk (0.7), sugarcane (0.5) and oil palm biomass (85.5) (Yacob, 2007). Oil Palm is the major source for biomass utilization and economically viable for conversion into biofuels such as methanol, ethanol, bio-oil and biodiesel. The sources of lignocellulosic material from oil palm include Empty Fruit Bunches (EFB), produced at the rate of 2.8 million tonnes per year on dry weight basis; oil palm frond (OPF), normally collected during replanting activity at 10.4 tonnes per hectare and estimated to be produced at 54.4 million tonnes per year from 2007–20; the trunks, obtained after 25 years during replanting, to supply 53.8% trunk fibers, 31.6% dried wood, and 14.5% bark and parenchyma (BFDIC, 2009). EFB contributes 30.5% of oil palm biomass produced in Malaysia (Shuit et al., 2009). It is renewable and accessible as it can be directly sourced at low cost from oil palm mill. EFB contains high amount of cellulose, hemicelluloses and lignin, and can replace sawdust powder of rubber plant, as good carbon sources for fermentation of POME (Saleh et al., 2012).

There are needs for further optimization of biomass-bioenergy co-generation in Malaysia and globally (Sumathi et al., 2008). The 9.7 hm^3 of ethanol produced annually is equivalent to 4.6 million tonnes of gasoline (Milbrandt and Overend, 2008), and oxygenated ethanol fuel is green with zero production of NO_X (Thiam and Bhatia, 2008). With high carbohydrate content, conversion of EFB to ethanol, syngas, butanol, bio-oil, hydrogen and biogas have added benefits with less environmental footprint (Geng, 2013). The approximately 20% palm oil production can be translated into 3.2 million tonnes of biodiesel which can compensate for 64% diesel consumption, with decreased dependency of 41% crude oil import (Milbrandt and Overend, 2008). There are, however issues related to food versus fuel that have to be addressed. Recycling and reuse of EFB for bioconversion has become pertinent in the light of environmental and socio-economic benefits to overcome unfavorable impacts of disposal treatment such as incineration and mulching. It is a potential feedstock for cellulose, compost, biosugar and bioethanol (Nazir, 2013). Traditionally, EFB is used as natural fuel for power and steam utilization in the palm oil mills. Straight burning of EFB however releases fine ash particles due to incomplete combustion. EFB conversion into biofuel requires critical steps including pre-treatment, hydrolysis, fermentation and distillation (Ishola et al., 2014). EFB has been converted to xylose, arabinose, mannose, galactose, sucrose and glucose, and the enzymatic fermentation has reportedly produce glucose at 0.43, 0.65 and 0.47 g^{-1}

and xylose at 0.26, 0.12 and 0.24 g⁻¹ dry weight basis using EFB, trunk and frond, respectively (Mohd et al., 2008). A total of 15 mg mL⁻¹ glucose could be fermented into 13.8% (w/w) ethanol which can be further improved by increasing the concentration of initial substrate. Infact, cellulosic ethanol production by EFB registers the highest ethanol titre of 48.54 g/L obtained through alkali pretreatment in a pilot scale reactor (Han et al., 2011). On dry weight basis, ethanol (L/tonne) can be produced from EFB (388), trunk (451) and frond (377) (Douglas et al., 2012). Aerobic and anaerobic co-digestion of EFB with microalgae could enhance biomethane production and POME treatment. Co-cultivation of *Nannochloropsis oculata* with EFB and POME produce the highest specific biogas production rate (1.13–1.14 m³ kg⁻¹ COD d⁻¹) and methane yield (4606–5018 mL CH₄ L⁻¹ POME ⁻¹), with equally high removal efficiency of COD (90–97%), BOD (84–98%) and TOC (65–80%) (Ahmad et al., 2014; Ahmad et al., 2015).

EFB is a potential source for bio-oil (Abdullah et al., 2011b). Bio-oil, typically known as brown liquids with pungent odor, is a high-density oxygenated liquid, produced from biomass which includes fragments of cellulose, hemicelluloses, lignin, and extractives. It is easy to use, transport and store, and can be used in diesel engines, turbines or boilers, in the manufacturing of specialty chemicals such as flavorings, and renewable resins and slow release fertilizers. Bio-oil is produced through special process known as fast pyrolysis. Gasification is the most suitable thermo-chemical route for EFB conversion to biofuels as it has the highest carbon conversion (>90%) and biofuel yield. However, due to the high viscosity and high water content of pyrolysis products, application of bio-oil as a biofuel is still challenging. Compared to other oil palm residues, such as oil palm kernel, EFB may not be a good candidate for solid fuels even after torrefaction pretreatment due to its high water content (Geng, 2013). Oil palm briquettes, made through briquetting process, may be better options for solid fuel. They are typically uniform with good physical properties such as easy to handle and feed and can be used as fuel in producing steam, district heating and electricity generation for larger commercial scale (Nasrin et al., 2008).

Biocomposites may involve combining the natural fiber and petroleum-originated non-biodegradable polymer or biodegradable polymer. Green composites refer to biocomposites which are derived from natural fibers and crop/bio-derived plastics that are more eco-friendly. Use of natural fibers as fillers in polymeric matrices confers several advantages over inorganic fillers especially in terms of renewability, low energy and cost, more deformability, biodegradability, combustibility, recyclability, more thermal and acoustic insulation properties, high specific strength, electrical resistance, lower density, less abrasiveness to treating apparatus and flexibility in use. The calorific value recovering at the end of life cycle is an individual advantage of natural fiber which is impossible for glass fibers (Shinoj et al., 2011). The natural fiber-reinforced plastic composites find applications in diverse fields such as aerospace, automotive parts, sports and amusement tools, boats, buildings and office products and machinery. Studies on the production and characterization of EFB-filled composites (both thermoset and thermoplastic) are concentrated on the behavior of the composites to different mechanical loading and characterization of physical, mechanical, thermal, electrical, and biodegradation properties. EFB-based composite properties are mainly related to matrix properties and the matrix selection is based on the required properties of the composites. Polypropylene (PP), natural rubber (NR), phenol formaldehyde (PF), polyvinyl chloride (PVC), polyurethane (PU), epoxy and polyester are among various matrices made into composites with oil palm fibers (Mahjoub et al., 2013). The purely extracted celluloses (PECs) extracted from EFB have been fabricated into PECs-PP composites by using injection-molding technique where the 25% PEC loading achieve the tensile strength of 26.7–27.3 MPa, without any addition of coupling agents. The extracted PECs are also successfully surface-engineered with acetate, oxalate and EDTA with different degree of substitution for Pb(II) removal where surface engineering with EDTA achieve 232.9–236.7 mg/g sorption. This suggests that the higher the number of carboxylic acids groups (ligand), the higher will be the metal sorption (Nazir, 2013).

Production of fiber board is one of the more emphasized industrial applications of oil palm wastes (Abdullah and Sulaiman, 2013). EFB and palm fibers can be used for medium density fiber-boards (MDF) and black-boards production. High-density fiberboard (HDF) terms hardboard is an engineered wood product. It is much harder and denser than MDF due to exploded wood fiber base. Another important consideration is that the polymeric insulated cables for high voltage engineering based on electrical treeing phenomenon have always been using synthetic retardants, with only limited numbers involving the use of organic materials (Jamil et al., 2012). EFB microfiller is a potential retardant to the beginning and growth of the electrical treeing phenomenon in polymeric materials for high voltage application (Ahmad et al., 2012). EFB, in actual facts, contains about 25% lignin and the hydrophobic behavior of lignin is suitable as coating materials

(Narapakdeesakul et al., 2013a; Narapakdeesakul et al., 2013b). EFB has also been the basic biochar sources in Malaysia with about 20 tonnes of biochar production daily (Sulaiman et al., 2011). Biochar is a stable organic matter applied to improve soil properties, and generated by heating the biomass to temperatures between 300 and 1000°C and low or zero oxygen concentrations. Carbonaceous materials including oil palm wastes, nutshells, wood, coir peat, and lignite can also be developed into activated carbon through either physical or chemical activation. Activated carbon has extremely porous and large surface area, making it suitable for adsorption and chemical reactions (Hameed et al., 2009). Conversion of lignocellulosic materials or EFB into Carbon Molecular Sieve (CMS) with precise and uniform size tiny pores for application as gas and liquid adsorbents is a promising application. The steps to obtain CMS from oil palm wastes including carbonization of the wastes, activation of the chars produced and pore modification of the activated carbons can be turned into economical, industrial-scale processes with medium technology requirement.

13.5.2 DELIGNIFICATION AND CELLULOSE EXTRACTION

Delignification is an important aspect of lignocellulosic biomass pre-treatment method. The type of biomass, the chemical agent, the reaction time and treatment temperature are factors affecting the efficacy of chemical treatments (Palamae et al., 2014). To increase the digestibility of EFB, different techniques have been reported such as the two-stage dilute acid hydrolysis (Millati et al., 2011), alkali pretreatment (Han et al., 2011), sequential dilute acid and alkali pretreatment (Kim et al., 2012), alkali and hydrogen peroxide pretreatment (Misson et al., 2009), sequential alkali and phosphoric acid pretreatment (Kim et al., 2012), aqueous ammonia (Jung et al., 2011), and solvent digestion (Abdullah et al., 2011b). Physico-chemical pretreatment such as ammonium fiber explosion (AFEX) (Lau et al., 2010), and super-heated steam (Bahrin et al., 2012) achieve hydrolysis efficiency of 90% and 66%, respectively. Alkali pretreatment is the most suitable method to prepare EFB for enzymatic hydrolysis and this is attributable to lignin degradation. Almost 100% lignin degradation is reportedly obtained when EFB is firstly treated with dilute NaOH, and subsequently with H_2O_2 (Misson et al., 2009). NaOH-pretreated EFB for bioethanol production has established the optimal conditions at 127.64°C at 22.08 min, using 2.89 mol/L NaOH.

Eco-friendly extraction of MCCs from EFB at low concentration of H_2O_2 (10% v/v) and formic acid (20% v/v) with ultrasound and heat assistance have been developed. Ultrasonic extraction carried out with H_2O_2 at 40 kHz and room temperature, yields 49% MCC with α-cellulose content of 91.3%, and crystallinity of 68.7%. Autoclave extraction with a mixture of H2O2 and formic acid at 120°C and further bleaching with H_2O_2 at 80°C, yields 64% of MCC with α-cellulose content of 93.7% and crystallinity of 70% (Nazir, 2013; Nazir et al., 2013).

EFB-pulp has been hydrolyzed with 2.5N HCl at 105°C for 30 min with constant agitation at 1:20 ratio of pulp to liquor. The reaction mixture was then filtered at room temperature, and washed repeatedly with distilled water, followed by 5% diluted NH_4OH (Mohamad Haafiz et al., 2013). The isolated MCC is a cellulose-I polymorph, with 87% crystallinity. In comparison, MCC isolated from jute using H_2SO_4 achieves 75% crystallinity (Jahan et al., 2011). The properties of cellulose nanowhiskers (CNW) from oil palm biomass-based MCCs are further investigated by comparing two techniques – MCC hydrolysis in 64% H_2SO_4 (96% purity) solution at 40°C for 60 min with strong agitation, and MCC swelling and separation into whiskers by using N, N-Dimethylacetamide (DMAc) (99% purity) with 0.5% lithium chloride (LiCl) (99% purity) as swelling agent, followed by ultrasonication over a period of 5 days (Mohamad Haafiz et al., 2014). The two different treatments achieve higher crystallinity of CNW at 84–88% as compared to 59% for the nanofibers from EFB by using H_2SO_4 (Fahma et al., 2010), 73.4% for CNW from mulberry (Li et al., 2009), and 69% for the nanocellulose from EFB by a chemo-mechanical technique (Jonoobi et al., 2011). Chemical swelling maintains the cellulose I structure while acid hydrolysis displays the coexistence of cellulose I and II allomorphs (Mohamad Haafiz et al., 2014). The high crystallinity is proposed due to the effect of strong acids which removes the amorphous regions (Li et al., 2009; Spagnola et al., 2012) but the greater challenge will be to translate these into industrial practices.

Ammonia fiber expansion (AFEX) pretreatment for cellulosic ethanol production from EFB achieves sugar yield close to 90% after enzyme formulation optimization. Post-AFEX size reduction is required to enhance the sugar yield possibly due to the high tensile strength (248 MPa) and toughness (2000 MPa) of palm fibers as compared to most cellulosic feedstock. The water extract from AFEX-pretreated EFB at 9% solids loading is highly fermentable where upto 65 g/L glucose can be fermented to ethanol within 24 h without nutrients supplement (Lau et al., 2010). A total glucose conversion rate of 86.3 /% is achieved using the Changhae Ethanol Multi Explosion (CHEMEX) facility with a cellulase loading of 50 FPU/g cellulose. Cellulase hydrolyzed EFB has resulted in 1.262 g/L ABE (acetone, butanol and ethanol) obtained in the medium containing 20 g/L sugar (Noomtim and Cheirsilp, 2011). Higher ABE yield of 3.47 g/L is obtained from treated EFB when compared to using a glucose-based medium and *Clostridium butyricum* EB6 at pH 6 (Ibrahim et al., 2012).

13.6 CONCLUSION AND FUTURE PERSPECTIVES

Pretreatment is a powerful tool to convert compact lignocelluloses into loosen or torn structures for penetration of chemical reagents or as template for microbial attack to digest and convert into desired product. Synergistic effects of physical and chemical treatments are often required to bring forth the desired results of cellulose yield. Investigations have now ventured into the realm of eco-friendly dissolution of lignocellulosic biomass especially with the application of ionic liquid and deep eutectic solvents. However, much work is needed to improve the economic aspects with less energy intensive process. The future lies in biorefinery concept and integrated utilization of lignocellulosic agro-wastes where bioenergy co-generation and waste remediation are put into practice whilst, extracting important biochemicals and biomaterials such as celluloses, hemicelluloses, lignin, organic acids and enzymes for various industrial applications.

KEYWORDS

- **Agro-wastes**
- **Bacterial celluloses**
- **Biomass**
- **Delignification**
- **Eco-friendly**
- **Fermentation**
- **Freeze explosion**

- Hemicellulose
- Hydrothermal pre-treatment
- Ionic liquid
- Lignocelluloses
- Lower plant cellulosic fibers
- Oil palm
- Ozonolysis
- Steam explosion
- Wet oxidation

REFERENCES

Abbott, A. P., Boothby, D., Capper, G., Davies, D. L., Rasheed, R. K. Deep eutectic solvents formed between choline chloride and carboxylic acids: versatile alternatives to ionic liquids. *J Am Chem Soc* 2004, 126, 9142–9147.

Abdullah, M. A., Nazir, M. S., Wahjoedi, B. A. Development of value-added biomaterials from oil palm agro-wastes. *2nd International Conference on Biotechnology and Food Sciences*, Bali, Indonesia, 2011a, pp. 1–3.

Abdullah, N., Sulaiman, F., Gerhauser, H. Characterization of oil palm empty fruit bunches for fuel application. *J Phys Sci* 2011b, 22(1), 1–24.

Abdullah, N., Sulaiman, F. The oil palm wastes in Malaysia. In *Biomass Now – Sustainable Growth and Use*, Darko, M. (Ed.), InTech Publishing: Rijeka, Croatia, 2013, pp. 75–100.

Abe, K., Yano, H. Comparison of the characteristics of cellulose microfibril aggregates isolated from fiber and parenchyma cells of Moso bamboo (*Phyllostachys pubescens*). *Cellulose* 2010, 17, 271–277.

Ahmad, A., Shah, S. M. U., Othman, M. F., Abdullah, M. A. Aerobic and anaerobic cocultivation of *Nannochloropsis oculata* with oil palm empty fruit bunch for enhanced biomethane production and palm oil mill effluent treatment. *Desalination and Water Treatment* 2015, 56(8), 2055–2065.

Ahmad, A., Shah, S. M. U., Othman, M. F., Abdullah, M. A. Enhanced palm oil mill effluent treatment and biomethane production by co-digestion of oil palm empty fruit bunches with *Chlorella* sp. *Canadian Journal of Chemical Engineering* 2014, 92, 1636–1642.

Ahmad, M. H., Jamil, A. A. A., Ahmad, H., Piah, M. A. M., Darus, A., Arief, Y. Z., Bashir, N. Oil palm empty fruit bunch as a new organic filler for electrical tree inhibition. *International Journal of Electronics and Electrical Engineering* 2012, 6(6), 213–218.

Alen, R. Structure and chemical composition of wood. In *Forest Products Chemistry*. Stenius, P. (Ed.), Fapet Oy: Helsinki, 2000, pp. 11–57.

Almustapha, S., Khan, A. A. A., Omar, A. A., Wahjoedi, B. A., Abdullah, M. A. Cellulose-modified carbon electrode for in situ lead detection. *Applied Mechanics and Materials* 2014, 625, 136–139.

Alvira, P., Tomás-Pejó, E., Ballesteros, M., Negro, M. J. Pretreatment technologies for an efficient bioethanol production process based on enzymatic hydrolysis: A review. *Bioresource Technology* 2010, 101, 4851–4861.

Andanson, J. M., Bordes, E., Devémy, J., Leroux, F., Pádua, A. A., Gomes, M. F. C. Understanding the role of co-solvents in the dissolution of cellulose in ionic liquids. *Green Chemistry* 2014, 16, 2528–2538.

Bahrin, E. K., Baharuddin, A. S., Ibrahim, M. F., Razak, M. N. A., Sulaiman, A., Abd-Aziz, S., Hassan, M. A., Shirai, Y., Nishida, H. Physicochemical property changes and enzymatic hydrolysis enhancement of oil palm empty fruit bunches treated with superheated steam. *BioResources* 2012, 7, 1784–1801.

Bak, J. S., Ko, J. K., Han, Y. H., Lee, B. C., Choi, I. G., Kim, K. H. Improved enzymatic hydrolysis yield of rice straw using electron beam irradiation pretreatment. *Bioresource Technology* 2009, 100, 1285–1290.

Balat, M., Balat, M., Kirtay, E., Balat, H. Main routes for the thermo-conversion of biomass into fuels and chemicals. Part 1: Pyrolysis systems. *Energy Conversion and Management* 2009, 50, 3147–3157.

Ballesteros, I., Ballesteros, M., Manzanares, P., Negro, M. J., Oliva, J. M., Sáez, F. Dilute sulfuric acid pretreatment of cardoon for ethanol production. *Biochemical Engineering Journal* 2008, 42, 84–91.

Banik, S., Bandyopadhyay, S., Ganguly, S. Bioeffects of microwave – A brief review. *Biosour Technol* 2003, 87, 155–159.

Beg, Q., Kapoor, M., Mahajan, L., Hoondal, G. S. Microbial xylanases and their industrial applications: a review. *Applied Microbiology and Biotechnology* 2001, 56, 326–338.

Behera, S., Arora, R., Nandhagopal, N., Kumar, S. Importance of chemical pretreatment for bioconversion of lignocellulosic biomass. *Renewable and Sustainable Energy Reviews* 2014, 36, 91–106.

BFDIC. Oil palm biomass. Beijing Forestry and Park Department of International Cooperation, Beijing, 2014, http://www.bfdic.com/en/Features/Features/79.html

Bhimte, N., Tayade, P. Evaluation of microcrystalline cellulose prepared from sisal fibers as a tablet excipient: A technical note. *AAPS Pharm Sci Tech* 2007, 8, E56–E62.

Bian, J., Peng, F., Peng, X. -P., Xu, F., Sun, R. -C., Kennedy, J. F. Isolation of hemicelluloses from sugarcane bagasse at different temperatures: Structure and properties. *Carbohydrate Polymers* 2012, 88, 638–645.

Bochek, A. M. Effect of hydrogen bonding on cellulose solubility in aqueous and nonaqueous solvents. *Russian Journal of Applied Chemistry* 2003, 76, 1711–1719.

Brink, D. L. Method of treating biomass material. *Solar Energy* 1996, 57, doi: 10.1016/s0038-092x(97)87966-2

Cardenas-Toro, F. P., Forster-Carneiro, T., Rostagno, M. A., Petenate, A. J., Maugeri Filho, F., Meireles, M. A. A. Integrated supercritical fluid extraction and subcritical water hydrolysis for the recovery of bioactive compounds from pressed palm fiber. *The Journal of Supercritical Fluids* 2014, 93, 42–48.

Carpita, N. C., Gibeaut, D. M. Structural models of primary-cell walls in flowering plants: consistency of molecular-structure with the physical-properties of the walls during growth. *Plant J* 1993, 3, 1–30.

Carvalheiro, F., Duarte, L. C., Gírio, F. M. Hemicellulose biorefineries: A review on biomass pretreatments. *Journal of Scientific and Industrial Research* 2008, 67, 849–864.

Chandel, A. K., Giese, E. C., Antunes, F. A., dos Santos Oliveira, I., da Silva, S. S. Pretreatment of sugarcane bagasse and leaves: unlocking the treasury of "Green currency." *Pretreatment Techniques for Biofuels and Biorefineries*, Springer: Berlin, 2013, pp. 369–391.

Chemat, F., Huma, Z., Khan, M. K. Applications of ultrasound in food technology: Processing, preservation and extraction. *Ultrasonics Sonochemistry* 2011, 18, 813–835.

Chen, W., Yu, H., Liu, Y., Hai, Y., Zhang, M., Chen, P. Isolation and characterization of cellulose nanofibers from four plant cellulose fibers using a chemical-ultrasonic process. *Cellulose* 2011, 18, 433–442.

Chiaramonti, D., Prussi, M., Ferrero, S., Oriani, L., Ottonello, P., Torre, P., Cherchi, F. Review of pretreatment processes for lignocellulosic ethanol production, and development of an innovative method. *Biomass and Bioenergy* 2012, 46, 25–35.

Curnow, O. J. Ionic liquids: Some of their remarkable properties and some of their applications. *Chemistry in New Zealand* 2012, 76, 118–122.

Da Costa, L. A. M., João, K. G., Morais, A. R. C., Bogel-Łukasik, E., Bogel-Łukasik, R. Ionic liquids as a tool for lignocellulosic biomass fractionation. *Sustainable Chemical Processes* 2013, 1(3), 1–31.

Diallo, A. O., Len, C., Morgan, A. B., Marlair, G. Revisiting physico-chemical hazards of ionic liquids. *Sep Purif Technol* 2012, 97, 228–234.

Disale, S. T., Kale, S. R., Kahandal, S. S., Srinivasan, T. G., Jayaram, R. V. Choline chloride 2 $ZnCl_2$ ionic liquid: an efficient and reusable catalyst for the solvent free Kabachnik–Fields reaction. *Tetrahedron Letter* 2012, 53, 2277–2279.

Elanthikkal, S., Gopalakrishnapanicker, U., Varghese, S., Guthrie, J. T. Cellulose microfibers produced from banana plant wastes: isolation and characterization. *Carbohydrate Polymers* 2010, 80, 852–859.

Fahma, F., Iwamoto, S., Hori, N., Iwata, T., Takemura, A. Isolation, preparation, and characterization of nanofibers from oil palm empty-fruit-bunch (OPEFB). *Cellulose* 2010, 17, 977–985.

Farias-Campomanes, A. M., Rostagno, M. A., Meireles, M. A. A. Production of polyphenol extracts from grape bagasse using supercritical fluids: Yield, extract composition and economic evaluation. *The Journal of Supercritical Fluids* 2013, 77, 70–78.

Faulon, J. L., Carlson, G. A., Hatcher, P. G. A three-dimensional model for lignocellulose from gymnospermous wood. *Organic Geochemistry* 1994, 21, 1169–1179.

Feng, L., Chen, Z. Research progress on dissolution and functional modification of cellulose in ionic liquids. *Journal of Molecular Liquids* 2008, 142(1), 1–5.

Fitz Patrick, M., Champagne, P., Cunningham, M. F., Whitney, R. A. A biorefinery processing perspective: treatment of lignocellulosic materials for the production of value-added products. *Bioresource Technology* 2010, 101, 8915.

Flauzino Neto, W. P., Silvério, H. A., Dantas, N. O., Pasquini, D. Extraction and characterization of cellulose nanocrystals from agro-industrial residue–soy hulls. *Industrial Crops and Products* 2013, 42, 480–488.

Foo, Y. N., Foong, K. Y., Yousof, B., Kalyana, S. A renewable future driven with Malaysian palm oil-based green technology. *Journal of Oil Palm and The Environment* 2011, 2, 1–7.

Foston, M., Ragauskas, A. J. Changes in lignocellulosic supramolecular and ultrastructure during dilute acid pretreatment of Populus and switch grass. *Biomass and Bioenergy* 2010, 34, 1885–1895.

Gabriele, G., Cerchiara, T., Salerno, G., Chidichimo, G., Vetere, M. V., Alampi, C., Gallucci, M. C., Conidi, C., Cassano, A. A new physical–chemical process for the efficient production of cellulose fibers from Spanish broom (*Spartium junceum* L.). *Bioresource Technology* 2010, 101, 724–729.

Galbe, M., Zacchi, G. Pretreatment: The key to efficient utilization of lignocellulosic materials. *Biomass and Bioenergy* 2012, 46, 70–78.

Geng, A. Conversion of oil palm empty fruit bunch to biofuels. In *Liquid, Gaseous and Solid Biofuels – Conversion Techniques*, Zhen, F. (Ed.), InTech Publishing: Croatia, 2013, DOI: 10.5772/53043.

Geoffrey, D. Fungal and bacterial biodegradation: white rots, brown rots, soft rots, and bacteria. In *Deterioration and protection of sustainable biomaterials, Vol. 1158, ACS Symposium Series*, American Chemical Society, USA, 2014, pp. 23–58.

Gericke, M., Fardim, P., Heinze, T. Ionic liquids – promising but challenging solvents for homogeneous derivatization of cellulose. *Molecules* 2012, 17, 7458–7502.

Ghali, A. E., Marzoug, I. B., Baouab, A. H. V., Roudesi, M. S. Separation and characterization of new cellulosic fibers from the *Huncus acutus* L. plant. *BioResources* 2012, 7, 2002–2018.

Gilberto, S., Julien, B., Alian, D. *Luffa cylinderica* as a lignocellulosic source of fiber, microfibrillated cellulose, and cellulose nanocrystals. *BioResources* 2010, 5, 727–740.

Gunderson, J. H., Elwood, H., Ingold, A., Kindle, K., Sogin, M. L. Phylogenetic relationships between chlorophytes, chrysophytes, and oomycetes. *Proceedings of the National Academy of Sciences* 1987, 84, 5823–5827.

Gunny, N., Anas, A., Arbain, D. Ionic liquids: green solvent for pretreatment of lignocellulosic biomass. *Advanced Materials Research* 2013, 701, 399–402.

Mood, H. S., Hossein Golfeshan, A., Tabatabaei, M., Salehi Jouzani, G., Najafi, G. H., Gholami, M., Ardjmand, M. Lignocellulosic biomass to bioethanol, a comprehensive review with a focus on pretreatment. *Renewable and Sustainable Energy Reviews* 2013, 27, 77–93.

Hall, D. A., Happey, F., Lloyd, P. F., Saxl, H. Oriented cellulose as a component of mammalian tissue. *Proceedings of the Royal Society of London Series B Biological Sciences* 1960, 151, 497–516.

Hameed, B. H., Tan, I. A. W., Ahmad, A. L. Preparation of oil palm empty fruit bunch-based activated carbon for removal of 2, 4, 6-trichlorophenol: Optimization using response surface methodology. *Journal of Hazardous Materials* 2009, 164(2), 1316–1324.

Han, M. H., Kim, Y., Kim, S. W., Choi, G. W. High efficiency bioethanol production from OPEFB using pilot pretreatment reactor. *Journal of Chemical Technology and Biotechnology* 2011, 86, 1527–1534.

Harmsen, P., Huijgen, W., Bermudez, L., Bakker, R. *Literature Review of Physical and Chemical Pretreatment Processes for Lignocellulosic Biomass.* Food and Biobased Research Institute: Wageningen, 2010.

Harmsen, P., Lips, S., Bakker, R. *Pretreatment of Lignocellulose for Biotechnological Production of Lactic Acid.* Food and Biobased Research Institute: Wageningen, 2013.

Haverty, D., Dussan, K., Piterina, A. V., Leahy, J. J., Hayes, M. H. B. Autothermal, single-stage, performic acid pretreatment of *Miscanthus* × *giganteus* for the rapid fractionation of its biomass components into a lignin/hemicellulose-rich liquor and a cellulase-digestible pulp. *Bioresource Technology* 2012, 109, 173–177.

Hayyan, M., Hashim, M. A., Hayyan, A., Al-Saadi, M. A., AlNashef, I. M., Mirghani, M. E. S., Saheed, O. K. Are deep eutectic solvents benign or toxic? *Chemosphere* 2013, 90, 2193–2195.

Hayyan, A., Mjalli, F. S., Al-Nashef, I. M., Al-Wahaibi, T., Al-Wahaibi, Y. M., Hashim, M. A. Fruit sugar-based deep eutectic solvents and their physical properties. *Thermochim Acta* 2012, 541, 70–75.

Hayyan, M., Mjalli, F. S., Hashim, M. A., Al-Nashef, I. M. A novel technique for separating glycerine from palm oil-based biodiesel using ionic liquids. *Fuel Process Technol* 2010, 91, 116–120.

Hendriks, A., Zeeman, G. Pretreatments to enhance the digestibility of lignocellulosic biomass. *Bioresource Technology* 2009a, 100, 10–18.

Hendriks, A. T. W. M., Zeeman, G. Pretreatments to enhance the digestibility of lignocellulosic biomass. *Bioresource Technology* 2009b, 100, 10–18.

Hendry, D. *Investigation of Supercritical Fluids for Use in Biomass Processing and Carbon Recycling.* University of Missouri: Columbia, 2012, 1–91.

Herrera, A., Téllez-Luis, S. J., Ramirez, J. A., Vázquez, M. Production of xylose from sorghum straw using hydrochloric acid. *Journal of Cereal Science* 2003, 37, 267–274.

Hsu, C. L., Chang, K. S., Lai, M. Z., Chang, T. C., Chang, Y. H., Jang, H. D. Pretreatment and hydrolysis of cellulosic agricultural wastes with a cellulase-producing *Streptomyces* for bioethanol production. *Biomass and Bioenergy* 2011, 35, 1878–1884.

Holm, J., Lassi, U. Ionic liquids in the pretreatment of lignocellulosic biomass. In *Ionic Liquids: Applications and Perspectives*, Kokorin, A. (Ed.), InTech Publishing: Croatia, 2011, pp. 545–560.

Hu, F., Ragauskas, A. Pretreatment and lignocellulosic chemistry. *Bioenergy Research* 2012, 5, 1043–1066.

Hu, Z., Wen, Z. Enhancing enzymatic digestibility of switch grass by microwave-assisted alkali pretreatment. *Biochemical Engineering Journal* 2008, 38, 369–378.

Ibrahim, M. F., Aziz, S. A., Razak, M. N. A., Phang, L. Y., Hassan, M. A. Oil palm empty fruit bunch as alternative substrate for acetone-butanol-ethanol production by *Clostridium butyricum* EB6. *Applied Biochemistry and Biotechnology* 2012, 166, 1615–1625.

Intanakul, P., Krairiksh, M., Kitchaiya, P. Enhancement of enzymatic hydrolysis of lignocellulosic wastes by microwave pretreatment under atmospheric pressure. *Journal of Wood Chemistry and Technology* 2003, 23, 217–225.

Ishola, M. M., Taherzadeh, M. J. Effect of fungal and phosphoric acid pretreatment on ethanol production from oil palm empty fruit bunches (OPEFB). *Bioresource Technology* 2014, 165, 9–12.

Isik, M., Sardon, H., Mecerreyes, D. Ionic liquids and cellulose: dissolution, chemical modification and preparation of new cellulosic materials. *International Journal of Molecular Sciences* 2014, 15, 11922–11940.

Jacqueline, I., Kroschwitz, A. S. *Encyclopedia of Chemical Technology, Vol. 20,* Wiley-Inter-Science: New York, 2001.

Jahan, M. S., Saeed, A., He, Z., Ni, Y. Jute as raw material for the preparation of microcrystalline cellulose. *Cellulose* 2011, 18, 451–459.

Jamil, A. A. A., Kamarol, M., Mariatti, M., Bashir, N., Ahmad, M. H., Arief, Y. Z., Muhamad, N. A. Organo-montmorillonite as an electrical treeing retardant for polymeric insulating materials. IEEE International Conference on Condition Monitoring and Diagnosis, 2012, pp. 237–240.

Jean-Luc, W., Olivier, B., Jean, P. M. *Cellulose Science and Technology.* EPFL Press: Lausanne, Switzerland, 2010.

Johar, N., Ahmad, I., Dufresne, A. Extraction, preparation and characterization of cellulose fibers and nanocrystals from rice husk. *Industrial Crops and Products* 2012, 37, 93–99.

Jonoobi, M., Khazaeian, A., Md Tahir, P., Azry, S. S., Oksman, K. Characteristics of cellulose nanofibers isolated from rubberwood and empty fruit bunches of oil palm using chemomechanical process. *Cellulose* 2011, 18, 1085–1095.

Jordan, D. B., Bowman, M. J., Braker, J. D., Dien, B. S., Hector, R. E., Lee, C. C., Mertens, J. A., Wagschal, K. Plant cell walls to ethanol. *Biochemical Journal* 2012, 442, 241–252.

Jung, Y. H., Kim, I. J., Han, J. I., Choi, I. G., Kim, K. H. Aqueous ammonia pretreatment of oil palm empty fruit bunches for ethanol production. *Bioresource Technology* 2011, 102, 9806–9809.

Kamphunthong, W., Hornsby, P., Sirisinha, K. Isolation of cellulose nanofibers from para rubberwood and their reinforcing effect in poly(vinyl alcohol) composites. *Journal of Applied Polymer Science* 2012, 125, 1642–1651.

Kargarzadeh, H., Ahmad, I., Abdullah, I., Dufresne, A., Zainudin, S., Sheltami, R. Effects of hydrolysis conditions on the morphology, crystallinity, and thermal stability of cellulose nanocrystals extracted from kenaf bast fibers. *Cellulose* 2012, doi. 10.1007/s10570–10012–19684–10576

Karimi, K., Shafiei, M., Kumar, R. Progress in physical and chemical pretreatment of lignocellulosic biomass. In *Biofuel Technologies*, Springer: Berlin, 2013, pp. 53–96.

Karp, S. G., Woiciechowski, A. L., Soccol, V. T., Soccol, C. R. Pretreatment strategies for delignification of sugarcane bagasse: a review. *Brazilian Archives of Biology and Technology* 2013, 56, 679–689.

Kim, S., Kim, C. H. Bioethanol production using the sequential acid/alkali-pretreated empty palm fruit bunch fiber. *Renewable Energy* 2013, 54, 150–155.

Kim, S., Park, J. M., Seo, J. W., Kim, C. H. Sequential acid-/alkali-pretreatment of empty palm fruit bunch fiber. *Bioresource Technology* 2012, 109, 229–233.

Klemm, D., Heublein, B., Fink, H. P., Bohn, A. Cellulose: fascinating biopolymer and sustainable raw material. *Angewandte Chemie* 2005, 44, 3358–3393.

Knez, Ž., Markočič, E., Leitgeb, M., Primožič, M., Hrnčič, M. K., Škerget, M. Industrial applications of supercritical fluids: A review. *Energy* 2014, 77, 235–243.

Krassig, H., Schurz, J. *Ullmann's Encyclopedia of Industrial Chemistry, 6 Edn.,* Wiley-VCH, Weinheim: Germany, 2002.

Kroon, M. C., Casal, M. F., Den Bruinhorst, A. V. Pretreatment of lignocellulosic biomass and recovery of substituents using natural deep eutectic solvents/compound mixtures with low transition temperatures. Patent WO 2013153203 A1, 2013.

Kubisa, P. Ionic liquids as solvents for polymerization processes – progress and challenges. *Progress in Polymer Science* 2009, 34(12), 1333–1347.

Kumar, P., Barrett, D. M., Delwiche, M. J., Stroeve, P. Methods for pretreatment of lignocellulosic biomass for efficient hydrolysis and biofuel production. *Industrial and Engineering Chemistry Research* 2009, 48, 3713–3729.

Laitinen, A. Supercritical fluid extraction of organic compounds from solids and aqueous solutions. Technical Research Centre of Finland, VTT: Finland, 1999, pp. 58–84.

Lau, M. J., Lau, M. W., Gunawan, C., Dale, B. E. Ammonia fiber expansion (AFEX) pretreatment, enzymatic hydrolysis, and fermentation on empty palm fruit Bunch fiber

(EPFBF) for cellulosic ethanol production. *Applied Biochemistry and Biotechnology* 2010, 162, 1847–1857.

Lee, H., Hamid, S., Zain, S. Conversion of lignocellulosic biomass to nanocellulose: structure and chemical process. *The Scientific World Journal* 2014, doi: 10.1155/2014/631013.

Leh, C. P., Rosli, W. D. W., Zainuddin, Z., Tanaka, R. Optimisation of oxygen delignification in production of totally chlorine-free cellulose pulps from oil palm empty fruit bunch fiber. *Industrial Crops and Products* 2008, 28, 260–267.

Lejeune, A., Deprez, T. *Cellulose: Structure and Properties, Derivatives and Industrial uses.* Nova Science Publisher Inc.: New York, 2010.

Li, M. F., Sun, S. N., Xu, F., Sun, R. C. Formic acid based organosolv pulping of bamboo (*Phyllostachys acuta*): comparative characterization of the dissolved lignins with milled wood lignin. *Chemical Engineering Journal* 2012, 179, 80–89.

Li, R., Fei, J., Cai, Y., Li, Y., Feng, J., Yao, J. Cellulose whiskers extracted from mulberry: A novel biomass production. *Carbohydrate Polymers* 2009, 76, 94–99.

Lin, L., Yan, R., Jiang, W., Shen, F., Zhang, X., Zhang, Y., Deng, S., Li, Z. Enhanced enzymatic hydrolysis of palm pressed fiber based on the three main components: cellulose, hemicellulose, and lignin. *Appl Biochem Biotechnol* 2014, 173, 409–420.

Lin, S. Y., Lin, I. S. Ullmann's Encyclopedia of Industrial Chemistry, 6 Edn., Wiley-VCH: Weinheim, Germany, 2002.

Long, Y. Biodegradable polymer blends and composites from renewable resources. John Wiley: New Jersey, 2009.

Lora, J. H., Glasser, W. G. Recent industrial applications of lignin: a sustainable alternative to non-renewable materials. *Journal of Polymers and the Environment* 2002, 10, 39–48.

Lu, B., Xu, A., Wang, J. Cation does matter: how cationic structure affects the dissolution of cellulose in ionic liquids. *Green Chemistry* 2014, 16, 1326–1335.

Lundqvist, J., Teleman, A., Junel, L., Zacchi, G., Dahlman, O., Tjerneld, F. Isolation and characterization of galactoglucomannan from spruce (*Picea abies*). *Carbohydr Polym* 2002, 48, 29–39.

Luque de, C. M. D., Priego, C. F. Analytical applications of ultrasound. Vol. 26, Elsevier: Netherlands, 2007.

Mahdy, A., Mendez, L., Ballesteros, M., González-Fernández, C. Autohydrolysis and alkaline pretreatment effect on *Chlorella vulgaris* and *Scenedesmus* sp. methane production. *Energy* 2014, 78, 48–52.

Malkapuram, R., Kumar, V., Negi, Y. S. Recent development in natural fiber reinforced polypropylene composites. *Journal of Reinforced Plastics and Composites* 2009, 28, 1169–1189.

Mahjoub, R., Bin Mohamad Yatim, J., Mohd Sam, A. R. A review of structural performance of oil palm empty fruit bunch fiber in polymer composites. *Advances in Materials Science and Engineering* 2013, doi: 10.1155/2013/415359.

Mandal, A., Chakrabarty, D. Isolation of nanocellulose from waste sugarcane bagasse (SCB) and its characterization. *Carbohydrate Polymers* 2011, 86, 1291–1299.

Martins, M., Teixeira, E., Corrêa, A., Ferreira, M., Mattoso, L. Extraction and characterization of cellulose whiskers from commercial cotton fibers. *Journal of Materials Science* 2011, 46, 7858–7864.

Mason, T. J., Paniwnyk, L., Lorimer, J. P. The uses of ultrasound in food technology. *Ultrasonics Sonochemistry* 1996, 3, S253–S260.

Milbrandt, A., Overend, R. P. *Survey of Biomass Resource Assessments and Assessment Capabilities.* APEC Energy Working Group, APEC Economies, 2008.

Millati, R., Wikandari, R., Trihandayani, E. T., Cahyanto, M. N., Taherzadeh, M. J. Niklasson, C. Ethanol from oil palm empty fruit bunch via dilute-acid hydrolysis and fermentation by *Mucor indicus* and *Saccharomyces cerevisiae*. *Agriculture Journal* 2011, 6, 54–59.

Misson, M., Haron, R., Kamaroddin, M. F. A., Amin, N. A. S. Pretreatment of empty palm fruit bunch for production of chemicals via catalytic pyrolysis. *Bioresource Technology* 2009, 100, 2867–2873.

Ming-Fei, L., Yong-Ming, F., Run-Cang, S., Xu, F. Characterization of extracted lignin of bamboo (*Neosinocalamus affinis*) pretreated with sodium hydroxide/urea solution at low temperature. *BioResources* 2010, 5, 1762–1778.

Mohamad Haafiz, M. K., Hassan, A., Zakaria, Z., Inuwa, I. M. Isolation and characterization of cellulose nanowhiskers from oil palm biomass microcrystalline cellulose. *Carbohydrate Polymers* 2014, 103, 119–125.

Mohamad Haafiz, M. K., Eichhorn, S. J., Hassan, A., Jawaid, M. Isolation and characterization of microcrystalline cellulose from oil palm biomass residue. *Carbohydrate Polymers* 93, 628–634.

Mohd, T., Amal, N., Ahmad, Z. F., Mohd, F. F., Ali, N., Hassan, O. The usage of empty fruit bunch (EFB) and palm pressed fiber (PPF) as substrates for the cultivation of *Pleurotus ostreatus*. *Jurnal Teknologi F* 2008, 49, 189–196.

Morán, J., Alvarez, V., Cyras, V., Vázquez, A. Extraction of cellulose and preparation of nanocellulose from sisal fibers. *Cellulose* 2008, 15, 149–159.

Mosier, N., Wyman, C., Dale, B., Elander, R., Lee, Y. Y., Holtzapple, M., Ladisch, M. Features of promising technologies for pretreatment of lignocellulosic biomass. *Bioresource Technology* 2005, 96, 673–686.

Nada, A. -A. M. A., Mohamed, Y. E.-K., El-Sayed, E. S. A., Fatma, M. A. Preparation and characterization of microcrystalline cellulose (MCC). *BioResources* 2009, 4, 1359–1371.

Narapakdeesakul, D., Sridach, W., Wittaya, T. Novel use of oil palm empty fruit bunch's lignin derivatives for production of linerboard coating. *Progress in Organic Coatings* 2013a, 76(7), 999–1005.

Narapakdeesakul, D., Sridach, W., Wittaya, T. Synthesizing of oil palm empty fruit bunch lignin derivatives and potential use for production of linerboard coating. *Songklanakarin Journal of Science and Technology* 2013b, 35(6), 705–713.

Narayanaswamy, N., Dheeran, P., Verma, S., Kumar, S. Biological pretreatment of lignocellulosic biomass for enzymatic saccharification. In *Pretreatment Techniques for Biofuels and Biorefineries*, Fang, Z. (Ed.), Springer: New York, 2013, pp. 3–34.

Nasrin, A. B., Ma, A. N., Choo, Y. M., Mohamad, S., Rohaya, M. H., Azali, A., Zainal, Z. Oil palm biomass as potential substitution raw materials for commercial biomass briquettes production. *Am J Appl Sci* 2008, 5(3), 179–183.

Nazir, M. S. Eco-friendly extraction, characterization and modification of microcrystalline cellulose from oil palm empty fruit bunches. PhD Thesis, Universiti Teknologi Petronas, Malaysia, 2013.

Nazir, M. S., Wahjoedi, B. A., Yussof, A. W., Abdullah, M. A. Eco-friendly extraction and characterization of cellulose from oil palm empty fruit bunches. *Bioresources* 2013, 8(2), 2161–2172.

Nieuwolt, S., Ghazalli, M. Z., Gopinathan, B. *Agro-Ecological Regions in Peninsular Malaysia*. Serdang: Selangor, Malaysia, 1982.

Ninomiya, K., Inoue, K., Aomori, Y., Ohnishi, A., Ogino, C., Shimizu, N., Takahashi, K. Characterization of fractionated biomass component and recovered ionic liquid during repeated process of cholinium ionic liquid-assisted pretreatment and fractionation. *Chemical Engineering Journal* 2015, 259, 323–329.

Noomtim, P., Cheirsilp, B. Production of butanol from palm empty fruit bunches hydrolysate by *Clostridium acetobutylicum*. *Energy Procedia* 2011, 9, 140–146.

Palamae, S., Palachum, W., Chisti, Y., Choorit, W. Retention of hemicellulose during delignification of oil palm empty fruit bunch (EFB) fiber with peracetic acid and alkaline peroxide. *Biomass and Bioenergy*, 2014, 66, 240–248.

Pasquini, D., Pimenta, M. T. B., Ferreira, L. H., Curvelo, A. A. S. Extraction of lignin from sugar cane bagasse and *Pinus taeda* wood chips using ethanol-water mixtures and carbon dioxide at high pressures. *The Journal of Supercritical Fluids* 2005, 36, 31–39.

Perasso, R., Baroin, A., Qu, L. H., Bachellerie, J. P., Adoutte, A. Origin of the algae. *Nature* 1989, 339, 142–144.

Perlack, R. D., Wright, L. L., Turhollow, A. F., Graham, R. L., Stokes, B. J., Erbach, D. C. In *Biomass as Feedstock for a Bioenergy and Bioproducts Industry: The Technical Feasibility of a Billion-Ton Annual Supply*. U.S. Department of Energy: Oak Ridge, TN, 2005.

Potumarthi, R., Baadhe, R. R., Bhattacharya, S. Fermentable sugars from lignocellulosic biomass: technical challenges. In *Biofuel Technologies – Recent Developments*, Gupta, V. K., Tuohy, M. G. (Eds.), Springer: Germany, 2013, pp. 3–27.

Ragauskas, A. J., Huang, F. Chemical pretreatment techniques for biofuels and biorefineries from softwood. In *Pretreatment Techniques for Biofuels and Biorefineries*, Fang, Z. (Ed.), Springer: New York, 2013, 151–179.

Ralph, J., Brunow, G., Boerjan, W. Lignins. In *eLS*, John Wiley and Sons Ltd.: New York, 2001.

Reddy, N., Yang, Y. Biofibers from agricultural byproducts for industrial applications. *Trends in Biotechnology* 2005, 23, 22–27.

Ries, M. E., Radhi, A., Keating, A. S., Parker, O., Budtova, T. Diffusion of 1-ethyl-3- methylimidazolium acetate in glucose, cellobiose, and cellulose solutions. *Biomacromolecules* 2014, 15, 609–617.

Rosa, M. F., Medeiros, E. S., Malmonge, J. A., Gregorski, K. S., Wood, D. F., Mattoso, L. H. C., Glenn, G., Orts, W. J., Imam, S. H. Cellulose nanowhiskers from coconut husk fibers: effect of preparation conditions on their thermal and morphological behavior. *Carbohydrate Polymers* 2010, 81, 83–92.

Rosa, S. M. L., Rehman, N., de Miranda, M. I. G., Nachtigall, S. M. B., Bica, C. I. D. Chlorine-free extraction of cellulose from rice husk and whisker isolation. *Carbohydrate Polymers* 2012, 87, 1131–1138.

Rostagno, M. A., Prado, J. M., Vardanega, R., Forster-Carneiro, T., Meireles, M. A. A. Study of recovery of fermentable sugars with subcritical water and carbon dioxide from palm fiber and grape seed. *Chemical Engineering Transactions* 2014, 37, 403–408.

Run Cang, S., Jeremy, T. Comparative study of lignin isolated by alkali and ultrasound-assisted alkali extraction from wheat straw. *Ultrasonics Sonochemistry* 2002, 9, 85–93.

Saleh, A. F., Kamarudin, E., Yaacob, A. B., Yussof, A. W., Abdullah, M. A. Optimization of biomethane production by anaerobic digestion of palm oil mill effluent using response surface methodology. *Asia-Pacific Journal of Chemical Engineering* 2012, 7, 353–360.

Sánchez, Ó. J., Montoya, S. Production of bioethanol from biomass: an overview. In *Biofuel Technologies – Recent Developments*, Gupta, V. K., Tuohy, M. G., Eds., Springer: Germany, 2013, pp. 397–441.

Sarkar, N., Ghosh, S. K., Bannerjee, S., Aikat, K. Bioethanol production from agricultural wastes: An overview. *Renewable Energy* 2012, 37, 19–27.

Sathitsuksanoh, N., George, A., Zhang, Y., Percival, H. New lignocellulose pretreatments using cellulose solvents: a review. *Journal of Chemical Technology and Biotechnology*, 2013, 88(2), 169–180.

Schell, D., Walter, P., Johnson, D. Dilute sulfuric acid pretreatment of corn stover at high solids concentrations. *Appl Biochem Biotechnol* 1992, 34–35, 659–665.

Scheller, H. V., Ulvskov, P. Hemicelluloses. *Annu Rev Plant Biol* 2010, 61, 263–289.

Shahbaz, K., Mjalli, F. S., Hashim, M. A., Al-Nashef, I. M. Eutectic solvents for the removal of residual palm oil-based biodiesel catalyst. *Sep Purif Technol* 2011, 81, 216–222.

Sheltami, R. M., Abdullah, I., Ahmad, I., Dufresne, A., Kargarzadeh, H. Extraction of cellulose nanocrystals from mengkuang leaves (*Pandanus tectorius*). *Carbohydrate Polymers* 2012, 88, 772–779.

Shinoj, S., Visvanathan, R., Panigrahi, S., Kochubabu, M. Oil palm fiber (OPF) and its composites: A review. *Industrial Crops and Products* 2011, 33(1), 7–22.

Shree, P. M., Anne-Sophie, M., Bruno, C., Claude, D. Production of nanocellulose from native cellulose – various options utilizing ultrasound. *BioResources* 2012, 7, 422–436.

Shuit, S. H., Tan, K. T., Lee, K. T., Kamaruddin, A. H. Oil palm biomass as a sustainable energy source: A Malaysian case study. *Energy* 2009, 34(9), 1225–1235.

Singh, B. S., Lobo, H. R., Shankarling, G. S. Choline chloride based eutectic solvents: magical catalytic system for carbon–carbon bond formation in the rapid synthesis of β-hydroxy functionalized derivatives. *Catal Commun* 2012, 24, 70–74.

Sjöström, E. *Wood Chemistry, Fundamentals and Applications*. Academic Press: New York, 1981, pp. 51–70.

Socha, A. M., Parthasarathi, R., Shi, J., Pattathil, S., Whyte, D., Bergeron, M., George, A., Tran, K., Stavila, V., Venkatachalam, S., Hahn, M. G., Simmons, B. A., Singh, S. Efficient biomass pretreatment using ionic liquids derived from lignin and hemicellulose. *Proceedings of the National Academy of Sciences of the United States of America* 2014, 111(35), E3587–E3595.

Solomon, T. W. G. *Organic chemistry*, 4th edn., John Wiley and Sons: New York, 1988.

Spagnola, C., Rodriguesa, F. H. A., Pereira, A. G. B., Fajardoa, A. R., Rubiraa, A. F., Muniz, E. C. Superabsorbent hydrogel composite made of cellulose nanofibrils and chitosan-graft–poly(acrylic acid). *Carbohydrate Polymers* 2012, 87, 2038–2045.

Stewart, D. Lignin as a base material for materials applications: Chemistry, application and economics. *Industrial Crops and Products* 2008, 27, 202–207.

Stewart, J. J., Akiyama, T., Chapple, C., Ralph, J., Mansfield, S. D. The effects on lignin structure of overexpression of ferulate 5-hydroxylase in hybrid poplar. *Plant Physiology* 2009, 150, 621–635.

Sulaiman, F., Abdullah, N., Gerhauser, H., Shariff, A. An outlook of Malaysian energy, oil palm industry and its utilization of wastes as useful resources. *Biomass and Bioenergy* 2011, 35(9), 3775–3786.

Sumathi, S., Chai, S. P., Mohamed, A. R. Oil palm as a source of renewable energy in Malaysia. *Renewable and Sustainable Energy Reviews* 2008, 12, 2404–2421.

Sun, J. X., Xu, F., Sun, X. F., Xiao, B., Sun, R. C. Physico-chemical and thermal characterization of cellulose from barley straw. *Polymer Degradation and Stability* 2005, 88, 521–531.

Sun, R., Hughes, S. Fractional extraction and physico-chemical characterization of hemicelluloses and cellulose from sugar beet pulp. *Carbohydrate Polymers* 1998, 36, 293–299.

Sun, X. F., Sun, R. C., Su, Y., Sun, J. X. Comparative study of crude and purified cellulose from wheat straw. *Journal of Agricultural and Food Chemistry* 2004, 52, 839–847.

Swatloski, R. P., Spear, S. K., Holbrey, J. D., Rogers, R. D. Dissolution of cellulose with ionic liquids. *Journal of the American Chemical Society* 2002, 124, 4974–4975.

Tadesse, H., Luque, R. Advances on biomass pretreatment using ionic liquids: An overview. *Energy and Environmental Science* 2011, 4, 3913–3929.

Taherzadeh, M. J., Karimi, K. Pretreatment of lignocellulosic wastes to improve ethanol and biogas production: A review. *International Journal of Molecular Sciences* 2008, 9, 1621–1651.

Thiam, L. C., Bhatia, S. Catalytic processes towards the production of biofuels in a palm oil and oil palm biomass-based biorefinery. *Bioresource Technology* 2008, 99, 7911–7922.

Torget, R., Teh-An, H. Two-temperature dilute-acid prehydrolysis of hardwood xylan using a percolation process. *Appl Biochem Biotechnol* 1994, 45–46, 5–22.

Uma Maheswari, C., Reddy, O. K., Muzenda, E., Guduri, B. R., Varada Rajulu, A. Extraction and characterization of cellulose microfibrils from agricultural residue – *Cocos nucifera* L. *Biomass and Bioenergy* 2012, 46, 555–563.

Vanholme, R., Demedts, B., Morreel, K., Ralph, J., Boerjan, W. Lignin biosynthesis and structure. *Plant Physiology* 2010, 153, 895–905.

Vázquez, M., Oliva, M., Téllez-Luis, S. J., Ramírez, J. A. Hydrolysis of sorghum straw using phosphoric acid: evaluation of furfural production. *Bioresource Technology* 2007, 98, 3053–3060.

Ventura, S. P. M., Marques, C. S., Rosatella, A. A., Afonso, C. A. M., Goncalves, F., Coutinho, J. A. P. Toxicity assessment of various ionic liquid families towards *Vibrio fischeri* marine bacteria. *Ecotoxicology and Environmental Safety* 2012, 76, 162–168.

Wanrosli, W. D., Rohaizu, R., Ghazali, A. Synthesis and characterization of cellulose phosphate from oil palm empty fruit bunches microcrystalline cellulose. *Carbohydrate Polymers* 2011, 84, 262–267.

Wanrosli, W. D., Zainuddin, Z., Lee, L. K. Influence of pulping variables on the properties of *Elaeis guineensis* soda pulp as evaluated by response surface methodology. *Wood Science and Technology* 2004, 38, 191–205.

Watanabe, T. Potential of cellulosic ethanol. In *Lignocellulose conversion: Enzymatic and microbial tools for bioethanol production*, Faraco, V., Ed., Springer Verlag: Berlin, 2013, pp. 1–20.

Wise, L. E., Murphy, M., D'Addieco, A. A. Chlorite hollocellulose, its fractionation and bearing on summative wood analysis and on studies on the hemicelluloses. *Paper Trade* 1946, 122, 35–42.

Wood, N., Ferguson, J. L., Nimal Gunaratne, H. Q., Seddon, K. R., Goodacre, R., Stephens, G. M. Screening ionic liquids for use in biotransformations with whole microbial cells. *Green Chemistry* 2011, 13, 1843–1851.

Xia, S., Baker, G. A., Li, H., Ravula, S., Zhao, H. Aqueous ionic liquids and deep eutectic solvents for cellulosic biomass pretreatment and saccharification. *RSC Advances* 2014, 4, 10586–10596.

Xu, W., Miller, S. J., Agrawal, P. K., Jones, C. W. Depolymerization and hydrodeoxygenation of switch grass lignin with formic acid. *Chem Sus Chem* 2012, 5, 667–675.

Yacob, S. Progress and challenges in utilization of oil palm biomass. In *Asian Science and Technology Seminar,* Japan Science and Technology Agency: Jakarta, Indonesia, 2007.

Yang, W., Li, P., Bo, D., Chang, H. The optimization of formic acid hydrolysis of xylose in furfural production. *Carbohydrate Research* 2012, 357, 53–61.

Yat, S. C., Berger, A., Shonnard, D. R. Kinetic characterization for dilute sulfuric acid hydrolysis of timber varieties and switch grass. *Bioresource Technology* 2008, 99, 3855–3863.

Yi, Z., Zhongli, P., Zhang, R. Overview of biomass pretreatment for cellulosic ethanol production. *International Journal of Agricultural and Biological Engineering* 2009, 2, 51–68.

Yoon, M., Choi, J., Lee, J. W., Park, D. H. Improvement of saccharification process for bioethanol production from *Undaria* sp. by gamma irradiation. *Radiation Physics and Chemistry* 2012, 81, 999–1002.

Yu, L. In *Biodegradable Polymer Blends and Composites from Renewable Resources.* John Wiley and Sons: Hoboken, New Jersey, USA, 2009.

Yuan, Z., Cheng, S., Leitch, M., Xu, C. Hydrolytic degradation of alkaline lignin in hot-compressed water and ethanol. *Bioresource Technology* 2010, 101, 9308–9313.

Yunus, R., Salleh, S. F., Abdullah, N., Biak, D. R. A. Effect of ultrasonic pre-treatment on low temperature acid hydrolysis of oil palm empty fruit bunch. *Bioresource Technology* 2010, 101, 9792–9796.

Zavrel, M., Bross, D., Funke, M., Büchs, J., Spiess, A. C. High-throughput screening for ionic liquids dissolving (ligno-) cellulose. *Bioresource Technology* 2009, 100, 2580–2587.

Zhang, H., Wu, J., Zhang, J., He, J. 1-Allyl-3-methylimidazolium chloride room temperature ionic liquid: A new and powerful nonderivatizing solvent for cellulose. *Macromolecules* 2005, 38, 8272–8277.

Zhang, Q., Benoit, M., De Oliveira, V. K., Barrault, J., Jérôme, F. Green and inexpensive choline-derived solvents for cellulose. *Decrystallization Chemistry* 2012, 18, 1043–1046.

Zhao, X. B., Wang, L., Liu, D. H. Effect of several factors on peracetic acid pretreatment of sugarcane bagasse for enzymatic hydrolysis. *Journal of Chemical Technology and Biotechnology* 2007, 82, 1115–1121.

Zhao, X. B., Wang, L., Liu, D. H. Peracetic acid pretreatment of sugarcane bagasse for enzymatic hydrolysis: a continued work. *Journal of Chemical Technology and Biotechnology* 2008, 83, 950–956.

Zhao, X. B., Liu, D. H. Fractionating pretreatment of sugarcane bagasse by aqueous formic acid with direct recycle of spent liquor to increase cellulose digestibility – the formiline process. *Bioresource Technology* 2012, 117, 25–32.

Zheng, Y., Zhao, J., Xu, F., Li, Y. Pretreatment of lignocellulosic biomass for enhanced biogas production. *Progress in Energy and Combustion Science* 2014, 42, 35–53.

Zhu, S. D., Wu, Y. X., Chen, Q. M., Yu, Z. N., Wang, C. W., Jin, S. W., Ding, Y. G., Wu, G. Dissolution of cellulose with ionic liquids and its application: a mini-review. *Green Chemistry* 2006, 8, 325–327.

CHAPTER 14

ANAEROBIC BIOREACTORS FOR THE TREATMENT OF CHLORINATED HYDROCARBONS

RICARDO ALFÁN-GUZMÁN, MATTHEW LEE, and
MICHAEL MANEFIELD

School of Biotechnology and Biomolecular Sciences, The University of New South Wales, Sydney, NSW, 2052, Australia

CONTENTS

14.1 INTRODUCTION

The fate and persistence of hazardous chemicals in the environment have been a concern for the last decades. Industrialization and extensive agricultural activities have led to their accumulation in the environment, becoming an evident risk for many ecosystems and human health, since most of them are known carcinogens. Among these substances, chlorinated hydrocarbons (also know as organochlorines) comprise a large spectrum of compounds that are of enormous industrial and economic importance because of their applications as insecticides, fungicides, solvents, hydraulic and heat-transfer fluids, cleaning agents, degreasers and intermediates for chemical syntheses (Reineke et al., 2011). They constitute the largest single group of compounds on the list of priority pollutants compiled by the U.S. Environmental Protection Agency (http://www.epa.gov/tio/download/remed/engappinsit-bio.pdf). A selection of chlorinated hydrocarbons is shown in Figure 14.1. Historically the management of these compounds, whether they are used as primary substances, or are waste products derived from other industrial processes, has resulted in the contamination of groundwater and soils all over the world. There exists therefore a strong need to develop novel techniques to reduce their environmental impact.

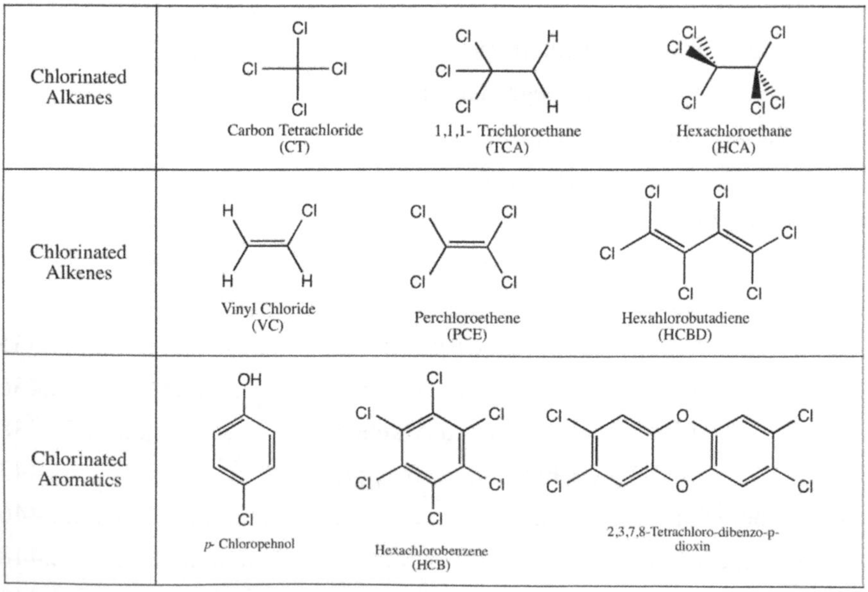

FIGURE 14.1 Chlorinated hydrocarbons.

Chlorinated hydrocarbons are volatile, posses varying solubility, and have long half-lives. These characteristics, among others, determine their fate and transport in the environment. The number of chlorine atoms in the molecule, directly affects their chemistry, as the number of substituted chlorines increases, so does the molecular weight and density, while vapor pressure and aqueous solubility decrease (http://www.epa.gov/tio/download/remed/engappinsitbio.pdf). Table 14.1 lists physical and chemical properties of some of the most commonly found chlorinated hydrocarbons. This selection can be found in either soil or water and some of them are manmade and some of them are generated from precursor manufacturing. Because of these properties they are highly stable and they resist abiotic and biological degradation in common naturally occurring environments. Further they display toxicity in acute doses and are classed as potential or proven carcinogens (US-EPA, 2000; Bhatt et al., 2006). This chapter will focus on biological technologies for *ex situ* destruction of organochlorines. Firstly, biodegradation mechanisms will be described. Secondly, biological reactors will be presented (different types, design and operational considerations). Finally, case studies of bioreactors treating chlorinated hydrocarbons under anaerobic conditions will be presented.

14.2 BIODEGRADATION MECHANISMS

There is a widely studied variety of microorganisms capable of undertaking biodegradation processes of chlorinated hydrocarbons, having as final end products less toxic compounds, or even reaching full mineralization. In this chapter only bacterial systems will be discussed. It is well known that the number and the position of the substituted chlorines determine the biological mechanism of degradation (McCarty, 1987). Compounds with a larger number of chlorines, such as PCE or HCB, are most likely degraded under anaerobic conditions by reductive processes, since they are highly oxidized. Less chlorinated compounds, like Dichloromethane (DCM) can be degraded under aerobic conditions (Distefano et al., 1991; Holliger and Schraa, 1994).

14.2.1 AEROBIC BIODEGRADATION

In the presence of oxygen, some organochlorines may be biodegraded to CO_2, H_2O and Cl^- either directly in an energy harvesting process or by

TABLE 14.1 Physical and Chemical Properties (modified from US-EPA, 2000)

Compound	Molecular weight (g/mole)	Aqueous solubility (mg/L)@25°C	Liquid density (g/mL) @4°C	Vapor pressure (mmHg) @25°C	Henry's Law dimension-less constant
Alkanes					
Carbon Tetrachloride (CT)	153.8	757	1.59	90	1.01
1,1,1-Trichloroethane (TCA)	133.4	1500	1.34	123	0.57
Chloroform (CF)	119.4	8200	1.48	151	0.122
Hexachloroethane (HCA)	236.7	0.05	2.09*	0.4**	0.115
Alkenes					
Perchloroethene (PCE)	165.8	150	1.62	17.8	0.579
Trichloroethene (TCE)	131.4	1100	1.46	57.9	0.335
Vinyl chloride (VC)	62.5	2670	Gas	2660	0.981
Hexachlorobutadiene (HCBD)	260.7	2–2.55	1.55	0.15	0.234
Aromatics					
Hexachlorobenzene (HCB)	284.7	0.01	2.004	1.09×10^{-5}	0.031

*@25°C, **22°C, 20°C.

cometabolism (Hartmans and De Bont, 1992). Direct oxidation occurs more often with less chlorinated compounds, like DCM or VC. This process yields carbon for cellular constituents and energy. On the other hand, highly chlorinated compounds, like PCE, are commonly oxidized fortuitously by cometabolism, in this case microorganisms obtain no benefit (Samprini, 1994). This section will only address direct oxidation.

14.2.2 DIRECT OXIDATION

It has been reported that some microorganisms can use chlorinated hydrocarbons, both aliphatic and aromatic, as sole carbon sources, while using molecular oxygen as electron acceptor in an energy yielding process (Hartmans and DeBont, 1992; Lee et al., 2000). For instance direct oxidation of VC by *Burkholderia cepacia* G4 or *Mycobacterium aurum* takes place inside the cell. In this process, chloroxirane, an unstable epoxide, is produced and it is known to be the limiting factor in VC biodegradation due to its toxicity. The breakdown of this epoxide is achieved by the enzyme epoxialkane coenzyme M transferase (Field, 2004); formic and oxiglycolic acids are also formed (Figure 14.2), finally full mineralization is reached (Hartmans and DeBont, 1992). Methylotrophs like *Methylophilus* sp. DM11, *Hyphomicrobium* sp. DM2 and *Methylobacterium* sp. DM4 are known for being capable of biodegrading DCM as a sole carbon source (Trotsenko and Torgonskaya, 2009). Liu and Chen (2009) demonstrated that under aerobic conditions, species like *Azospirillum* and *Alcaligenes* are able to use HCB as a sole carbon source.

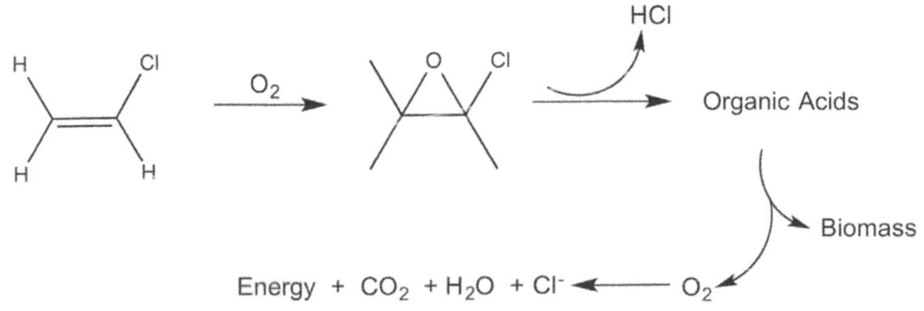

FIGURE 14.2 Aerobic biodegradation of vinyl chloride (modified from Kielhorn, 2000).

14.2.3   ANAEROBIC BIODEGRADATION

In the anaerobic mode of degradation the organochlorine serves as electron acceptor instead of O_2. Reductive dechlorination generally involves the sequential replacement of a chlorine atom on a chlorinated hydrocarbon with a hydrogen atom (Figure 14.3) and has been observed to occur both directly and cometabolically. In direct dechlorination, the mediating bacteria use the chlorinated hydrocarbons as an electron acceptor in energy-producing redox reactions. Cometabolic mechanisms occur under iron reducing, sulfate reducing or methanogenic environments when bacteria incidentally dechlorinate organochlorines in the process of using another electron acceptor to generate energy. Dehalofermantation is another well-known mechanism. During this process, bacteria use chlorinated compounds as sole carbon and energy sources in the absence of an exogenous electron acceptor (US-EPA, 2000; Smidt and de Vos, 2004; Reineke et al., 2011; Lee et al., 2012).

Biological reductive dechlorination is a type of anaerobic respiration in which a chlorinated compound is used as a terminal electron acceptor. Reductive dechlorination enables the conservation of energy via electron transport phosphorylation (ATP formation), this process may occur through different reaction-mechanisms; *hydrogenolysis*, in which one chlorine atom is removed and replaced by hydrogen or by *dehaloelimination* which implies the removal of two adjacent chlorines, leading to the formation of an additional carbon-carbon bond (Holliger et al., 1990; Middeldorp et al., 2010). The process is also known as *organohalide respiration*. Hydrogen is commonly used an electron donor in such cases and is typically supplied indirectly through the fermentation of organic substrates like lactate, acetate, pyruvate or methanol (Smidt and de Vos, 2004). For *organohalide*

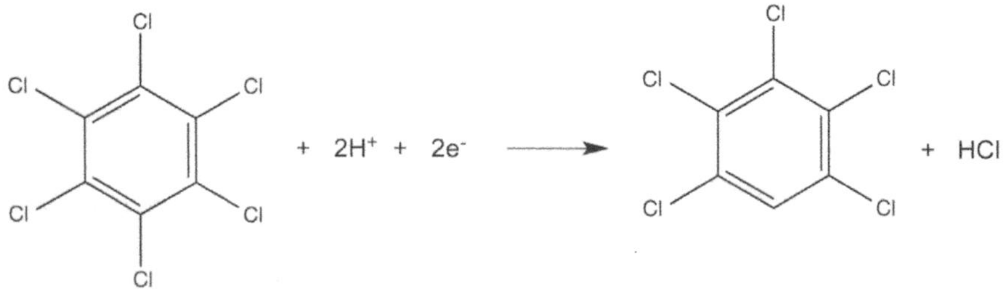

FIGURE 14.3   Reductive dechlorination reaction of Hexachlorobenzene.

respiration to be thermodynamically favorable, redox potential must be sufficiently low, -300 mV to -500 mV (Suthersan, 2001). Based on Gibb's free energy values, *organohalide respiration* is more energetically favorable than some of the other anaerobic respiratory processes. For instance, the $\Delta G^{0'}$ for dehalogenated HCB using hydrogen as electron donor is -494.3 kJmol^{-1}, which is more than the $\Delta G^{0'}$ for sulfate reduction (-153 kJmol^{-1}) or methanogenesis (-131 kJmol^{-1}) (Thaur et al., 1977).

Bacterial organohalide respiration is the most important process in detoxification of chlorinated compounds under anaerobic conditions. Organohalide respiring bacteria (ORB) known to date belong to the low G-C Gram positive bacteria (*Desulfitobacterium* and *Dehalobacter*) and the proteobacteria *Desulfomonile*, *Desulfuromonas*, *Dehalospirillum* and *Dehalococcoides* (*Chloroflexi* phyla). New organohalide respiring microorganisms, which form a physiologically and phylogenetically coherent group designated *Anaeromyxobacter delogenans* and represent the first taxon in the *Myxobacteria* capable of anaerobic growth, have also been isolated (Reineke et al., 2011; Hug, 2013). The keystone in *organohalide respiration* is a group of enzymes named reductive dehalogenases (RD). These are single-polypeptide membrane-associated anaerobic enzymes that are synthesized as preproteins with a cleavable twin arginine translocation (TAT) peptide signal (Bisaillon et al., 2010). Their molecular weight ranges between 50 to 65 kDa and they possess a monomeric corrinoid dependent structure, usually containing a cobalamin cofactor, like cyanocobalamin (Vitamin B12), which is also a nutritional requirement for the growth of ORB. It is assumed that these corrinoids are located at the active site of the enzymes based on the ability of cobalamins to catalyse dehalogenation in the presence of low-potential electron donors (Lee et al., 2010; Reineke et al., 2011; Schipp et al., 2013), as shown on Figure 14.4. Two iron-sulfur clusters have also been identified in these enzymes, excepting the chlorobenzoate-reductive dehalogenase of *Desulfomonile tiedjei* (Copley, 2003; Reineke et al., 2011; Wagner et al., 2013). Whilst the presence of Vitamin B12 in the media or the environment is crucial for ORB, not all ORB can synthesize the whole molecule or a precursor *de novo*, therefore they need other species to produce it or if grown as pure cultures, exogeneous addition is required. For instance, *Dehalococcoides mccartyi* CBDB1 isolates require low concentrations of this vitamin (<10 nM) to reductively dechlorinated HCB, using only acetate as a carbon source and H$_2$ as an electron donor (Adrian et al., 2000; Hölscher et al., 2003).

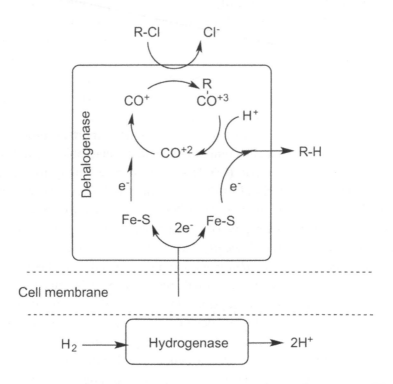

FIGURE 14.4 Oraganohalide respiration in *Dehalospirillum multivorans* (modified from Middeldorp et al., 2010).

14.3 ANAEROBIC BIOREACTOR ENGINEERING

Contaminated soil or groundwater can be remediated *in situ*, by direct exogenous addition of nutrients like reduced organic carbon, nitrogen or phosphate or any other essential compound to stimulate growth of indigenous bacteria (bioestimulation) or by addition of non-indigenous, actively growing, specialized microbial strains (bioaugmentation). *In situ* remediation is more often used for contaminated groundwater. Some advantages of *in situ* remediation are that water or soils can be treated without the need of being pumped out, excavated or transported elsewhere, thereby reducing operational costs. Nevertheless, it requires longer periods of time and distribution or diffusion of nutrients is never uniform. On the other hand, *ex situ* treatments like biopiles (excavated contaminated soils mixed with amendments and placed on a treatment area) or bioreactors, are easier to control and monitor and homogenous distribution of nutrients can be achieved leading to shorter operation times (http://www.epa.gov/oust/pubs/tums.htm). This section will focus in describing different types of bioreactors.

For a long time bioreactors have been used for a wide variety of purposes, from the fermentation of carbohydrates to produce wine or beer in antiquity to the more recent production of monoclonal antibodies. Such systems have also been used in the environmental field, mostly in the treatment of wastewaters. In this section some of the most commonly used anaerobic bioreactors will be presented. Anaerobic treatment processes are characterized by a lower energy demand due to the absence of aeration, slower microbial growth rates, lower chemical oxygen demand (COD) removal, longer start-up periods, high alkalinity, low sludge production and biogas production (Stephenson et al., 2000). Bioreactors for the treatment of wastewaters or for the destruction of specific toxic compounds need to maintain higher biomass densities, which can then provide greater resistance to any inhibitory substances in the influent, thus, loading rates are primarily dictated by the concentration of active biomass (Stephenson et al., 2000; Liao et al., 2006). Apart from temperature, cell density and pH, among others basic parameters, there is a wide range of features that are considered during design and operation. These include:

- Reactor volume (V_r). Total reactor volume V_r:[L or m^3].
- Working volume (V_o). Reactors are never operated to their maximum capacity, because mechanical agitation increases the liquid height in the reactor, gas production increases the pressure inside the vessel or there might be foam formation (Doble et al., 2004; Kennes et al., 2009). Therefore, bioreactors are almost always operated, by convention, at 70% of the Vr. V_o:[L or m^3].
- Flow Rate (Q). Volume of fluid, which passes through a given area per unit of time. Flow rate determines the hydraulic retention time, directly affects mass transfer, mixing patterns and granule formation and may increase shear stress on the cells (Villadsen et al., 2011; Taweel et al., 2012). Q:[m^3 h^{-1}].
- Hydraulic Retention Time (HRT). This parameter is a measure of the average length of time that a soluble or solid compound remains in a reactor. Its value is calculated by dividing the reactor volume (V_r) by the inlet flow rate (Q), HRT:[hours or days] (Villadsen et al., 2009).

HRT is extremely important since it affects the reaction between the microorganisms and the target compounds and influences sedimentation time, especially in reactors like Upflow Anaerobic Sludge Blankets (UASB) (Lettinga et al., 1997; von Sperling, 2007; Lettinga, 2010).

- Organic Loading Rate (OLR). OLR is defined as the weight of organic matter per day applied over volume of the reactor. This parameter is important for the operation of digesters and systems based on sedimentation or granulation, since it determines the biological activity that will take place in the system (Babaee and Shayagan, 2011); OLR:[kg L^{-1} d^{-1}].
- Removal Rate. Complete of partial transformation of a target compound, expressed as percentage.
- Steady state. This is one of the most important concepts for bioreactor operation. A system reaches the steady state when all the state variables (entropy, temperature, pressure) are constant. Specifically, in biological systems, this is achieved when cells grow at a constant rate and culturing parameters also remain constant (dissolved gases, nutrient consumption, biodegradation rates, pH, etc.). The operation of bioreactors for biodegradation purposes must be performed once this state has been achieved. However, in anaerobic systems, particularly if working with a complex microbial community, this might imply longer operation times (compared to aerobic processes), since anaerobic bacteria are known to posses slower doubling times and growth rates may vary from species to species in a diverse community (Crécy et al., 2007; Villadsen et al., 2011).
- Feeding regime. Bioreactors can be operated under three different kinds of feeding regimes: (a) Batch mode, where the volume of the reactor remains constant, inoculum, media and other reagents are added at the beginning of the process; (b) Fed-batch, under this regime all components of the process are added at the beginning, but throughout the process, any of these can be added as required, raising the volume as a consequence; finally, (c) Continuous mode, under this condition, media is fed to the system at the rate as it exits, keeping the volume constant (Stoner, 1993; Villadsen et al., 2011).

Common anaerobic bioreactor configurations are designed to obtain good mixing and biomass separation. This can be achieved by filtration, sedimentation and digestion before returning the clarified or treated liquid to the bioreactor; use of anaerobic filters or upflow clarification or the use sequential reactors systems (Stephenson et al., 2002). Some of the most utilized anaerobic reactor configurations are briefly described below.

14.3.1 EXAMPLES OF ANAEROBIC BIORFACTORS

Continuous stirred-tank reactor (CSTR): In this steady state system, reactants and products are continuously added, mixed and withdrawn. As reactants enter the tank, they are immediately diluted, which in many cases favors the desired reaction or decreases the impact of toxic byproducts (Figure 14.5). Fast mixing also allows easy control over exothermic reactions. Mechanical (propellers) or hydraulic (air spargers) agitation is required to achieve uniform composition (Stoner, 1993). CSTRs are considered ideal systems since all the components in the tank are homogenously distributed, excluding the formation of death zones (regions of low or no mixing) or hot spots (regions with higher temperatures) (Stoner, 1993; Coulson, 1994). The use of CSTRs, besides the already listed advantages, provides a source for yielding large volumes of steady state cells or proteins. In the environmental field, toxic or recalcitrant compounds, like TCE or PCE, can be reductively

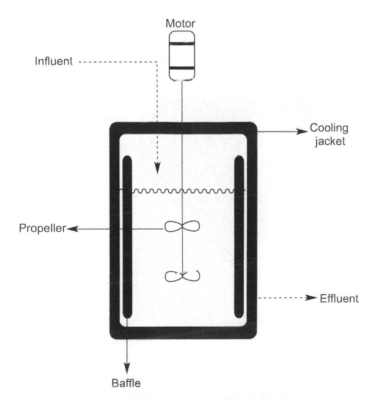

FIGURE 14.5 Continuous stirred-tank reactor (CSTR).

dechlorinated in CSTRs by continuously maintaining low concentrations of these compounds. Whilst these systems provide low operating costs (unless cooling jackets are needed) and are easy to clean, CSTRs have the lowest conversion rates (Coulson, 1994; Delgado et al., 2014).

Upflow Anaerobic Sludge Blanket (UASB): Industrial scale UASBs have been used for more than two decades in the treatment of municipal wastewater. The main characteristics of this reactor type are the upflow feeding and the formation of a flocculent or granular blanket with good sedimentation capacity, where biological degradation takes place (Figure 14.6). This system keeps the biomass in the reaction compartment, where the sludge is developed, the influent comes in through the bottom of the reactor causing

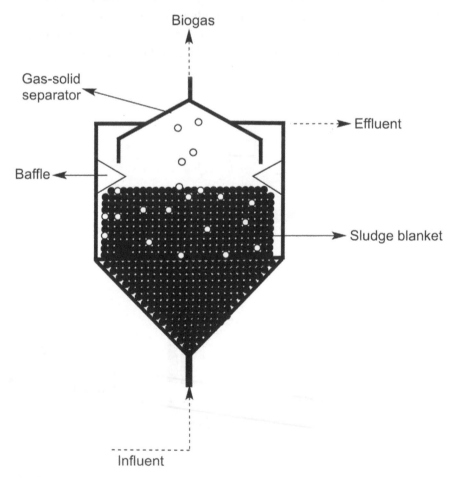

FIGURE 14.6   Upflow anaerobic sludge blanket (UASB).

recirculation of the biomass, this recirculation is achieved by sedimentation of the solids in the settling compartment, followed by return by simple gravity to the reaction compartment (Lettinga et al., 1991; von Sperling, 2007). In UASB processes, activated sludge is fed together with the waste with no prior sedimentation needed. The upflow inlet forms small granules or particles that remain in suspension and provide a large surface on which organic matter can attach and undergo biodegradation (Arceivala and Asolekar, 2007). Since these systems are operated continuously, there is no mechanical mixing required. UASB reactors usually posses a gas/solid separator, which allows the biogas formed to exit the tank freely. In wastewater treatment this biogas is often methane. The liquid effluent exits the reactor from the top of the tank where it can undergo further treatment or be recirculated to the tank, if needed, increasing HRT. These reactors have also been used in the treatment of toxic compounds, like organochlorines for long periods of time maintaining stable dechlorination rates (Sponza, 2001; Basu and Asolekar, 2012). These bioreactors do not need cooling jackets or any other system to keep the temperature constant and can be operated at room temperatures 18–20°C. As mentioned before, mixing in the system is provided by the constant upflow feeding, therefore no mechanical or hydraulic mixing systems are needed, keeping construction and operation costs low. Some disadvantages are the low toxicity resistance, since biomass can be slowly washed out of the system, thus low HRT most be avoided, nevertheless high HRT may lead to higher capital costs; start up times and recovery from stressed conditions can be protracted (Lettinga et al., 1991; Bal and Dhagat, 2001; Arceivala and Asolekar, 2007).

Membrane Bioreactor (MBR): The use of MBRs is often required under extreme conditions like high temperatures or salinity, where biofilm, floc or granule formation cannot be achieved or when full biomass retention must be assured (Lettinga et al., 1997). The aim of MBRs is to concentrate either biomass or sludge to increase contact time between the target compound and the microorganism, thus increasing biodegradation. To accomplish this, micro or ultra filtration processes are applied (Szentgyörgyi and Bakó, 2010), hydrophobic membranes used for this purpose often have pore sizes around 0.1 μm, ensuring biomass retention, producing a substantially clarified effluent or permeate. For anaerobic biodegradation processes, gas transfer, especially H_2, is extremely important, since it serves as an electron donor for reductive processes. These systems can provide this gas directly to the biomass attached to its surface guaranteeing 100% transfer (Judd, 2006).

There are two different MBR configurations. In one configuration the membrane is located inside the reactor and the permeate exits the membrane while still being inside the reactor. This process operates at lower fluxes (F=Flow rate/time), but requires higher membrane surface area. In another configuration, known as an external loop MBR, the membrane is located outside the tank (Figure 14.7). This configuration is capable of elevated flow and higher trans-membrane pressure, and allows an easier clean up of the system (Szentgyörgyi and Bakó, 2010). MBRs face several operational hurdles like organic fouling, which is a typically caused by accumulation of colloidal material and bacteria on the membrane, requiring high liquid velocities across the membrane or agitation systems to minimize fouling.

Anaerobic filters: Support media is used to provide a resistant surface to which microorganism can attach. In this process the inlet flow runs upwards through a column packed with these materials (glass beads, polyurethane, granular activated carbon, sand, foam polymers, granite, etc.). Pollutants are then dissolved and then absorbed by the biofilm (Figure 14.8). Packaging materials are chosen based on the extent of their surface area, void volumes and resistance to compaction (Aizpuru et al., 2005; Kennes et al., 2009). These systems are more often used in treatment of gas or extremely volatile compounds; therefore, the gas stream must be dissolved in water to be

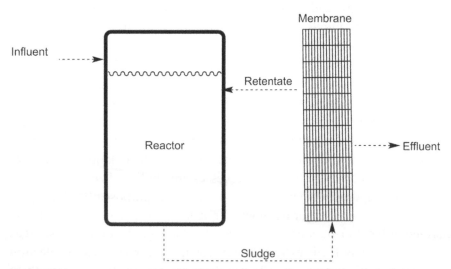

FIGURE 14.7   External-loop membrane bioreactor (modified from Szentgyörgyi and Bakó, 2010).

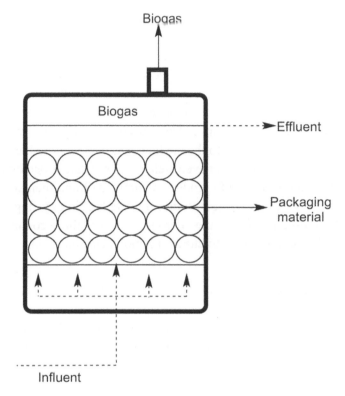

FIGURE 14.8   Biofilter.

accessible to the microorganisms present in the biofilm. This can be achieved through direct circulation of the contaminated gas stream or by dissolving the inlet stream in a liquid phase located outside the reactor, known as *bio-washing* (Devinny, 1998; Kennes et al., 2009). High gas-liquid contact surfaces, reaction control, transference rate between the biomass and the pollutant, low start-up costs and relatively easy operation are some advantages of these systems. Disadvantages are low recovery time after fluctuations, difficult control over parameters such as pH and clogging as a result of biomass development (Aizpuru et al., 2005; Kennes et al., 2009).

14.3.2   REACTORS FOR THE ANAEROBIC BIODEGRADATION OF CHLORINATED HYDROCARBONS: CASE STUDIES

In this section some examples of different types of bioreactors operated anaerobically to affect reductive dechlorination will be considered. Most of them are modifications of the bioreactors described in section 14.3.1 including changes in the direction of the inlet flow, different packing materials and ways of delivering electron donors to the system, among others. It is important to emphasize that the majority of these systems have only been tested on laboratory-scale and only few of them have used pure cultures as inoculum. Most of them provide a fermentable organic compound, like lactate or methanol, as a carbon and energy source. No cobalamins, vitamins or any other cofactor is added exogenously to the medium, suggesting possible commensal interactions among bacteria or even archaea. For instance, bacteria present in microbial communities could produce Vitamin B12 and then supply it to ORB (Figure 14.9). In only a few studies, molecular biology fingerprinting tools have been used to identify the microorganisms in these communities in an attempt to link community structure with bioreactor function.

14.3.2.1   Chlorinated Alkanes

In 2002, Shlötelburg and collaborators conducted studies on reductive dechlorination of 1,2-Dichloropropane (DCP) using a mixed culture

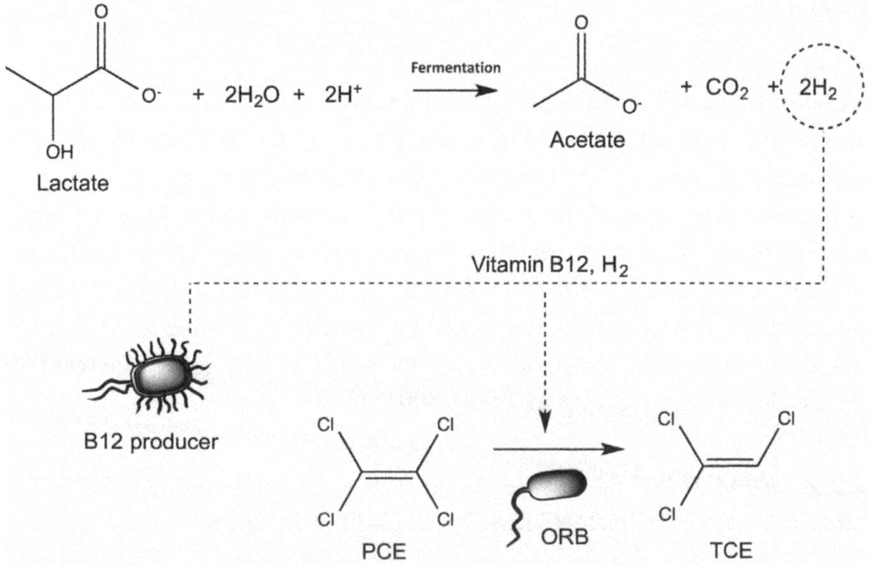

FIGURE 14.9   Suggested interaction mechanism in mixed cultures during the reductive dechlorination of organochlorines.

taken from sediments in a location close to a sewer of a propyleneoxide-producing plant. Microorganisms capable of dechlorinating DCP were enriched and immobilized on 1 cm³ polyurethane foam cubes and introduced into a 5 L stainless steel fluidized-bed reactor (biofilter) containing mineral salt medium with 7 mg L⁻¹ DCP and 50 mg L⁻¹ methanol, HRT 24 h, 30°C and pH 6.2. The system was operated continuously for 4 months and it was proven that 95% of the DCP was transformed to propane during the operation. Microbial community analysis was also performed, showing abundance of *Dehalococcoides* and *Dehalobacterrestrictus*-like bacterial strains, known for the dechlorination of TCE. Reductive dechlorination of 1,1,1-TCA and CT was tested using a packed bed reactor in the form of a column packed with CT contaminated sediments (Niemet and Semprini, 2005). In this study, the bioreactor was operated initially in batch mode and subsequently continuously, with and without the addition of nitrate as an additional electron acceptor. Benzoate and acetate were fed as carbon sources/electron donors in separate experimental sets. The system was kept at 22°C, pH varied between 7.6 and 7.9. Synthetic groundwater solution amended with 250 μg L⁻¹ CT and 1,1,1-TCA was fed at a rate of 340 mL d⁻¹, HRT varied between 0.5 and 2 days. Batch results showed removal of 91% of both TCA and CT when the system was fed with these two compounds as the only electron acceptors and benzoate was the electron donor. On the other hand, continuous operation of the system led to less comprehensive removal of the target pollutants. This study stated that CT removal was linked to biological processes; nevertheless, no clear evidence of this was shown. 1,1,2-TCA biodegradation under methanogenic conditions was evaluated in a 12.5 L acrylic UASB reactor, operated continuously with 5 different HRTs (12, 18, 24, 30, 36 hours). The reactor was inoculated with 5 L of anaerobic granular sludge from another UASB reactor, used to treat chlorinated aromatic compounds. The system was fed with artificial wastewater containing 1,1,2-TCA in different concentrations (5–40 mg L⁻¹), acetate as carbon source, initial ORL was 2 kg COD m⁻³d⁻¹ and at pH 7.6. A control reactor with no 1,1,2-TCA was also set up. This study showed the importance of HRT in dechlorination activity driven by microbial communities under granulation conditions. It was observed that 1,1,2-TCA removal decreased from 99.8% to 96.5% when HRT was reduced from 36 to 12 hours. Importantly, with lower HRT the accumulation of 1,1,2-TCA and Dichloroethane (DCA) was observed (Basu and Asolekar, 2012).

Cyclic chlorinated alkane biodegradation has also been studied in bioreactors. In 2006, Quintero and coworkers assessed the potential use of stirred-tank bioreactors to degrade Hexachlorocyclohexane (HCH) isomers contained in soil slurry cultures. A 5 L reactor equipped with a single turbine propeller was used. The system was inoculated with 400 g of soil spiked with 100 mg kg^{-1} of each HCH isomer (α, β, γ and δ), sodium bicarbonate was used as a pH buffer and sodium sulphide as reducing agent. A mixture of volatile fatty acids (VFA; acetic, propionic and butyric 4:1:1) was fed as a source of carbon and energy. The reactor was operated at 30°C at pH 7 and 350 rpm for 1 year. Sludge, HCH concentration and carbon source (VFA replaced for starch) were modified throughout the experiment duration. The results of this experiment demonstrated that when the system was operated with a sludge concentration of 8 g L^{-1} (as volatile suspended solids, VSS), 25–100 mgHCH kg^{-1} and starch (2 gCOD L^{-1}) as carbon source, α- and β-HCH were completely degraded within 10 days. On the other hand, 90% of γ- and δ-HCH was removed after 50 days of incubation, in this case, biodegradation products like dichlorobenzenes were detected. No microbial community analysis was performed.

14.3.2.2   Chlorinated Alkenes

Popat and Dehusses (2009) performed lab-scale batch trials in a 2.2 L anaerobic biofilter using a mixed culture containing several *Dehalococcoides* strains, among other bacteria, for the treatment of a vapor stream containing TCE. Sodium lactate, was fed 10 times in excess of the stoichiometric requirement for production of hydrogen necessary for complete dechlorination, temperature was kept at 25°C. The effect of pH was evaluated by changing the concentration of buffers in the medium (KH_2PO_4, K_2HPO_4 and $NaHCO_3$). After 200 days of operation, 90% transformation was achieved, with *cis*-DCE, VC and ethene produced as the main end products. Formation on any of these was directly related to pH variations. At pH 8.3 *cis*-DCE formation was predominant, while at 6.85–6.9 ethene production increased. In another study focusing on TCE reduction a 2 L UASB reactor, commonly used in treatment of wastewater, was inoculated with sludge and fed with glucose as a carbon and energy source and TCE as electron acceptor. Over more than 230 days of continuous operation at 37°C and a HRT of 7 h, 90% TCE and 94% COD removal was observed at loading rates as high as 160 mg

TCE L^{-1}d^{-1} and 14 g L^{-1}d^{-1} (Sponza, 2001). UASB reactors are known for the formation of granules, consisting of a diverse group of microorganisms in sludge providing increased resistance to xenobiotics (Lettinga et al., 1991). This demonstration revealed UASB systems to be a suitable option, not only for the treatment of wastewater, but also for toxic compounds such as organochlorines.

Fathepure and Tiedje (1994) evaluated the role of *Desulfomonile tiedjei* DCB-1 in the dechlorination of both PCE and 3-chlorobenzoate (3-CB) in a biofilm column. The configuration of the system included a 50×2.5 cm glass column packed with 0.3 cm diameter glass beads as bacterial support. The column was equipped with a water jacket for maintaining the biofilm at 35°C. The reactor reached steady state after 4 months of continuous operation. To maintain anaerobic conditions, sodium sulphide was added (125 mg L^{-1}). The maximum observed dechlorination rates of PCE and 3-CB fed at 6 μM and 1 nM were 2.0 and 414 μmol L h^{-1}, respectively. This corresponds to a PCE consumption rate of 3.7 nmol h^{-1} (mg of protein)$^{-1}$ [88.9 μmol (g of protein)$^{-1}$d^{-1}]. The rate of PCE dechlorination increased from 2.0 to 10.3 μmol L^{-1} h^{-1} when the influent PCE was increased from 6 to 120 μM, respectively. Concentrations of 60 μM and above decreased reactor performance. PCE was mainly converted to TCE and *cis*- and *trans*-DCE at all the tested flow rates. Vinyl chloride (VC) was not detected, thus suggesting dechlorination of PCE to non-chlorinated products, but no ethene formation was reported. Chlorinated alkanes such as 1,1,2-TCA (10.7 μM) and CF (7.54 μM) were also tested. These were converted to 1,2 DCA and Dichlormethane (DCM) at 37% and 68.8%, respectively.

In 2008, Hwu and Lu performed experiments testing biodegradation of PCE in UASB reactors, focusing on finding an optimum HRT that could lead to higher removal rates. In a 2 L glass reactor, 700 mL of anaerobic sludge from a food-processing waste treatment system were added. Two HRT were tested (1 and 4 days) and the reactor was operated continuously in the dark at room temperature, with a PCE loading rate of 3 mg L^{-1} d^{-1}. Lactate and sucrose were provided as carbon sources (2:1 ratio based on COD) at a loading rate of 3125 mgCOD L^{-1} d^{-1}. Dechlorination of PCE was evaluated by liquid effluent and headspace gas analysis. After almost 40 days of operation, it was shown that 4 days was the optimum HRT value, since PCE removal increased from 51% to 87%, when compared to HRT of 1 day. Biotransformation rate values also changed significantly from 10.5±2.3 to 21.3±3.7 μmol d^{-1}, for HRTs of 4 and 1 days respectively. Higher HRT showed that ethene was the main breakdown product, also with low methane production. The presence of *Dehalococcoides*-like cells was detected in the granules formed in the reactor using Fluorescence *In Situ* Hybridization (FISH) with a 16S rRNA-targeted oligonucleotide. When HRT time was higher, so was the number of *Dehalococcoides*-like species. This study concluded that higher HRT increases the contact between microorganisms and chlorinated ethenes, therefore, increasing *organohalide respiration*.

Almost all the systems described, rely on fermentation of organic compounds like lactate or glucose to provide H_2 to ORB in order to reductively dechlorinate chlorinated hydrocarbons, nevertheless, there are systems that directly provide H_2 by sparging, H_2-generating electrodes or by diffusion through a gas transfer membrane (Chung et al., 2007), however, exogenous addition of this gas implies high operational and safety risks, due to the flammability of H_2. In 2007, Chung, Brown and Rittmann, conducted a study on TCE reductive dechlorination in a H_2-based denitrifying membrane-biofilm reactor (MBfR). In this type of bioreactor (Figure 14.10), hydrogen is directly delivered as an electron donor by diffusion through a membrane wall. Bacteria living on the surface of an outer support, oxidize H_2, the generated electrons are then used by ORB to reduce chlorinated compounds (Chung et al., 2007; Martin and Nerenberg, 2012).

The design of this reactor included a glass tube containing a bundle of hollow-fiber membranes, a composite bubble-less gas transfer membrane, where H_2 flowed through. Inoculation was made with a mixed culture previously exposed to perchlorate and high nitrate concentrations. TCE was fed through the influent at a rate of 0.02 mL min^{-1} and other electron acceptors (nitrate and sulfate) were also fed in the influent. After more than 120 days of continuous circulation, approximately 93% of TCE was transformed to VC. DNA extractions, PCR and clone libraries was performed to determine the species in the microbial community; such analysis showed the presence of *Dehalococcoides mccartyi* strains CBDB1, FL2 and BAV1, among other species, including denitrifiers. Interestingly, no acetate (only carbon source used by *D. mccartyi* strains) was added to the influent, suggesting that the autotrophic reactions involving CO_2 and H_2 must have produced at least traces of acetate to sustain *D. mccartyi* growth (Chung et al., 2007).

Bioreactors are not only designed in the hope of being utilized on-site, they are also used to produce ORB biomass to be applied in contaminated sites for bioaugmentation purposes. In 2014, Delgado and coworkers used a lab-scale continuous stirred-tank reactor (CSTR) to grow a TCE respiring

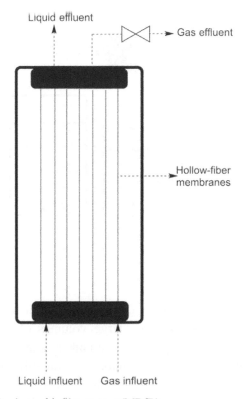

FIGURE 14.10 Membrane-biofilm reactor (MBfR).

consortium, containing *Dehalococcoides maccartyi* and *Geobacter*. In this reactor configuration, biomass floats freely in the medium with no attachment to any support surface. In this study, the system consisted of a series of three reactors, as shown in Figure 14.11. The first vessel contained fresh non-inoculated medium which was pumped to the second reactor that contained the consortium (both magnetically stirred). Finally the third vessel served as a collecting system from the effluent.

The minimal salt medium was fed at a 4 day HRT and amended with 3 mM TCE, 2 mM lactate, 15 mM methanol, 5 mL L^{-1} ATTC® MD-VS™ vitamin supplement and 500 mg L^{-1} Vitamin B12. A mixture of cysteine and sodium sulphide was used as a reducing agent, sodium bicarbonate was used as a buffer (pH 6.5 to 7.5) and temperature was kept at 30°C. No H$_2$ was added. Over 120 days, reductive dechlorination of TCE was sustained at influent concentrations of 1 and 2 mM, with 97% converted to ethene. The *Dehalococcoides* biomass concentration reached 10^{12} cells L^{-1}

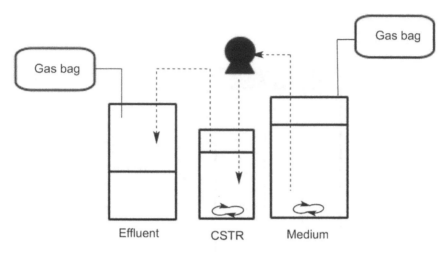

FIGURE 14.11   CSTR system (modified from Delgado et al., 2013).

(Delgado et al., 2014). Vainberg, Condee and Steffan in 2009 conducted a scaling up experiment to produce large volumes of *Dehalococcoides*-containing consortia. In this study stirred-tank bioreactors were used for batch culturing of three different commercially available consortia for three sites contaminated with chlorinated solvents, PCE or TCE. The scaling up process was carried out in three steps and finally, cells were harvested by membrane filtration and stored at 4°C. *Dehalococcoides* cell numbers were determined using quantitative PCR (Figure 14.12):

- Reactor I. A 20 L stirred-tank reactor (with propeller) was steam sterilized and flushed with N_2 to maintain anaerobic conditions before inoculation with 2 L of the corresponding consortium. The tank was filled up to 16–18 L with minimal salt medium. After inoculation, 10% yeast extract solution and PCE or TCE (10 mg L^{-1}) were added. The system was kept at 30°C, agitated at 100 rpm and pH was maintained between 6.4 and 7.2 by NaOH additions and sparging N_2 to remove dissolved CO_2. PCE/TCE were added constantly when the concentration of cDCE was reduced to 1–3 mg L^{-1}. When the cultures reached an optical density of 1 at 550 nm, they were transferred to Reactor II.
- Reactor II. In this 750 L stirred-tank bioreactor (also with a propeller), the consortium grown on Reactor I served as seed for this system. The final volume of medium was 550 L. Operational parameters such as pH (no NaOH was utilized) and temperature were the same. Agitation velocity was decreased to 60 rpm. Sodium lactate was fed continuously

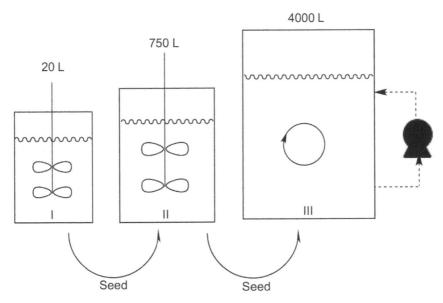

FIGURE 14.12 Scaling up bioreactors for the production of *Dehalococcoides*-containing consortium.

at a flow rate of 0.02–0.04 mL h^{-1} L_{medium}^{-1}. PCE/TCE addition was kept at a rate of 0.9–1.2 mL h^{-1} L_{medium}^{-1}. After 13–15 days of operation, the culture in this reactor reached 10^{10}–10^{11} *Dehacoccoides* L^{-1}.

- Reactor III. A 4000 L (4 m^3) reactor (with a working volume of 3200 L) was inoculated with the content of Reactor II. This system was not agitated by an impeller, instead, a centrifugal pump was used to recirculate the medium throughout the culturing process. Substrate feeding and other operational parameters were the same as in Reactor II. After 25 days of operation, the final biomass concentration in Reactor III, was 10^{11}–10^{12} *Dehalococcoides* L^{-1} and dechlorinating activity was approximately 80 mg PCE h^{-1} g dry weight^{-1}.

14.3.2.3 Chlorinated Aromatics

Anaerobic biodegradation of chlorinated aromatics has been conducted using packed columns with sand or sediments. Bosma et al. (1998) studied the biodegradation of chlorinated benzenes on packed columns with Rhine River sediments under methanogenic conditions. This study revealed that Hexachlorobenzene (HCB), Pentachlorobenzene (PCB), Tetrachlorobenzene

(TeCB) and Trichlorobenzene (TCB) were broken down to 1,3- and 1,4-Dichlorobenzene (DCB); when 1,3- and 1,4-DCB were tested, respectively, Monochlorobenzene (MCB) was the only end product. Finally when MCB tests were conducted no dechlorination was observed. No microbial community analysis was performed. Several bioreactor configurations exist offering a wide range of advantages in the treatment of wastewaters containing recalcitrant compounds (Bolaños et al., 2001). Horizontal Anaerobic Immobilized Bioreactors (HAIBs) are an example of these technologies. These bioreactors are characterized by the predominant plug-flow regime, which allows the development of different microbes as a function of substrate availability and consumption; therefore, intermediate metabolites resulting from degradation in the first regions of the reactor can be degraded further down (Bolaños, 2001; de Nardi, 2002).

Damainovic et al. (2009) designed a HAIB (Figure 14.13) to assess its potential to dechlorinate Pentachlorophenol (PCP). Two 0.3 L reactors were packed with cubic polyurethane foam matrices containing immobilized anaerobic sludge. Both reactors were fed continuously with glucose, acetic

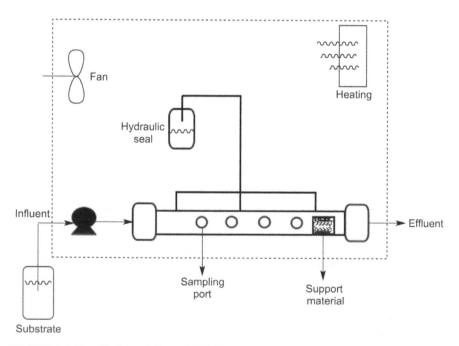

FIGURE 14.13   Biodegradation of PCP in a HAIB reactor (modified from Damainovic et al., 2009).

and formic acid as substrates in an equivalent rate of 1.1–1.7 kgCOD m^{-3}d^{-1} and PCP 0.05 to 2.59 mg L^{-1} d^{-1}. HRT varied between 18 and 14 in each reactor. The system was kept inside a temperature controlled chamber at 30°C, pH ranged from 8.4 to 8.9. A removal efficiency of 98% was achieved with Dichlorophenol (DCP) as the only chlorophenol detected. No absorption of PCP or its breakdown products was detected in the polyurethane matrices.

As mentioned, HRT is one of the most important parameters in the successful biodegradation of xenobiotics. Degradation of 4-Chlorophenol (CP) in a UASB reactor under methanogenic conditions was studied. The aim of the research was the optimization of 4-CP removal through variations in HRT. To achieve this goal, 3 L of anaerobic sludge were used to inoculate a 12.5 L acrylic reactor. The sludge was not previously exposed to the 4-CP or any other congener. The experiments were carried out in four different reactors, containing the same amount of inoculum. In each reactor, a different HRT was tested (6,8, 12 and 16 h). Variations in HRT were achieved by changing the influent flow rate, which led to medium to high loading rates (2–5.4 kg m^{-3} d^{-1}). The system was fed with 4-CP 40 mg L^{-1} and sodium acetate as a carbon source at 3 mg L^{-1}. Each reactor was operated continuously under pseudo-steady state for 25–30 days. A control reactor without 4-CP was also set up. Experiments concluded HRT of 12 h was optimal to achieve the highest 4-CP removal rates at 88.3%. Degradation of 4-CP was confirmed via stoichiometric release of Cl ions and *ortho* mode of ring cleavage. As expected, methanogenic activity was 1.2-fold higher in the control reactor (Majumder and Gupta, 2008). As mentioned before, UASB reactors are widely used in the treatment of wastewater and some organochlorines. Second generation UASBs contemplate modification that increase their efficiency, for instance Recycling Upflow Fixed Bed reactors (R-UFB) increase microbial mass stability and provide uniform stability of nutrients, pH and temperature when they are operated in batch recycle mode (Malina and Pohland, 1992).

Pagano et al. (1995) confirmed that the use of a R-UFB led to significant dechlorination of Aroclor 1248 spiked sediments. A 6 L column was packed with sand and operated at 30°C (Figure 14.14). Sanitary landfill leachate was used as carbon source, since it contains a vast amount of volatile fatty acids. The microbial community present in the sediments was used as a source of dechlorinating bacteria. After 91 days of batch recycle mode, 23% of Aroclor was dechlorinated, the highest removal was observed within 7 weeks, biphenyl (a liver toxin) was the main breakdown product. The microbial community was not identified.

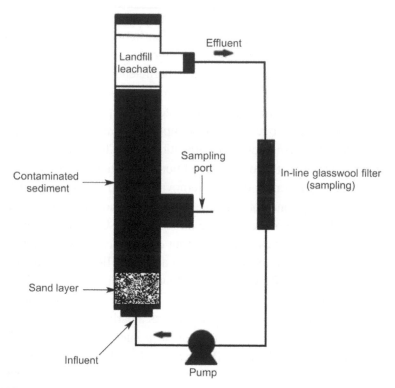

FIGURE 14.14   Recycling upflow fixed bed bioreactor (modified from Pagano, 1995).

14.4   CONCLUSION

Most of the systems presented in this chapter reported removal efficiencies above 90%, making them suitable options for on-site treatment of chlorinated hydrocarbons. Selection of any of these reactors needs careful consideration based on factors like toxicity of the parent organochlorines, as well as of the daughter products, mixing and diffusion, among others. For suspended-growth systems, like CSTR, organlochlorines toxicity determines the success of the process. Low concentrations of these compounds need to be fed to avoid biomass losses and decrease of removal rates, prolonging operation times for highly contaminated sites. On the other hand, immobilized cell systems, like UASB or biofilters, provide higher biomass yields, which lead to higher removal rates, and toxicity does not affect biomass in the same extent as in suspended-growth systems. Nevertheless, these reactors face limitation with low contaminant diffusion, temperature gradients and non-homogenous mixing. MBRs as well as immobilized cell reactors, provide toxicity tolerance, better gas diffusion and easier temperature control, however they possess the highest operational costs.

pH plays one of the most important roles in the operation and success of bioreactors for the treatment of organochlorines. Some of the case studies presented showed that pH values above 8.5 during the reductive dechlorination of PCE could lead to the formation of cis-DCE or VC, both well-known carcinogens. Therefore implementation of pH monitoring and regulation is crucial to maximize contaminant load reduction. At the same time, most of these reactors were kept at 30°C, which is generally the optimum temperature for ORB growth, as well as for dehalogenases activity. Heating jackets or controlled temperature chambers are required and in terms of bigger scale reactors this could imply higher operational costs. The effect of HRT on the efficiency of the processes has been demonstrated, with low HRT providing insufficient contact time between microorganisms and the target compound, on the other hand longer HRT can increase biomass exposure to toxic metabolites.

Another important factor to consider for successful operation of anaerobic bioreactors, is the nature of the inoculum. Almost all of these systems used mixed microbial communities, with the advantages that no exogenous supply of H_2 (electron donor) or vitamin B12 (reductive dehalogenase cofactor) are needed. At the same time, this may also lead to competition between a wide range of bacteria and ORB for carbon sources like acetate, or even for H_2, which could impact the ORB biomass density and would make the process harder to control. The use of pure cultures, like *Dehalococcoides*, *Dehalobacter* or other ORB, may increase the efficiency and control of the processes, but implies exogenous addition of the previously mentioned compounds, which means higher operational risks and costs.

The aim of many of the described anaerobic bioreactors was only to test the potential use of this technology to remediate soil, water or air streams contaminated with organochlorines. In this regard, most of them were successful, even without reaching 100% removal. However, partial reductive dechlorination in other systems led to low removal rates (<30%) and to the formation of other, equally or more, toxic compounds like VC, biphenyl or monochlorobenzene, making the efficiency of these processes questionable, especially for extremely recalcitrant compounds like HCB, Polychlorinated Biphenyls or CF. Finally, it is important to note that all these reactors were designed and operated on lab-scale, where processes can be more easily controlled and monitored. Further work is required before implementing these systems on bigger scales to treat larger volumes of contaminated soils or water. Inappropriate selection of a bioreactor may lead to high operation and design costs, lower removal efficiencies and prolonged operation times. Future advances will come from bold interactions between process engineers and microbiologists.

KEYWORDS

- **Anaerobic biodegradation**
- **Bioreactors**
- **Chlorinated hydrocarbons**
- **Organohalide respiring bacteria**
- **Reductive dechlorination**

REFERENCES

Adrian, L., Szewzyk, U., Wecker, J., Görisch, H. Bacterial dehalorespiration with chlorinated benzenes. *Nature* 2000, 480, 580–583.

Aizpuru, A., Malhautier, L., Fanlo, J. Biofiltration of a mixture of volatile organic compounds on granular activated carbon. *Biotech Bioeng* 2005, 83(4), 479–488.

Arceivala, S., Asolekar, S. R. In *Wastewater Treatment for Pollution Control and Reuse*. Tata McGraw-Hill: New Delhi, 2007, pp. 173–199.

Babaee, A., Shauegan, J. Effect of organic loading rates on production of methane from anaerobic digestion on vegetables waste. *Wor Ren Ener Congr*, Sweden, 2011.

Bal, A. S., Dhagat, N. N. Upflow anaerobic sludge blanket reactor – a review. *Indian J Environ Heal* 2001, 43(2),1–82.

Basu, D., Asolekar, S. Performance of a UASB reactor in the biotreatment of 1,1,2-trichloroethane. *Jour Environ Sci Heal* 2012, 47(2), 267–273.

Bhatt, P. K., Mudliar, S., Chakrabarti, T. Biodegradation of chlorinated compounds - a review. *Critical Reviews in Environ Sci Tech* 2007, 37, 165–198.

Bisaillon, A. B., Lepine, F., Deziel, E., Villemur, R. Identification and characterization of a novel CprA reductive dehalogenase specific to highly chlorinated phenols from *Desulfitobacterium hafniense* strain PCP–1. *Appl Environ Microbiol* 2010, 76, 7536–7540.

Bosma, N. P., van der Meer, J. R., Schraa, G., Tros, M. E., Zehn, A. J. B. Reductive dechlorination of all trichloro- and dichlorobenzene isomers. *FEMS Microbiol Ecol* 1998, 53, 223–229.

Bolaños, M. L. R., Varesche, M. B. A., Zaiat, M., Foresti, E. Phenol degradation in horizontal-flow anaerobic immobilized biomass (HAIB) reactor under mesophilic conditions. *Water Sci Technol* 2001, 44, 167–174.

Chung, J., Krajmalnik-Brown, R., Rittmann, B. Bioreduction of trichloroethene using a hydrogen-based membrane biofilm reactor. *Environ Sci Technol* 2007, 42(2), 477–483.

Crécy de, E., Metzagar, D., Allen, D., Pénicaud, M., Lyons, B., Hansen, J. Development of a novel continuous culture device for experimental evolution of bacterial population. *Appl Microbiol Biotech* 2007, 77, 489–496.

Coulson, J. M. *Chemical and Biochemical Reactors and Process Control*. Elsevier: Oxford, 1994, pp. 71–102.

Copley, S. D. Aromatic dehalogenases: Insights into structures, mechanisms and evolutionary origins. In *Dehalogenation: Microbial Processes and Environmental Applications*. Häggablom M. M., Bossert, I. D. (Eds.), Kluwer Academic Publishers: USA, 2003, pp. 51–98.

Distefano, T., Gossett, J., Zinder, S. Reductive dechlorination of high concentrations of tetrachloroethene to ethene by an anaerobic enrichment culture in the absence of methanogenesis. *Appl Environ Microbiol* 1991, 57, 2287–2292.

Devinny, J. Monitoring biofilters used for air pollution control. *Pract Period Haz Tox Radioact* 1998, 2(2), 78–85.

Damainovic, M. H. R. Z., Moraes, E. M., Zaiat, M., Foresti, E. Pentachlorophenol (PCP) dechlorination in horizontal-flow anaerobic immobilized biomass (HAIB) reactors. *Bioresour Technol* 2009, 100, 4361–4367.

Delgado, A. G., Fajardo-Williams, D., Popat, S. C., Torres, C. I. Successful operation of continuous reactors at short retention time results in high-density, fast-rate *Dehalococcoides* dechlorinating cultures. *Appl Microbiol Biotech* 2014, 98, 2729–2737.

de Nardi, I., Varesche, M. B. A., Zaiat, M., Foresti, E. Anaerobic degradation of BTEX in a packed-bed reactor. *Water Sci Tech* 2002–45, 175–180.

Doble, M., Kruthiventi, A. K., Gikar, V. G. Biotransformation and Bioprocesses. CRC Press: New York, 2004, pp. 169–191.

Fathepure, B. Z., Tiedje, J. M. Reductive dechlorination of tetrachloroethylene by a chlorobenzoate-enriched biofilm reactor. *Environ Sci Tech* 1994, 28(4), 746–752.

Field, J. Biodegradability of chlorinated solvents and related chlorinated aliphatic compounds. *Reviews in Environ Sci Biotech* 2004, 3(3), 185–254.

Hartmans, S. Aerobic vinyl chloride metabolism in *Mycobacterium aurum* L1. *Appl Environ Microbiol* 1992, 58(4), 1220–1226.

Holliger, C., Schraa, G., Stams, A., Zehnder, A. Reductive dechlorination of 1,2-dichloroethane and chloroethane by cell suspensions of methanogenic bacteria. *Biodeg* 1990, 1, 253–261.

Holliger, C., Schraa, G. Physiological meaning and potential for application of reductive dechlorination by anaerobic bacteria. *FEMS Microbiol Rev* 1994, 15, 297–305.

Hölscher, T., Görisch, H., Adrian, L. Reductive dehalogenation of chlorobenzene congeners in cell extracts of *Dehalococcoides* sp. strain CBDB1. *Appl Env Microbiol* 2003, 69(5), 2999–3001.

Hug, L. A., Maphosa, F., Leys, D., Löffler, F. E., Smidt, H., Edwards, E. A., Adrian, L. Overview of organohalide-respiring bacteria and a proposal for a classification system for reductive dehalogenases. *Phil Trans R Soc B* 2013, 368, 20120322.

Hwu, C. S., Lu, C. J. Continuous dechlorination of tetrachloroethene in an upflow anaerobic sludge blanket reactor. *Biotech Lett* 2008, 30, 1589–1593.

Judd, S. The MBR Book: principles and applications of membrane bioreactors for water and wastewater treatment. Elsevier Science: Oxford, 2006, pp. 55–76.

Kennes, C., Eldon, R., Veiga, M. Bioprocesses for air pollution control. *J Chem Biotech* 2009, 84,1419–1436.

Kielhorn, J., Melber, C., Wahnschaffe, U., Aitio, A., Mangelsdorf, I. Vinyl chloride: still a cause of concern. *Environ Heal Persp* 2000, 108(7), 579–588.

Lee, P. K. H., Cheng, D., Hu, P., West, K. A., Dick, G. J., Brodie, E. L., Andersen, G. L., Zinder, S. H., He, J., Álvarez-Cohen, L. Comparative genomics of two newly isolated *Dehalococcoides* strains and an enrichment using a genus microarray. *ISME J* 2011, 5, 1014–1024.

Lee, M., Cord-Ruwisch, R., Manefield, M. A process for the purification of organochlorine contaminated activated carbon: Sequential solvent purging and reductive dechlorination. *Wat Resear* 2010, 44, 1580–1590.

Lee, M., Low, A., Zemb, O., Koening, J., Michaelsen, A., Manefield, M. Complete chloroform dechlorination by organochlorine respiration and fermentation. *Environ Microbiol* 2012, 14(4), 883–894.

Lettinga, G., Hulshoff, P. L. W. UASB process design for various types of wastewaters. *Water Sci Techn* 1991, 24, 87–107.

Lettinga, G., Field, J., van Lier, J., Zeeman, G., Hulshoff, L. W. Advanced anaerobic wastewater treatment in the near future. *Water Sci Tech* 1997, 35(10), 5–12.

Lettinga, G. The route of anaerobic waste (water) treatment toward global acceptance. In *Environmental Anaerobic Technology: Applications and New Developments,* Fang, H. H. P. (Ed.), Imperial College Press: Singapore, 2010, pp. 1–12.

Lioa, B. Q., Kraemes, J. T., Bagley, D. M. Anaerobic membrane bioreactors: Applications and research directions. *Crit Rev Environ Sci Techn* 2006, 36, 489–530.

Liu, T., Chen, Z., Shen, Y. Aerobic biodegradation of hexachlorobenzene by an acclimated microbial community. *Int J Environ Pollut* 2009, 2(3), 235–244.

Majumder, P. S., Gupta, S. K. Degradation of 4-chlorophenol in UASB reactor under methanogenic conditions. *Bioresour Tech* 2008, 99, 4168–4177.

Malina, J. F., Pohland, F. G. Design of anaerobic processes for the treatment of industrial and municipal wastes. CRC Press: USA, 1992, pp. 61–70.

Martin, K. J., Nerenberg, R. The membrane biofilm reactor (MBfR) for water and wastewater treatment: principles, applications and recent developments. *Bioresour Tech* 2012, 122, 83–94.

McCarty, P. L. Breathing with chlorinated solvents. *Science* 1997, 276, 1521–1522.

Middeldorp, P. L., Pas, B., Eekert, M., Kengen, S., Schraa, G., Stams, A. Anaerobic microbial reductive dehalogenation of chlorinated ethenes. *Bioremed J* 1999, 3(3), 151–169.

Pagano, J. J., Scrudato, R. J., Roberts, R. N., Bemis, J. C. Reductive dechlorination of PCB-contaminated sediments in an anaerobic bioreactor system. *Environ Sci Tech* 1995, 29, 2584–2589.

Popat, S. C., Deshusses, M. A. Reductive dehalogenation of trichloroethene vapors in an anaerobic biotrickling filter. *Environ Sci Tech* 2009, 43, 856–7861.

Quintero, J. C., Moreira, M. T., Lema, J. M., Feijoo, G. An anaerobic bioreactor allows the efficient degradation of HCH isomers in soil slurry. *Chemosphere* 2006, 63, 1005–1013.

Rieneke, W. M. C., Kaschabek, S., Pieper, D. Chlorinated hydrocarbon metabolism. *Encyclop Life Sci* 2011, DOI: 10.1002/9780470015902.a0000472.pub3

Samprini, L. *In situ* bioremediation of chlorinated solvents. *Environ Heal Perps* 1994, 103(5), 101–105.

Schipp, C. J., Marco-Urrea, E., Kublik, A., Seifert, J., Adrian, L. Organic cofactor in the metabolism of *Dehalococcoides mccartyi* strains. *Phil Trasn R Soc B* 2013, 368, doi: 10.1098/rstb.2012.0317

Schlötelburg, C., von Wintzingerode, C., Hauck, R., von Wintzingerode, F., Hegemann, W., Göbel, U. B. Microbial structure of an anaerobic bioreactor population that continuously dechlorinates 1,2-dichlorpropane. *FEMS Micrbio Ecol* 2002, 39, 229–237.

Smidt, H., de Vos, W. M. Anaerobic microbial dehalogenation. *Annu Rev Microbiol* 2004, 58, 43–73.

Sponza, D. T. Rapid granulation and sludge retention for tetrachloroethylene removal in an upflow anaerobic sludge blanket reactor. *Biotech Lett* 2001, 23, 1209–1216.

Stephenson, T., Judd, S., Jefferson, B., Brindle, K. Membrane bioreactors for wastewater treatment. IWA Publishing: London, 2000, pp. 9–95.

Stoner, D. *Biotechnology for the Treatment of Hazardous Waste.* CRC Press: USA, 1993, pp. 1, 45.

Szentgyörgyi, E., Bakó, K. Anaerobic membrane bioreactors. *Hung J Indust Chem* 2010, 38(2), 181–185.

Suthersan, S. Natural and enhanced remediation systems. CRC Press: USA, 2001, pp. 51–58.

Taweel-Al, A. M., Shah, O., Aufderheide, B. Effect of mixing on microorganism growth in loop bioreactors. *Int J Chen Eng* 2012, doi: 10.1155/2012/984827

Thauer, R. K., Jungermann, K., Decker, K. Energy conservation in chemotrophic anaerobic bacteria. *Bacterio Rev* 1977, 41, 100–180.

Trotsenko, Y., Torgonskaya, M. The aerobic degradation of dichloromethane: Structural-functional aspects (a review). *Appl Biochem Microbiol* 2009, 45(3), 261–276.

Vainberg, S., Condee, C. W., Steffan, R. J. Large-scale production of bacterial consortia for remediation of chlorinated solvent-contaminated groundwater. *J Ind Microbiol. Biotechnol.* 2009, 36, 1189–1197.

Villadsen, J., Nielsen, J., Lidén, G. *Bioreaction Engineering Principles.* Springer: New York, 2011, pp. 66–76.

von Sperling, M. Waste stabilization ponds. IWA Publishing: London, 2007, pp. 46–66.

Wagner, A., Kleinsteuber, S., Sawers, G., Smidt, H., Lechner, U. Regulation of reductive dehalogenase gene transcription in *Dehalococcoides mccartyi*. *Phil Trasn R Soc B* 2013, 368, 20120317.

INDEX

Milton Keynes UK
Ingram Content Group UK Ltd.
UKHW031138141024
449569UK00024B/1248

9 781774 635827